Electrical and Electronic Measurements and Instrumentation

Electrical and Electronic Measurements and Instrumentation

G. N. Srinivas
Professor, JNTUHCEH,
JNTUH College of Engineering,
Hyderabad.

S. Narasimha
Professor,
TKR College of Engineering & Technology,
Hyderabad.

BSP **BS Publications**
A unit of **BSP Books Pvt. Ltd.**
4-4-309/316, Giriraj Lane,
Sultan Bazar, Hyderabad - 500 095

Electrical and Electronic Measurements and Instrumentation *by G. N. Srinivas and S. Narasimha*

Published by:

BSP BS Publications
A unit of **BSP Books Pvt., Ltd.**
4-4-309/316, Giriraj Lane, Sultan Bazar,
Hyderabad - 500 095
Phone : 040 - 23445600, 23445688
e-mail : info@bspbooks.net

ISBN: 978-93-8681-992-5 (Hardbound)

Dedicated to

Our Parents

Preface

This book have been designed as a study material for Electrical, Electronics and Instrumentation students studying in various universities. This book attempt to provide simple explanations about measurements and instrumentation, throughout the book chosen examples (solved problem) and bits are presented with detailed explanations.

The chapters in the book are arranged in a proper sequence i.e., Electrical, Electronics Measurements and Instrumentation. That permits each topic to build upon earlier studies, which is important in understanding the concept.

The chapter 1 includes the discussion of various types of measuring instrumentals, classifications, operations and applications.

The chapter 2 explains about the instrument transforms their use in power measurement and also discussed about power factor meters, frequency meters and synchro scopes.

The detail analysis of 3-phase and single-phase power, Energy measurements, errors and compensation and extension of ranges, comparision is included in chapter 3 and the chapter 4.

The chapter 5 dedicated to the discussion of various types of potentiometers.

The chapter 6 explain the classification of resistance and measurement methods and applications.

The chapter 7 is dedicated to A.C bridges used for the measurement of basic electrical parameters R, L and C.

The chapter 8 explain various magnêtic measurement methods.

The electronic instruments in a new chapter 9 as its use has become more common place in recent years discussed about the digital voltmeters and chapter 10 dual trace oscilloscope is also included in this edition. These digital instruments are gaining wide acceptance in electronic instrumentation.

The chapter 11 includes the discussion of various types of transducers and of applications.

The chapter 12 is dedicated to the Introduction of the display devices and signal generators classifications and applications.

The chapter 13 includes the discussion of the cathode ray oscilloscope and its applications.

We will be profusely grateful to the readers for pin pointing any lapses and errors. We could continue to welcome constructive criticism and fruitful suggestions for further improvement of the book.

Finally, we wish to thanks the publisher and the entire team of BSP for bringing out this book.

-Authors

Contents

CHAPTER 2

Instrument Transformers

CHAPTER 3

Measurement of Power

CHAPTER 4

Measurement of Energy

CHAPTER 6

Resistance Measurements

CHAPTER 7

A.C Bridges

CHAPTER 8

Magnetic Measurements

CHAPTER 9

Electronic Instruments

CHAPTER 10

Dual Trace Oscilloscope

CHAPTER 11

Transducers

CHAPTER 12

Display Devices and Signal Generators

CHAPTER 13

Cathode Ray Oscilloscope

CHAPTER 1

Measuring Instruments

1.1 Introduction

The fundamental quantities of electrical engineering such as current, voltage, power, energy, power factor and frequency have to be measured with the help of instruments for the purpose of computing the system efficiency and stability. The instruments which are designed to measure these quantities are called measuring instruments. The size and capacity of the instrument vary depending on the magnitude of measurement but the basic working principle is the same.

1.2 Classification of Measuring Instruments

Measuring instruments are classified as:

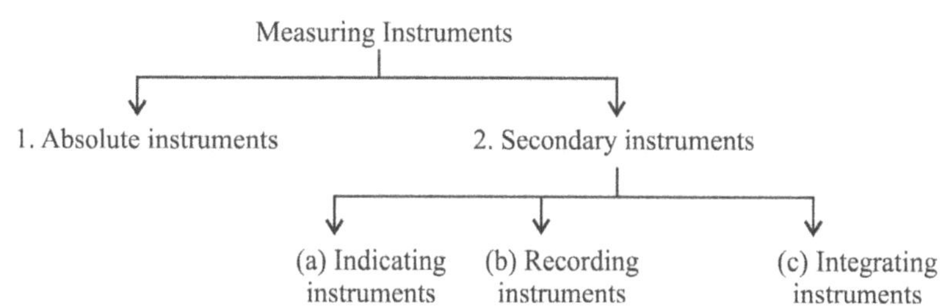

1.2.1 Absolute Instruments

These are those which give the quantity to be measured in terms of instrument constant and its deflection and requires no comparison with any others standard instruments.

Example: Tangent galvanometer and Rayleigh's current balance.

1.2.2 Secondary Instruments

These are the instruments which give directly the value of the quantity being measured by the amount of deflection which is pre-calibrated by comparison with absolute instruments or one which has already been calibrated. Such instruments are designed to serve mainly the three purposes.

1. Indicate the instantaneous value being measured and record it.
2. Indicate the instantaneous value of the quantities being measured.
3. Measure the different quantities and integrate to give a unstable result.

1.2.2.1 Indicating Instruments

These are the instruments which indicate the magnitude of the instantaneous values being measured by means of pointer over a calibrated scale.

The indication of pointer also change with respect to time.

Example: Ammeter, voltmeter, wattmeter, frequency meter, power factor meter etc.

1.2.2.2 Recording Instruments

These instruments not only read the instantaneous values but also record continuously the magnitude of the quantity on a paper over a period of time.

The moving system of the instrument carries an inked pen (pointer) which rest lightly on a graph or chart. That is moved at a low speed and uniform, in a direction perpendicular to that of the deflection of the pointer.

These instruments are generally used in power house.

1.2.2.3 Integrating Instruments

These are measure the total quantity of electricity detected over a period of a time.

Example: Energy meter (Ampere-hour and watt-hour meters) etc.

1.3 Essentials of Indicating Instruments

In most of the indicating instruments three distinct torque are required for operation. They are:

1. Deflecting torque T_d (or) deflection system
2. Controlling torque T_c (or) control system
3. Damping torque (or) damping system

The above systems use one of the following effect, produced by current (or) voltage, produce deflecting torque.

1. ***Thermal effect:*** The current to be measured is passed through small element which heats it. The temperature converted into e.m.f using thermocouple producing current again.
2. ***Induction effect:*** Eddy currents are produced due to magnetic flux (alternating). This producers force to move non magnetic disc. Induction principle is used for AC quantities.
3. ***Magnetic effect:*** Force is developed when current carrying conductor placed in magnetic field force. F = BIL.
4. ***Electrostatic effect:*** When two plates are charged, electrostatic forces are developed. This force is used to displace one plate. The displacement is directly proportional to the voltage which displacement is calibrated to magnitude of voltage
   - Suitable for voltmeters only.

1.4 Deflecting System

The deflecting torque cause the moving system of the instrument to move from its initial zero position. The magnitude of the deflecting torque depends upon the magnitude of the measurable quantity. The torque is produced by use of any one of the effects of electric current such as magnetic, electro-magnetic, heating, electrostatic etc. The method of production of the deflecting torque and its relation to the measurable quantity depends upon the type of instruments. As long as the instrument is connected in the circuit and measurable quantity is present in the instrument operating mechanism, the deflecting torque is constant and continuous and so the pointer will try to rotate as a motor if not controlled. Hence there exists a necessity to control the movement of the pointer such that it comes to rest at the deflected position. This torque is called controlling torque. The action of the deflecting torque for moving system containing pointer is shown in Fig. 1.1.

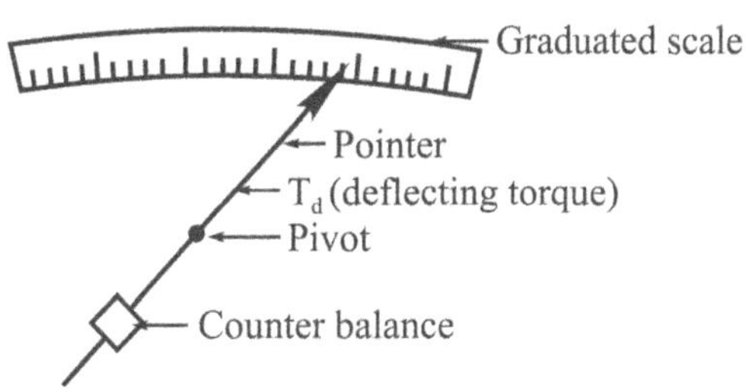

Fig. 1.1 Deflecting system

Effects of indicating Instruments

Effects used	Type of measurements	Applications
Thermal	Current, Voltage on both AC and DC	Ammeters, voltmeters
Magnetic effect	Current, Voltage, power and energy AC and DC	Ammeter, voltmeter, watt meters, integrating meters etc.
Electro-static effect	Voltmeter on both AC and DC system	Voltmeter
Chemical effect	Ampere hours in DC system	Voltmeter
Electromagnetic Induction effect	Voltage, Current, power, energy, in AC system	Voltmeter, Ammeter, wattmeter, energy meter

1.5 Controlling System

The magnitude of the movement of the moving system would be somewhat indefinite under the influence of deflecting torque unless some controlling methods existed. This method opposes the deflecting torque and increase in deflection of the moving system. Thus limits the movement and ensures that the magnitude of the deflection is always the same for a given value of quantity to be measured. Under the influence of the methods (Technique), the pointer will return to its zero position.

Controlling system has mainly two technique (or) methods. Those are

Controlling system

1. Spring control
2. Gravity control

1.5.1 Spring Control

A helical hair spring usually made of Phosphor Bronze is attached to the moving system in such away that its one end is fixed to the moving system while the outer end to a rigid body. As the angle of deflection increases, the stress in the spring increases due to which it exerts a torque on the moving mechanism opposite to the motion of the moving mechanism. This torque is proportional to the angle of deflection.

$$T_c \propto \theta$$

$$I \propto \theta \qquad [\text{when at } T_d = T_c]$$

$$\boxed{T_c = \frac{Edt^3}{i2l}\theta \text{ kg-m}}$$

E → Young modulus of the spring material kg/m

d → Depth of the Flat spring

t → thickness in meters

l → length of the strip

θ = angle of deflection

The main essential requirements of spring material

1. It should be non-magnetic
2. It should have low-resistance
3. It should be cheap and durable
4. Its temperature coefficient of resistance should be low
5. It should not get subjected to appreciable fatigue during operation.

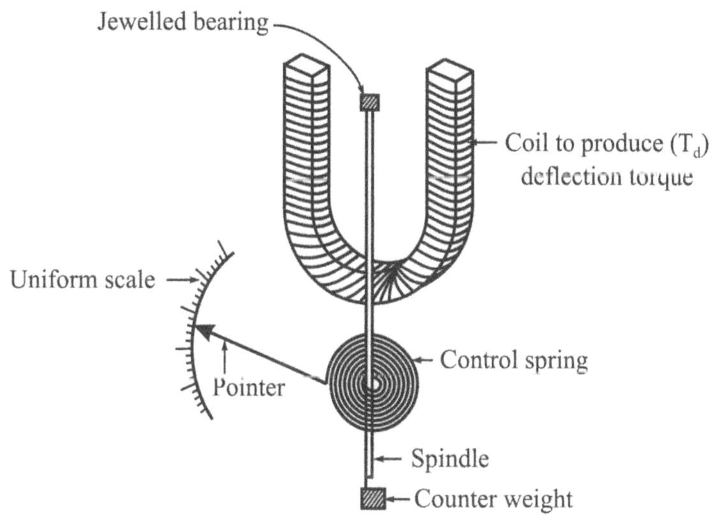

Fig. 1.2 Spring control

1.5.2 Gravity Control

In this gravity control system a small amount of weight is attached to the moving system in such a way that it gets activated at the time of deflection only and produce a T_c in some proportion to the deflection. In the Fig. 1.3(a) shown the zero position of the pointer, the controlling torque is '0'. This position is 'A' of the weight in Fig. 1.3(b). If the system deflects, the weight position also changes, as in Fig. 1.3(b).

$T_c = w \sin\theta \times l$

w → the weight of control arm

$l \rightarrow$ length of arm

$\theta \rightarrow$ angle of deflection

w and l is constant for a set design

$$T_c \propto \sin\theta$$

At deflected positions

$$T_d = T_c \qquad [\because T_d \propto I]$$

$$\therefore \qquad \boxed{I \propto \sin\theta}$$

Thus the deflection is proportional to current (i.e., Quantity to be measured).

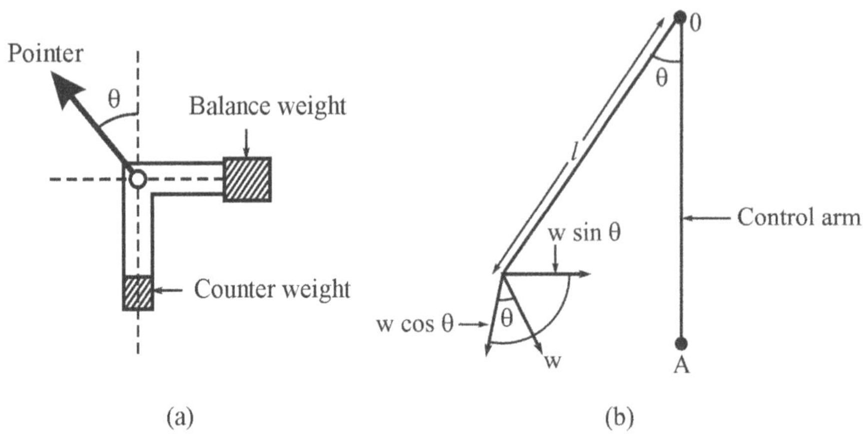

Fig. 1.3 Gravity control

1.5.3 Comparison of Controlling Systems

Spring control	Gravity control
1. The scale in uniform	1. The scale in non uniform
2. The readings can be taken very accurately	2. The readings cannot be taken accurately
3. Very popularly used in most of the Instruments	3. Rarely used for indicating portable instruments
4. Simple, rigid but costlier compared to gravity control	4. Simple, cheap but delicate
5. The system need not be in vertical position	5. The system must be in vertical position only.
6. The controlling torque is proportional to θ.	6. The controlling torque is proportion to sin θ.
7. Controlling torque is fixed.	7. Controlling torque can be varied

1.6 Damping Torque

Damping force or torque is also necessary to avoid oscillation of the moving system about its final deflected position owing to the inertia of the moving parts and to bring the moving system to rest in its deflected position quickly.

If the instrument is **under damped** (Fig. 1.4), the moving system will oscillate about its final position and take some time to come to rest in its steady position. If the instrument is **over damped** the moving system will becomes slow and sluggish. When the degree of damping is such that the pointer moving system, quickly to its deflected position without oscillation. This damping is said to be **critical damped** and the instrument is said to be 'dead beat' in practice. To obtain best result, the damping is adjusted to the value slightly less than the critical value.

The damping torque must operate only while the moving system of the instrument is actually moving and always oppose its motions. It must not effect the steady deflection produced by the deflecting torque.

The damping torque must be obtained by

(a) Air friction

(b) Fluid friction

(c) Eddy current methods

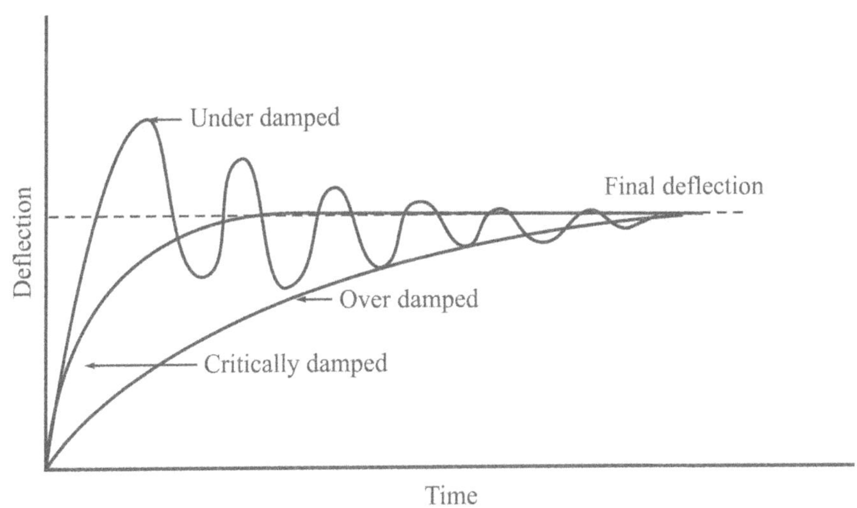

Fig. 1.4 Damping system

1.6.1 Air Friction Damping

This method uses a light aluminium piston to the moving system shown in Fig. 1.5. The piston moves inside the air chamber with least clearance along with the movement of the pointer. The clearance between the piston and sides of the chamber is kept least and uniform. As the piston moves inside the chamber rapidly, the air inside the closed space is compressed and the pressure so developed opposes the motion of the piston. If the piston moves out of the chamber rapidly the pressure in the closed space decrease than normal and the pressure on the open side of the piston would be greater than that of the opposite side. In the process of equilising the pressure motion of the piston is again opposed. The only care that must be taken for efficient damping is that the arm carrying the piston is not bent or the piston does not touch the walls of the chamber else it would lead to serious error in reading.

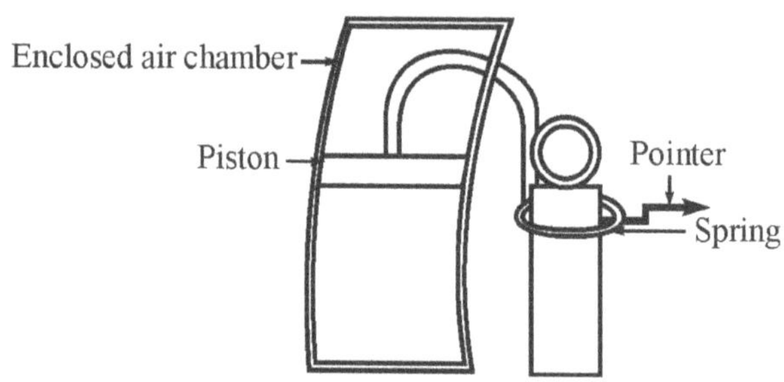

Fig. 1.5 Air friction damping

1.6.2 Fluid Friction Damping

This method is similar to air friction damping, only air is replaced by fluid (Fig. 1.6). The fluid is opposing motion. Damping force due to fluid is greater than air, due to more viscosity.

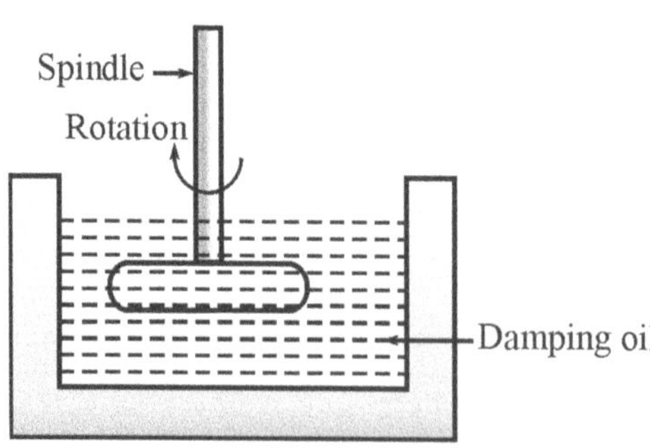

Fig. 1.6 Fluid friction damping

Disadvantage of fluid damping

1. The instrument has to be used only in vertical position.
2. Oil creeps out if the sealing is not perfect.

1.6.3 Eddy Current Damping

This method is mainly based on Faraday's law and Lenz's law. When a conductor moves in magnetic field cut's the flux, emf induced in it and direction of this emf is to oppose the cause producing it. This current interacts with the magnetic field to produce an electromagnetic torque which opposes the motion. This torque is proportional to the strength of the magnetic field and the 'current' produced. The current is proportional to emf which in turn is proportional to velocity of the conductor. Thus if the strength of the magnetic field is constant, the torque is proportional to velocity of the conductor. This method is the most effective.

1.7 Permanent Magnet Moving Coil (PMMC)

These instruments are universally used for DC measurements. The basic working principle of a PMMC instrument is the same at that D' Arsonval Galvanometer but it slightly differs from the D' Arsonval Galvanometer in construction.

A permanent magnet type moving coil instrument is shown in Fig. 1.7. It consists of permanent powerful magnet with soft Iron pole pieces. A cylindrical Iron core is mounted between the two poles of the magnet giving very narrow air gap in which the sides of a pivoted light rectangular coil lines. The rectangular coil is wound of many turns from fine wire on light aluminium or copper former and acts as moving elements. The current is used into and out of the coil by means of phosphor bronze hair springs provided at both ends. The springs also provide the controlling torque. When the current to be measured is passed through the coil, a deflecting torque is produced on account of reaction of the permanent magnetic field with the coil magnetic field. The direction of deflecting torque can be determined by applying "Fleming left hand rule".

If the current in amperes flowing, the coil of turns N and length l and B is the flux density in tesla in the air gap, then deflecting force

$$F \propto B\, i\, l\, N \quad \text{Newton}$$

If 'r' is the mean distance in meters of the wires forming the sides of the coil from the axis of rotation, then deflecting torque

$$T_d = F \times 2r = 2\, B\, i\, l\, Nr \quad \text{N-m}$$

In general, torque

T = Ampere turn on coil × Area of coil × Flux density in the air gap

From above expression it is obvious that if flux density B in the air gap is constant, then deflecting torque

$$T_d \propto i$$

Since such instruments are spring controlled

$\therefore$ controlling torque $T_c \propto \theta$

Since in steady deflection position $T_c = T_d$

$$\therefore \qquad \theta \propto i$$

Hence, such instrument have *Uniform* scale. Eddy currents are induced in the aluminium on which the coil is wound. Thus the *Eddy current damping* is provided.

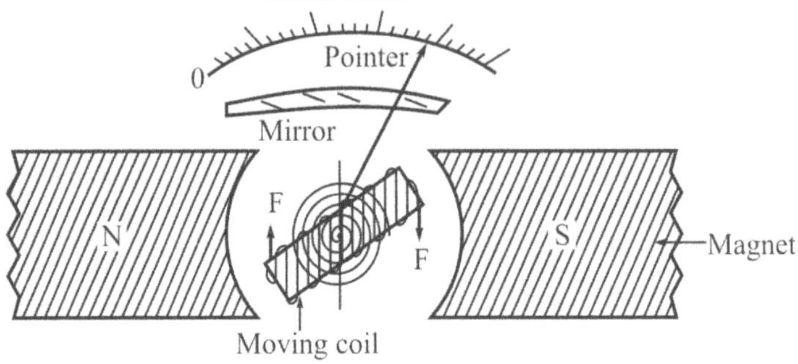

Fig. 1.7 Permanent magnet moving coil

1.7.1 Advantages

(i) Uniform scale
(ii) Low power consumption because driving power is small (25μW to 200μW)
(iii) No hysteresis loss as the former is of copper or aluminium
(iv) High torque/weight ratio
(v) Very effective and reliable eddy current damping
(vi) No effect of stray magnetic flux as instance polarised or unidirectional field is employed
(vii) Range can be extended with shunts or multipliers

1.7.2 Disadvantages

(i) These cannot be used for A.C measurements
(ii) Friction and temperature might introduce errors in the case of other instruments
(iii) These are costlier in comparison with moving iron instrument because of delicate construction and the necessary accurate machining and asserably of various parts

(iv) The using of control spring and of permanent magnets might cause errors which can be considerably obviated by careful choice of material and preparing during manufacture.

1.7.3 Torque Equation

The Torque for a moving coil instrument is obtained for the basic law of the electromagnetic torque. The deflecting torque is given by,

$$T_d = NB\,l\,dT \qquad \text{or} \qquad T_d = NBAI$$

$$T_d = GI$$

where $G = NBA = \text{constant}$

where,

T_d = deflecting torque in N – m

B = flux density in air gap. Wb/m^2

A = effective coil area. m^2

N = number of tuns of the coil

I = current in the moving coil, amperes

The spring control provides a controlling torque and is proportional to the angular deflection of the pointer.

$$T_c = K\theta$$

where, T_c = Controlling torque

K = Spring constant, Nm/rad (or) Nm/deg

θ = angular deflection

For final steady deflection

$$T_c = T_d$$

∴ Final steady state position

$$T_c = T_d \text{ or } \qquad GI = K\theta$$

$$\theta = \left(\frac{G}{K}\right)I$$

∴ Current $$I = \left(\frac{K}{G}\right)\theta$$

As the deflection is directly proportional to the current passing in the meter, we get a uniform scale for the instrument.

1.8 D' Arsonval Galvanometer

The construction of a D' Arsonval Galvanometer is shown in Fig. 1.8. The description of different parts is given below.

1. **Moving Coil:** It is the current carrying element. It is either rectangular or circular in shape and consists of a number of turns of fine wire. This coil is suspended so that it is free to turn about its vertical axis of symmetry. It is arranged in a uniform, radial, horizontal magnetic field in the air gap between pole pieces of a permanent magnet and iron core. The iron core is spherical in shape if the coil is circular but is cylindrical if the coil is rectangular. The iron core is used to provide flux path of low reluctance and therefore to produce strong magnetic field for the coil to move in. This increases the deflecting torque and hence the sensitivity of the galvanometer. The length of air gap is about 1.5 mm. In some galvanometers, the iron core is omitted resulting in decreased value of flux density and the coil is made narrower to decrease the air gap. Such a galvanometer is less sensitive but its moment of inertia is smaller on account of its reduced radius and consequently a short-periodic time.

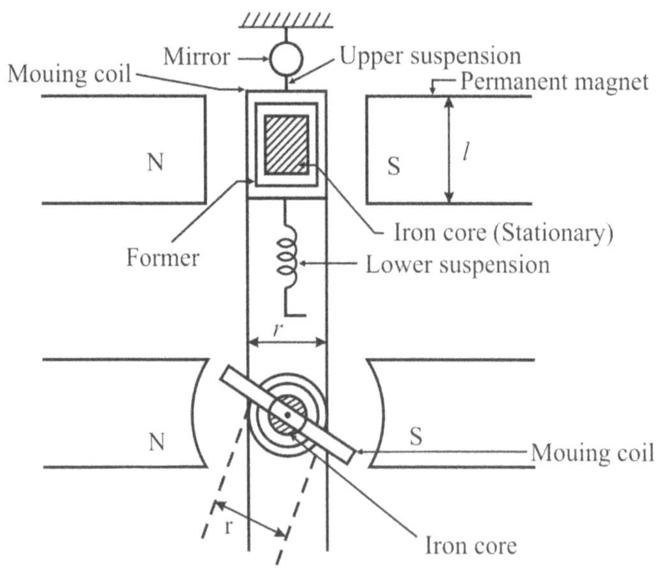

Fig. 1.8 D'Arsonval galvanometer

2. **Damping:** There is a damping torque present owing to production of eddy current in the metal former on which the coil is mounted. Damping is also obtained by

connecting a low resistance across the galvanometer terminals. Damping torque depends upon the resistance and we can obtain critical damping by adjusting the value of resistance.

Iron core: It is spherical if coil is circular and cylindrical if coil is rectangular. It is basically used to provide low reluctance path to the magnetic flux and to produce strong magnetic field, thus ensures higher deflecting torque and better sensitivity of the galvanometer. The air gap is about 1/16 inches i.e., about 1.5 mm. If small movement of inertia is necessary, the iron core can be omitted but it decreases the sensitivity.

3. **Suspension:** The coil is supported by a flat ribbon suspension which also carries current to the coil. The other current connection in a sensitive galvanometer must be levelled carefully so that the coil hangs straight and centrally without rubbing the poles or the sag to iron cylinder. Some portable galvanometers which do not require exact levellinge have "taut suspensions" consisting of straight flat strips kept under tension far at the both top and at the bottom.

   The upper suspension consists of gold or copper wire of nearly 0.0125 or 0.025 mm diameter rolled into the form of a ribbon. This is not very strong mechanically, so that the galvanometers must be handled carefully without jerks. Sensitive galvanometers are provided with coil clamps to take the strain from suspension while the galvanometer is being moved.

Induction: The suspension carries a small mirror upon which a beam of light it costs. The beam of light is reflected on to a scale upon which the deflection is measured. This scale is usually about 1 meter away from the instrument, although ½ meter may be used for greater compactness.

Damping: The damping is eddy current damping. The eddy currents developed in the metal former on which oil is mounted, are responsible to produce damping torque. For effective damping a low resistance is connected across the galvanometer terminals. By adjusting the value of this resistance damping can be changed and critical damping can be achieved.

Zero adjustment: A torsion head is provided for the adjustment of the coil position and zero setting.

Indication: The suspension carries a small mirror upon which a beam of light is cast through a glass window in the outer brass case surrounding the instrument. The beam of light is reflected on the scale. The scale is usually 1 m away from the mirror.

1.8.1 Torque Equation

The various parameters involved in torque equation are;

r = width of coil in meters

l = length of coil measured along vertical axis in m.

N = Number of turns of coil

B = Flux density in air gap in Wb/m^2 or Tesla

i = current through coil in A

K = Spring constant or restoring constant in Nm/rad

α (alfa) = angle between plane of coil and direction of magnetic field

A = area of coil in $m^2 = l \times r$

θ_f = Final steady state deflection of coil in rad

F = Force on each side of coil $= NB\, il \sin\alpha$ N

T_d = Deflecting torque $= F \times d = NB\, il \sin\alpha\, r$

$\therefore\ T_d = NBi\, A \sin\alpha$

As the field is radial in nature, $\alpha = 90^\circ$ hence $\sin\alpha = 1$

$\therefore\ T_d = NBiA = Gi = Gi$ Nm

where, $G = NBA$ = Galvanometer constant.

The restoring torque provided by the spring is directly proportional to the final deflection of the coil.

$$\therefore \qquad T_c = K\,\theta_f$$

For final steady state position of coil. $T_d = T_c$

$$Gi = K\,\theta_f$$

$$\theta_f = \frac{Gi}{K}$$

The scale is calibrated in mm. The scale is at a distance of 1 m from the mirror as shown in the Fig. 1.9.

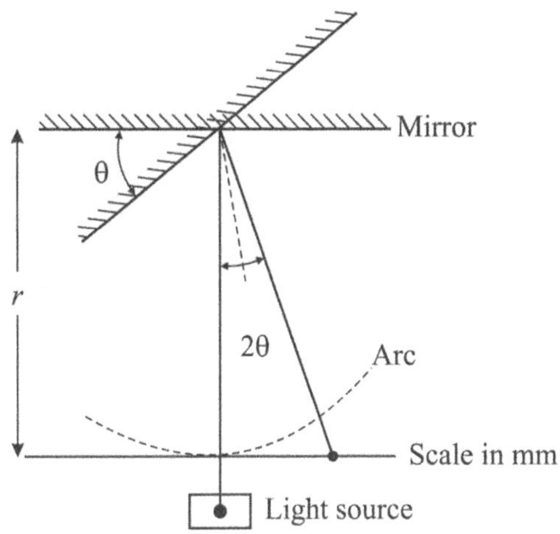

Fig. 1.9 Measurement of deflection in mm

For small deflection, the radius of arc and angle of turning decide the deflection; the angle through which the beam get reflected is $2\theta_f$, if mirror is turned through θ_f.

$\therefore$ d in mm $= 2\theta_f \times r$

$$\therefore \quad d = \frac{2Gir}{K} \text{ mm}$$

Note: Usually r = 1 m = 1000 mm for the galvanometer.

1.9 Intrinsic Constants of Galvanometer

The various intrinsic constants of galvanometer are

1.9.1 Constant of Inertia (J)

The inertia of the system opposes the motion. Thus inertia produces a retarding torque given by

$$T_i = J\frac{d^2\theta}{dt^2}$$

J = Constant of inertia about axis of rotation in kg-m^2

where $\frac{d^2\theta}{dt^2}$ = angular acceleration

θ = Deflection at any time t

1.9.2 Control Constant

The elasticity of the suspension is proportional to the displacement which produces controlling torque. This is required to bring the moving system back to the original position

$$T_c = K\theta$$

K = Control constant or restoring constant in Nm/rad

1.9.3 Displacement Constant (G)

The constant G defined in torque equation of a galvanometer is called displacement constant.

$$G = NBA = NBl \times r \quad \text{Nm/A}$$

1.9.4 Damping Constant

Another torque retarding the motion is friction in air and elastic hysteresis in the suspension. It is assumed to be proportional to the angular velocity of the moving system.

$$T_d = D\,\frac{d\theta}{dt}$$

D = Damping constant in Nm/rad s^{-1}

1.10 Dynamic behaviour of Galvanometer

The dynamic behaviour of galvanometer is analysed through its equation of motion.

Thus $$J\frac{d^2\theta}{dt^2} + D\frac{d\theta}{dt} + K\theta = Gi$$

This is second order differential equation governing the galvanometer motion. The solution of this equation has two parts.

1. Complementary function (C.F)
2. Particular integral (P.I)

The C.F represents the *transient* behaviour while P.I represents the *steady state* condition i.e., final deflection of the moving system.

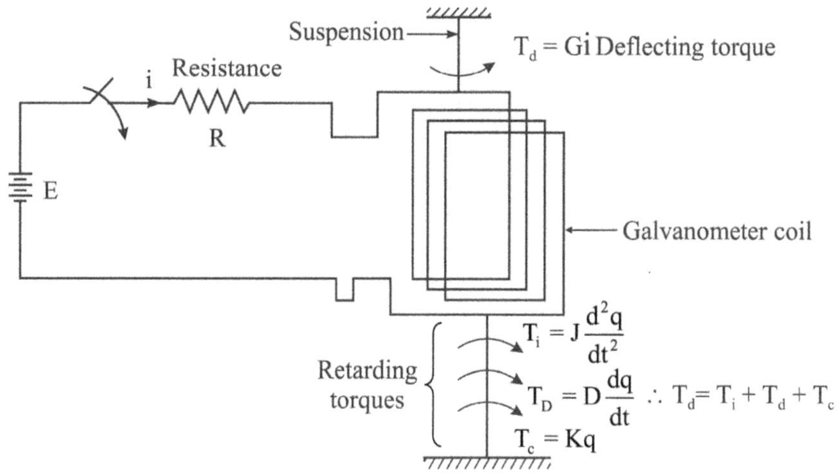

Fig. 1.10 Torque acting in galvanometer motion

The behaviour of system before it achieve the steady state is transient behaviour when transient behaviour dies out, the system achieves final steady state position.

The auxiliary equation of above differential equation is obtained as,

$$Jm^2 + D_m + K = 0$$

The roots of this equation are

$$m_1 = \frac{-D + \sqrt{D^2 - 4JK}}{2J}$$

$$m_2 = \frac{-D - \sqrt{D^2 - 4JK}}{2J}$$

Hence the solution has two exponential terms of powers m_1 and m_2

$$\therefore \qquad \theta = Ae^{m_1 t} + Be^{m_2 t} \text{ (C.F)}$$

where A, B = Constants.

Now when steady state is reached then $\frac{d\theta}{dt}$ and $\frac{d^2\theta}{dt^2}$ are zero as moving system attains a final steady state position of θ_f. Using in equation Gi = $J\frac{d^2\theta}{dt} + D\frac{d\theta}{dt} + K\theta$, we get P.I as

$$K\theta_f = Gi \qquad \text{i.e., } \theta_f = \frac{Gi}{K} \text{ as derived earlier}$$

Thus the complete solution in the addition of C.F and P.I.

$$\theta = \underbrace{A\,e^{m_1 t} + B\,e^{m_2 t}}_{\text{Transient term}} + \underbrace{\theta_f}_{\text{Steady state}}$$

Now the transient terms may be purely exponential or oscillatory which depends on the nature of roots m_1 and m_2. This defines the various damping conditions of the system.

1.11 Undamped Motion

The motion existing when damping is made zero i.e., D = 0 is called undamped motion of the system. The roots m_1 and m_2 are purely imaginary with zero real part for this case.

Note: The oscillations with zero damping are natural oscillations without opposition, having highest frequency. This frequency of undamped oscillations is called natural frequency of oscillations and denoted as ω_n.

$$\omega_n = \sqrt{\frac{K}{J}} \text{ rad/s} \qquad [\text{putting } D = 0 \text{ in } \omega_d]$$

$$\alpha = \tan^{-1}\left[\sqrt{\frac{4JK - D^2}{D}}\right] = \tan^{-1}\alpha = 90^\circ, \ [D = 0]$$

Thus the solution of undamped motion is

$$\theta = \theta_f\left[1 - \frac{2\sqrt{JK}}{2\sqrt{JK}}e^0 \sin\left(\omega_n t + 90^\circ\right)\right]$$

$$\therefore \qquad \theta = \theta_f\,[1 - \cos\omega_n t]$$

These are the oscillations with constant frequency and amplitude about the final position θ_f. Such oscillations are called sustained oscillations.

These oscillations are shown in the Fig. 1.11.

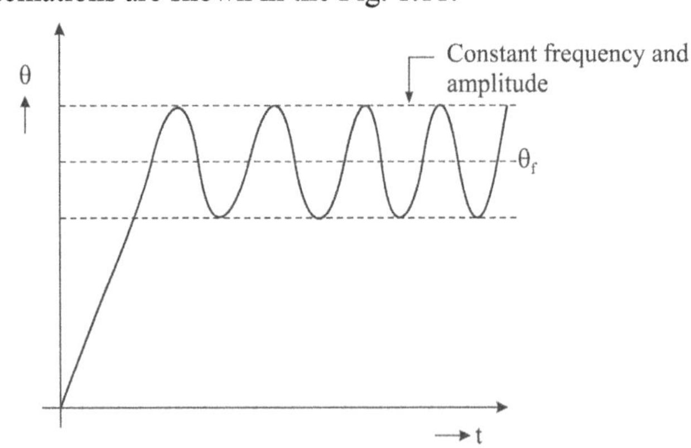

Fig. 1.11 Undamped motion of galvanometer

1.12 Underdamped Motion

Both m_1 and m_2 are complex conjugates of each other having negative real part.

$$D^2 - 4KJ < 0 \quad \text{Underdamped}$$

Note: The transient behaviour is damped oscillations i.e., oscillations of decreasing amplitude. After sometime amplitude becomes zero and system achieves steady state.

The roots are imaginary and complex conjugates of each other represented as

$$m_1, m_2 = \frac{-D \pm \sqrt{(-1)^2\left[4JK - D^2\right]}}{2J} \quad \text{but } \sqrt{-1} = J$$

$$\therefore \quad m_1, m_2 = -\frac{D}{2J} \pm j\sqrt{4JK - D^2} = -\alpha \pm J\omega d \qquad .....(1.1)$$

where $\alpha = \dfrac{D}{2J}$ and $\omega_d = \dfrac{\sqrt{4JK - D^2}}{2J}$

Thus the solution becomes

$$\theta = A\,e^{(-\alpha + j\omega_d)t} + B\,e^{(-\alpha - j\omega_d t)} + \theta_f \qquad .....(1.2)$$

$$\therefore \quad \theta = e^{-\alpha t}\left[A\,e^{+j\omega_d t} + B\,e^{-j\omega_d t}\right] + \theta_f$$

But $\quad e^{+j\theta} = \cos\theta + j\sin\theta, \quad e^{-j\theta} = \cos\theta - \sin\theta$

$$\theta = e^{-\alpha t}\left[A\,(\cos\,\omega_d t + j\,\sin\,\omega_d t) + B(\cos\,\omega_d t - j\,\sin\,\omega_d t)\right] + \theta_f$$

$$\therefore \quad \theta = e^{-\alpha t}\left[(A+B)\cos\,\omega_d\,t + j(A-B)\sin\omega_d t\right]$$

$$\theta = e^{-\alpha t}\left[P\,\cos\,\omega_d + Q\sin\omega_d t\right] \qquad .....(1.3)$$

where $\quad P = A + B \quad$ and $\quad Q = j(A - B)$

Let $\quad \theta = F\,e^{-\alpha t}\sin(\omega_d t + \alpha) \qquad .....(1.4)$

$$\theta = F\,e^{-\alpha t}\left[\sin\,\omega_d t\,\cos\,\alpha + \cos\,\omega_d t\,\sin\,\alpha\right] \qquad .....(1.5)$$

Comparing equation (1.3) and equation (1.5), $P = \sin\alpha$ and $Q = \cos\alpha$

Hence $\quad \alpha = \tan^{-1}\dfrac{P}{Q} \quad$ and $\quad F = \sqrt{P^2 + Q^2}$

Thus the final equation for θ underdamped case is

$$\theta = F\, e^{-\frac{D}{2J}t} \sin\left(\omega_d t + \alpha\right) + \theta_f \qquad \text{.....(1.6)}$$

where $\omega_d = \dfrac{\sqrt{4JK - D^2}}{2J}$ = damped frequency of oscillations in rad/s

The nature of such oscillations is shown in the Fig. 1.12.

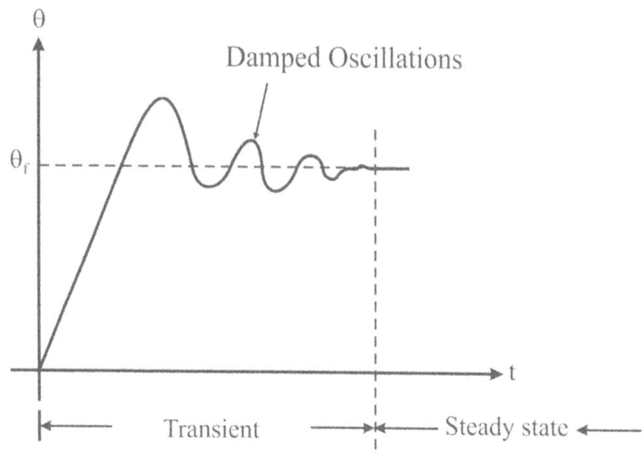

Fig. 1.12 Underdamped motion of galvanometer

The constant F and α are to be obtained from initial conditions

Obtaining F and α:

To obtain F and α i.e., P and Q use initial condition i.e.,

at $t = 0$ and $\theta = 0$ in the equation (1.6) ,

$$\therefore \qquad 0 = F \sin(\alpha) + \theta_f$$

$$\therefore \qquad \sin\alpha = -\frac{\theta_f}{F} = P \qquad \text{.....(1.7)}$$

Differentiating equation (1.6) with the

$$\frac{d\theta}{dt} = F\left(-\frac{D}{2J}\right)e^{-\frac{D}{2J}t} \sin\left(\omega_d t + \alpha\right) + F\, e^{-\frac{D}{2J}} + \omega_d \cos\left(\omega_d t + \alpha\right)$$

But at $t = 0, \quad \dfrac{d\theta}{dt} = 0$

$$\therefore \qquad 0 = -\left(\frac{D}{2J}\right)^{(F)} \sin\alpha + F\,\omega_d \cos\alpha \qquad .....(1.8)$$

$$\therefore \qquad \tan\alpha = \omega_d \times \frac{2J}{P} = \frac{\sqrt{4JK - D^2}}{D} \times \frac{2J}{D}$$

$$\therefore \qquad \tan\alpha = \frac{\sqrt{4JK - D^2}}{D} \qquad .....(1.9)$$

Using (1.7) and (1.8)

$$0 = -\left(\frac{D}{2J}\right)(F)\left(-\frac{\theta_f}{F}\right) + F\omega_d \cos\alpha$$

$$\cos\alpha = -\frac{D\theta_f}{2JF\,\omega_d} = Q \qquad .....(1.10)$$

But $\sin^2\alpha + \cos^2\alpha = 1$

$$\therefore \qquad \left(-\frac{\theta_f}{F}\right)^2 + \left(-\frac{D\theta_f}{2J\,F\,\omega_d}\right)^2 = 1$$

$$F = -\,\theta_f\sqrt{\frac{4J^2\omega_d^{\,2} + D^2}{4J^2\omega_d^{\,2}}} \qquad .....(1.11)$$

Using $\quad \omega_d = \dfrac{\sqrt{4JK - D^2}}{2J}$ in equation (1.11)

$$\therefore \qquad F = -\,\theta_f\left[\frac{2\sqrt{JK}}{\sqrt{4JK - D^2}}\right] \qquad .....(1.12)$$

Using equation (1.9) and (1.11) in equation (1.6)

$$\theta = -\,\theta_f\left[\frac{2\sqrt{JK}}{\sqrt{4JK - D^2}}\right]e^{-\frac{D}{2J}t}\sin(\omega_d t + \alpha) + \theta_f$$

$$\therefore \qquad \theta = \theta_f\left[1 - \frac{2\sqrt{JK}}{\sqrt{4JK - D^2}}\right]e^{-\frac{D}{2J}t}\sin(\omega_d t + \alpha) \qquad .....(1.13)$$

The angular frequency is ω_d hence

$$f_d = \frac{\omega_d}{2\pi} = \frac{1}{2\pi}\frac{\sqrt{4JK - D^2}}{2J}\ \text{Hz} \qquad .....(1.14)$$

T_d = time period

$$T_d = \frac{1}{f_d} = 2\pi\left[\frac{2J}{\sqrt{4JK - D^2}}\right] \quad .....(1.15)$$

1.13 Critically Damped Motion

For critically damping, the roots m_1 and m_2 are equal, real and negative. Thus

$$D^2 - 4JK = 0 \quad \text{i.e.,} \quad D^2 = 4JK \text{ and}$$

$$D = 2\sqrt{JK} \quad \text{for critical damping}$$

$$m_1 = m_2 = -\frac{D}{2J} \text{ and } D = 2\sqrt{JK}$$

Note: For this case, the transient response is not critically damped oscillatory but purely exponential such that pointer attains the steady position of very quickly.

As the response is not oscillatory, equation (1.13) is not applicable for this case, the solution for the critically damped case is,

$$\theta = \theta_f + e^{-\frac{D}{2J}t}[A + Bt] \quad ......(1.16)$$

Using $t = 0, \theta = 0$

$$0 = \theta_f + A \quad .....(1.17)$$

Differentiating (1.16) and using $t = 0$, $\frac{d\theta}{dt} = 0$,

$$\frac{d\theta}{dt} = -\frac{D}{2J}e^{-\frac{D}{2J}t}[A + Bt] + e^{-\frac{D}{2J}t}B$$

$$0 = -\frac{D}{2J}A + B \quad .....(1.18)$$

$$A = -\theta_f \text{ and } B = -\frac{D\theta_f}{2J}$$

$$\theta = \theta_f\left[1 - e^{-\frac{D}{2J}t}\left(1 + \frac{D}{2J}t\right)\right] \quad .....(1.19)$$

The value of damping constant for the critical damping is denoted by D_c and $D_c = 2\sqrt{JK}$.

Thus,

$$\frac{D}{2J} = \frac{2\sqrt{JK}}{2J} = \sqrt{\frac{K}{J}} = \omega_n$$

Using in equation (1.19)

$$\theta = \theta_f\left[1 - e^{-\omega_n t}(1 + \omega_n t)\right] \quad .....(1.20)$$

1.14 Overdamped Motion

The amount of damping is mathematically measured by defining a ratio of actual damping D and critical damping D_c. This is called a *damping ratio* and denoted by a greek letter ξ. This is also called *relative damping.*

$$\xi = \frac{D}{D_c} = \frac{D}{2\sqrt{JK}} = \text{Damping ratio} \qquad .....(1.21)$$

Thus when the actual damping is more than the damping for critical case, the motion is called overdamped and the roots m_1 and m_2 are real, *unequal* and negative.

Note: For overdamped case, the transient motion is purely exponential and nonoscillatory. The pointer attains final position of exponentially, taking more time than that of critical damping.

More the value of damping, the pointer response is slow and sluggish, taking more time to attain the final position.

Note: For critical damping $\xi = 1$ while for overdamped case the damping ratio $\xi > 1$.

The solution for overdamped motion interms of ξ is given as,

$$\theta = \theta_f \frac{\xi + \sqrt{\xi^2 - 1}}{1 + 2\sqrt{\xi^2 - 1}} e^{-\omega_n t\left(\xi - \sqrt{\xi^2 - 1}\right)} - \frac{\xi - \sqrt{\xi^2 + 1}}{2\sqrt{\xi^2 - 1}} e^{-\omega_n t(\xi + \sqrt{\xi^2 - 1})} \qquad .....(1.22)$$

The pointer motion for critically damped and the overdamped case is shown in the Table 1.1.

Table 1.1 Types of galvometer motion

	Range of ξ	Motion	Nature of roots m_1 and m_2	Response
1.	$\xi = 0$	Undamped	Purely complex with zero real part	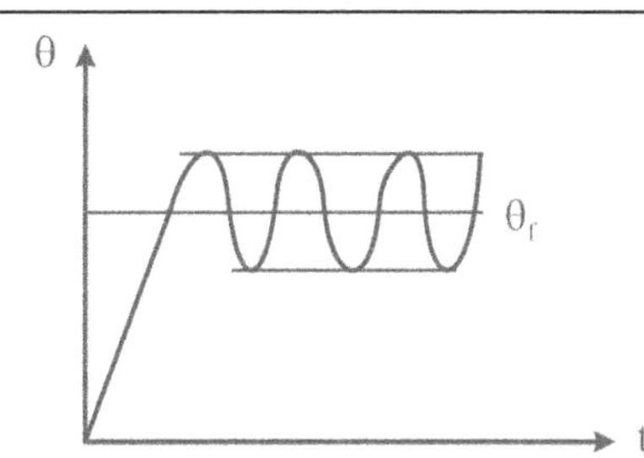 Pure oscillations

Table 1.1 *Contd...*

	Range of ξ	Motion	Nature of roots m_1 and m_2	Response
2.	$0 < \xi < 1$	Underdamped	Complex conjugate with negative real part	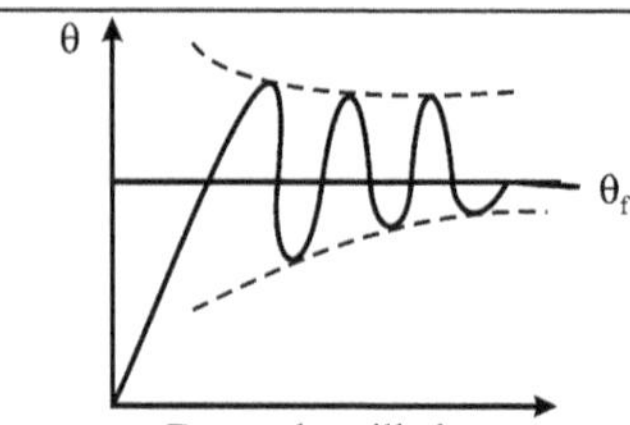 Damped oscillations
3.	$\xi > 1$	Overdamped	Real, unequal, negative	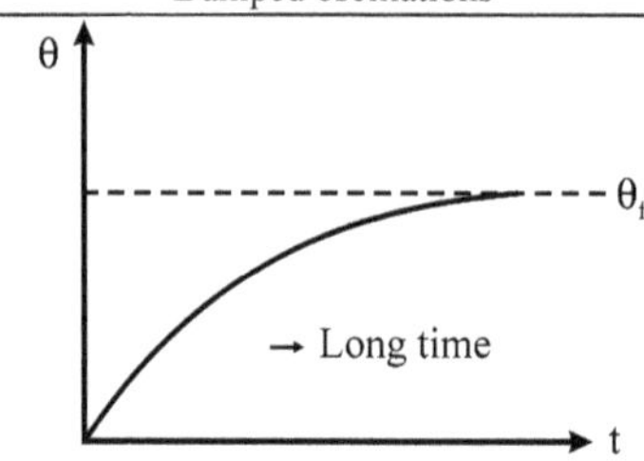 Purely exponential but very slow
4.	$\xi = 1$	Critically damped	Real, equal, negative	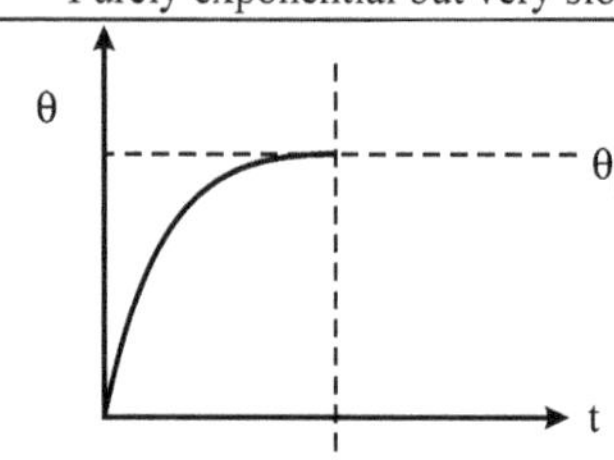 Purely exponential and quick

Key point: As this motion is slow, practically overdamping is avoided in the instruments. The critical damping is preferred practically for the instruments.

1.15 Logarithmic Decrement

Consider the underdamped galvanometer motion as shown in the Fig. 1.13.

The amount by which the pointer exceeds its final position θ_f, during first attempt is called the *First overshoot.*

This is maximum in all the over shoots existing in the transient period.

$\therefore$ θ_1 = maximum deflection at $t = t_1$

First overshoot $= \theta_1 = \theta_f$ .....(1.23)

Thus at $t = t_1$, the deflection is maximum equal to θ_1. According to maxima theorem, the time t_1 at which deflection is maximum, must satisfy $\left.\frac{d\theta}{dt}\right|_{t=t_1} = 0$.

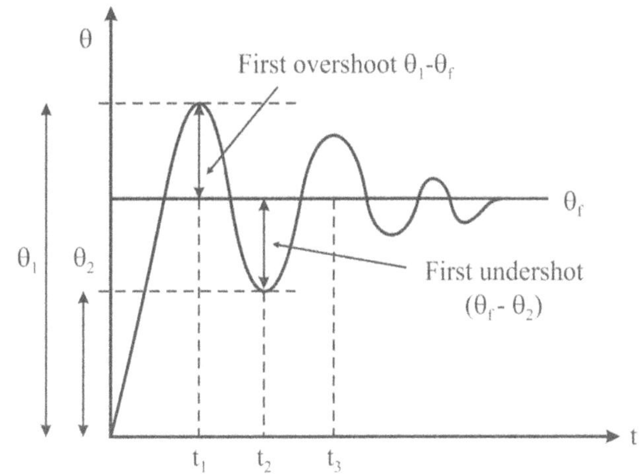

Fig. 1.13 Overshoot and undershoot in motion

Rearranging equation

$$\theta = \theta_f\left[1 - \frac{2\sqrt{JK}}{4JK - D^2} e^{-\frac{D}{2J}t} \sin(\omega_d t + \alpha)\right]$$

interms of ξ as,

$$\left[\theta = \theta_f\left[1 - \frac{1}{\sqrt{1-\xi^2}} e^{-\xi\omega_n t} \sin(\omega_d t + \alpha)\right]\right] \qquad .....(1.24)$$

where $\omega_d = \omega_n\sqrt{1-\xi^2},\ \alpha = \tan^{-1}\left[\frac{\sqrt{1-\xi^2}}{\xi}\right]$

At $t = t_1$, $\frac{d\theta}{dt} = 0$

i.e., $$\frac{d\theta}{dt} = \frac{-\theta_f}{\sqrt{1-\xi^2}}\left\{(-\xi\omega_n)e^{-\xi\omega_n t}\sin(\omega_d t + \alpha) + e^{-\xi\omega_n t}\omega_d\cos(\omega_d t + \alpha)\right\} = 0$$

As ξ, ω_d and ω_n are not zero, the above equation gets satisfied by,

$$(-\xi\omega_n)\sin(\omega_d t + \alpha) + \omega_d\cos(\omega_d t + \alpha) = 0.$$

$$\therefore \quad \tan(\omega_d t + \alpha) = \frac{\omega_d}{\xi\omega_n} = \frac{\omega_n\sqrt{1-\xi^2}}{\xi\omega_n} = \frac{\sqrt{1-\xi^2}}{\xi} = \tan\alpha$$

This equation is satisfied when $\omega_d t = n\pi$ because $\tan(n\pi + \theta) = \tan\theta$.

$$\therefore \quad t = \frac{n\pi}{\omega_d}$$

For first overshoot $n = 1$, $t = t_1$

$$\therefore \quad t_1 = \frac{\pi}{\omega_d} = \frac{\pi}{\omega_n\sqrt{1-\xi^2}} \qquad .....(1.25)$$

Putting in equation 1.24

$$\therefore \quad \theta_1 = \theta_f[1 + e^{-\pi\xi/\sqrt{1-\xi^2}}] \qquad .....(1.26)$$

Note that $\sin\,(\omega_d t_1 + \alpha)\big|_{t_1=\frac{\pi}{\omega_d}} = +\sin\alpha = \sqrt{1-\xi^2}$

$$\therefore \quad \text{First overshoot} = \theta_1 - \theta_f = \theta_f\,(e^{-\pi\xi/1-\xi^2}) \qquad .....(1.27)$$

$$\text{At} \quad n = 2 + t_2 = \frac{2\pi}{\omega_d}$$

$$\therefore \quad \theta_2 = \theta_f\left[1 - e^{-2\pi/1-\xi^2}\right]$$

$$\therefore \text{ first undershoot} = \theta_f - \theta_2 = \theta_f\left(e^{-2\pi\xi/1-\xi^2}\right) \qquad .....(1.28)$$

Taking ratio of first over and undershoots

$$\frac{\theta_1 - \theta_f}{\theta_f - \theta_2} = \frac{e^{-\pi\xi/\sqrt{1-\xi^2}}}{e^{-2\pi\xi/\sqrt{1-\xi^2}}} = e^{+\pi\xi/\sqrt{1-\xi^2}} \qquad .....(1.29)$$

$$\therefore \quad ln\left[\frac{\theta_1 - \theta_f}{\theta_f - \theta_2}\right] = \frac{\pi\xi}{\sqrt{1-\xi^2}} \qquad .....(1.30)$$

The natural logarithm of the ratio of two successive swings is called logarithmic decrement and denoted by λ.

$$\therefore \quad \lambda = ln\left[\frac{\theta_1 - \theta_f}{\theta_f - \theta_2}\right] = \frac{\pi\xi}{\sqrt{1-\xi^2}} \qquad .....(1.31)$$

$$\text{Now,} \quad T_d = \frac{2\pi}{\omega_d} = \frac{2\pi}{\omega_n\sqrt{1-\xi^2}}$$

While $$T_o = \frac{2\pi}{\omega_n}$$

$$\therefore \quad \frac{T_o}{T_d} = \sqrt{1-\xi^2} \qquad \text{.....(1.32)}$$

$$\therefore \quad \lambda = \pi\xi\left(\frac{T_d}{T_o}\right) \qquad \text{......(1.33)}$$

where T_o = time period corresponding to ω_n.

Thus equation $\theta = \theta_f\left[1 - \frac{2\sqrt{JK}}{\sqrt{4JK-D^2}} e^{-\frac{D}{2J}t} \sin(\omega_d t + \infty)\right]$ can be expressed as,

$$\theta = \theta_f\left[1 - \frac{T_d}{T_o} e^{-2\pi\xi t/T_o} \sin\left(\frac{2\pi t}{T_d} + \sin^{-1}\frac{T_o}{T_d}\right)\right] \qquad \text{.....(1.34)}$$

$$\theta = \theta_f\left[1 - \frac{\omega_n}{\omega_d} e^{-\omega_d \lambda t/\pi} \sin\left(\omega_d t + \sin^{-1}\frac{\pi\xi}{\lambda}\right)\right] \qquad \text{.....(1.35)}$$

1.16 Settling Time

The time required by the pointer to achieve the steady state when the complete transient behaviour dies out is called settling time and denoted as T_s.

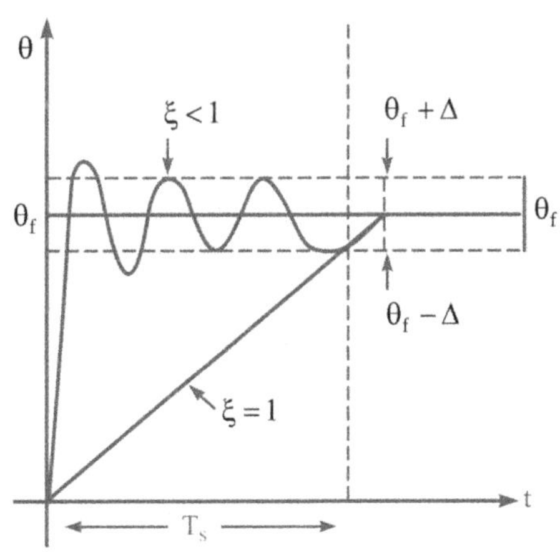

Fig. 1.14 Setting time

Practically as the exponential term is present in the equation, and is kept between 0.6 to 1 and motion is underdamped hence oscillatory. While a particular band is defined about θ_f denoted by $\pm \Delta$. When transient oscillations decrease and enter into $\theta_f \pm \Delta$ zone and remain thereafter. Within this interval then it is said that the pointer has achieved the steady state. This is shown in the Fig. 1.14.

For $\pm 2\%$ band defined from accuracy point of view,

$$T_s = \frac{4}{\xi\omega_n}$$

This is possible for ξ between 0.6 to 1.0 hence practically ξ is designed between this range.

1.17 Effect of External Resistance on Damping

The damping opposition to the motion by dissipating the energy of radiation. In galvanometer the damping is provided by two types.

1.17.1 Mechanical Damping

This is due to the friction present in the mechanical of the pointer. This is not very significant. The damping torque produced due to such mechanical objects is given by,

$$T_m = D_m \frac{d\theta}{dt}$$

where D_m = Mechanical damping constant

1.17.2 Electromagnetic Damping

This is effective damping than the mechanical damping. It is produced due to reduced effect when coil moves in a magnetic field. Thus when coil moves in a magnetic fields,

(i) The eddy currents are induced in the metal former.

(ii) The emf is induced in coil which circulates through coil

These two effects cause damping called electromagnetic damping.

Let R = Resistance of galvanometer circuit= $R_g + R_x$

where R_g = resistance of galvanometer coil

R_x = External resistance connected for damping

When coil rotates, emf is induced in it which is given by,

$$e = 2N \times Blv$$

where $$V = \frac{r}{2}\omega = \frac{r}{2}\frac{d\theta}{2t} = \text{linear velocity}$$

∴ $$e = 2NBl\frac{r}{2}\frac{d\theta}{dt} = NBA\frac{d\theta}{dt} \quad [l \times r = \text{Area A}]$$

but $$G = NBA$$

∴ $$e = G\frac{d\theta}{dt}$$

∴ $$i = \frac{e}{R} = \frac{G}{R}\frac{d\theta}{dt}$$

The torque produced due to this current flowing through the coils is

$$T_{coil} = N \times Bil \times r = NBA\, i = Gi$$

∴ $$T_{coil} = G \times \frac{G}{R}\frac{d\theta}{dt} = \frac{G^2}{R}\frac{d\theta}{dt} = D_{coil}\frac{d\theta}{dt}$$

where D_{coil} = Damping constant of coil circuit

∴ $$\left[D_{coil} = \frac{G^2}{R}\right]$$

Now let us find damping due to the metal former.

It consists of one strip i.e., N = 1.

∴ $$T_f = Bil \times r = BAi$$

Now $$BA = \frac{G}{N} \quad \text{and} \quad i = \frac{G}{N\,R_f}\frac{d\theta}{dt}$$

as N = 1 for former

where R_f = Resistance of former.

∴ $$T_i = \frac{G}{N} \times \frac{G}{N\,R_f}\frac{d\theta}{dt} = \frac{G^2}{N^2R_f}\frac{d\theta}{dt} = D_{former}\frac{d\theta}{dt}$$

where D_{former} = Damping constant of former

∴ $$\left[D_{former} = \frac{G^2}{N^2R_f}\right]$$

Total electromagnetic damping = $(D_{coil} + D_{former})\dfrac{d\theta}{dt}$

$$\therefore \qquad T_e = \left[\frac{G^2}{R} + \frac{G^2}{N^2 R_f}\right]\frac{d\theta}{dt}$$

$$T_e = De\frac{d\theta}{dt}$$

$$D_e = \frac{G^2}{R} + \frac{G^2}{N^2 R_f}$$ = (G^2/N^2Rf) damping constant due to electromagnetic damping

Total damping due to both effect is,

$$T_D = T_m + T_e = [D_m + D_e]\frac{d\theta}{dt} = D = \frac{d\theta}{dt}$$

where $D = D_m + D_e$

1.18 Critical Resistance for Damping

The mechanical damping is very small and can be neglected.

$$\therefore \qquad D = D_e = \frac{G^2}{R} + \frac{G^2}{N^2 R_f}$$

Practically, the damping due to metal former is also very small.

$$\therefore \qquad D = \frac{G^2}{R}$$

For the critical damping, $\xi = 1 \quad R = R_c$ and $D = D_c = 2\sqrt{JK}$

$$\therefore \qquad 2\sqrt{JK} = \frac{G^2}{R_c}$$

$$\therefore \qquad R_c = G^2/2\sqrt{JK}$$

$$\therefore \qquad R_x = R_c - R_g = \frac{G^2}{2\sqrt{JK}} - R_g$$

This is the value of external resistance required to adjust damping to the critcal damping. It is called external critical damping resistance (ECDR).

1.19 Sensitivity of Galvanometer

The sensitivity of galvanometer can be defined with respect to current, voltage or resistance of galvanometer. Thus there are three sensitivities associated with a galvanometer.

1.19.1 Voltage Sensitivity

It is defined as the deflection obtained in scale divising per unit voltage impressed on the galvanometer.

$$S_v = \frac{V}{i\,R_g} \text{ m / V} \qquad .....(1.36)$$

1.19.2 Current Sensitivity

It is defined as the deflection obtained per unit current.

$$\therefore \qquad S_i = \frac{\theta_f}{i} = \frac{G_i}{Ki} = \frac{G}{K} \text{ rad/H} \qquad .....(1.37)$$

Practically the value of current and deflection are very small, hence the current sensitivity is expressed in mm/μA.

$$\therefore \qquad S_i = \frac{d}{i} \quad \text{but} \quad d = 2\theta_f; \qquad r = \frac{2G_i}{K} r$$

$$\therefore \qquad S_i = \frac{2Gr}{K} \text{ m/A}$$

As usually r = 1 m = 1000 mm,

$$S_i = \frac{2000G}{K \times 10^6} = \frac{G}{500K} \text{ mm/μA}$$

$$\text{But} \qquad d = \frac{2Gir}{K} = \frac{2000Gi}{K} \text{ in mm for } r = 1 \text{ m}$$

$$\therefore \qquad S_v = \frac{2000Gi}{iKR_g \times 10^6} \text{ mm/μV}$$

$$\therefore \qquad S_v = \frac{G}{500\,K\,R_g} \text{ mm/μV} \qquad .....(1.38)$$

1.19.3 Megohm Sensitivity

It is the resistance of the circuit in megohm so that the deflection is one scale division when one volt is impressed in the galvanometer.

$$\therefore \qquad S_0 = \frac{d}{i \times 10^{-6}} \text{ M}\Omega \text{ / scale division} \qquad .....(1.39)$$

But $$d = \frac{2000\ Gi}{K}\ \text{mm for } r = 1\ \text{m},$$

∴ $$S_0 = \frac{G}{500\ K} M\Omega / \text{mm} \qquad .....(1.40)$$

For high sensitivity, G must be larger and K should be small. As G = NBA, to get high sensitivity, coil must be having more N, more cross-section area and must be placed in high flux density magnetic field. Hence the coil has large number of turns of fine were as area of cross-section cannot be increased beyond limits. The K can be decreased by using small stiffness constants springs.

1.20 Errors in Moving Coil Instruments

The common errors are accountable due to change in temperature, friction, mechanical unbalance, and operational. These are discussed in detail below.

1.20.1 Temperature Error

Apart from the change in room temperature, a change in temperature in the coil and in the spring effects the instrument reading as a whole. Increase in temperature of the coil increases its resistance, due to which the current decrease proportionately resulting in decrease in deflection. When the instrument is used as voltmeter, resistance also increases in the associated high resistance connected in series. With the effect that the deflection of the coil decreases, besides, increase in temperature of the control spring decreases stress in it. With the result that reading shows an error. To some extent, the error is minimised by winding the series resistance coil of material of very small temperature co-efficient of resistance. Hence the instrument has a temperature range of working.

1.20.2 Mechanical Unbalance

The efficiency of the moving system is high only when the weight of the moving system is well balanced and acts on that Jewel disturbed, bearing axially. Under any circumstance of use if the balance is destroyed, the friction between the pivot and the jewel bearing dominates, resulting decrease in deflection. Besides an unbalanced coil will have sluggish movement due to which the torque produced is also affected. The effect of these is the decrease in deflection. To overcome this the following precautions are taken during use.

1. The instrument should not be given any jerk during use.
2. As far as possible the instruments should be used in a position which keeps the spindle vertical.
3. The instrument should not be operated near a heater or other heating appliances.

1.20.3 Frictional Error

The mechanical frictional force acting at the pivots oppose the deflecting force. At the time of initial graduation of the sale this is taken into account but as time elapses, the frictional surface gets effected and friction increases due to which errors in the deflection results. These errors are minimised by adopting a moving system of light construction, mounted on jewelled bearing and keeping the moving system dust proof.

1.20.4 Observational Error

This error is due to misreading of the scale/parallax in reading and error estimation. Errors due to parallax is eliminated by providing a mirror below the pointer under the slit provided in the scale.

1.20.5 Ageing

Constant use of meters and ageing affects the stress in the control spring due to which the controlling torque decreases and the deflection increases. This amounts to failure of control mechanism which is rectified by replacing the control spring with new one and the instrument recalibrated.

A meter is said to have 100% accuracy when the reading of the pointer and the actual reading are equal. The error in the meter is said to be zero. If the error exists, the level of accuracy falls to that extent. Normally for a good design, the error percentage lies between $\pm$ 0.05% to $\pm$ 5% for full scale deflection.

An instrument which takes least current for the full scale deflection is said to be more sensitive. The ampere turns of the moving coil being constant for a set design, the least current for full scale deflection can be obtained by having higher number of turns. Another point to be kept in mind is the torque weight ratio. The least weight of the moving system will have higher torque therefore, increasing the number of turns means increase in weight. Hence, there should be a compromise between the number of turns in the moving coil and torque weight ratio, for an economical result. Sensitivity is normally measured in ohm/volt of full scale deflection. Higher is the value, higher is the sensitivity of the instrument. A 20000 ohms/volt instrument is more sensitive than a 1000 Ω/volt meter.

1.21 Moving Iron Instrument

The most commonly used ammeter and voltmeters in laboratories and switch board at commercial frequencies are moving iron instrument. These instruments are cheap, robust, and can be manufactured to reasonable accuracy. The operating principle is simple. It consists of a vane of soft iron of high permeability steel which forms the moving system of the instrument along with the pointer, spring, pivots and damping device. The vane is so situated that it gets attracted or repelled from by a magnetic force in a solenoid carrying current. While doing so, it deflects proportional to the quantity being measured. Thus there are two general types of moving iron instruments.

1. Attraction type
2. Repulsion type

The attraction type instruments operate on the principle of operation of single piece of soft iron into a magnetic field and repulsion type instruments operate on the principle of repulsion of two adjacent iron pieces magnetised by the same magnetic field. In both types of these instruments, the current to be measured or a definite fraction of the current to be measured proportional to the voltage to be measured it passed through a coil of wire. This current carrying coil sets up the necessary field. A certain number of ampere-turns is required for the operation of the instrument and this number can be made up either having a few turns and large current vice-versa. The instrument to be used as an ammeter is provided with a coil of few turns of thick wire in order to have low resistance and carry large current, and that to be used as a voltmeter is provided with a coil of large number of turns of fine wire in order to have high resistance and draw as small current as possible.

1.21.1 Attraction (or single-iron) Type M.I Instrument

Section of view of attraction type moving iron instrument is shown in Fig. 1.16. It consists essentially of a coil or solenoid and oval shaped iron pivoted in such a way that it can move in or out of the solenoid. To this iron a pointer is attached so that it may deflect along with the moving iron over a graduated scale. The iron is made of sheet metal specially shaped to give a scale as nearly uniform as possible. When the current to be measured (or a definite fraction of the current to be measured or propogational to the voltage to be measured) is passed through the solenoid, a magnetic field is set up inside the solenoid which in turn magnetises the iron. Thus the iron is attracted into the coil, causing the spindle and the pointer to rotate. The design of the shape of the iron is largely by trial. Such instruments normally have spring control an pneumatic damping. In Fig. 1.15 gravity control attraction type instrument is shown. This has been superceded by spring control instruments. These are better than the gravity control instruments as these can be used in any position. Hence M.I instrument can be used for both AC & DC measurements. Due to square law response, the scale of the moving iron instrument is non-uniform.

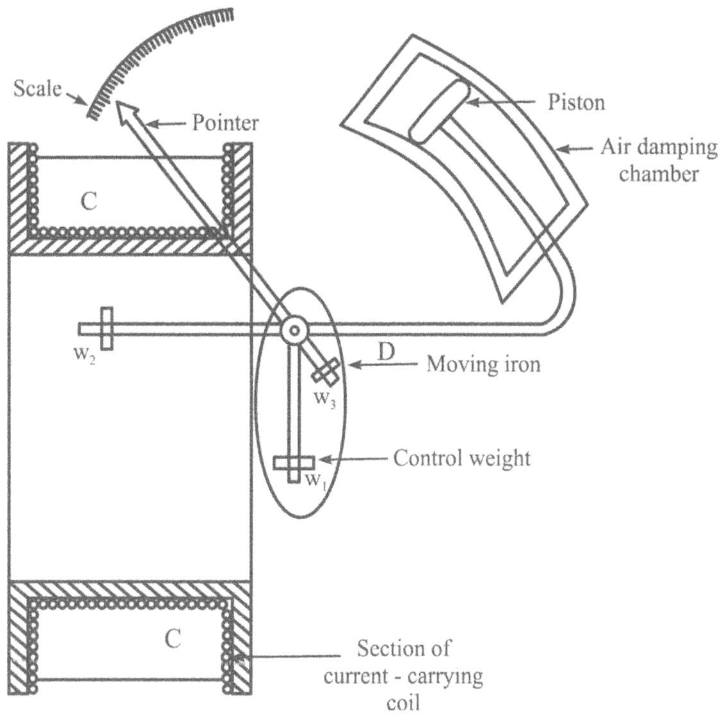

Fig. 1.15 Moving iron attraction type instrument

1.21.2 Torque Equation of Moving Iron Instruments

Consider a small increment in current supplied to the coil of the instrument (Fig. 1.16). Due to this current let $d\theta$ be the deflection under the deflecting torque T_d, due to such deflection some mechanical work will be done,

$$\therefore \quad \text{mechanical work} = T_d \, d\theta$$

There will be a change in the energy stored in the magnetic field due to the change in inductance. This is because the vane tries to occupy the position of minimum reluctance hence the force is always in such a direction so as to increase the inductance of coil. The inductance is inversely proportional to the reluctance of the magnetic circuit of coil.

Let I = Initial current

L = Instrument inductance

θ = Deflection

dI = Increase in current

$d\theta$ = Change in deflection

dL = Change in inductance

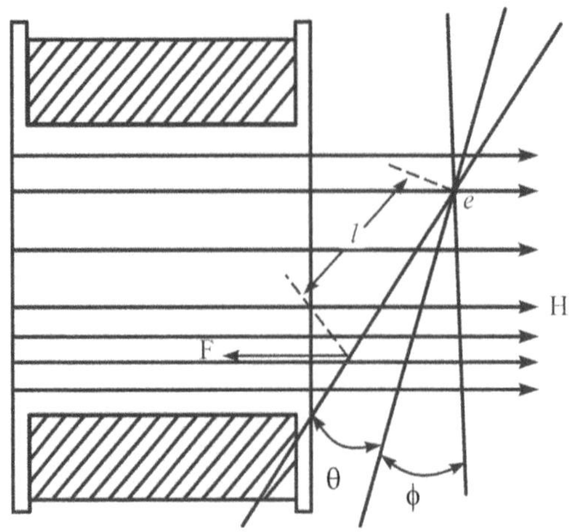

Fig. 1.16

In order to effect an increment dI in the current, there must be an increase in the applied voltage given by,

$$e = \frac{d(LI)}{dt} = I\frac{dL}{dt} + L\frac{dI}{dt} \quad \text{as both I and L are changing}$$

The electrical energy supplied is given by,

$$eIdt = \left(I\frac{dL}{dt} + L\frac{dI}{dt}\right)Idt$$

$$= I^2dL + IL\,dI$$

The stored energy increases from $\frac{1}{2}LI^2$ to $\frac{1}{2}(L + dL)(I + dI)^2$

Hence the change in the stored energy is given by,

$$= \frac{1}{2}(L + dL)(I + dI)^2 - \frac{1}{2}LI^2$$

Neglecting higher order terms, this becomes,

$$IL\,dI + \frac{1}{2}I^2dL$$

The energy supplied is nothing but increase in stored energy plus the energy required for mechanical work done.

$$\therefore \quad I^2dL + ILdI = ILdI + \tfrac{1}{2}I^2dL + T_d.d\theta$$

$$T_d.d\theta = \frac{1}{2} I^2 dL$$

$$\therefore \quad \boxed{T_d = \frac{1}{2} I^2 \frac{dL}{d\theta}}$$

While the controlling torque is given by,

$$T_c = K\theta$$

where K = Spring constant

$$K\theta = \frac{1}{2} I^2 \frac{dL}{d\theta} \quad \text{under equilibrium}$$

$$\therefore \quad \theta = \frac{1}{2} \frac{I^2}{K} \frac{dL}{d\theta}$$

Thus the deflection is proportional to the square of the current through the coil and the instrument gives square law response.

Ranges:

1. Ammeter from about 0.02 A to 0.800 A maximum without current transformer.
2. Voltmeters from about 0.1 V to 0.800 V maximum without potential transformer.

1.22 Moving Iron Repulsion Type Instrumcnt (Double-Iron) Type

These instruments have two vanes inside the coil, the one is fixed and other is movable. When the current flows in the coil, both the vanes are magnetised with like polarities induced on the same side. Hence due to the repulsion of like polarities, there is a force of repulsion between the two vanes causing the movement of the moving vane. The repulsion type instruments are the most commonly used instruments.

The two different designs of repulsion type instruments are:

1. Radial vane type and 2. Co-axial vane type

1.22.1 Radial Vane Repulsion Type Instrument

The Fig. 1.17 shows the radial vane repulsion type instrument. Out of the other moving iron mechanisms, this is the most sensitive and has most linear scale. The two vanes are a radial strips of iron. The fixed vane is attached to the coil, while the movable vane is attached to the spindle and suspended in the induction field of the coil. The needle of the instrument is attached to this vane.

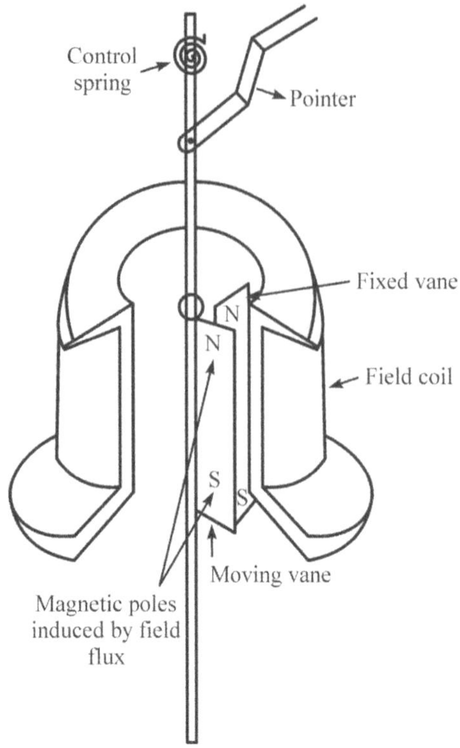

Fig. 1.17 Radial vane repulsion type instrument

Even though the current through the coil is alternating there is always repulsion between the like poles of the fixed and the movable vane, hence the deflection of the pointer is always in the same direction. The deflection is effectively proportional to the actual current and hence the scale is calibrated directly to read amperes or volts. The calibration is accurate only for the frequency for which it is designed because the impedance is different for different frequencies.

1.22.2 Concentric Vane Repulsion Type Instrument

The Fig. 1.18 shows concentric vane repulsion type instrument. The instrument has two concentric vanes. One is attached to the coil frame rigidly while the other can rotate coaxially inside the stationary vane. Both the vanes are magnetised to the same polarity due to the current in the coil. Thus the movable vane rotates under the repulsive force. As the movable vane is attached to the pivoted shaft, the repulsion results in a rotation of the shaft. The pointer deflection is proportional to the current in the coil. The concentric vane type instrument is moderately sensitive and the deflection is proportional to the square of the current through coil. Thus the instrument is said to have square law response. The scale of the instrument is non-uniform in nature. Thus whatever may be the direction of the current in the coil, the deflection in the moving iron instruments is in the same direction. Hence moving iron instruments can be used for both A.C and D.C measurements. Due to square law response, the scale of the moving iron instrument is non-uniform.

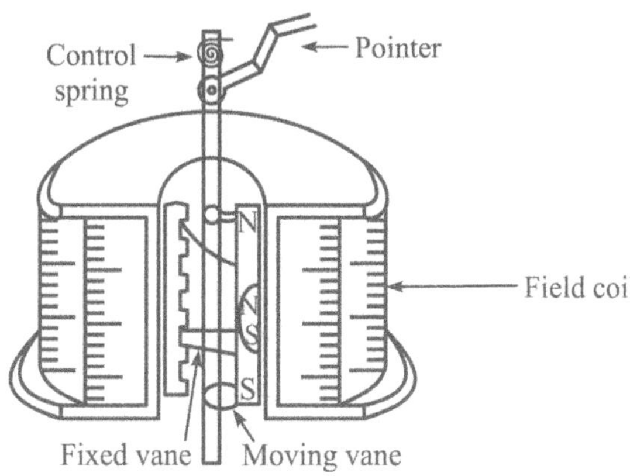

Fig. 1.18 Concentric vane repulsion type instrument

1.23 The advantages and disadvantages of M.I. Instrument

1.23.1 Advantages

1. It can be used on both A.C. and D.C supply
2. The instrument is robust, simple in design and free from maintenance
3. It possesses high starting torque
4. There are no current carrying parts in the moving system hence these meters are extremely rugged and reliable
5. These are capable of giving good accuracy. Modern moving iron instruments have a d.c error of 2% or less
6. It can withstand momentary overloads
7. It can give reasonable accuracy within the limits of both precision and industrial grades

1.23.2 The main disadvantages are

1. Scale is not uniform throughout
2. There are serious errors due to hysteresis, frequency changes and stray magnetic fields
3. Due to the non linearity of B-H curve, the deflecting torque is not exactly proportional to the square of the current

4. The stiffness of the spring decreases with increase in temperature
5. Change in frequency of operation causes serious error.
6. Power consumption at low voltage is high

1.24 Errors in M.I Instrument

There are two types of errors which occur in moving iron instrument. Errors which occur with both A.C and D.C and the other which occur only with A.C only.

Errors with both D.C and A.C

1.24.1 Hysteresis Error

This error occurs as the value of flux density is different for the same current for ascending and descending values. The value of flux density is higher for descending value of current (and voltage) than for ascending values. This error can be minimized by making the iron parts small so that they demagnetize themselves quickly. Another matched is to work the iron parts at low values of flux density so that the hysteresis effects are small.

Hysteresis may produce a 2 to 3% error. With the use of nickel iron alloys with narrow hysteresis loops, the error may be brought down to less than 0.05%.

1.24.2 Stray Magnetic Fields

The error due to stray magnetic fields (fields other than the operating magnetic field) may be appreciable. As the operating magnetic field is weak (about 0.006 to 0.0075 ub/m^2 at full scale deflection) and hence can be easily distorted. Such errors depend upon the direction of the stray magnetic field relative to the field of the instrument. These errors can be minimized by using an iron case or a thin iron shield over the working parts.

1.24.3 Temperature Error

The effect of temperature changes on moving iron instrument arises chiefly from the temperature coefficient of spring. The error may be 0.02% per °C change in temperature. In voltmeters, error are caused due to self-heating of coil and series resistance. The temperature of the coil may increase by 10 to 20°C for a power consumption of 1 W. The resistance increases (by about 4 to 8%), causing a decrease in current for a given voltage. This produces a decreased deflection. Therefore the series resistance should be made a material like manganin which has a small temperature co-efficient. The value of series resistance should be very large as compared with the coil resistance in order to minimize errors due to self-heating. In the case of switch board instruments, the series resistance is about 10 times the coil resistance.

1.25 Errors with A.C only

1.25.1 Frequency Error

Changes in frequency may cause errors due to changes of reactance of the working coil and also due changes of magnitude of eddy current set up in the metal parts of instrument.

1.25.2 Reactance of Instrument Coil

The change of reactance of the instrument coil is important in case of voltmeters where an additional resistance R_s is used in series with the instrument coil. Let the resistance and inductance of the instrument coil be R and L. Then the current I in the instrument coil for a given applied voltage 'V' is given by:

$$I = \frac{V}{\sqrt{(R+R_S)^2 + \omega^2 L^2}}$$

The deflection of the M.I voltmeter depends upon the current through the coil. Therefore the deflection for a given voltage will be less at high frequencies than at low frequencies. To some extent compensation to this type of error is possible by connecting a capacitor 'C' across the series resistance R_s as shown in Fig. 1.19.

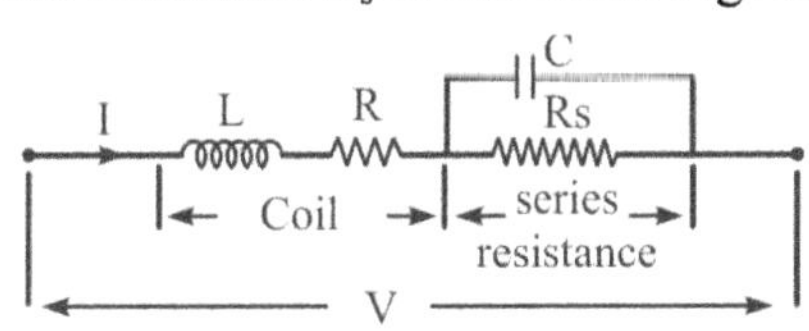

Fig. 1.19 Frequency compensation for M.I voltmeters

The idea of shunting the series resistance is to make the circuit behave like a pure resistance so that the frequency changes have no effect on the readings of the instrument. Thus the compensated instrument will have a p.f nearly equal to unity. As the resistance of the meter, R is considerably smaller than the series multiplier resistance R_s, it can be assumed that it is sufficient to ensure that the magnitude Z of the impedance of the circuit formed by L, R_s and C has a value R_s when used on A.C

Now $$Z = j\omega L + \frac{R_s}{1 + j\omega C R_s} = j\omega L + \frac{R_s - j\omega C R_s^2}{1 + \omega^2 C^2 R_s^2}$$

$$Z^2 = R_s^2 (1 - \omega^2 C^2 R_s^2)^2 + \omega^2 (L - C R_s^2)^2$$

This must equal R_s^2 in order that the a.c calibration at all frequencies and d.c calibration is the same.

$$R_s^2 = R_s^2 (1 - \omega^2 C^2 R_s^2)^2 + \omega^2 (L - C R_s^2)^2$$

$$L^2 - 2LC\,R_s^2 - C^2 R_s^4 = 0 \qquad \text{(or)} \qquad L = 2.41\,C\,R_S^2$$

$$C = \frac{1}{2.41}\frac{L}{R_s^2} = 0.41\frac{L}{R_s^2}$$

$$\therefore \quad C = 0.41\frac{L}{R_s^2}$$

It should be understood that the above analysis is valid for limited range of frequency which in practical cases is up to 125 Hz.

1.25.3 Eddy Current Error

When instrument is used for a.c measurement, the eddy currents are produced in the iron parts of the instrument. The eddy current affects the instrument current causing the change in the deflecting torque. This produces the error in the meter reading. As eddy currents are frequency dependent, frequency change cause eddy current error.

1.25.4 Hysteresis Error

Due to hysteresis effect, the flux density for the same current while ascending and descending values is different. While descending, the flux density is higher and while ascending it is lesser. So meter reads higher for descending value of current or voltage. So remedy for this is to use smaller iron parts which can demagnetise quickly or to work with lower flux densities.

1.26 Comparison between moving Coil and moving Iron Instrument

	Moving Coil		Moving Iron
1.	Coil moves in the magnetic field	1.	Soft iron moves in the magnetic field
2.	Deflecting torque is proportional to current	2.	Deflecting torque is proportional to square of the current
3.	Damping is provided by eddy current	3.	Damping is provided by air friction
4.	Spring controlled instrument	4.	Gravity controlled instrument
5.	Controlling torque is proportion to the angle of deflection	5.	Controlling torque is proportion to sin θ
6.	Scale is uniform	6.	Non-uniform scale, cramped at the beginning and at end
7.	Delicate, sensitive and accurate	7.	Robust, reasonable accurate
8.	Costly	8.	Cheap
9.	Low power consumption	9.	Power consumption is higher than moving coil
10.	It is used in D.C circuit	10.	It can be used in both A.C and D.C circuit
11.	Can be used as voltmeter, Ammeter, Galvanometer and ohmmeter	11.	Can be used as ammeter, voltmeter and wattmeter.

1.27 Basic D.C Ammeter

The basic d.c ammeter is nothing but a D′Arsonval galvanometer. The coil winding of a basic meter is very small and light and hence it can carry very small currents. So as mentioned earlier, for large currents, the major part of current is required to be by passed using a resistance called "shunt". It is shown in the Fig. 1.20.

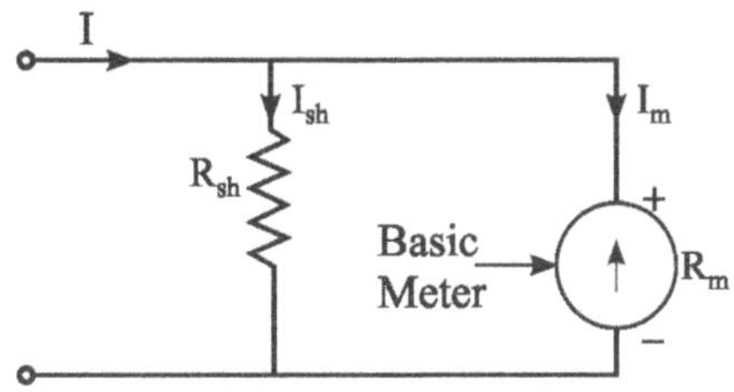

Fig. 1.20 Basic d.c ammeter

The shunt resistance can be calculated as:

Let R_m = internal resistance of coil

R_{sh} = shunt resistance

I_m = full scale deflection current

I_{sh} = shunt current

I = total current

$I = I_{sh} + I_m$

As the two resistances R_{sh} and R_m are in parallel, the voltage drop across them is same.

$$\therefore \quad I_{sh} R_{sh} = I_m R_m \qquad \therefore \quad R_{sh} = \frac{I_m R_m}{I_{sh}}$$

but $I_{sh} = I - I_m$

$$\therefore \quad R_{sh} = \frac{I_m R_m}{I - I_m} \qquad \therefore \quad R_{sh} = \frac{R_m}{(m-1)}$$

where, $m = \frac{I}{I_m}$

The m is called "multiplying power" of the shunt and defined as the ratio of total current to the current through the coil. It can be expressed as,

$$m = \frac{I}{I_m} = 1 + \frac{R_m}{R_{sh}}$$

The shunt resistance may consist of a constant temperature resistance wire within the case of the meter or it may be external shunt low resistance.

Thus to increase the range of ammeter 'm' times, the shunt resistance required is $\frac{1}{(m-1)}$ times the basic meter resistance. This is nothing but "extension of ranges of an ammeter".

1.28 Multi Range Ammeters

The range of the basic d.c ammeter can be extended by using number of shunts and a selector switch. Such a meter is called "Multi range ammeter" and shown in the Fig. 1.21.

R_1, R_2, R_3 and R_4 are four shunts. When connected in parallel with the meter, they can give four different ranges I_1, I_2, I_3 and I_4. The selector switch S is multi position switch, having low contact resistance and high current carrying capacity. The make before break type switch is used for the range changing. If the ordinary switch is used while range changing, the switch remains open and full current passes through the meter. The meter may get damaged due to such high current. So make before break switch is used. The design of such switch is so that it makes contact with next terminal before completely breaking the contact with previous terminal. The multi range ammeters are used for the ranges up to 50 A. While using the multi range ammeter, highest range should be used first and thus the current range should be decreased till good upscale reading is obtained. All the shunts are very precise resistance and hence cost of such multi range ammeter is high.

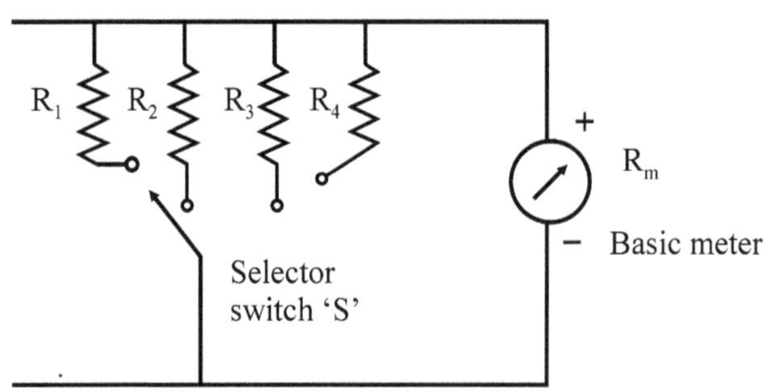

Fig. 1.21 Multi-range ammeter

The mathematical analysis of basic d.c ammeters equally applicable to such multi range ammeter. Thus,

$$R_1 = \frac{R_m}{m-1}$$

$$R_2 = \frac{R_m}{m_2 - 1} \text{ at so on}$$

where m_1, m_2, m_3 .... are the multiplying powers for the currents I_1, I_2, I_3 ....

1.29 The Aryton Shunt or Universal Shunt

We have seen that in multi range ammeter a make before break switch is a must. The aryton shunt or the universal shunt eliminates the possibility of having a meter without a shunt. The meter with the aryton shunt is shown in the Fig. 1.22.

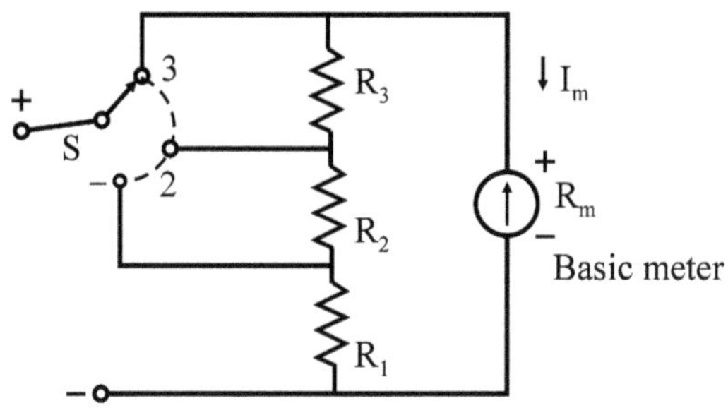

Fig. 1.22 Ammeter with aryton shunt

The selector switch S, selects the appropriate shunt required to change the range of the meter. When the position of the switch is at "1", then the resistance R_1 is in parallel with the series combination of R_2, R_3 and R_m, hence current through the shunt is more than the current through the meter, thus protecting the basic meter. When the switch is in the position '2', then the series resistance of R_1 and R_2 are in parallel with the series combination of R_3 and R_m; the current through the meter is more than through the shunt in this position. In the position '3' the resistance R_1, R_2 and R_3 are in series and acts as the shunt. In this position, the maximum current flows through the meter. This increases the sensitivity of the meter.

The voltage drop across the two parallel branches is always equal.

Thus, $$I_{sh} R_{sh} = I_m R_m$$

but in position '1', R_1 is in parallel with $R_2 + R_3 + R_m$

$$\therefore \quad I_1 [R_1] = I_m [R_2 + R_3 + R_m] \quad .....(1.41)$$

where, I_1 is the first range required

In position 2, $R_1 + R_2$ is in parallel with $R_3 + R_m$

$$\therefore \quad I_2 (R_1 + R_2) = I_m (R_3 + R_m) \quad .....(1.42)$$

where, I_2 is the second range required

In position 3, $R_1 + R_2 + R_3$ is in parallel with R_m.

$$\therefore \quad I_3 (R_1 + R_2 + R_3) = I_m R_m \quad .....(1.43)$$

where, I_3 is the third range required.

The current range I_3 is the minimum while I_1 is the maximum range possible. Solving the equations (1.41), (1.42) and (1.43) the required Aryton shunt can be designed.

1.30 Requirement of a Shunt

1. The shunt resistance should be stable and cast out with time.
2. The shunt resistance should not carry currents which will cause excessive temperature rise.
3. The temperature co-efficient of shunt and the meter should be low and should be as equal as possible.
4. The type of material used to join the shunts should have low thermo dielectric voltage drop i.e., the soldering of joints should not cause a voltage drop.
5. The resistance should have low thermal electromotive force with copper.
6. Due to the soldering, the values of resistance should not be changed.

The "MANGANIN" is usually used for the shunts of d.c instruments while the constant is useful for the shunt of a.c instruments.

1.31 Precautions to be taken while using an Ammeter

1. While using multi range ammeter, first use the highest current range and then decrease the current range until sufficient deflection is obtained. So to increase the accuracy, finally select the range which will give the reading near full scale deflection.
2. The polarities must be observed correctly. The opposite polarities deflect the pointer in opposite direction against the mechanical stop and this may damage the pointer.
3. As the ammeter resistance is very low, it should never be connected across any source of e.m.f. Always connect an ammeter in series with the load.

1.32 Basic D.C Voltmeter

The basic D.C. voltmeter is nothing but a PMMC D' Arsonval galvanometer. The resistance is required to be connected in series with the basic meter to use it as a voltmeter. This series resistance is called a multiplier. The main function of the multiplier is to limit the current through the basic meter so that the meter current does not exceed the full scale deflection value. The voltmeter measures the voltage across the two points of a circuit or a voltage across a circuit component. The basic d.c. voltmeter is shown in Fig. 1.23.

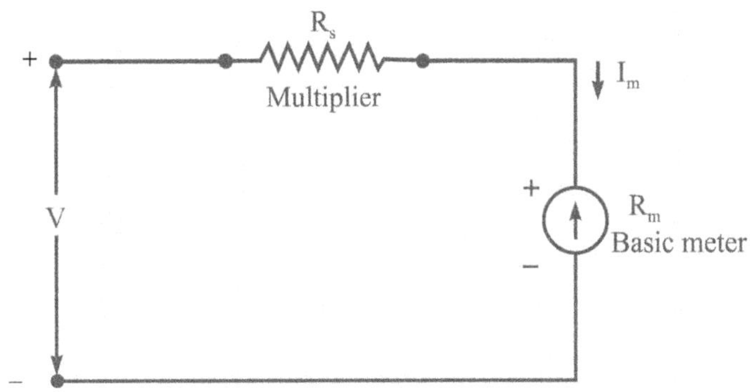

Fig. 1.23 Basic D.C voltmeter

The voltmeter must be connected across the two points or a component to measure the potential difference, with the proper polarity.

The multiplier resistance can be calculated as,

Let R_m = Internal resistance of meter

R_s = series multiplier resistance

I_m = full scale deflection current

V = full range voltage to be measured

From Fig. 1.23

$$\therefore \quad V = I_m (R_m + R_s)$$

$$V = I_m R_m + I_m R_s$$

$$\therefore \quad I_m R_s = V - I_m R_m$$

$$\therefore \quad R_s = \frac{V}{I_m} - R_m$$

The "multiplying factor" for multiplier is the ratio of full range voltage to be measured and the drop across the basic meter.

Let v = drop across the basic meter = $I_m R_m$

$$m = \text{multiplying factor} = \frac{V}{v} = \frac{I_m (R_m + R_s)}{I_m R_m}$$

$$\therefore \quad m = 1 + \frac{R_s}{R_m}$$

Hence multiplier resistance can also be expressed as,

$$R_s = (m - 1) R_m$$

Thus to increase the range of voltmeter 'm' times, the series resistance required is (m – 1) times the basic meter resistance. This is nothing but "extension of ranges of a voltmeter".

1.32.1 Multirange Voltmeter

The range of the basic d.c voltmeter can be extended by using number of multipliers and a selector switch. Such a meter is called "multi range voltmeter" and is shown in the Fig. 1.24.

The R_1, R_2, R_3 and R_4 are the four series multipliers.

When connected in series with meter, they can give four different voltage range as V_1, V_2, V_3 and V_4. The selector system 'S' is multi position switch by which the required multiplier can be selected in the circuit.

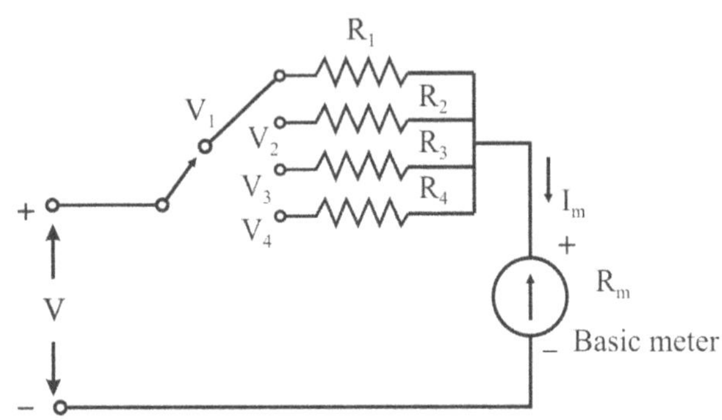

Fig. 1.24 Multirange voltmeter

The mathematical analysis of basic d.c voltmeter is equally applicable for such multi-range thus,

$$R_1 = \frac{V_1}{I_m} - R_m,\ R_2 = \frac{V_2}{I_m} - R_m \text{ and so on}$$

1.32.2 Practical Multirange Voltmeter

In this arrangement (Fig. 1.25) the multipliers are connected in a series string, the connections are brought out from the junctions of the resistance. The selector switch is used to select the required voltage range.

When the switch S is at position V_1, $R_1 + R_2 + R_3 + R_4$ acts as a multiplier resistance. When the switch S is at position V_4 is the lowest voltage range while 'V_1' is the maximum voltage range.

More practical arrangement of multiplier resistance is shown in the Fig. 1.25.

The multiplier resistance can be calculated as:

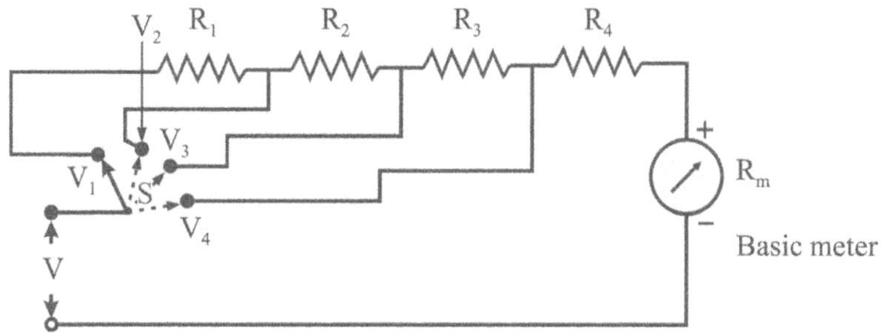

Fig. 1.25

In position V_4, the multiplier is R_4 only, the total resistance of the circuit is say RT.

$$\therefore \quad R_T = \frac{V_4}{I_m} \qquad R_4 = R_T - R_m \qquad .....(1.44)$$

In position V_3, the multiplier is $R_3 + R_4$

$$R_T = \frac{V_3}{I_m}$$

$$R_3 + R_4 = R_T - R_m \qquad R_3 = R_T - (R_m + R_4) \qquad .....(1.45)$$

on position V_2, the multiplier is $R_2 + R_3 + R_4$

$$\therefore \quad R_T = \frac{V_2}{I_m}, \qquad \therefore \quad (R_2 + R_3 + R_4) = R_T - R_m$$

$$\therefore \quad R_2 = R_T - (R_m + R_3 + R_4) \qquad .....(1.46)$$

In position V_1, the multiplier is $R_1 + R_3 + R_4$ $\therefore$ $R_T = \frac{V_1}{I_m}$

$$\therefore \quad R_1 + R_2 + R_3 + R_4 = R_T - R_m$$

$$R_1 = R_T - (R_m + R_2 + R_3 + R_4) \qquad .....(1.47)$$

using the equations (1.44), (1.45), (1.46) and (1.47) multipliers can be designed. The advantage of this arrangement is that the multiplier except R_4 have standard resistance value and can be obtained commercially in precision tolerances. The first resistance i.e., R_4 only is the resistance having special value and must be manufactured specially to meet the circuit requirements.

1.33 Sensitivity of Voltmeters

In a multi range voltmeter, the ratio of the total resistance R_T to the voltage range remains same. This ratio is nothing but the reciprocal of the full scale deflection current of the meter i.e., $1/I_m$. This value is called "sensitivity" of the voltmeter.

Thus the sensitivity of the voltmeter is defined as

$$S = \frac{1}{\text{full scale deflection current}}$$

$$S = \frac{1}{I_m} \frac{\Omega}{V} \frac{K\Omega}{V} \quad \text{(ohm/V or kilo ohm/V)}$$

Note: The sensitivity range is specified on the meter dial and it indicates the resistance of the meter for a one volt range.

The internal resistance of the voltmeter is not the same in each of its ranges. The higher is the range of the voltmeter, greater is its internal resistance. Internal resistance of a voltmeter can be obtained from its sensitivity as,

Internal resistance of voltmeter = maximum voltage (range) × sensitivity in Ω / V

The sensitivity is useful in calculating the resistance of a multiplier in d.c. voltmeter.

Consider the practical multi range voltmeter circuit shown in the Fig. 1.26.

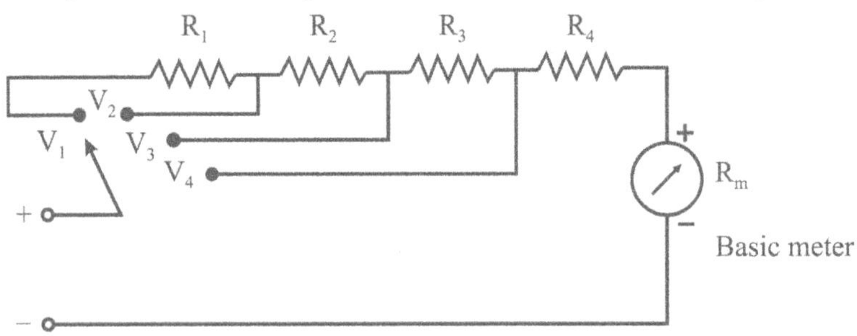

Fig. 1.26 Practical multi range voltmeter

S = sensitivity rating in Ω/V

R_m = Internal resistance of basic meter or coil.

Then the multiplier resistance can be obtained as

$$R_1 = SV_1 - (R_m + R_2 + R_3 + R_4)$$

$$R_2 = SV_2 - (R_m + R_3 + R_4)$$

$$R_3 = SV_3 - (R_m + R_4)$$

$$R_4 = SV_4 - R_m$$

where V_1, V_2, V_3, and V_4 are the required voltage ranges.

[**Note:** this method is called the "sensitivity method" of calculating the multiplier resistance.]

1.34 Loading Effect

While selecting a meter for a particular measurement, the sensitivity rating is very important. A low sensitive meter may give the accurate reading in low resistance circuit but will produce totally inaccurate reading in high resistance circuit.

The voltmeter is always connected across the two parts between which the potential difference is to be measured. It is connected across a low resistance then as voltmeter in high, most of the current will pass through a low resistance and will produce the voltage drop which will be nothing, but the true reading, but if the voltmeter is connected across the high resistance then due to two high resistance in parallel, the current will divide almost equally through two paths. Thus the meter will record the voltage drop across the high resistance which will be much lower than the true reading.

Thus the low sensitivity instrument when used in high resistance circuit gives a lower reading than the true reading. This is called "LOADING EFFECT" of the volt meter. It is mainly cause due to low sensitivity instruments.

1.35 Requirements of a Multiplier

1. The change in their resistance with temperature should be small.
2. They should be non-inductively wound for a.c. meters.
3. Their resistance should not change with time.

Commonly used resistance materials for construction of a multiplier are mangenin and constantan.

1.36 Precautions to be taken while using a Voltmeter

The following precautions must be taken while using a voltmeter:

1. While using the multi range voltmeter, first use the highest range and then decrease the voltage range until the sufficient deflection is obtained.
2. Take care of the loading effect. The effect can be minimised by using high sensitivity voltmeters.
3. The polarities must be observed correctly. The wrong polarities deflect the pointer in the opposite direction against the mechanical stop and this may damage the pointer.
4. The voltmeter resistance is very high and it should always be connected across the circuit or component whose voltage is to be measured.

1.37 Ammeter and Voltmeter

The meters which connected in series with the circuit whose current is to be measured are called "ammeter". The power loss in ammeter is $I^2 R_a$. Where R_a is ammeter resistance. To have low power loss, ammeter resistance must be very low.

The meters which are connected in parallel with the circuit whose voltage is to be measured are called "voltmeters". The power loss in voltmeter is V^2/R_v where R_v is voltmeter resistance. To have low power loss, voltmeter resistance must be very high.

The construction and working principle of both the meters is same. Both are basically current sensing devices but they have following difference.

Ammeter		Voltmeter	
1.	It is a current measuring device which measures current through circuit	1.	Which measure potential difference between the two points of a circuit.
2.	Always connected in series with circuit.	2.	Always connected in parallel with the circuit.
3.	The resistance is very small.	3.	The resistance is, very high.
4.	Deflection torque is produced by current to be measured directly.	4.	Deflecting torque is produced by a current which is proportional to the voltage to be measured.

Electrostatic Instruments

Basically electrostatic instruments are all are voltmeters. Practically such instruments may be used for measurement of current and power but both the types of measurements require measurement of voltage across a known independence. The main advantage of such instruments is the measurement of high voltage in both A.C and D.C circuits without any errors due to eddy current losses and hysteresis.

Force and Torque equation of Electrostatic Instruments

The electrostatic instruments is based on the operation principle that there exists a force between the two plates with opposite charges; this force can be obtained using the principle that the mechanical work done is equal to the stored energy if there is a relative motion of Plates, the principle of electrostatic voltmeter is as shown in the Fig. 1.27.

Consider two plates A and B where Plate A is fixed and plate B is movable. The plates are oppositely charged and are restrained by a spring connected to the fixed point. Let potential difference of 'V' volt be applied to the plates; then a force of attraction F-newton exists between them. Plate B moves towards A until this force is balanced by that of the spring. The capacitance between the plate is then C-farad and the stored energy E is

$$E = \frac{1}{2}CV^2 \text{ Joule} \qquad .....(1.48)$$

When applied voltage increases by dV, the current flowing through capacitance also changes and θt is given by.

$$i = \frac{dQ}{dt} = \frac{d}{dt}(CV) \qquad [\because Q = CV] \qquad .....(1.49)$$

$$I = C\frac{dV}{dt} + V\frac{dC}{dt}$$

The input electrical energy is given by

$$E_i = VI\ dt$$

$$= V \times \left(C\frac{dV}{dt} + V\frac{dC}{dt}\right)dt$$

$$= CV\ dV + V^2\ dC \qquad .....(1.50)$$

and also to the change in applied voltage by value dV, the capacitance increase by dC because plate B moves towards a fixed plate A. Which decreases the distance of separation between two plates increasing net capacitance.

The new energy stored is given by

$$E' = \left(\frac{1}{2}\ C + dC\right)(V + dV)^2 \qquad .....(1.51)$$

The change in stored energy is given by

$$E' - E = \frac{1}{2}(C + dC)(V + dV)^2 - \frac{1}{2}CV^2$$

$$= \frac{1}{2}(C + dC)\left(V^2 + 2V\ dV + dV^2\right) - \frac{1}{2}CV^2$$

$$= \frac{1}{2}CV^2 + CV\,dV + \frac{1}{2}CdV^2 + \frac{1}{2}V^2dC + 2VdVdC + \frac{1}{2}dCdV^2 - \frac{1}{2}CV^2$$

Neglecting higher order terms of small quantities such as dC and dV, we get

$$E' - E = \frac{1}{2}V^2\ dC + CV\,dV \qquad ......(1.52)$$

from the law of conservation of energy we have,

Input energy = Increment in stored energy + mechanical work done

$$Ei = (E' - E) + F\,dx$$

$$CV\,dV + V^2\ dC = \left(\frac{1}{2}V^2dC + CV\,dV\right) + (Fdx)$$

$$F\,dx = CV\,dV + V^2dC - \frac{1}{2}V^2dC - CV\,dV$$

$$F\,dx = \frac{1}{2}V^2dC$$

The force of attraction between plates is

$$\boxed{F=\frac{1}{2}V^2\frac{dC}{dx}} \qquad \text{.....(1.53)}$$

From eq. 1.53 we know that the force of attraction is directly proportional to the square of the applied voltage V.

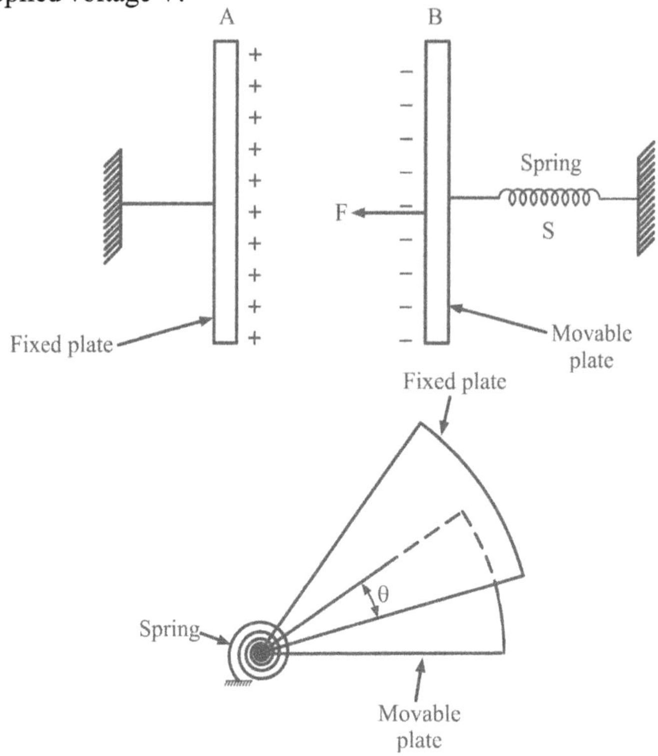

Fig. 1.27 Principle of electrostatic voltmeter

From the above theory can be extended to the rotational motion, with the angular deflection θ in place of the linear displacement x.

$$\therefore \qquad T_d = \frac{1}{2}V^2\frac{dC}{d\theta}$$

In case of the meter uses the spring control with torsional spring constant K then

$$T_c = K\theta$$

where K = Spring constant

θ = deflection.

but $T_d = T_C$ for steady position ......(1.54)

$$\frac{1}{2}V^2\frac{dC}{d\theta} = K\theta$$

$$\therefore \quad \boxed{\theta = \frac{1}{2k}V^2\frac{dC}{d\theta}} \quad .....(1.55)$$

Since

1. The deflection is proportional to the square of voltage to be measured. The scale is non uniform which is compressed at the lower end. (square low response)
2. This instrument can be used for A.C and D.C measurements as the deflection is proportional to the square of the voltage to be measured.

1.38 Type of Electrostatic Voltmeters

Operating Principle

The operating principle of an electrostatic instrument depends on the force of attraction between two or more electrically charged conductors between which a potential difference in maintained and this force gives rise to a deflection torque. The electrostatic mechanism resembles a variable capacitor. Where the force existing between the two parallel plates is a function of the potential difference applied to them.

Basically there are two types of Electrostatic instruments.

1. Quadrant type (10 kV to 20 kV)
2. Attracted – disc type (above 20 kV)

1.38.1 Quadrant - Type Electrostatic Voltmeter

Quadrant – Type electrometers are used for the measurement of voltage up to 10 kV to 20 kV.

Construction and working: It consists essentially of two sets of mutual plates, one movable and the other fixed, the former being of very light construction (e.g., of aluminium). The movable plate, together with the end of the spiral spring, is attached to the spindle carrying the pointer of the instrument (Fig. 1.28). These two plates (fixed & movable) constitute a capacitor whose capacitance changes as the pointer moves on the scale.

When the voltage to be measured is applied between the fixed and movable plates. The plates acquire opposite charges that are proportional to the potential difference or the voltage. The electric field set up between the plates causes the movable plate field and pointer to move to the right until the deflecting force is balanced by a spring or other restoring force. The deflecting torque is directly proportional to the square of the applied voltage and thus, these instruments measure r.m.s voltages.

The deflecting torque of this type of instrument is very small, unless the applied voltage is extremely large.

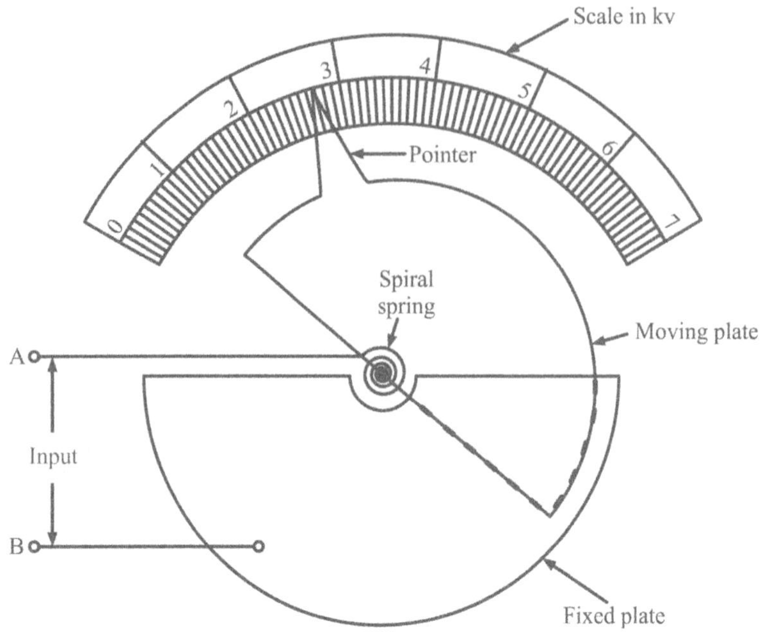

Fig. 1.28 Quadrant-type electrostatic voltmeter

The force on the plates may be increased by using a greater number of quadrants. Such an instrument known as Kelvin's multicellular electrostatic voltmeter. There are two type of the electrical connections in the quadrants electrometer,

1. Heterostatic connection
2. Idiostatic Connection.

1.38.2 Heterostatic Connection

This connection is shown in Fig. 1.29. A high tension battery is used for charging the needle to a potential considerably above that of the quadrants to which the negative of the voltage to be measured is connected.

In this connection, the quadrants are connected together in diagonally opposite pairs. The moving vane i.e needle in positively charged due to battery the deflecting force due to top and bottom quadrants on movable needle cancels each other on both sides. The only deflecting force responsible is force of attraction between left quadrant and right moving sector and force of repulsion between right quadrant and left moving sector.

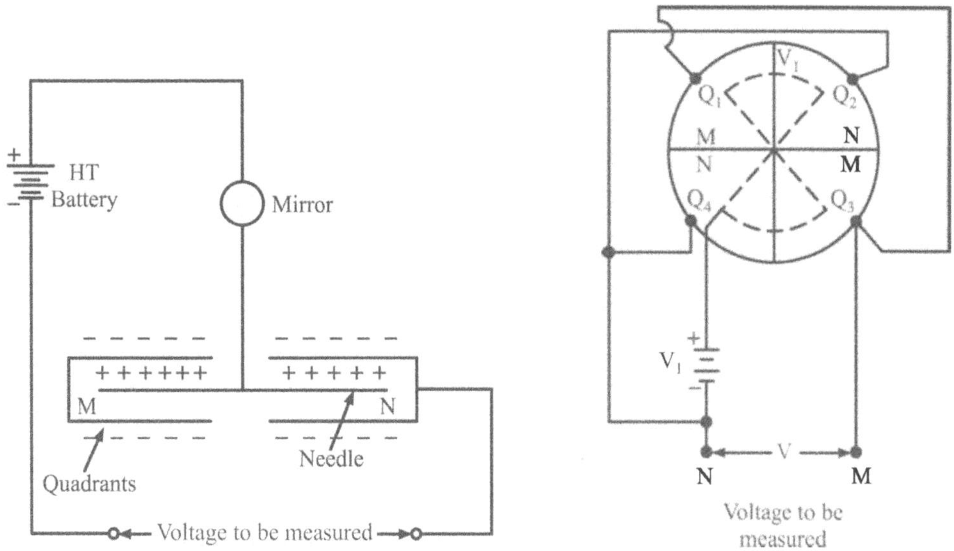

Fig. 1.29 Heterostatic connection

1.38.3 Theory of Heterostatic Connection

In order to develop torque equation for the heterostatic connection consider only one half portion of the needle with two quadrants adjacent to it. This is shown in the Fig. 1.30.

The needle is considered as a sector of circle with radius r. Now this arrangement of two quadrants with needle exactly in between, resembles the two capacitors placed side by side. At equilibrium position as needle is placed symmetrically, the capacitance C_1 and C_2 are equal. But when needle rotates, the value of one capacitor becomes greater than other

Let V_1 = Potential of needle

V = voltage being measured

V_A = Potential of quadrant Q_1

V_B = Potential of quadrant Q_2

C_1 = Capacitance of left hand capacitor

C_2 = Capacitance of Right hand Capacitor

d = Distance of needle from either top or bottom plates of quadrants

For a parallel capacitor

Capacitance $$C_1' = \frac{\varepsilon A}{d} \quad .....(1.56)$$

where ε = permittivity of medium

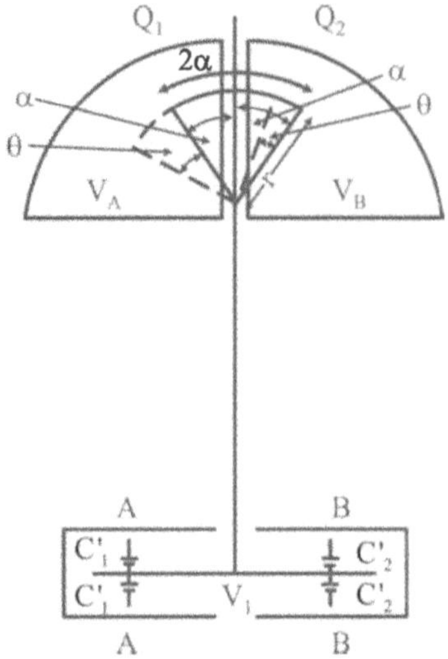

Fig. 1.30 Position of needle with some deflection

$$A = \text{Area of Vane} = \frac{r \times (\alpha + \theta) \times r}{2} = \frac{1}{2} r^2 (\alpha + \theta)$$

$$C_1' = \frac{\varepsilon \left(\frac{1}{2} (\alpha + \theta) r^2 \right)}{d}$$

$$\therefore \quad C_1' = \frac{\varepsilon r^2 (\alpha + \theta)}{2d} \qquad \text{.....(1.57)}$$

This is because the needle spans an angle $(\alpha + \theta)$ under quadrant Q_1

Capacitance $$\boxed{C_1 = 2C_1' = \frac{\varepsilon}{d} (\alpha + \theta) r^2 = \frac{\varepsilon r^2 (\alpha + \theta)}{d}} \qquad \text{.....(1.58)}$$

Similarly, capacitance $$\boxed{C_2 = 2C_2' = \frac{\varepsilon r^2 (\alpha - \theta)}{d}} \qquad \text{.....(1.59)}$$

The energy stored in $C_1 = \frac{1}{2} C_1 (V_1 - V_B)^2$

The energy stored in $C_2 = \frac{1}{2} C_2 (V_1 - V_A)^2$

Total energy stored in,

$$\boxed{w = \frac{1}{2} C_1 (V_1 - V_B)^2 + \frac{1}{2} C_2 (V_1 - V_A)^2} \qquad \text{.....(1.60)}$$

Let 'T_θ' be the torque in the position θ then for an infinitesimal charge dθ of the needle, the work done in moving system $T_\theta\, d\theta$. This work done is equal to the increase in the stored energy dw.

$$\therefore \qquad T_\theta\, d\theta = dw$$

$$T_\theta = \frac{dw}{d\theta}$$

$$\therefore \qquad T_\theta = \frac{d}{d\theta}\left[\frac{1}{2}C_1(V_1 - V_B)^2 + \frac{1}{2}C_2(V_1 - V_A)^2\right]$$

$$T_\theta = \frac{1}{2}(V_1 - V_B)^2\frac{dC_1}{d\theta} + \frac{1}{2}(V_1 - V_A)^2\frac{dC_2}{d\theta} \qquad \text{.....(1.61)}$$

$$\frac{dC_1}{d\theta} = \frac{d}{d\theta}\left[\frac{\varepsilon r^2(\alpha + \theta)}{d}\right] = \frac{\varepsilon r^2}{d}$$

$$\frac{dC_2}{d\theta} = \frac{d}{d\theta}\left[\frac{\varepsilon r^2(\alpha - \theta)}{d}\right] = -\frac{\varepsilon r^2}{d}$$

Using in the eq. 1.61

$$T_\theta = \frac{1}{2}(V_1 - V_B)^2 \times \frac{\varepsilon r^2}{d} - \frac{1}{2}(V_1 - V_A)^2\frac{\varepsilon r^2}{d}$$

$$T_\theta = \frac{\varepsilon r^2}{2d}\left[(V_1 - V_B)^2 - (V_1 - V_A)^2\right] \qquad \text{......(1.62)}$$

But the medium is air hence $\varepsilon = \varepsilon_0$

$$\therefore \qquad T_0 = \frac{\varepsilon_0 r^2}{2d}\{(V_A - V_B)[2V_1 - (V_B + V_A)]\} \qquad \text{.....(1.63)}$$

The above expression is obtained considering only two quadrants and half needle hence for all four quadrants the deflecting torque will be doubled.

$$\therefore \qquad \boxed{T_\theta = \frac{\varepsilon_0 r^2}{d}\{[(V_A - V_B)(2V_1 - (V_B - V_A)]\}} \qquad \text{.....(1.64)}$$

Note: The torque is positive only when

$2V_1 > (V_B + V_A)$. Now V is the Potential to be measured and is equal to $V_A - V_B$

$$\therefore \qquad T_\theta = \frac{\varepsilon_0 r^2}{2d}\{V[2V_1 - (V_B + V_A]\}$$

If potential of needle V_1 is very large compared to voltage to be measured than,

$$T_\theta = \frac{\varepsilon_0 r^2}{d} \times V \times 2V_1$$

$$\boxed{T_\theta \propto 2VV_1} \text{ for heterostatic with } V_1 \text{ large}$$

1.38.4 Idiostatic Connection

This connection generally used in commercial instrument. In this type of connection needle is connected to any one of the pairs of quadrant as shown in the Fig. 1.31 directly without external voltage.

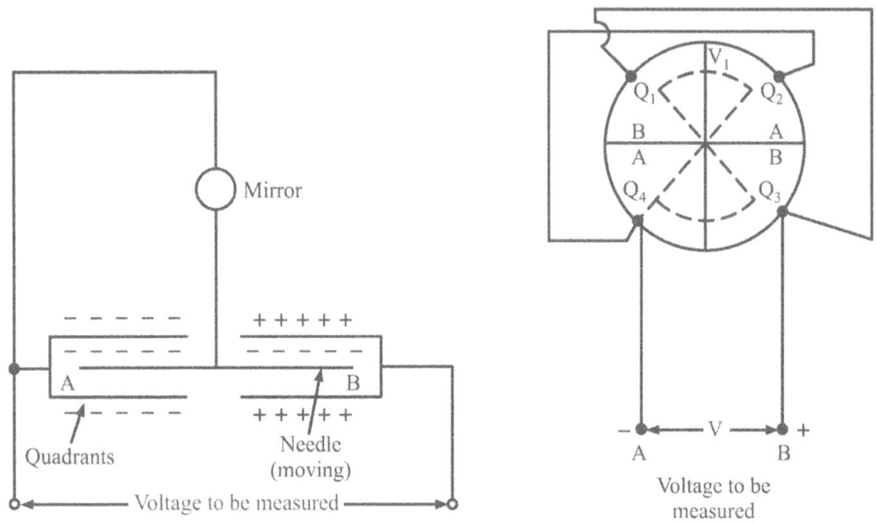

Fig. 1.31 Idiostatic connection

The moving needle is negatively charged, the left hand quadrant is negatively charged and the right hand quadrant is positively charged; the force of attraction on needle due to top and bottom parts of right hand quadrant cancel each other. So there is no motion of needle due to right hand quadrant, similarly the force of repulsion on needle due to top and bottom parts of left hand quadrant also cancel each other.

The right hand positively charged quadrant attracts the part of the needle near to left hand quadrant while the left hand negatively charged quadrant repels the part of the needle to right hand quadrant. This rotates the needle and hence the pointer.

1.38.5 Theory of Idiostatic Connection

For idiostatic connection, the external voltage applied to the needle $V_1 = 0$ V and the potential of quadrant nothing but the voltage V_1 which is applied to the needle directly.

$$\therefore \quad V_1 = V_A$$

While the voltage to be measured is $V = V_B - V_A$

Thus using these value in the expression of T_θ,

$$T_\theta = \frac{\varepsilon_0 r^2}{d}\{(V)[2V_A - V_A - V_B]\} = \frac{\varepsilon_0 r^2}{d}\{(V)(-V)\}$$

$$\boxed{T_\theta = \frac{\varepsilon_0 r^2}{d} V^2}$$

The negative sign is neglected as it indicates the direction of rotation opposite to that which has been assumed.

$$\boxed{T_\theta \propto V^2} \quad \text{For Idiostatic}$$

Note: As torque is proportional to square of the applied voltage the scale is non uniform for idiostatic connection, note that as deflecting torque is proportional to the square of the voltage to be measured. The idiostatic connection is used for A.C measurements.

1.39 Kelvin Multicellular Voltmeter

It is one of the most important commercial form of an electrostatic voltmeter. It is basically a quadrant electrometer with large number of needles and only one quadrant. Basically it is used for a voltage range of 100 to 1000 volts. By modifying basic voltmeter it is possible to measure voltage of the range of 40 V only

Thus to obtain a very high force for very small voltage, a large number of cells are used in the instruments hence it is called multicellular.

The Kelvin multi cellular voltmeter shown in Fig. 1.32.

In this type of instrument, the fixed system consists of a cellular structure composed of two sets of triangular metal plates which are fixed to brass supports. The moving system consists of an equal number of aluminium vanes mounted on a spindle and suspended by a phosphor-bronze wire. The upper end of the wire is connected to an elliptical spring (coach spring) which is mounted on a torsion head for zero adjustments. The spring is provided for protection against accidental fracture of suspension due to vibration and other such factor. A safety sleeve is placed to cone in contact with a stop in the event of any sudden jerk which would otherwise tend to break the suspension. A pointer attached to the moving system deflects on the scale. The controlling torque is provided by the tension in the phosphor-bronze wire when the vanes are deflected from the zero position.

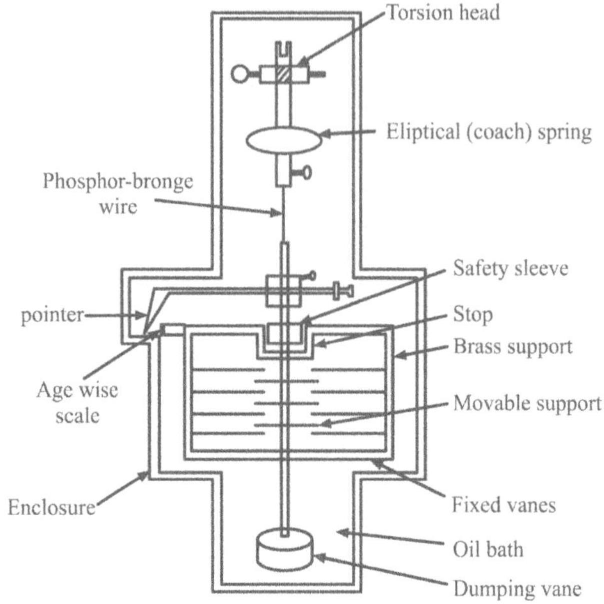

Fig. 1.32 Kelvin's multicellular electrostatic voltmeter

The damping is provided by a vane immersed in an oil dashpot.

The deflection torque for a *n* cell instrument

$$T_d = n \times \text{torque of one cell}$$

$$\therefore \quad \boxed{T_d = n\frac{\varepsilon}{d}r^2 V^2}$$

1.40 Attracted Disc Electrostatic Voltmeter

The attracted disc type instruments are generally used for the measurement of voltages above 20 kV. The system consists of two plates such that one plate can move freely while other is fixed. Both the plates are perfectly insulated from each other. The voltage to be measured is applied across the plates as a supply voltage as shown in the Fig. 1.33. Due to the supply voltage, electrostatic field gets produced which develops a force of attraction between the two plates. Due to the force of attraction the movable plate gets deflected. In this mechanism the controlling torque is provided by a spring.

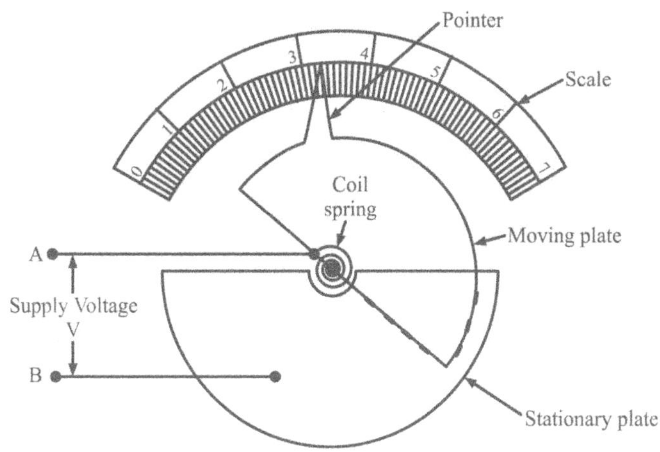

Fig. 1.33 Attracted disc type electrostatic instrument

1.41 Theory of Attracted Disc Type Voltmeter

As we have seen already that the force of attraction F between the two parallel plates with potential difference V between them is given by, $F = \frac{1}{2}V^2 \frac{dC}{dx} N$

Let A be the area of each plate and $\varepsilon = \varepsilon_0 \varepsilon_r$ is the permittivity of the medium, Let d be the distance between the two plates

Then the capacitance C is given by,

$$C = \frac{A\varepsilon}{d} \text{ farads} \qquad .....(1.65)$$

Differentiating eq. 1.65 with respect to x,

$$\frac{dC}{dx} = -\frac{A\varepsilon}{d^2} \qquad .....(1.66)$$

The negative sign indicates that with decrease in distance of separation d, capacitance C increase. So neglecting negative sign, the force of attraction can be rewritten as.

$$\boxed{F = \frac{1}{2}V^2\left(\frac{A\varepsilon}{d^2}\right)N} \qquad .....(1.67)$$

The potential difference V between the two plates is given by

$$\boxed{V = \sqrt{\frac{2Fd^2}{A\varepsilon}} \text{ Volts}} \qquad .....(1.68)$$

Thus this instrument gives as absolute determination of voltage as it is given in terms of force and linear dimensions. The deflecting force in adequate only when the voltage to be measured is high. For avoiding the errors due to corona effect, special construction is necessary to ensure good insulation. The superior dielectric strength of a high vacuum is used in modern instruments which enables to get more force for a given voltage with very small clearance between the plates.

1.42 Kelvin Absolute Electrometer

Construction

It consists of two discs one moving and other fixed. The moving disc is carried by a spring and is suspended from a micrometer head as shown in Fig. 1.34. The moving ring to reduce the fringing effects. The zero setting of the instrument is done with the help of an optical sighting.

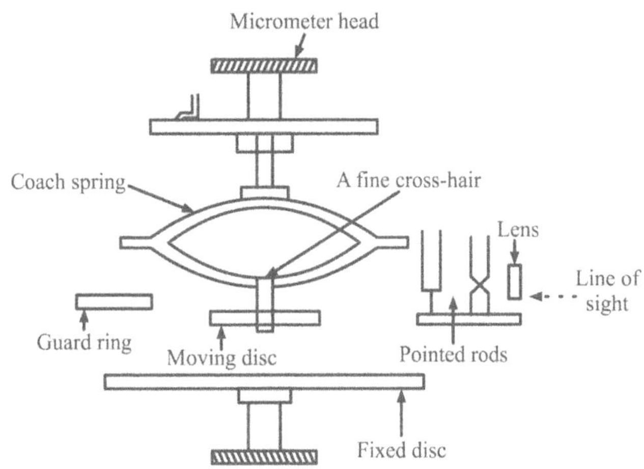

Fig. 1.34 Kelvin absolute electrometer

Operation

The voltage to be measured is applied between the two discs. The moving disc is attracted downwards and is brought back to zero position by turning the micrometer head. The displacement is measured by the micrometer which is calibrated in terms of force. The voltage is then determined in terms of the force and the dimensions of the instrument.

Theory

Let

d = distance between plates : M

A = Area of Plates : M^2

$= \text{Area of moving plate} + \frac{1}{2}$ area of air gap between moving plate and guard ring.

ε = Permittivity of dielectric : F/m

V = voltage being measured : V.

F = Force between disc: N

$$\text{Force } F = \frac{1}{2}V^2\frac{dC}{dx} N$$ [from Force eqn instrument] electro static

Let A be the area of each plated and e = ε_0 ε_r is permittivity of the medium. Let 'd' be the distance between the two plates. Then the capacitance C is given by.

$$C = \frac{A\varepsilon}{d} \text{ farads} \qquad \text{.....(1.69)}$$

Differentiating eq. 1.69 with respect to x,

$$\frac{dC}{dx} = -\frac{A\varepsilon}{d^2} \qquad \text{.....(1.70)}$$

The negative sign indicates that with decrease in distance of separation d, capacitance C increase. So neglecting negative sign, the force of attraction can be rewritten as

$$\boxed{F = \frac{1}{2}V^2\left(\frac{A\varepsilon}{d^2}\right)N} \qquad \text{.....(1.71)}$$

The potential difference V between the two plates in given by

$$\boxed{V = \sqrt{\frac{2Fd^2}{A\varepsilon}} \text{ volts}} \qquad \text{......(1.72)}$$

This instruments gives an absolute determination of potential difference as the potential difference is given in terms of force and linear dimensions. The deflecting force is adequate only when the voltage to be measured is high. For avoiding the errors due to corona effect, special construction is necessary to ensure good insulation. The superior dielectric strength of a high vacuum is used in modern instruments which enable to get more force for a given voltage with very small clearance between the plates.

The disadvantage of this instrument is that when the voltage being measured is small a few hundred (volt the two discs) should be very near together in order to get an appreciable force, in such cases the measurement of distance between the plates is difficult to carry out. The solution lies in increasing the voltage between fixed and moving plates by using hetero static connections.

1.43 Advantages and Disadvantages of Electrostatic Instruments

The various advantages of electrostatic instruments are:

1. They can be manufactured with a very high accuracy
2. They can be used on either AC or DC and over a fairly large range frequencies
3. The instrument may be calibrated with DC and yet the calibration would be valid for AC also since the deflection is independent of the waveform of the applied voltage
4. Since no iron is used for their construction. They are free from hysteresis, eddy current losses and temperature errors.
5. Their power losses in negligible
6. They are unaffected by stray magnetic field although they have to be guarded against any stray electrostatic field.
7. They can be used up to 100 kHz frequency without any serious loss of accuracy.
8. Once the discs are charged, no more current is drawn from the circuit and the instruments represents infinite impendence.

Disadvantages

1. The scale is not uniform
2. These are large in size, bulky and not very robust
3. They are not suitable for low voltage measurements
4. These are expensive instruments.

Limitations of Electrostatic Instruments

1. Low voltage Voltmeter are liable to friction errors.
2. Since the deflection torque is proportional to the square of the voltage, their scales are not uniform although some uniformity can be obtained by suitably shaping the quadrants of the Instruments.
3. Their use is limited to certain special application particularly in A.C. circuits of relatively high voltage where the current taken by other instruments would result in erroneous indications.
4. They are inherently, Laboratory-type rather than industrial-type instruments.
5. They are expensive and are not likely to be durable.

1.44 Extension of Range of Electrostatic Voltmeter

The range of various instruments can be extended using multipliers. Similarly the range of electrostatic instruments is also extended using multipliers. The multipliers used for electrostatic instruments are of two types:

1. Resistance Potential divider
2. Capacitance Multipliers

1. ***Resistance Potential Divider:*** By use of resistance potential divider for extending the range of an electrostatic Instrument is shown in Fig. 1.35.

Here,

R = Total resistance of potential divider

V = Voltage to be measured

r = the resistance whose voltage drop is applied to an electrostatic voltmeter

v = the voltage across an electro static voltmeter

C = the capacitance of an electrostatic voltmeter.

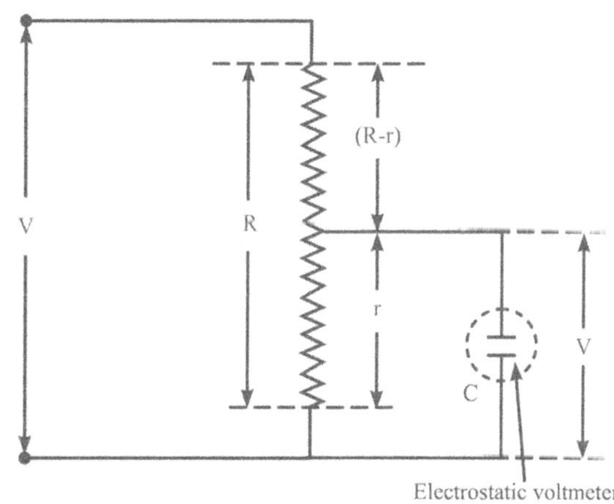

Fig. 1.35 Resistance potential divider

The resistance r and capacitance C forms a parallel circuit and the equivalent impedance is

$$Z = r \| -j X_C \qquad \text{where} \qquad X_C = \frac{1}{w_C}$$

$$= \frac{r \times \left(\frac{-j}{\omega C}\right)}{r - \frac{j}{\omega C}} = \frac{-jr}{r\omega C - j} = \frac{-jr \times j}{(r\omega C - j) \times j}$$

$$Z = \frac{r}{1 + jr\omega C} \qquad \text{.....(1.73)}$$

Thus the equivalent independence across the voltage V is,

$$Z_T = R - r + Z$$

$$= R - r + \frac{r}{1 + jr\omega C} = \frac{(R-r)(1+jr\omega C)+r}{(1+jr\omega C)}$$

$$\therefore \quad Z_T = \frac{R + j\omega rC(R-r)}{(1+jr\omega C)} \quad .....(1.74)$$

The factor by which voltage is charged due to potential divider is called its multiplying power and given by

$$m = \frac{V}{v} = \frac{Z_T}{z} = \frac{\frac{R + j\omega rC(R-r)}{(1+jr\omega C)}}{r/1 + jr\omega C}$$

$$\therefore \quad m = \frac{R + j\omega rC(R-r)}{r} = \frac{R}{r} + j\omega C(R-r) \quad .....(1.75)$$

The numerical value of multiplying power m is

$$\boxed{m = \sqrt{\left(\frac{R}{r}\right)^2 + \omega^2 C^2 (R-r)^2}} \quad .....(1.76)$$

If ω, 'C' and 'r' are very small then $w^2C^2r^2 \leq 1$ and can be neglected.

$$\boxed{m = \frac{R}{r}\sqrt{1 + \frac{\omega^2 C^2 r^2 (R-r)^2}{R^2}} \approx \frac{R}{r}} \quad ......(1.77)$$

This can be easily analyzed

1. At high voltages, the accuracy is very less due to stray capacitance effects.
2. When used for A.C measurements, it should be wound non inductively and capacitor leakage resistance must be high.
3. At high voltage, the power losses and wastage is excessive, high cost.

Thus this method is not suitable for high voltages but used for D.C measurements as capacitance potential divider cannot be used for D.C. circuits.

2. ***Capacitance Multipliers:*** The capacitance multiplier method is nothing but the use of capacitance potential divider. There are two methods of connecting capacitor for potential division.

   ***Method* I:** In this method a single capacitor is connected in series with the voltmeter and the voltmeter to be measured and applied across the combination as shown in Fig. 1.36.

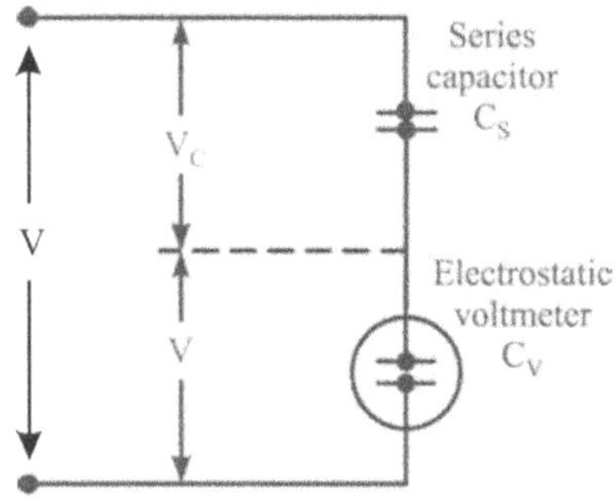

Fig. 1.36 Capacitor multiplier method-I

C_s = Series Capacitor
C_V = Capacitor of Voltmeter
v = Voltage across voltmeter
V = Voltage to be measured

The total capacitance across the supply is,

$$C_t = \frac{C_S \times C_V}{C_S + C_V} \quad C_S \text{ and } C_V \text{ in series.}$$

The total impedance across the supply is

$$Z_t = \frac{1}{j\omega C_t} = \frac{C_S + C_V}{j\omega C_S C_V} \qquad \text{......(1.78)}$$

The impedance of voltmeter is, $Z = \frac{1}{j\omega C_V}$ .....(1.79)

Thus the multiplying power of the multiplier is

$$m = \frac{V}{v} = \frac{Z_t}{Z} = \frac{\frac{C_S + C_V}{j\omega C_s C_V}}{1/j\omega C_V}$$

$$m = Cs + Cv/Cs = \boxed{m = \frac{C + C_V}{C} = 1 + \frac{C_V}{C_S}} \qquad \text{.....(1.80)}$$

1. The voltmeter capacitor varies with the deflection of the moving needle hence the voltmeter must be calibrated along with the series multiplier capacitor.
2. To have high value of multiplying power, the voltmeter capacitor must be high.

Method II: In many practical cases a set of capacitors connected in series across the voltage to be measured is used. The voltmeter is connected across one of the suitable capacitors as shown the Fig. 1.37.

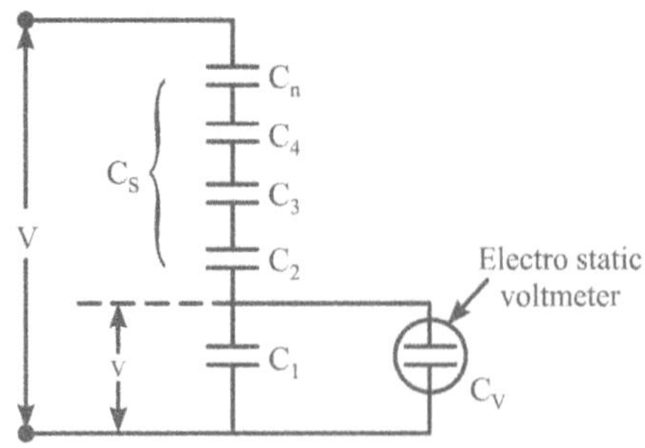

Fig. 1.37 Capacitor multiplier-II

The capacitors C_1 and C_V are in parallel hence their resultant is $C_1 + C_V$.

While C_2, C_3, C_4---C_n are in series and their equivalent is C_S where,

$$\frac{1}{C_S} = \frac{1}{C_1} + \frac{1}{C_2} + \ldots\ldots + \frac{1}{C_n}$$

Thus C_3 and $(C_v + C_1)$ are in series hence the resultant capacitor across the voltage 'V' is

$$C_t = C_s\,(C_1 + C_v)/C_s + C_2 + C_v$$

$$Z_t = \frac{1}{jwct} = \frac{C_s + C_1 + C_v}{jwc_s\,(C_1 + C_v)}$$

while across the voltage 'V' the capacitance is $C_1 + C_v$ hence the impendence is, $z = \frac{1}{jw(C_1 + C_v)}$ the multiplying power is,

$$m = \frac{Z_t}{Z} = \frac{V}{v} = \frac{\dfrac{C_S + C_1 + C_V}{jwC_j(C_1 + C_V)}}{\dfrac{1}{jw(C_1 + C_v)}} = m = 1 + \left[\frac{C_1 + C_v}{C_s}\right]$$

If C_1 is large with respect to C_v, then there is no appreciable change in the multiplying power alongwith the deflection of the pointer.

Problems with Solutions

1. The coil of a moving coil voltmeter is 40 mm × 30 mm wide and has 100 turns wound on it. The control spring exerts a torque of 0.25×10^{-3} N-m when the deflection is 50 divisions on the magnetic field in the air gap is 1 wb/m^2. Estimate the resistance that must be put in series with the coil to give 1V/division. Resistance of voltmeter is 10,000 ohms.

Solution:

Given data:

Number of turns, N = 100

Controlling torque $T_c = 0.25 \times 10^{-3}$ N-m

Deflection, θ = 50 scale divisions

Dimensions of the coil = 40 mm × 30 mm (l × d)

Flux density in the air gap, B = 1 wb/m^2

Required voltage per division = 1 V / division

Resistance of voltmeter, $R_m = 10{,}000\ \Omega$

Series resistance R_{se} = ?

- The deflecting torque is

$$\begin{aligned} T_d &= NBldI \quad \text{N-m} \\ &= 100 \times 1 \times 40 \times 10^{-3} \times 30 \times 10^{-3} \times I \\ &= 0.12\ I\ \text{N-m} \end{aligned}$$

- The controlling torque is

$$\begin{aligned} T_c &= k\theta \\ &= 0.25 \times 10^{-3}\ \text{N-m} \end{aligned}$$

- At equilibrium or steady state conditions, the controlling torque is equal to deflecting torque

$$T_c = T_d$$

$$0.25 \times 10^{-3} = 0.12\ I$$

$$\therefore \qquad I = \frac{0.25 \times 10^{-3}}{0.12} = 2.083 \times 10^{-3}\ \text{A}$$

- Therefore current at full scale deflection, I = 2.083 mA
- Let, the resistance to be added in series with the coil be 'R_{se}'
- The resistance of the voltmeter in circuit = $(R_{se} + R_m)\Omega$
- The voltage across the instrument = current at full scale division × resistance of the voltmeter in circuit.

$$\begin{aligned} I(R_m + R_{se}) &= 2.083 \times 10^{-3}(10000 + R_{se}) \\ &= 20.83 + 2.083 \times 10^{-3} R_{se} \end{aligned}$$

This produces a deflection of 50 divisions

$$\therefore \quad \text{Volts per division} = \frac{\text{voltage across instrument}}{\text{no. of divisions}}$$

$$= \frac{20.83 + 2.083 + 10^{-3} R_{se}}{50} \text{ V / division}$$

$$\therefore \quad \frac{20.83 + 2.083 + 10^{-3} R_{se}}{50} = 1$$

$$20.83 + 2.083 \times 10^{-3} R_{se} = 50$$

$$\therefore \quad R_{se} = 14000\ \Omega$$ **Ans.**

∴ A resistance of 14000 Ω or 14 kΩ must be added in series with the coil to obtain 1V/division

2. In a measurement of resistance by the substitution method, a standard resistance of 100 kΩ is used. The galvanometer has a resistance of 2000 Ω and given the following deflections.

(i) With unknown resistance, 46 divisions

(ii) With standard 'R', 40 divisions

Find value of the unknown resistance

Solution:

Given data:

Standard 'R', S = 100 kΩ

Galvonometer 'R', G = 2 kΩ

Deflection with standard 'r', $\theta_1 = 40$ division

Deflection with unknown 'r', $\theta_2 = 46$ division

Unknown resistance, x = ?

The deflection of the galvanometer is directly proportional to the I passing through the circuit and, I is inversely proportional to the resistance of the circuit.

$$\because \quad \theta < I < \frac{1}{R}$$

From substion method, analysis, $\dfrac{\theta_1}{\theta_2} = \dfrac{X+G}{S+G}$

$$X + G = (S+G)\frac{\theta_1}{\theta_2}$$

∴ unknown 'R', $X = (S+G)\dfrac{\theta_1}{\theta_2} - G$

$$= (100\ k\Omega + 2\ k\Omega)\frac{40}{46} - (2000)$$

$$\mathbf{= 86.69\ k\Omega}$$ **Ans.**

3. In the moving coil voltmeter, Resistance = 10 kΩ, dimensions of coil = 30 mm × 30 mm, no.of turn of coil = 100, flux density in the air gap = 0.08 wb/m^2 and spring constant = 3 × 10^{-6} N-m per degree. Find the deflection produced by a voltage of 200 V.

Solution:

Given data:

$R = 10\ k\Omega$ $A = 30 \times 30 \times 10^{-6} m^2$

$N = 100$

$B = 0.08$ wb/m^2, $k = 3 \times 10^{-6}$ N-m / deg.

$V = 200V$

Find $\theta = ?$

$$\Rightarrow T_d = NBI(L \times d)$$

$$= NBIA$$

But $I = \dfrac{V}{R} = \dfrac{200}{10K} = 20\ mA$

∴ $T_d = 100 \times 0.08 \times 30 \times 30 \times 10^{-6} \times 20 \times 10^{-3}$ .....(1)

$$= 144 \times 10^{-6}\ N\text{-}m$$

Controlling torque $T_c = k\theta$

$T_c = 3 \times 10^{-6}\theta$ .....(2)

For equilibrium condition

eq. (1) = eq. (2)

∴ $T_d = T_c \Rightarrow 144 \times 10^{-6} = 3 \times 10^{-6} \times \theta$

$$\theta = \frac{144 \times 10^{-6}}{3 \times 10^{-6}}$$

∴ $\boldsymbol{\theta = 48^o}$ **Ans.**

4. In a moving coil instrument, the coil is wound with 50 turns in square section former of mean length 4 cm long each side. The coil charge in a uniform radial field of 0.08 wb/m^2. Find the torque produced in it.

Solution:

Torque = ?

$F = NBil$ Newton

$$\text{Torque} = \text{Force} \times \text{Width}$$
$$= NBil \times \text{width}$$
$$= NBil^2 \text{ N-m}$$
$$= 50 \times 0.08 \times 10 \times 10^{-3} \times 0.04^2$$
$$\mathbf{T = 6.4 \times 10^{-4} \text{ N-m}}$$ **Ans.**

5. Calculate the shunt required to extent the range up to 10A when it gives full scale deflection with a current of 50 mA. The resistance of the coil is 0.08 Ω.

Solution:

Total current to be measured = 10 A

Instrument current = 50 mA = 0.05 A

$$\text{Multiplying power} = \frac{I}{I_m} = \frac{10}{0.05} = 200$$

$$\text{Resistance of shunt} = R_s = \frac{R_m}{M-1}$$

$$= \frac{0.08}{200-1} = \frac{0.08}{199} = \mathbf{4 \times 10^{-4}\ \Omega}$$ **Ans.**

6. Moving coil instrument whose resistance is 2.5 Ω gives full scale deflection with a current of 1 mA. This instrument is to be used with a manganin shunt to extent its range to 100 mA. Calculate the error caused by a 10 °C rise in temperature when,

(i) Copper moving coil is connected directly

(ii) A 75 Ω manganin resistance is used in series with the instrument moving coil.

The temperature coefficient of copper is 0.004/°C and that of manganin is 0.00015 / °C

Solution:

Given data:

Resistance of instrument $R_m = 25\ \Omega$

Full scale deflecting current, $I_m = 1$ mA

New range with manganin shunt, I = 100 mA

Series resistance $R_{se} = 75\ \Omega$

Coefficient of temperature of copper, $\alpha_c = 0.004 /^{\circ}$ C

Coefficient of temperature of manganin, $\alpha_m = 0.00015 /^{\circ}$ C

Find

Error caused by 10 °C rise in temperature = ?

(i) When copper moving coil is connected directly across the manganin shunt as shown in the Fig. (a).

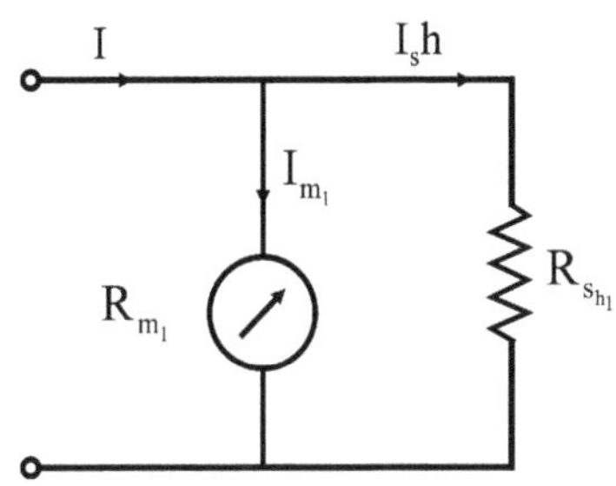

Fig. (a)

$$\text{Multiplying factor of shunt } m = \frac{I}{I_m} = \frac{100}{1} = 100$$

$$\therefore \text{ Resistance of shunt } R_{sh} = \frac{R_m}{m-1} = \frac{25}{100-1} = 0.2525\ \Omega$$

The relation between the temperature and resistance is given by

$$R' = R(1+\alpha T)$$

where α = Temperature coefficient at 10 °C rise in temperature

Instrument resistance

$$R_m' = R_{m_1}(1+\alpha_c T)$$

$$= 25(1+0.004\times 10) = 26\ \Omega$$

Shunt resistance

$$R_{sh_1} = R_{sh_1}(1+\alpha_m T)$$

$$= 0.2525(1+10\times 0.00015)$$

$$= 0.2529\ \Omega$$

∴ Current through meter for 100 mA in the main circuit will be

$$I_{m_1} = I \times \frac{R'_{sh}}{R'_{sh} + R'_{sn}} = 100 \times \frac{0.2529}{26 + 0.2529} = 0.9633 \text{ mA}$$

∴ % error will be $\dfrac{I_{m_1} - I_m}{I_m} \times 100 = \dfrac{0.9633 - 1}{1} \times 100 = -3.67\%$

The negative sign indicates that the meter is reading low.

(ii) When a 75 Ω manganin resistance is used in series with the M.C of the instrument as shown in Fig. (b).

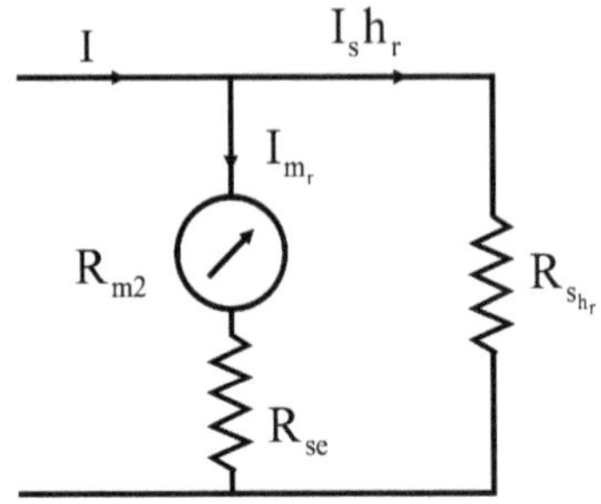

Fig. (b)

Total resistance of the meter circuit

$$R_{m_2} = R_{m_1} + R_{se}$$

$$R_{m_2} = 25 + 75 = 100\ \Omega$$

∴ Shunt resistance $R_{sh_2} = \dfrac{R_{m_2}}{m - 1}$

$$= \frac{100}{100 - 1} = \frac{100}{99}$$

$$= 1.0101\ \Omega$$

At 10 °C rise in temperature

The total resistance of instrument circuit

$$R_T = R_{m_1}(1 + \alpha_c T) + T_{se}(1 + \alpha_m T)$$

$$= 25(1 + 0.004 \times 10) + 75(1 + 0.00015 \times 10)$$

$$= 26 + 75.1125 = 101.1125\ \Omega$$

Shunt resistance $R'_{sh_2} = R_{sh_2}(1+\alpha_m T)$

$= 1.0101\ (1 + 0.00015 \times 10)$

$= 1.0116\ \Omega$

∴ Current through meter for 100 mA in the main circuit will be

$$I_{m_2} = I \times \frac{R'_{sh_2}}{R_T + R'_{sh_2}} = 100 \times \frac{1.0116}{101.1125 + 1.0116}$$

$= 0.9906$ mA

∴ Percentage error will be

$$\frac{I_{m_2} - I_m}{I_m} \times 100 = \frac{0.9906 - 1}{1} \times 100$$

$= -0.94\%$ **Ans.**

Hence the percentage error declines from 3.67% to 0.94% by the addition of series manganin resistor to the moving coil.

7. A moving coil instrument gives a full scale deflection of 10 mA, when the potential difference across its terminals is 100 mV. Calculate

(i) The shunt resistance for a full scale deflection corresponding to 100 A.

(ii) The series resistance for full scale reading with 1000 V

The power dissipation in each case

Solution:

Given data:

Full scale deflection current $I_{fs} = 10$ mA

Voltage applied $V_m = 100$ mV

(i) For full scale deflection corresponding to 100 A

To find Shunt resistance R_{sh} = ? Power dissipation P_{M1} = ?

(ii) For full scale reading with 1000 V

To find Series resistance R_{se} = ? Power dissipation P_{m2} = ?

(i) Fig. (a) shows the circuit connection to get a full scale deflection corresponding to 100 A

Meter resistance $(R_m) = \frac{V_m}{I_{fs}} = \frac{100 \times 10^{-3}}{10 \times 10^{-3}} = 10\ \Omega$

∴ $R_m = 10\ \Omega$

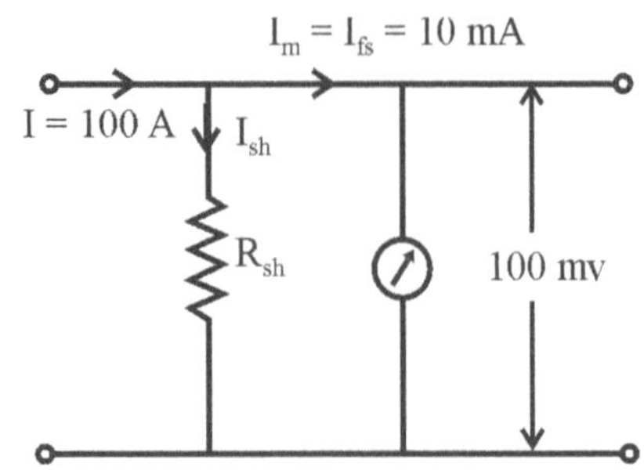

Fig. (a)

Voltage across R_{sh},

$$V_{sh} = V_m = 100 \text{ mV}$$

$$V_m = I_{sh} R_{sh}$$

$$I_{sh} R_{sh} = 100 \times 10^{-3}$$

$$(I - I_m) R_{sh} = 100 \times 10^{-3}$$

$$(100 - 10 \times 10^{-3}) R_{sh} = 100 \times 10^{-3}$$

$$R_{sh} = \frac{100 \times 10^{-3}}{99.99} = 1.0001 \times 10^{-3} \Omega$$

$$\therefore \quad R_{sh} = 1.0001 \text{ m}\Omega$$

Power dissipation of the meter

$$P_{m_1} = V_m \times I$$

$$= 100 \times 10^{-3} \times 100$$

$$\therefore \quad P_{m_1} = 10 \text{ W}$$

(ii) Fig. (b) shows the circuit arrangement for full scale reading with 1000 V.

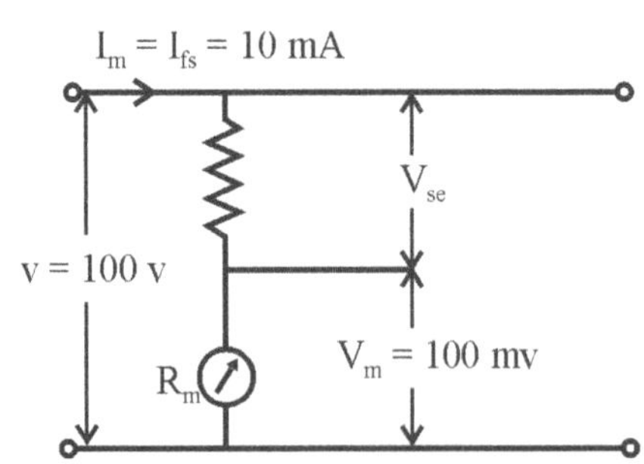

Fig. (b)

Voltage across R_{se},

$$V_{se} = V - V_m$$

$$I_m R_{se} = 1000 - (100 \times 10^{-3})$$

$$10 \times 10^{-3} \times R_{se} = 999.9$$

$$R_{se} = \frac{999.9}{10 \times 10^{-3}} = 99990\ \Omega$$

$$\therefore \quad R_{se} = 99.99\ k\Omega$$

The power dissipation of the meter

$$P_{m_2} = V \times I_m$$

$$\therefore \quad P_{m_2} = 1000 \times 10 \times 10^{-3} = 10\ W$$

8. The measured value of a resistor is 105 Ω, where as its true value is 100 Ω. Determine the relative error.

Solution:

Measured value, A_m = 105 Ω

True value, A = 100 Ω

Absolute error, $\epsilon_0 = A_m - A = 105 - 100$

$$= 5\ \Omega$$

Relative error, $\epsilon_r = \frac{\epsilon_0}{A} = \frac{5}{100} = 0.05$

Percentage error = 5

9. In a moving coil instrument, the moving coil consists of 100 turns wound on a square former coil length 3 cm. The flux density in the air gap is 0.06 T. Calculate the turning movement acting on the coil when carrying a current of 12 mA.

Solution:

Given data:

Number of turns on the coil N = 100

Mean length of the coil, l = 3 cm = 0.03 m

Flux density in the air gap B = 0.06 T

Current flowing through the coil i = 0.012 A

The deflecting force F = BilN

$$= 0.06 \times 0.012 \times 0.03 \times 100 = 0.00216\ N$$

Distance between the centres of the sides of coil

$$2r = \text{width of the former} = 0.03 \text{ m}$$

Turning movement acting on the coil = $F \times 2r$

$$= 0.00216 \times 0.03$$

$$= 648 \times 10^{-7} \text{ N-m}$$ **Ans.**

Note: The flux density may be also given as 0.06 ωb/m^2 (1T = 1ωb/m^2)

10. The inductance of a moving iron instrument is given by inductance $L = (10 + 5\theta - \theta^2)\ \mu H$. Where θ is the deflection in radian from zero position. The spring constant is 12×10^{-6} Nm/rad. Estimate the deflection for a current of 5 amps.

***Solution*:**

Given data:

Inductance $L = \left(10 + 5\theta - \theta^2\right)\ \mu H$

Spring constant $K = 12 \times 10^{-6}$ Nm/rad

Current I = 5 A

Deflection θ = ?

The rate of change of inductance with deflection

$$\frac{dL}{d\theta} = 5 - 2\theta\ \mu H / \text{rad}$$

We know that in case of M.I instruments,

Deflection,

$$\theta = \frac{1}{2} \times \frac{I^2}{K} \times \frac{dL}{d\theta}$$

$$\theta = \frac{1}{2} \times \frac{(5)^2}{12 \times 10^{-6}} \times (5 - 2\theta) \times 10^{-6}$$

$$\theta = 1.0416(5 - 2\theta)$$

$$\theta = 5.20833 - 2.0833\ \theta$$

$$3.0833\ \theta = 5.20833$$

$$\theta = 1.689 \text{ rad or}$$

∴ $$\boldsymbol{\theta = 96.78^{\circ}}$$ **Ans.**

11. The following data refers to a moving coil voltmeter: resistance = 10.000 Ω, dimensions of coil = 30 mm × 30 mm, number of turns of coil = 100, flux density in the air gap = 0.08 ωb/m^2, spring constant = 3 × 10^{-6} N-m per degree. Find deflection produced by a voltage of 200 V.

Solution:

Given data

$R = 10.000\ \Omega$

$A = 30 \times 30 \times 10^{-6}\ m^2$

$N = 100$

$B = 0.08\ \omega b/m^2$

$K = 3 \times 10^{-6}$ N-m / degree

$V = 200$ V

$\theta = ?$

$$T_d = N \times B \times (L \times d) \times I = NBAI \qquad [\because A = L \times d]$$

$$I = \frac{V}{R} = \frac{200}{10.000} = 20\ mA$$

$$\therefore \quad T_d = 100 \times 0.08 \times 30 \times 30 \times 10^{-6} \times 20 \times 10^{-3} = 1.44 \times 10^{-4}\ N-m \qquad .....(1)$$

Controlling torque $T_c = k\theta, T_c = 3 \times 10^{-6}\theta$ .....(2)

For equilibrium condition eqn. (1) = eqn. (2)

$$\therefore \quad T_d = T_c$$

$$1.44 \times 10^{-4} = 3 \times 10^{-6} \times \theta$$

$$\theta = \frac{144 \times 10^{-6}}{3 \times 10^{-6}}$$

$$\therefore \quad \theta = 48'$$ **Ans.**

12. A moving coil milli-voltmeter has a resistance of 200 Ω and the full scale deflection is reached when a potential difference of 100 mV is applied across the terminals. The moving coil has effective dimensions of 30 mm × 25 mm and is wound with 100 turns. The flux density in the air gap is 0.2 ωb/m^2. Determine the control constant, if the final deflection is 100° and a suitable diameter of copper wire for coil winding if 20% of the total instrument resistance is due to the coil winding. Resistivity of copper is 1.7 × 10^{-8} Ω-m.

Solution:

Given that

Meter resistance $R_m = 200\ \Omega$

Potential difference corresponding to full scale deflection $V_{fsd} = 100$ mV

Moving coil dimensions = $(l \times d) = 30\text{ mm} \times 25\text{ mm}$

Number of turns, N = 100

Flux density in air gap, $B = 0.2\ \omega b / m^2$

Final (or) full scale deflection $\theta_{fsd} = 100^\circ$

Coil winding resistance, R_c = 20% of R_m

Resistivity of copper, $\rho = 1.7 \times 10^{-8}\ \Omega$-m

Control constant, K = ? Diameter of copper wire, D = ?

The deflecting torque at full scale deflection

$$T_d = NBld\ I_{fed} = NBld\ \frac{V_{fsd}}{R_m}$$

$$= 100 \times 0.2 \times 30 \times 10^{-3} \times 25 \times 10^{-3} \times \frac{100 \times 10^{-3}}{200}$$

$$= 7.5 \times 10^{-6}\ \text{N-m}$$

At equilibrium or steady state, deflection torque is equal to control torque

$T_c = T_d$

$K\theta_{fsd} = 7.5 \times 10^{-6}$

$$K = \frac{7.5 \times 10^{-6}}{100} = 7.5 \times 10^{-8}\ \text{N} - \text{m} - \text{deg}$$

$$= \frac{7.5 \times 10^{-8}}{\frac{\pi}{180}} = 4.297 \times 10^{-6}\ \text{N} - \text{m} / \text{rad}$$

The resistance of coil winding is given as

$$R_c = 20\%\text{ of }R_m$$

$$= \frac{20}{100} \times 200 = 40\ \Omega$$

But $$R_c = \frac{NeL_{mt}}{a}$$

where, L_{mt} = Length of mean turn

$$= 2(l + d)$$

$$= 2(30 + 25) = 110 \text{ mm}$$

$$a = \frac{NeL_{mt}}{T_c} = \frac{100 \times 1.7 \times 10^{-8} \times 110 \times 10^{-3}}{40}$$

$$= 4.675 \times 10^{-9} \text{ m}^2$$

Also, we know that,

Area of cross-section, $a = \frac{\pi D^2}{4}$

$$D = \sqrt{\frac{4a}{\pi}} = \sqrt{4 \times \frac{4.675 \times 10^{-9}}{\pi}}$$

$$= 7.715 \times 10^{-5} \text{m}$$

$$= 0.077 \text{ mm}$$

∴ Control constant,

$$K = 7.5 \times 10^{-8} \text{ N-m/deg (or) } 4.297 \times 10^{-6} \text{ N-m/rad}$$

and diameter of copper wire. D = 0.077 mm

13. A PMMC instrument has a full scale deflection of 90° for a current of 2 A. The deflecting torque in a PMMC instrument is directly proportional to current in the moving coil. Find the value of current required for a deflection of 30° if the instrument is

1. Sprind control
2. Gravity controlled

Solution:

Given data:

$$\theta_1 = 90^\circ$$

$$I_1 = 2 \text{ A}$$

$$\theta_2 = 30^\circ$$

$$T_d \propto I$$

1. In case of spring controlled

   In a spring controlled instrument, the controlling torque is proportional to deflection i.e. $T_c = k\theta$, $T_c \propto \theta$

From the given data, $T_d \propto I$

For equilibrium position $T_d = T_e$, $\theta \propto I$

$$\frac{90^\circ}{30^\circ} = \frac{2}{I_2} \qquad \because \left[\frac{\theta_1}{\theta_2} = \frac{I_1}{I_2}\right]$$

$$\therefore I_2 = \frac{30 \times 2}{90^\circ} = \frac{60}{90} = 0.67\text{A}$$

0.67 A **Ans.**

2. In case of gravity controlled

   For gravity controlled instruments, the controlling torque is proportional to sine of the angle of deflection, that is,

$$T_d \propto \sin\theta \quad T_d \propto I$$

$$\therefore \sin\theta \propto I, \quad \frac{\sin\theta_1}{\sin\theta_2} = \frac{I_1}{I_2}$$

$$\frac{\sin 90^\circ}{\sin 30^\circ} = \frac{2}{I_2}$$

$$\therefore I_2 = \frac{\sin 30 \times 2}{\sin 90^\circ}$$

$$I_2 = \frac{0.5 \times 2}{1}$$

$I_2 = 1\text{A}$ **Ans.**

14. A basic D′Arsonval meter movement with an internal resistance $R_m = 100\ \Omega$ and a full scale current of $I_m = 1$ mA is to be converted into a multirage D.C voltmeter with ranges of 0-10 V, 0-50 V, 0-250 V, 0-500 V. Find the value of various resistance using the potential divider arrangement.

***Solution*:**

Given data:

$$R_m = 100\ \Omega, I_m = 1\ \text{mA}$$

Find $R_1 = ?,\ R_2 = ?,\ R_3 = ?,\ R_4 = ?$

$\therefore$ Voltage across the meter movement

$$V = I_m R_m = 1 \times 10^{-3} \times 100 = 100\ \text{mV}$$

∴ Voltage multiplying factors

$$M_1 = \frac{10}{100 \times 10^{-3}} = 100$$

$$M_2 = \frac{50}{100 \times 10^{-3}} = 500$$

$$M_3 = \frac{250}{100 \times 10^{-3}} = 2500$$

$$M_4 = \frac{500}{100 \times 10^{-3}} = 5000$$

Let R_1, R_2, R_3, R_4 be the values of resistance for the voltmeter with ranges 0-10, 0-5, 0-250, 0-500, respectively.

$$R_1 = (M_1 - 1)R_m = (100 - 1) \times 100 = 9900\Omega = 9.9\ K\Omega$$
$$R_2 = (M_2 - M_1)R_m = (500 - 100) \times 100 = 40\ K\Omega$$
$$R_3 = (M_3 - M_2)R_m = (2500 - 500) \times 100 = 200\ K\Omega$$
$$R_4 = (M_4 - M_3)R_m = (5000 - 2500) \times 100 = 250\ K\Omega$$

Review Questions

1. How the various measuring instruments are classified? What are the basic requirements of any measuring instruments?
2. State the various effects with which deflecting torque is produced?
3. State the difference between spring control and gravity control methods to produce control Torque?
4. Explain various methods of providing damping torque in an indicating instrument.
5. What are the essentials of indicating instruments?
6. Write a short note on controlling system.
7. Write a short note on Damping systems.
8. Write the advantages and Disadvantages of PMMC.
9. Derive the Torque equation of D′ Arsonval galvanometer?
10. Explain Dynamic behaviour of Galvonometer?
11. Explain – Intrinsic constants of D′ Arsenal galvanometer.

12. Explain – under damped, over damped, critically damped motions in D'Arsonval galvanometer?
13. Explain about logarithmic decrement in galvanometer.
14. Explain effect of External resistance on Damping and derive the value of critical resistance for damping.
15. Write a short not on sensitivity of galvanometer?
16. Explain briefly errors in moving coil instruments?
17. Derive the Torque equation of Moving Iron Instruments?
18. Explain with a neat sketch about 1. Radial vane repulsion 2. Concentric vane Repulsion Type Instruments?
19. What are the advantages and disadvantages of M.I. Instruments?
20. State and explain each in briefly errors in M.I. Instrument
21. State the comparison between moving coil and Moving Iron Instrument
22. What are the requirements of a shunt? The aryton shunt
23. What are the precautions to be taken while using as an ammeter
24. Write short note on 1. Basic DC. Ammeter 2. Basic D.C voltmeter
25. What are the precautions to be taken while a volt meter?
26. What is loading effect?
27. Describe the construction and working of PMMC instrument
28. Derive Torque equation for PMMC.
29. State the errors in PMMC instruments.
30. What are requirements of shunts and multiplyers?
31. Write a short note on (i) Kelvin multicellular voltmeter (ii) Kelvin absolute electrometer.
32. List the advantages and Disadvantages of electrostatic instruments.
33. Explain attracted-disc type electrometer with neat diagram.
34. Explain principle of electrostatic instrument.
35. Explain briefly quadrant type electrometer.

Quiz Questions

1. The most effective damping is

   (a) fluid friction (b) eddy current

   (c) pneumatic (d) electromagnetic

2. A 0-100 V range voltmeter has sensitivity 1000 Ω/volt. Its internal resistance is:
   (a) 1000 Ω (b) 10,000 Ω
   (c) 100 kΩ (d) 1000 kΩ
3. The moving iron instruments can be used for measuring
   (a) direct current and voltages (b) AC current and voltages
   (c) both (a) and (b) (d) radio frequency current
4. The meter that is suitable for only direct current measurement is:
   (a) MI type (b) hot-wire
   (c) electro dynamic type (d) PMMC
5. In MI instrument, the deflecting torque is proportional to
   (a) I^3 (b) 'I' (c) $I^{3/2}$ (d) I^2
6. Indicating Instruments measure:
   (a) peak value (b) RMS value
   (c) average value (d) none
7. In MC ammeter scale on the dial is
   (a) cramped at the beginning (b) cramped in middle
   (c) cramped at end (d) uniform throughout
8. The following meter reads DC 'voltage' and current accurately.
   (a) moving iron (b) PMMC
   (c) dynamometer (d) none
9. The moving coil in a dynamometer wattmeter is connected
   (a) across the supply (b) in series with the fixed coil
   (c) in series with the load (d) across the load
10. A dynamometer type wattmeter responds to time
    (a) average value of active power (b) average of reactive power
    (c) peak value of active power (d) peak value of reactive power
11. Two 100 μA full scale PMMC meters are employed to construct a 10 V and a 100 V full scale voltmeter. These meters will have figures of merit (sensitises)
    (a) 100 kΩ/V and 10 kΩ/V (b) 100 kΩ/V and 1 kΩ/V
    (c) 10 kΩ/V and 100 kΩ/V (d) 10 kΩ/V and 1 kΩ/V

12. A 0-10 mA PMMC ammeter reads 4 mA in a circuit. Its bottom control spring shapes suddenly, the meter will now read nearly.
    (a) 10 mA (b) 8 mA
    (c) 2 mA (d) zero
13. The scale of moving iron type instrument
    (a) cramped at the beginning (b) cramped at end
    (c) cramped in the middle (d) uniform throughout
14. In a moving coil instrument the scale on dial is
    (a) uniform throughout (b) cramped at the beginning
    (c) cramped at middle (d) cramped at the end
15. For repulsion type MI instruments
    (a) air friction (b) pneumatic
    (c) eddy correct (d) fluid friction
16. Which of the following instrument has the best accuracy
    (a) moving coil (b) moving iron
    (c) hot wire (d) thermal
17. The most effective damping is
    (a) fluid friction (b) electromagnetic
    (c) pneumatic (d) eddy current
18. A 25% error voltmeter is used to measure supply voltage of 100 V (DC). The meter reads
    (a) 125 V (b) 75 V
    (c) 100 V (d) 150 V
19. The kwh meter can be classified as an .... instrument
    (a) deflecting (b) digital
    (c) recording (d) indicating
20. The damping type used in moving iron instruments
    (a) air friction (b) pneumatic
    (c) eddy current (d) fluid friction
21. The disadvantage of permanent magnet moving coil instrument is
    (a) high power consumption
    (b) high cost relative to moving iron instrument

(c) low torque / weight ratio
(d) absence of effective and efficient eddy current damping

22. When an ac voltage is applied to a PMMC voltmeter
(a) the meter gets damped (b) meter reading is zero
(c) pointer will oscillate to and fro (d) the pointer will not move at all

23. Error which does not result in moving iron instrument for both ac and dc measurement
(a) stray magnetic field error (b) hysterisis error
(c) eddy current error (d) temperature error

24. The power consumption of PMMC instruments is typically about
(a) 0.25 W to 2 W (b) 25 μ W to 200 μW
(c) 0.25 mW to 2 mW (c) None

25. In an induction type of meter, maximum torque is produced when the phase angle, between the two fluxes is
(a) 0° (b) 45°
(c) 60° (d) 90°

26. In induction type of instrument
(a) $T_d \propto I^2$ (b) $T_d \propto T$
(c) $T_d \propto \sin\theta$ (d) $T_d' \propto \sqrt{I}$

27. The main problem with bar-graph meters is that
(a) they are not very sensitive
(b) they are unstable
(c) they cannot give very precise readings
(d) they can display only peak values

28. The meter movement in an illumination meter directly measures
(a) current (b) power
(c) voltage (d) energy

29. Ammeter shunts are useful because
(a) they prevent overheating of the meter movement
(b) they make a meter more physically rugged
(c) they allow for measurement of large currents
(d) they increase meter sensitivity

30. Suppose a certain current in a galvanometer causes the compass needle to deflect by 20 degree, then this current is doubled while the polarity stays the same. The angle of the needle deflection will
    (a) reverse direction (b) stay the same
    (c) increase (d) decrease
31. A hot-wire ammeter
    (a) Can measure ac as well as dc
    (b) Registers current changes very fast
    (c) Can indicate very low voltages
    (d) Measures electrical energy
32. Which of the following types of instrument is an integrating instrument
    (a) P.F meter (b) energy meter
    (c) Watt meter (d) frequency meter
33. Direct method is used to measure
    (a) Length (b) temperature
    (c) Pressure (d) voltage
34. Which instrument has the lowest resistance?
    (a) ammeter (b) voltmeter
    (c) frequency meter (d) negger
35. The best material of use for standard resistor is
    (a) manganin (b) aluminium
    (c) nichronme (d) platinum
36. Which of the following is absolute instrument
    (a) power facter-meter (b) tangent galvanometer
    (c) megger (d) frequency meter
37. The pointer in instrument is made up of
    (a) steel (b) iron
    (c) aluminum (d) copper
38. Ampere is one of the
    (a) supplementary unity (b) derived units
    (c) base units (d) units used to measure change

39. In moving system instruments the damping force is used to
    (a) brings the pointer to rest (Zero) position
    (b) move the pointer from its zero position
    (c) uniform movement of pointer
    (d) none of the above

Questions and Answers

1. What is meant by measurement?

Ans. Measurement is an act or the result of comparison between the quantity and a predefined standard.

2. Mention the basic requirements of measurement.

Ans. The standard used for comparison purpose must be accurately defined and should be commonly accepted. The apparatus used and the method adopted must be provable.

3. What are the 2 methods for measurement?

Ans. Direct method and Indirect method.

4. Explain the function of measurement system.

Ans. The measurement system consists of a transducing element which converts the quantity to be measured in an analogous form. The analogous signal is then processed by some intermediate means and is then fed to the end device which presents the results of the measurement.

5. Define Instrument.

Ans. Instrument is defined as a device for determining the value or magnitude of a quantity or variable.

6. List the types of instruments.

Ans. The 3 types of instruments are:

1. Mechanical Instruments
2. Electrical Instruments and
3. Electronic Instruments.

7. Classify instruments based on their functions.

Ans. Indicating instruments

Integrating instruments

Recording instruments

8. Give the applications of measurement systems.

Ans. The instruments and measurement systems are used for Monitoring of processes and operations, Control of processes and operations, experimental engineering analysis.

9. Why calibration of instrument is important?

Ans. The calibration of all instruments is important since it affords the opportunity to check the instrument against a known standard and subsequently to errors in accuracy.

10. Name the different essential torques in indicating instruments.

Ans. Deflecting torque

Controlling torque

Damping torque

11. Name the types of instruments used for making voltmeter and ammeter.

Ans. PMMC type

Moving iron type

Dynamometer type

Hot wire type

Electrostatic type

Induction type.

12. State the advantages of PMMC instruments

Ans. Uniform scale.

No hysterisis loss

Very accurate

High efficiency.

13. State the disadvantages of PMMC instruments

Ans. Cannot be used for ac measurements

Some errors are caused by temperature variations.

14. How the range of instrument can be extended in PMMC instruments.

Ans. In ammeter by connecting a shunt resister

In voltmeter by connecting a series resister.

15. State the advantages of Dynamometer type instruments

Ans. Can be used for both dc and ac measurements.

Free from hysterisis and eddy current errors.

16. **State the advantages of Moving iron type instruments**

Ans. Less expensive

Can be used for both dc and ac

Reasonably accurate.

17. **State the advantages of Hot wire type instruments**

Ans. Can be used for both dc and ac

Unaffected by stray magnetic fields

Readings are independent of frequency and waveform.

18. **In a gravity controlled instrument, the defection angle is proportional to**

Ans. Sine-inverse of measureand

20. The preferred damping condition for indicating instruments is

Ans. A damping coefficient of 0.8 to 1

22. Two helical springs are used in a D′Arsonval meter movement because

Ans. It compensates for temperature changes

23. In the circuit, the voltage across the 500 K resistor is exactly 10 V. If a voltmeter with a sensitivity of 20 K/V is used to measure the voltage between point A and B, what are the readings indicated on its 50 V and 5 V range?

Ans. 8, 2.86 V

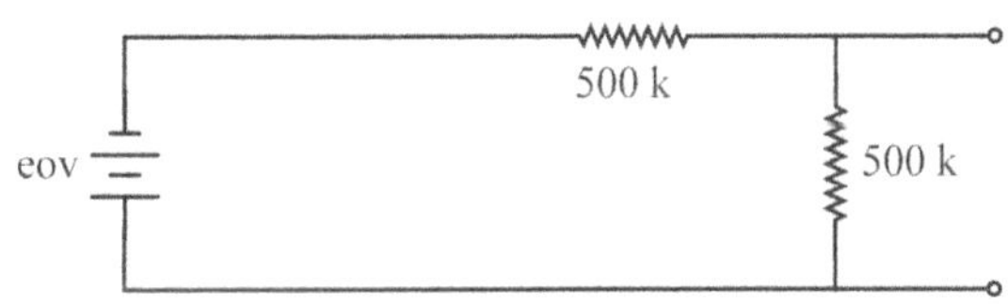

24. In the fig. shown below A_1, A_2, A_3 are ideal ammeter and, if A_1 and A_3 read 5 A, 13 A respectively, then reading of A_2 will be

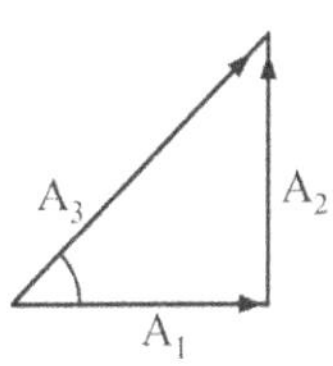

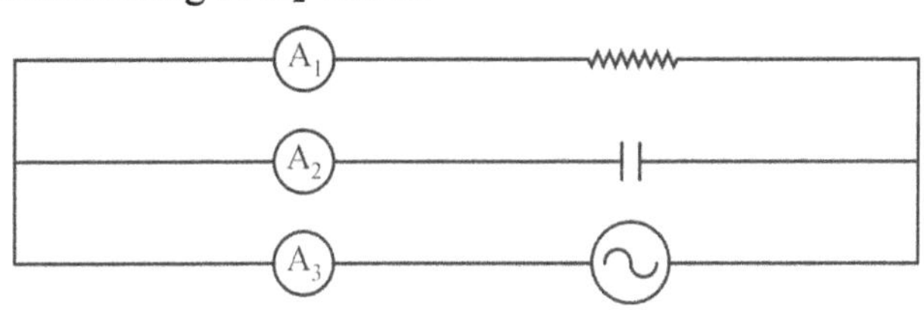

Ans. 12 A

25. Ammeter is connected in ________ with the supply

Ans. Series

26. Voltmeter is connected in ________ with supply

Ans. Parallel

27. Moving coil instrument use ________ effect

Ans. Magnetic

28. The most efficient damping is ________ density

Ans. Eddy Current

29. In PMMC instrument the scale is ________

Ans. Linear

30. The deflecting torque of a moving Iron instrument is proportional to ________

Ans. $(\text{current})^2$

31. Tangent galvanometer is example for ________ instrument

Ans. Absolute

32. Static error is a ________________________________

Ans. The difference between the measured value and true value of quantity.

33. ______________refers to the degree of closeness conformiting to the true value of quantity under measurement.

Ans. Accuracy

34. Accuracy is defined as ______________

Ans. The nearness of the indicated value to the true value of the quantity being measured

35. Sensitivity expressed in terms of _____________

Ans. V/cm

36. A voltmeter contain a ________ resistance in series.

Ans. High

37. In a moving iron meter, the deflection to torque is proportional to

Ans. Square of the current through the coil

38. Why is a MISC meter not recommended for d.c. measurement

Ans. The error is high due to hysteresis effect

39. A moving-iron voltmeter has a full-scale of 100 V. This meter's lowest non-zero marking would normally be

Ans. 20 V

40. The preferred damping condition for indicating instruments is

Ans. A damping coefficient of 0.8 to 1

41. The material most preferred for control spring is

Ans. Al

42. A moving coil instrument is used an ohmmeter. The indicating scale of the meter will be

Ans. Hyperbalic

43. The basic principle required to be satisfied for a.c/d.c measurement by a meter is

Ans. Deflection $\propto$ $(\text{current})^2$

44. In the circuit the voltage across the 500 KΩ resistor in exactly 10 V. If a voltmeter with a sensitivity of 20 K/V is used to measure the voltage between point A and B, what are the readings indicated on its 50 V and 5 V range?

Ans. 8, 2.86 V

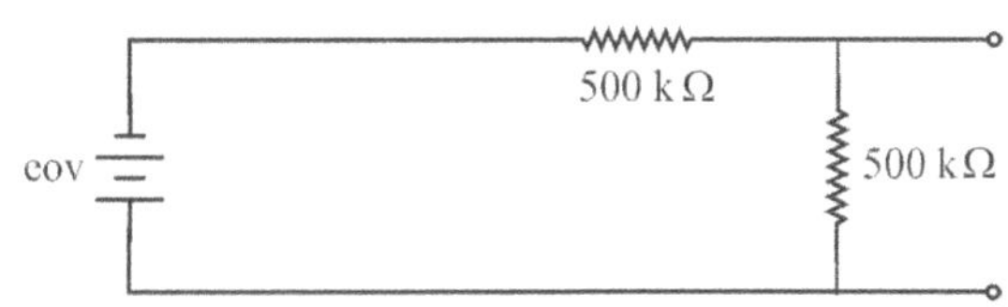

45. The current through the galvanometer and the direction of current in

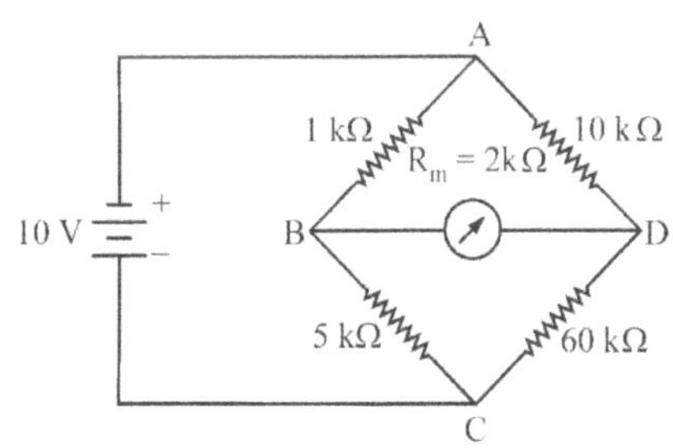

Ans. 21 μ A, DB

$$\text{Ex:}\quad Ig = \frac{E_{TH}}{R_{TH} + R_M} \Rightarrow R_{TH} = \frac{5}{6}K + \frac{600}{70}K = 9.4\ K,$$

$$E_{TH} = 10\left(\frac{1}{6} - \frac{10}{70}\right) = 10\left(\frac{1}{42}\right),\ I_g = \frac{10}{42} + (9.4\ K + 2\ K) = 21\ \mu\ A$$

46. In the fig. shown A_1, A_2, A_3 are ideal ammeters and, if A_1 and A_3 read 5 A, 13 A respectively, then reading of A_2 will be

Ex:

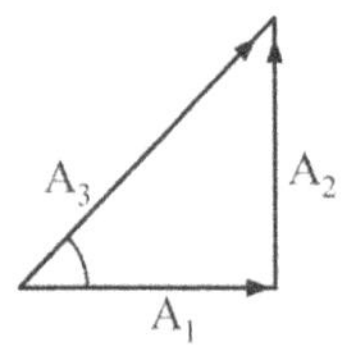

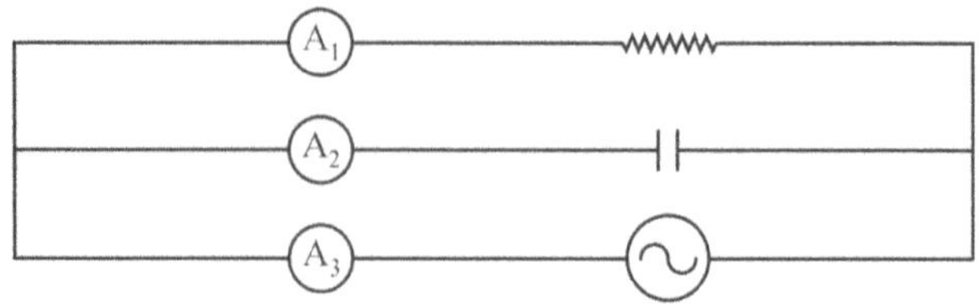

Ans. 12 A

47. Which one of the following quantities has the same dimension in both electromagnetic and electrostatic systems 1. current 2. electric energy 3. electric power. Select the correct answer using the codes given

Ans. 2 and 3

Match the Following

1. Match the decimal multiples and submultiples

(a)	tera	(p)	10^{12}
(b)	atto	(q)	10^{-18}
(c)	fernto	(r)	10^{-15}
(d)	nano	(s)	10^{-9}
(e)	pico	(t)	10^{-12}

2. Match the following

(a)	magnetic effect	(p)	integrating meter
(b)	heating effect	(q)	ammeters
(c)	chemical effect	(r)	D.C ampere-hour meter
(d)	electrostatic effect	(s)	voltmeter
(e)	electromagnetic induction	(t)	AC ammeter Effect

3. Match the following

(a)	thermocouple meter	(p)	a.c/d.c
(b)	moving Iron meter	(q)	a.c/d.c
(c)	moving coil meter	(r)	d.c only
(d)	induction meter	(s)	a.c only

4. Match List-I with List-II and select correct answer

List-I (Instrument)		List-II (Example)	
(a)	indicating	(p)	energy meter
(b)	analog	(q)	tangent galvanometer
(c)	integrating	(r)	ammeter
(d)	absolute	(s)	moving pointer

5. Match the following

(a)	moving coil instrument	(p)	Td $\propto$ I^2
(b)	moving iron instrument	(q)	Td $\propto$ $\sin\theta$
(c)	spring control	(r)	moving coil
(d)	gravity control	(s)	moving iron
(e)	air friction damping	(t)	$T_d \propto I$
		(u)	$T_c \propto \theta$

6. Match the following current and its effects

(a)	ampere	(p)	light bulb
(b)	milli ampere	(q)	mild shock
(c)	kilo ampere	(r)	welding
(d)	micro ampere	(s)	sensitive meters
(e)	pico ampere	(t)	sensitive galvanometer

7. Specification of voltage to be measured Type of instruments suitable

(a)	0 – 10 mV from a source of internal resistance of 1 m Ω	(p)	PMMC
(b)	thermo e.m.f ranging up to 5 mV from a thermo couple	(q)	electronic
(c)	supply voltage of 230 V 50 Hz	(r)	moving iron
(d)	r.m.s values of a voltage containing d.c and ripples at 50 Hz and harmonics	(s)	thermal

Quiz Questions Key

1. (b)	2. (c)	3. (c)	4. (d)	5. (b)
6. (b)	7. (d)	8. (b)	9. (a)	10. (a)
11. (a)	12. (d)	13. (a)	14. (a)	15. (b)
16. (c)	17. (d)	18. (b)	19. (c)	20. (b)
21. (a)	22. (b)	23. (b)	24. (c)	25. (c)

26. (d)	27. (c)	28. (a)	29. (a)	30. (c)
31. (a)	32. (b)	33. (a)	34. (a)	35. (a)
36. (b)	37. (c)	38. (c)	39. (a)	

Match the Following Key

1. a-s, b-p, c-t, d-r, e-q
2. a-s, b-p, c-t, d-q, e-r
3. a-p, b-q, c-r, d-s
4. a-r, b-s, c-p, d-q
5. a-t, b-p, c-u, d-q, e-s
6. a-r, b-p, c-t, d-q, e-s
7. a-p, b-q, c-r, d-s

CHAPTER 2

Instrument Transformers

2.1 Introduction

The limitations of extension of instrument range in the conventional instrument transformer serve two important purposes in measurement.

1. It help to increase the range of the instrument beyond the value that is feasible by use of shunts and multipliers.
2. It insulates the instrument from the high voltage high current line.

There are mainly two types of instrument transformer's

(a) Current Transformer (C.T) (b) Potential Transformer (P.T)

The C.T is better suited than shunt for increasing the range of the ammeter. The current transformer produces the same instrument reading regardless of the constants of the instrument within reasonable limits. In the same way the P.T is a better substitute to the series resistance for the same advantages as enunciated in connection with C.T. These are used in METERING and PROVIDING PROTECTION in power system from over current, over voltage.

2.2 Current Transformer

The heavy alternating currents which can not be sensed through normal Ammeter and current coils of watt meters, energy meters can easily be measured by use of C.T along low range instrument.

A transformer is a device which consists of two windings called primary and secondary. It transfers energy from one side to another with suitable change in the level of voltage or current. a C.T basically has a primary coil of one or more turns of heavy cross-sectional area. In some C.Ts the bar carrying high current may act as a primary, this is connected in series with the line carrying high currents.

The secondary of the transformer is made up of a large number of turns of fine wire having small cross-sectional area. This is usually rated for 5 A. This is connected to the coil of normal range Ammeter.

The general representation of a C.T as shown in the Fig. 2.1.

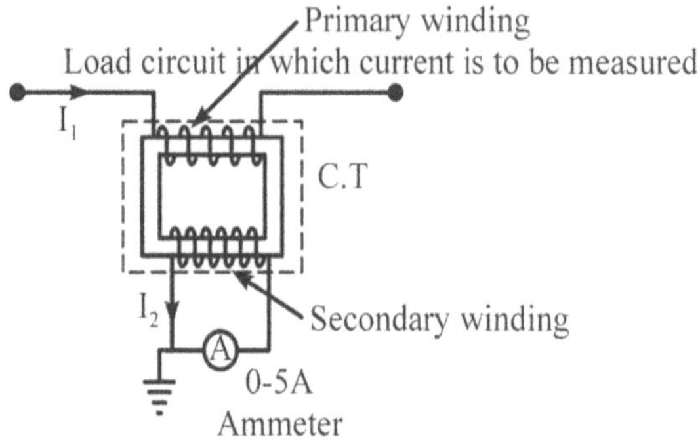

Fig. 2.1 Current transformer

2.2.1 Constructional Details of Current Transformer

The main importance lies in minimising the exciting ampere turns to keep the ratio and phase angle error small. For this purpose, the core material should have low reluctance and small iron loss the core material for wound type is Metal, or nickel-iron alloy containing copper which has the property of very high permeability, low loss and small retentivity is used in C.T.

(a) ***Winding:*** In order to reduce leakage reactance, the windings should be close together. Usually copper wire is used in the secondary and copper strip is used in the primary. The dimensions of which depends on the primary current. The windings are carefully designed to withstand forces at the time of short circuit current in the system. The bar primary ring core construction is generally used.

(b) ***Insulation:*** The primary and secondary windings are insulated by tape and varnish for lower line voltages but for high voltages of 6.6 kV and above, it is immersed in oil or compound filled. In the former case, the windings are wound over one another on cylinders of bakelite or similar material and the transformer is mounted in a short steel tank filled with transformer oil whereas in the latter case it enclosed in a tank which is filled up solid with insulating compound. In this case cooling is poor. Sometimes for oil immersed type, porcelean tube is also used to insulate primary and secondary. Portable CTs can also be obtained by actually winding through the core aperture, such CTs can carry high current range.

2.2.2 Working Principle

The C.T are basically step up Transformer i.e., stepping up a voltage from primary to secondary. Thus the current reduces from primary to secondary. So from current point of view, these are step down transformers, stepping down the current value considerably from primary to secondary.

Let N_1 = Number of turns of primary

N_2 = Number of turns of secondary

I_1 = Primary current

I_2 = Secondary current

For a transformer $$\frac{I_1}{I_2} = \frac{N_2}{N_1}$$

As N_2 is very high compared to N_1, the ratio I_1 to I_2 is also very high for current transformer. Such a current ratio is indicated for representing the range of current transformer.

Working on this current ratio and the meter reading on the secondary, the actual high line current flowing through the primary can be obtained.

Example: A 250:5, current transformers is used along with an ammeter. If ammeter is 3. A, estimate the line current?

Solution:

Given data: $$\frac{I_1}{I_2} = \frac{250}{5}$$

But ammeter as in secondary, $I_2 = 2.7$ A

$$\frac{I_1}{3} = \frac{250}{5}; \; I_1 = 150 \text{ A}$$

∴ line current is 150 A

2.2.3 Construction of Current Transformers

There are mainly two types of constructions used for the C.T.

Those are:

1. Wound type
2. Bar type

2.2.4 Wound Type

In wound type current trasformer construction, the primary is wound for more then one full turn, on the core. The construction is shown in Fig. 2.2.

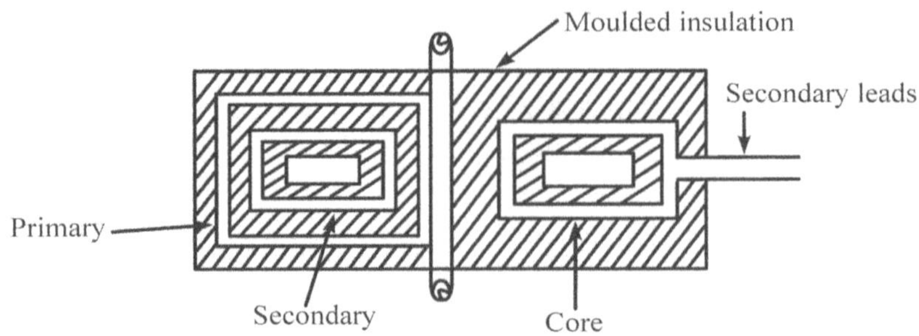

Fig. 2.2 Wound type current transformer

In a low voltage wound type current transformer the secondary winding is wound on a bakelite former. The heavy primary winding is directly wound on the top of the secondary winding with a suitable insulation in between the two. Otherwise the primary is wound completely separately and then taped with suitable insulating material and assembled with the secondary on the core.

The three forms of core used commonly. In Fig. 2.3(a), the rectangular form is built up of two L-punchings. The windings are placed in the shorter limbs with the primary wound over the secondary. In Fig. 2.3(b), gives considerable protection to the windings but is more difficult to build up than the other forms. The windings are placed on the central limb. In Fig. 2.3(c), the ring form is very commonly used when the primary current is large. The secondary winding is uniformly distributed around the ring and the primary winding is a single bar. This type of construction is very robust and has advantage of joint-less core and very small leakage reactance.

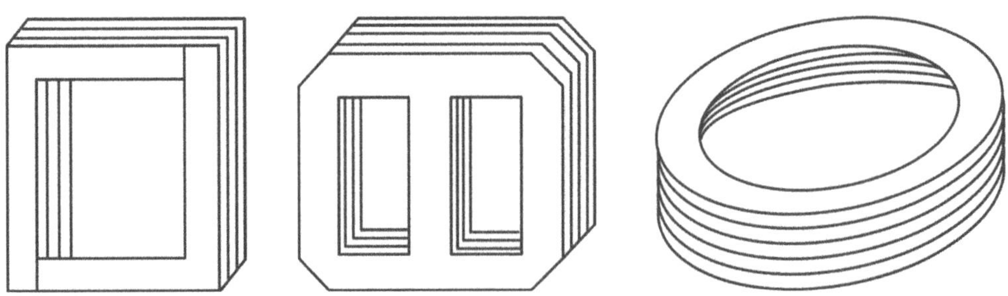

(a) (b) (c)

Fig. 2.3 Forms of cores

2.2.5 Bar Type Current Transformer

In bar type current transforms, the primary winding is nothing but a bar of suitable size. The construction is shown in Fig. 2.4.

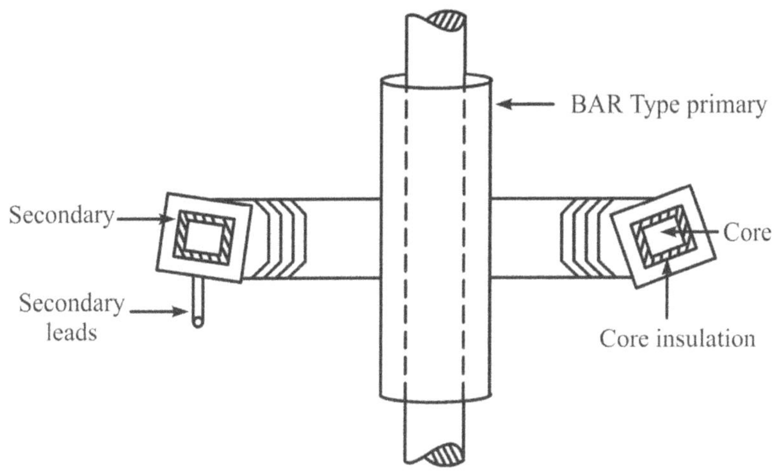

Fig. 2.4 Bar type current transformer

The insulation on the bar type primary is bakelized paper tube or a resin directly moulded on the bar. Such bar type primary is integral part of the current Transformer.

The core and the secondary windings are same in bar type transformer. The stampings used for the laminations in C.T must have high cross-sectional area than the ordinary transformer. Due to this, the reluctance of the interleaved corners remains as low as possible hence the corresponding magnetizing current is also small.

The windings are placed very close to each other so as to reduce the leakage reactance. To avoid the corona effect in bar type Transformer, the external diameter of the tube is kept large. The windings are so designed that without damage, they can withstand short circuit forces which may be caused due to short circuit in the circuit in which the C.T is inserted.

For small line voltages, the tape and varnish are used for insulation. For line voltage above 7 KV, the oil immersed or compound filled C.Ts are used.

2.2.6 Theory of Current Transformers

Consider the equivalent circuit of a C.T as shown in Fig. 2.5 with load.

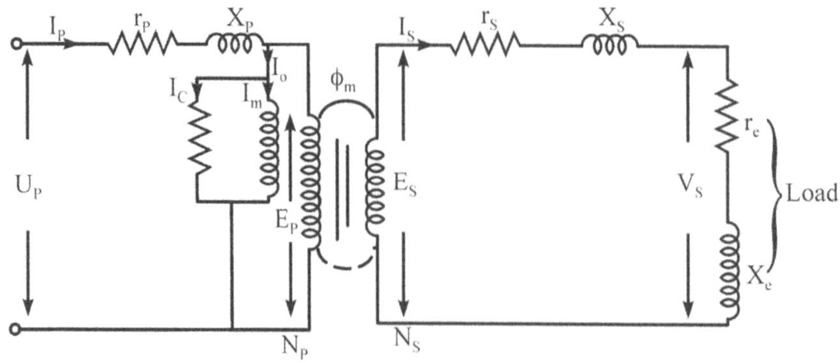

Fig. 2.5 Equivalent circuit of current transformer

Here,

r_p = Resistance of primary winding

X_p = Reactance of primary winding

r_s = Resistance of secondary winding

X_s = Reactance of secondary winding

r_e = Resistance of external burden of secondary

X_e = Reactance of external burden of secondary

E_P = Induced voltage in primary

E_S = Induced voltage in secondary

V_s = Secondary terminal voltage

I_P = Primary current

I_s = Secondary current

I_o = No load current (or) exciting current

I_c = Core loss component of I_o [$I_o \cos \phi$]

I_m = Magnetising component of I_o [$I_o \sin \phi$]

ϕ = Working flux of transformer

δ = Angle between E_s and I_s

θ = Phase angle of transformer

Δ = Phase angle of load or burden i.e., $r_e + jX_e$

$$= \tan^{-1} \frac{X_e}{r_e}$$

$\propto$ = Angle between I_o and working flux ϕ.

2.2.7 Derivation of Actual Ratio

Consider three phasor diagram of the transformer with a lagging p.f load as shown in Fig. 2.6(a).

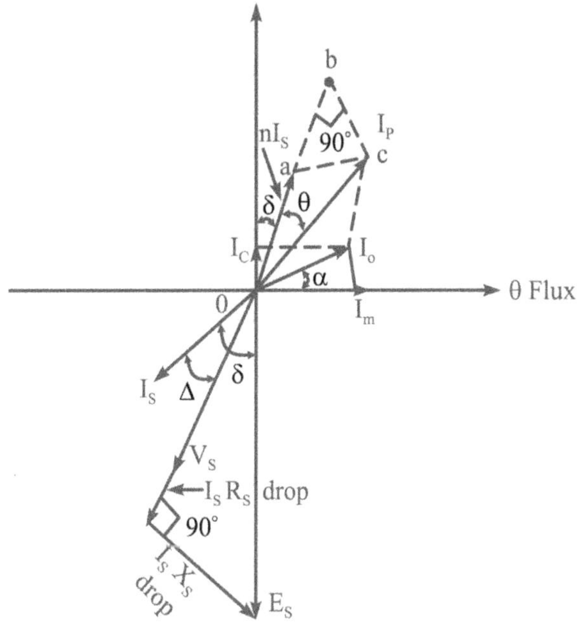

Fig. 2.6(a) Phasor diagram of current transformer

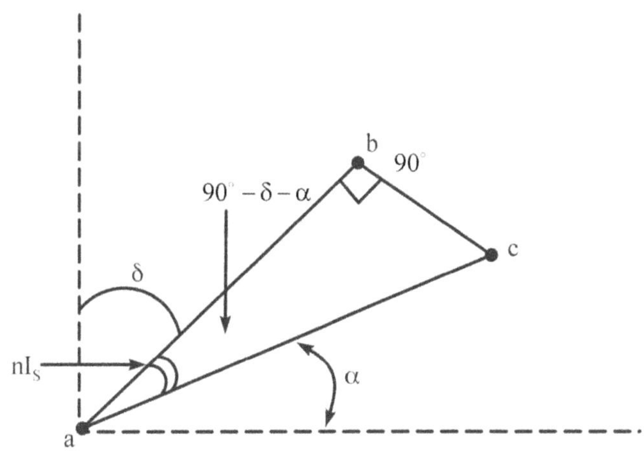

Fig. 2.6(b) A section of phasor diagram of a current transformer

Consider $\angle bac$ as shown in Fig. 2.6(b).

$\angle bac = 90° - \delta - \alpha$, $ac = I_o$, $oa = nI_s$, $oc = I_p$

$$\therefore \quad bc = ac \sin(90^\circ - \delta - \alpha) = I_o \sin[90^\circ - (\delta + \alpha)]$$

$$= I_o \cos(\delta + \alpha)$$

$$\therefore \quad ab = ac \cos(90^\circ - \delta - \alpha) = Io \cos[90^\circ - (\alpha + \delta)]$$

$$= I_o \sin(\delta + \alpha)$$

From right angle triangle Obc,

$$(Oc)^2 = (Ob)^2 + (bc)^2$$

$$= (Oa + ab)^2 + (bc)^2$$

$$= [nI_s + I_o \sin(\delta + \alpha)]^2 + [I_o + \cos(\delta + \alpha)]^2$$

$$= n^2 I_s^2 + 2nI_s I_o \sin(\delta + \alpha) + I_o^2 \sin^2(\delta + \alpha) + I_o^2 \cos^2(\delta + \alpha)$$

$$I_p = \sqrt{n^2 I_s^2 + 2n\, I_s\, I_o \sin(\delta + \alpha) + I_0^2} \quad \text{.....(2.1)}$$

$$\therefore \text{ actual ratio} = R = \frac{I_p}{I_s} \qquad \text{... As per definition}$$

$$\therefore \quad R = \frac{\sqrt{n^2 I_s^2 + 2n\, I_s\, I_o \sin(\delta + \alpha) + I_0^2}}{I_s} \quad \text{.....(2.2)}$$

Practically for properly designed Transformer, $I_o << nI_s$

$$\therefore \quad R = \frac{\sqrt{n^2 I_s^2 + 2n\, I_s\, I_o \sin(\delta + \alpha) + I_0^2 \sin^2(\delta + \alpha)}}{I_s}$$

Since, $I_0^2 = I_0^2 \sin^2(\delta + \alpha)$

$$\therefore \quad R = \frac{nI_s + I_o \sin(\delta + \alpha)}{I_s} = n + \frac{I_o}{I_s} \sin(\delta + \alpha) \quad \text{.....(2.3)}$$

This is approximate value of actual ratio but practically very close to actual result. The eq. in (2.3) can be expanded as,

$$R = n + \frac{I_o}{I_s}[\sin\delta\cos\alpha + \cos\delta\sin\alpha]$$

But, $I_o \cos\alpha = I_m$ and $I_o \sin\alpha = I_c$

$$R = n + \frac{I_m}{I_s}\sin\delta + \frac{I_c}{I_s}\cos\delta$$

$$\mathbf{R} = n + \frac{I_m \sin\delta + I_c \cos\delta}{I_s} \quad \text{.....(2.4)}$$

Note: The δ is positive for logging p.f load, while negative for leading p.f load.

2.3 Derivation of Phase Angle (θ) of Transformer

The phase angle 'θ' is defined as the angle between reversed secondary current phasor. That is reflected secondary current phasor and the primary current.

Sign convention:

1. θ is positive [if reflected secondary current leads primary current]
2. θ is negative [if secondary current lags primary current]

From phasor diagram (Fig. 2.6(a))

$$\therefore \qquad \theta = nI_s \wedge I_p$$

$$\tan\theta = \frac{bc}{ob} = \frac{bc}{oa+ab} = \frac{I_o\cos(\delta+\alpha)}{nI_s + I_o\sin(\delta+\alpha)}$$

∴ Now $\tan\theta \approx \theta$ as θ is very small.

$$\therefore \qquad \theta = \frac{I_o\cos(\delta+\alpha)}{nI_s + I_o\sin(\delta+\alpha)} \text{ radians} \qquad .....(2.5)$$

but $I_o << nI_s$ hence neglecting from denominator,

$$\theta = \frac{I_o\cos(\delta+\alpha)}{nI_{IS}} = \frac{I_o[\cos\delta\cos\alpha - \sin\delta\sin\alpha]}{nI_s}$$

$$\therefore \qquad \theta = \frac{I_m\cos\delta - I_c\sin\delta}{nI_s} \text{ radians} \qquad .....(2.6)$$

θ in degrees:

$$\theta = \frac{180^\circ}{\pi}\left[\frac{I_m\cos\delta - I_c\sin\delta}{nI_s}\right] \text{ degrees} \qquad .....(2.7)$$

2.4 Errors in Current Transformer

For an instrument transformer it is necessary that the transformation ratio must be exactly equal to turns ratio and phase of the secondary terms (voltage and current) must be displaced by exactly 180° from that of the primary terms (voltage and current). Two types of errors affect these characteristics of an instrument transformer, which are:

1. Ratio error 2. Phase angle error

2.4.1 Ratio Error

In the Current transformer, ratio $\frac{I_2}{I_1} = \frac{N_1}{N_2}$, but actually it is not so. The current ratio is not equal to turns ratio because of magnetizing and core loss components of the exciting current. It also gets affected due to the secondary current and p.f. The load current is not a constant fraction of the primary current. Similarly in case of P.T, the ratio $\frac{V_2}{V_1}$ is also not exactly equal to $\frac{N_2}{N_1}$ due to the factor mention above. Thus the transformer ratio is not constant but depends on the load current, p.f of load and exciting current of the transformer. Due to this fact, large error is introduced in the measurements done by the instrument transformers. Such error is called RATIO ERROR.

$$\therefore \qquad \% \text{ Ratio error} = \frac{\text{Nominal ratio} - \text{actual ratio}}{\text{Actual ratio}} \times 100$$

$$\therefore \qquad \% \text{ Ratio error} = \frac{K_n - R}{R} \times 100$$

2.4.2 Phase Angle Error

In the power measurement, it is a must that the phase of secondary current is to be displaced by exactly 180° from that of primary current for C. T, while the phase of secondary voltage is to be displaced by exactly 180° from that of primary voltage for P.T but actually it is not so. The error introduced due to this fact is called phase angle error. It is denoted by angle 'θ' by which phase difference between primary are secondary is different from 180°.

$$\therefore \qquad \theta = \frac{180}{\pi}\left[\frac{I_m \cos\delta - I_c \sin\delta}{nI_s}\right] \text{ degrees}$$

Approximate result: In practice, the loads, are inductive and δ is positive, and very small.

$$\therefore \qquad \sin\delta \approx 0 \text{ and } \cos\theta \approx \mathbf{1}$$

[Hence, the equation 2.4 and 2.7 becomes]

$$R = n + \frac{I_c}{I_s}$$

$$\theta = \left(\frac{180}{\pi}\right)\left(\frac{I_m}{nI_s}\right) \text{ degrees}$$

[if in terms of I_p, these can be written using $n = \frac{I_p}{I_s}$]

$$\therefore \qquad R = n + \frac{nI_c}{I_p}$$

$$\theta = \left(\frac{180}{\pi}\right)\left(\frac{I_m}{I_p}\right) \text{ degree}$$

2.5 Causes of Errors in Current Transformer

In an ideal Transformer, the actual transformation ratio would be equal to the turns ratio and the phase angle would be zero. However, as a result of physical limitions inherent in electric and magnetic circuit of the Transformer, these are departure from the ideal and consequently there are errors caused. The reasons are:

(i) There is some exciting mmf required by the primary winding to produce flux and, therefore, the transformer draws a magnetizing current I_m.

(ii) The transformer input must have a component which supplies the core losses (eddy current and hysterons) and I^2R losess of transformer windings due to flow of current I_o.

Loss component I_e is required to feed the losses associated with the flux and also the associated copper loss in winding due to flow of I_o.

(iii) The flux density in the core is not a linear function of the magnetizing force i.e., the transformer core becomes saturated.

(iv) There is always a magnetic leakage and consequently the primary flux linkages are not equal to the secondary flux linkages.

2.6 Characteristic of Current Transformer

1. Effect of p.f of secondary circuit

(a) ***Ratio error:*** For all inductives loads, δ is positive hence $\sin(\delta + \alpha)$ is positive hence actual Ratio 'R' is always greater than the turns ratio n. For capacitive burdens, δ is negative and R is less than the turns ratio.

(b) ***Phase angle error:*** When load in inductive and δ is small positive then θ is positive, but as δ approaches 90° as load becomes highly inductive then θ becomes negative. For capacitive load 'δ' is negative and θ is always positive.

2. Effect of change in burden on secondary

The secondary winding current burden increases means volt-ampere rating increases. Due to increased secondary current, secondary flux increase which induce more voltage on secondary. These both I_m and I_c increase to keep flux constant. Due to this, errors also increase. Thus more secondary burden means more errors.

(a) ***Effect of change in frequency:*** If frequency is increased at constant voltage then the flux and flux density decreases. Thus there is reduction in I_m and I_c and hence errors also get reduced.

(b) ***Effect of change in I_p:*** The I_p and I_s are directly related as I_p changes, I_s also changes. For low values of I_p, I_o is dominating thus I_m and I_c are also dominating parts of I_p. Thus errors are higher. As I_p increase, part of I_o becomes insignificant from I_p hence I_c and I_m also are less significant compared to I_p i.e., I_s. Thus errors are less.

2.7 Why Current Transformer of Secondary should not be open?

It is very important that the secondary of C.T should not be kept open; either it should be shorted or must be connected in series with a low resistance coil such as current coils of watt meter, coil of ammeter etc. If it is left open, then current through secondary becomes zero hence the ampere turns produced by secondary which generally oppose primary ampere turns becomes zero. As there is no counter m.m.f unopposed, primary m.m.f produce high flux in the core. This produce excessive core losses, heating the core beyond limits. Similarly hearvy e.m.f will be induced on the primary and secondary side. This may damage the insulation of the winding, this is danger from the operator point of view as well. It is usual to ground the C.T on the secondary side to avoid a danger of shock to the operator.

Hence never open the secondary winding circuit of a C.T which its primary winding is energised. Thus most of the C.T have a short circuit link or a switch at secondary terminals. When the primary to be energised, the short circuit link must be closed so that there is no danger of open circuit secondary.

2.8 Potential Transformer (PT)

Potential Transformer are used to measure high voltages by mean of low voltage voltmeter or to emergise the pressure coil of watt meter or other protective equipment like current breaker used in high voltage system. There is no basic difference in the operation of current transformer and potential transformer except the load current in the potential transformer is of the order of exciting current and the load impedance on the secondary is mostly resistance. In a potential transformer, accuracy of the voltage ratio, minimising the voltage drop are two important points of consideration. This is done to avoid ratio and phase error.

2.8.1 Constructional Details

In order to reduce the errors in potential transformer the reluctance of transformer core should be a small as possible and the flux density in the core should be fairly about 1 wb/m^2. The exciting current can then be made small. By placing the two windings close together with proper insulation, the reactance of the potential transformer can be kept small. Turns compensation can be applied more successfully than current transformer owing to the large number of primary turns. The actual ratio of the potential transformer is greater than the turns ratio 'n' but can be made as near as nominal ratio by adjusting the primary and secondary turns so that 'n' is less than nominal ratio.

Generally these transformer are oil immersed. Adequate insulation is provided in the form of porcelain or other type of bushings for the primary terminals. With regards to core material and type of winding, it is similar to a Current Transformer except the values of ampere turns of the primary and secondary. The connection diagram of a potential transformer is given in Fig. 2.7. Usually the primary winding is fitted with a fuze.

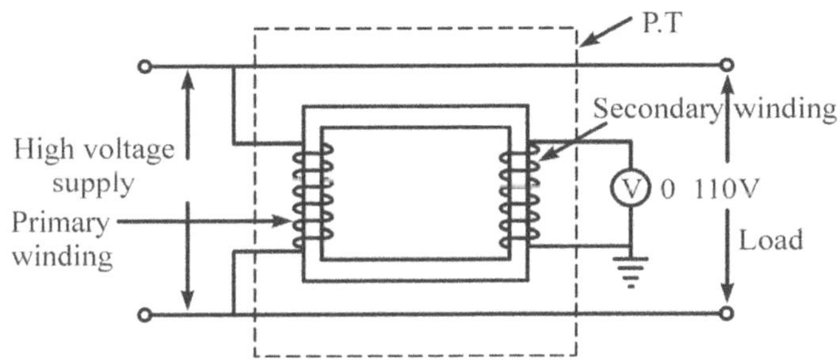

Fig. 2.7 Potential transformer

2.8.2 Theory of Potential Transformer

The loading of P.T is very small in practice hence exciting current I_o is of the order of I_s i.e., secondary winding current while in a normal potential transformer I_o is very small compared to I_s. The equivalent circuit of P.T is shown in the Fig. 2.8

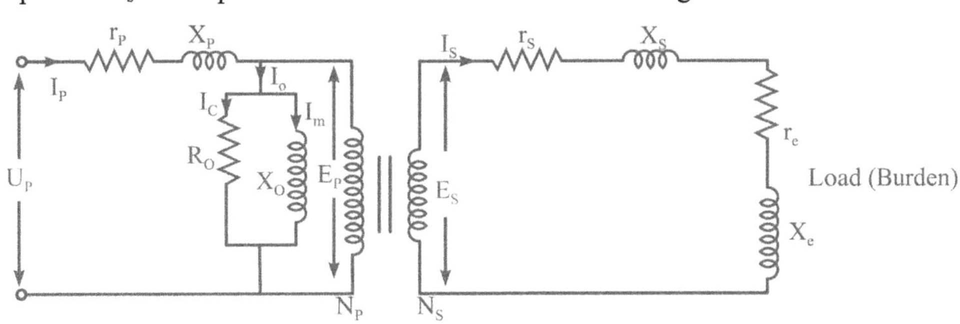

Fig. 2.8 Equivalent circuit of potential transformer

The various symbols are

E_p = Primary induced voltage
E_s = Secondary induced voltage
V_p = Primary applied voltage
V_s = Secondary terminal voltage
N_p = Primary turns
N_s = Secondary current
I_p = Primary current
I_s = Secondary current
ϕ = Working flux
I_m = Magnetising component of I_o
I_c = Core loss component of I_o
r_s X_s = Resistance and reactance of secondary winding
r_p X_p = Resistance and reactance of primary winding
r_e X_e = Resistance and reactance of burden
Δ = Phase angle of secondary load current

$$= \tan^{-1}\frac{X_e}{r_e}$$

For P.T $\qquad n = \dfrac{N_p}{N_s} = \dfrac{E_p}{E_s}$

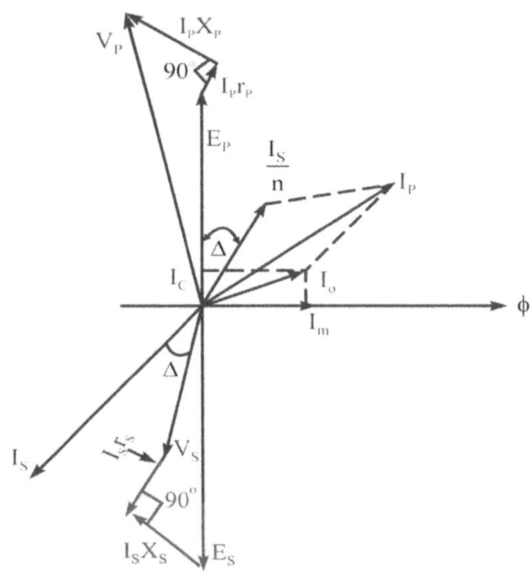

Fig. 2.9 Phasor diagram of a Potential Transformer

$I_p r_p$ is parallel to I_p | $I_s r_s$ is parallel to I_s

$\overline{E_p} + \overline{I_p r_p} + \overline{I_p X_p} = \overline{V_p}$ | $\overline{V_s} + \overline{I_s r_s} + \overline{I_s X_s} = \overline{E_s}$

2.8.3 Derivation of Actual Ratio

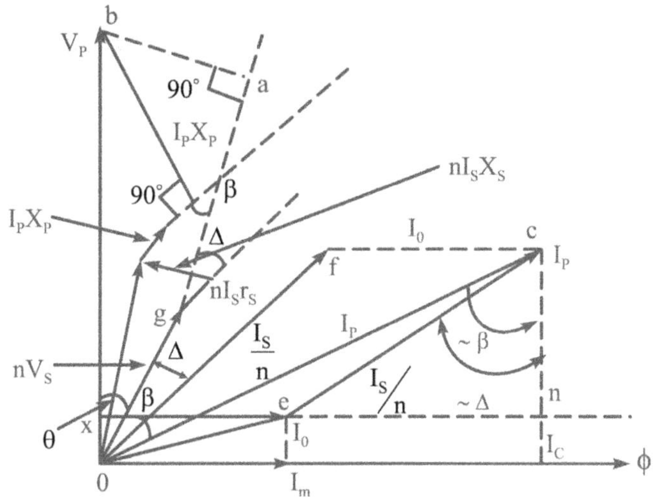

Fig. 2.10 Phasor diagram referred to primary

$$Oe = I_0, \quad Oc = I_p, \quad Of = \frac{I_s}{n}, \qquad Og = n\,V_s, \qquad Ob = V_p$$

θ = Phase angle of transformer = $V_p \wedge V_s$ reversed

Δ = phase angle of secondary load = $n\,V_s \wedge \dfrac{I_s}{n}$

β = Phase angle between I_p and I_s reversed

Oa = is n V_s extended and ba is perpendicular drawn from b on V_s extended

$$\therefore \quad Oa = Ob\cos\theta = V_p\cos\theta \qquad .....(2.8)$$

Now Oa is made up to various components.

$$Oa = n\,V_s + n\,I_s\,r_s\cos\Delta + n\,I_s\,X_s\sin\Delta + I_p\,r_p\cos\beta + I_p\,X_p\sin\beta \qquad ......(2.9)$$

Equating 2.8 and 2.9

$$\therefore \quad V_p\cos\theta = n\,V_s + n\,I_s\,(r_s\cos\Delta + X_s\sin\Delta) + I_p\,(r_p\cos\beta + X\sin\beta) \qquad .....(2.10)$$

For potential transformer, the load is nothing but a voltmeter hence θ is very small hence both V_p as well as nV_s can be assumed to be perpendicular to ϕ.

Thus approximately, $\angle Ocd \approx \beta$ and $\angle ecd \approx \Delta$

$$\therefore \quad cd = I_p \cos\beta = ch + hd = \frac{I_s}{n}\cos\Delta + I_c \qquad .....(2.11)$$

and
$$I_p \sin\beta = I_m + \frac{Is}{n}\sin\Delta \qquad .....(2.12)$$

As θ is very very small, $\cos\theta \approx 1$

$$\therefore \quad V_p \cos\theta = V_p \qquad .....(2.13)$$

Using in 2.10, all the results of equations (2.11), (2.12) & (2.13)

$$V_p = n\,V_s + n\,I_s\,(r_s \cos\Delta + X_s \sin\Delta) + r_p\left[\frac{I_s}{n}\cos\Delta + I_c\right] + X_p\left[I_m + \frac{I_s}{n}\sin\Delta\right]$$

$$\therefore V_p = nV_s + \frac{I_s}{n}\cos\Delta\left(n^2 r_s + r_p\right) + \frac{I_s}{n}\sin\Delta\left(n^2 X_s + X_p\right) + I_c r_p + I_m X_p \qquad .....(2.14)$$

Now $n^2 r_s + r_p = R_{1e}$ = equivalent resistance referred to primary

$n^2 X_s + X_p = X_{1e}$ = equivalent reactance referred to primary

$$\therefore \quad V_p = n\,V_s + \frac{I_s}{n}\,[R_{1e}\cos\Delta + X_{1e}\sin\Delta] + I_c r_p + I_m X_p \qquad .....(2.15)$$

Thus the actual ratio is,

$$R = \frac{V_p}{V_s} = n + \frac{\frac{I_s}{n}\,[R_{1e}\cos\Delta + X_{1e}\sin\Delta] + I_c r_p + I_m X_p}{V_s} \qquad .....(2.16)$$

The result also can be derived in terms of parameter referred to secondary. The equation of V_p can be written as

$$V_p = n\,V_s + \frac{n^2}{n} I_s \cos\Delta\left[r_s + \frac{r_p}{n^2}\right] + \frac{n^2}{n} I_s \sin\Delta\left[X_s + \frac{X_p}{n^2}\right] + I_c r_p + I_m X_p$$

But
$$r_s + \frac{r_p}{n^2} = R_{2e} \text{ = equivalent resistance referred to secondary}$$

$$X_s + \frac{X_p}{n^2} = X_{2e} \text{ = equivalent reactance referred to secondary}$$

$$V_p = n\,V_s + n\,I_s\,\left[R_{2e}\cos\Delta + X_{2e}\sin\Delta\right] + I_c r_p + I_m X_p \qquad .....(2.17)$$

$$\therefore \quad R = \frac{V_p}{V_s} = n + \frac{nI_s\left[R_{2e}\cos\Delta + X_{2e}\sin\Delta\right] + I_c r_p + I_m X_p}{V_s} \qquad .....(2.18)$$

2.8.4 Derivation of Phase Angle 'θ'

Referred to the phase diagram in Fig 2.10

$$\tan\theta = \frac{ab}{oa} = \frac{I_pX_p\cos\beta - I_pV_p\sin\beta + nI_sX_s\cos\Delta - nI_sr_s\sin\beta}{nV_s + nI_sr_s\cos\Delta + nI_sX_s\sin\Delta + I_pr_p\cos\beta \times I_pX_p\sin\beta}$$

In the expression of oa, the term other than nV_s are very small and can be neglected. Similarly as θ is very small tan θ ≈ θ.

$$\therefore \quad \theta = \frac{I_pX_s\cos\beta - I_pr_p\sin\beta + nI_sX_s\cos\Delta - nI_sr_s\sin\Delta}{nV_s}$$

$$= \frac{X_p\left[\frac{I_2}{n}\cos\Delta + I_c\right] - r_p\left[I_m + \frac{I_s}{n}\sin\Delta\right] + nI_sX_s\cos\Delta - nI_sr_s\sin\Delta}{nV_s}$$

$$= \frac{I_s\cos\Delta\left(\frac{X_p}{n} + nX_s\right) - I_s\sin\Delta\left(\frac{r_p}{n} + nr_s\right) + I_cX_p - I_mr_p}{nV_s}$$

$$= \frac{\frac{I_s\cos\Delta}{n}\left(X_p + n^2X_s\right) - \frac{I_s\sin\beta}{n}\left(r_p + n^2r_s\right) + I_cX_p - I_mr_p}{nV_s}$$

Now $\quad r_p + n^2 r_s = R_{1e}$ and $X_p + n^2 X_s = X_{1e}$

$$\theta = \frac{I_s}{n}\cos\Delta X_{1e} - \frac{I_s}{n}\sin\Delta R_{1e} + I_cX_p - \frac{I_mr_p}{nV_s}$$

$$\theta = \frac{I_s}{n}\left(X_{1e}\cos\Delta - R_{1e}\sin\Delta\right) + I_cX_p - \frac{I_mr_p}{nV_s} \text{ radians} \qquad \text{.....(2.19)}$$

In terms of quantities referred to secondary,

$$R_{2e} = \frac{R_{1e}}{n^2} \quad \text{and} \quad X_{2e} = \frac{X_{1e}}{n^2}$$

$$\frac{\theta = nI_s\left(X_{2e}\cos\Delta - R_{2e}\sin\Delta\right) + I_cX_p - I_mr_p}{nV_s}$$

$$\therefore \quad \theta = \frac{I_s}{n}\left(X_{2e}\cos\Delta - R_{2e}\sin\Delta\right) + \frac{I_cX_p - I_mr_p}{nV_s} \text{ radians} \qquad \text{.....(2.20)}$$

It can be noted that the phase angle θ is treated positive then V_s reversed i.e., nV_s leads the primary winding voltage V_p. The θ is treated negative when nV_s lags the primary winding voltage V_p.

Once R and θ are obtained then the errors in potential transformers are,

$$\therefore \quad \% \text{ ratio error} = \frac{K_n - R}{R} \times 100$$

and $\therefore$ Phase angle error $= \theta$ radians

2.9 Characteristic of Potential Transformers

1. **Effect of change in burden on secondary:** The change in burden on secondary changes the secondary current and secondary VA. As the secondary current increases primary current increases, this increases in the various drops for constant V_p; V_s decreases. This increases the transformer ratio R. Thus ratio error increase as the secondary current i.e., burden increase. Due to increases in voltage drop, V_p advances while V_s retards in phase thus the phase angle between V_p and V_s increases and becomes more negative as the secondary current increases.

2. **Effect of power factor of secondary circuit:** The p.f of secondary circuit depends on the p.f of the secondary burden. As p.f decreases, angle Δ increases. This shifts I_p towards I_o, as seen from the phasor diagram. Thus V_p and V_s becomes almost in phase with E_p and E_s respectively. thus V_p becomes more compared to E_p but as V_p is constant as supply voltage, the result is the reduction in E_p. Similarly the voltage V_s also reduces compared to E_s hence from equation 2.18 it is clear that the R increases as V_s decreases due to decrease in the p.f of secondary circuit while V_s advances in phase and V_p retards hence negative phase angle reduces as lagging p.f of secondary decreases.

3. **Effect of primary voltage:** The primary voltage is practically constant and there is hardly any significant change in it hence there is no significant change in the errors.

4. **Effect of change in frequency:** If frequency increases at constant voltage then the flux and flux density decreases. This reduces the values of I_m, I_c and hence I_o. This tends to reduce the errors. But increase in frequency increases the leakage reactance increasing the voltage drops. This increases the ratio and ratio error. Thus effects are opposite to each other and the resultant effect depends on relative values of I_o and the leakage reactance. While the phase angle increases due to increases in leakage reactance as well as decrease in I_o.

2.10 Errors in Potential Transformer

Increase in secondary burden increases the phase angle and the ratio error due to increase in resistance and reactance drop in the primary and secondary. A change in p.f from unity to lagging of the secondary increases ratio error; decreases phase angle. Reduction in frequency increases the core flux with corresponding increase in the exciting current and reduction in reactances drops.

Increases in exciting current increases ratio error and reduction in reactance drop decreases ratio error.

2.11 Application of Current Transformer and Potential Transformer

C.T and P.T have a wide application in high voltage and high current power system in the measurements of current, voltage and power. Besides, it is also used in the protection circuit as in the case of merz price protection and circuit breaker.

2.12 Advantages and Disadvantages of Instrument Transformer

Advantages

1. The rating of low range meter can be fixed irrespective of value of high voltage or current to be measured.
2. The normal range volt meter and ammeter can be used along with these transformers to measure high voltage and current.
3. These can be used for operating many type of protecting devices such as relays or pilot lights.
4. These transformer isolate the measurement from high voltage and current circuit. This ensures safety of the operator and makes the handling of the equipments very easy and safe.

Disadvantages

The only disadvantage of these instrument transformer is that they can be used only for A.C circuit and not for D.C circuit.

2.13 Difference between Instrument and Power Transformers

Instrument Transformer

1. Mainly used to extend the range of the instruments while measuring parameters like:

   Voltage, Current, Power and Energy etc.

2. They are basically step down transformer and used along with devices such as protective relays, indicators etc.
3. They are required to transform very small power as their loads are generally delicate moving elements of the instruments.
4. As the power Output is very small as loads are light hence heating is not severe.
5. Accuracy is the main consideration while designing and phase angle errors to minimum. Cost is the second consideration.

   Example are C.T and P.T

Power Transformer

1. Mainly used to change the voltage levels in a Power System
2. They can be used to step up or step down the voltage.
3. They are required to transform huge amount of power to the load.
4. As they handle large power, the heat dissipation is the major consideration and cooling arrangement is necessary.
5. The cost is the main consideration in the design while efficiency and regulation are the second considerations.

   Example: Distribution Transformers and Transmission Transformer

2.14 Power Factor Meter (p.f meter)

To measure the current, voltage and power in a A.C circuit, its p.f can be calculated from the relationship. i.e., $P = VI \cos \phi$

$$\therefore \qquad \cos \phi = \frac{P}{VI} \qquad [\cos \phi = \text{p.f of circuit}]$$

This method of determining the p.f of an electrical circuit is however, of low accuracy, has a number of disadvantages and is rarely used in practice. It is obviously desirable to have an instantaneous indication of the p.f of an A.C. circuit, especially where, p.f is varying continuously, without having resource to mathematical calculations of readings of several instruments. P.f meters indicate directly by a single reading of the circuit to which they are connected. The accuracy obtained with the use of p.f meter is sufficient for most purposes other than high precision testing. The basic construction of p.f meters is similar to a watt meter. There are mainly two types of p.f meters.

1. Electrodynamometer type and
2. Moving iron type

These instruments are described in details below.

2.14.1 Electrodynamometer Type Power Factor Meter

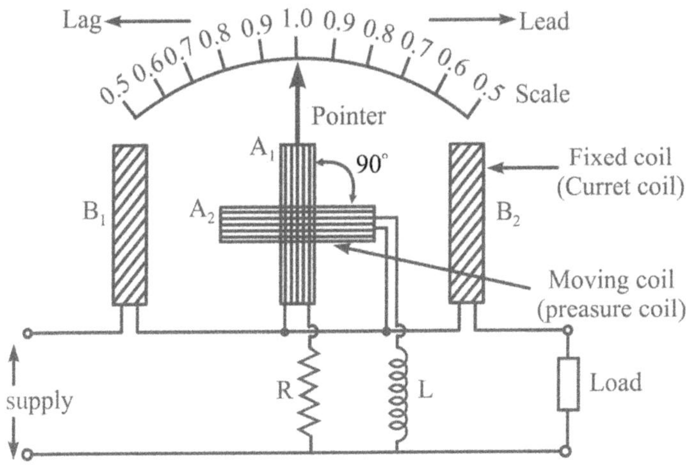

Fig. 2.11 Single phase electrodynamometer type power factor meter

Working Principle: It consists of two fixed coils B_1 B_2 as shown in Fig. 2.11 carrying current in the circuit under test and the magnetic field of these coils are proportional to the main current. Between the two fixed coils and within the field is pivoted to moving coil A_1 and A_2 placed at an angle of 90° apart. These coils move together and carry a pointer to indicate the p.f over a graduated scale. The dimensions and the number of turns of the two coils are same and produce equally strong magnetic field 90° displaced in space and time when equal current having a time phase difference of 90° are passed through them. The resistor 'R' and inductor 'L' are connected in series to these two coils and are so adjusted that these currents are equal when the frequency is normal. At other values of frequency, the two currents differ since the coil A_2 is dependent upon frequency while coil A_1 is independent of frequency.

Assuming that the time phase difference between the two currents in A_1 and A_2 is exactly by 90° i.e., time phase difference as zero and also the field of two fixed coils are uniform as in the direction of arrow. The torque on each coil for a given coil current is maximum when the coil is parallel to the field (Fig. 2.12) along XX. Now if the p.f angle is ϕ, the coil takes up a position of equilibrium displaced by 'θ' from the vertical and the torque due to the two coils are equal and opposite. Since the current in coil A_1 is in phase with system voltage and the field in which it moves is proportional to the system current, then coil A_1 movement is 'watt meter' movement displaced by $(90 - \theta)$ from the maximum torque position.

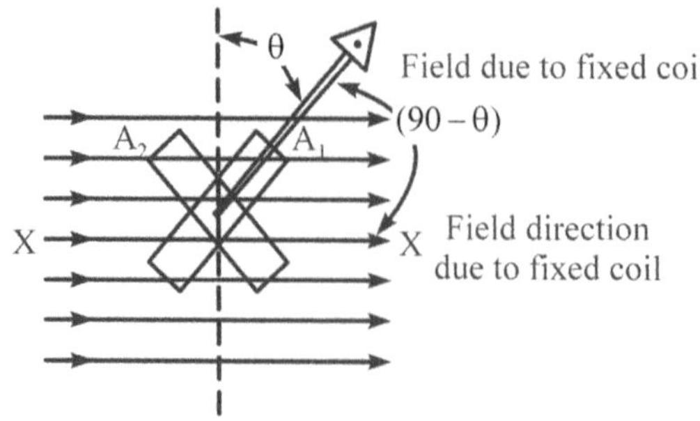

Fig. 2.12

θ = Angle of deflection

ϕ = Power factor angle

$\therefore$ The torque of $A_1 = T_{A1} = K_1$ VI $\cos\phi \cos(90 - \theta)$

where, K_1 is a constant.

Similarly since the current in coil A_2 lags 90° on the system voltage and coil B is a sine meter movement displaced θ from the maximum torque position.

$\therefore$ torque in $A_2 = T_{A2} = K_2$VI $\sin\phi \cos\theta$ [$\because K_2$ = constant]

At the final deflected position

$$T_{A1} = T_{A2}$$

$\therefore$ $$K_1 VI \cos\phi \cos(90^\circ - \theta) = K_2 VI \sin\phi \cos\theta$$

or $$\cos(90^\circ - \theta) = \sin\phi \cos\theta$$

or $$\frac{\cos(90^\circ - \theta)}{\cos\theta} = \frac{\sin\phi}{\cos\phi}$$

or $$\frac{\sin\theta}{\cos\theta} = \tan\theta$$

or $$\tan\theta = \tan\phi \quad \therefore \quad \boxed{\theta = \phi}$$

Therefore the angular position taken by the coils is equal to the system phase angle whether it is leading or lagging side. Connection diagram of single phase dynamometer type p.f meter is shown in Fig. 2.11.

2.15 Moving Iron Power Factor Meter

There are mainly two types of moving iron power factor meters,

1. Rotating field type
2. Alternating field type

The advantages of moving iron power factor meter over the dynamometer type are:

1. These instruments are simple and robust in construction, they are comparatively cheap also.
2. A scale extends over 360°.
3. The working forces in moving iron are larger.
4. All coils in moving iron are fixed so no ligaments are required.

Disadvantages

The calibration of these instruments is appreciably affected by variations in supply frequency, voltage and waveform.

But due to the losses in the iron parts, the accuracy of moving iron power factor meters is much less than electrodynamometer type.

2.15.1 Rotating Field Type Moving Iron Power Factor Meter

This type of meter is used for balanced type of loads with the help of three coils A_1, A_2 and A_3 these coils are supplied from three phase system through CTs. Coil B is connected at the centre of the system of coils as shown in Fig. 2.13(a) & (b) and is connected in series with a resistance across two lines of the supply. A sector shaped iron rod is pivoted at the centre with sector shaped iron pieces I_1 and I_2 at its ends. Damping vanes and a pointer is fixed on the spindle with no controlling force. The moving system carries an alternating flux. It does not tend to rotate continuously but sets in a definite position which depends upon the relative phases of the current in coil B which is in phase with voltage and current in coil A_3. The deflection of the system is approximately equal to the angle of the phase displacement between current and voltage in three phase circuit. Due to the presence of iron in the instrument and variation of reactance of coil B with the variation of supply frequency, the error is introduced in the reading. Hence, it is necessary to calibrate the instrument at normal frequency.

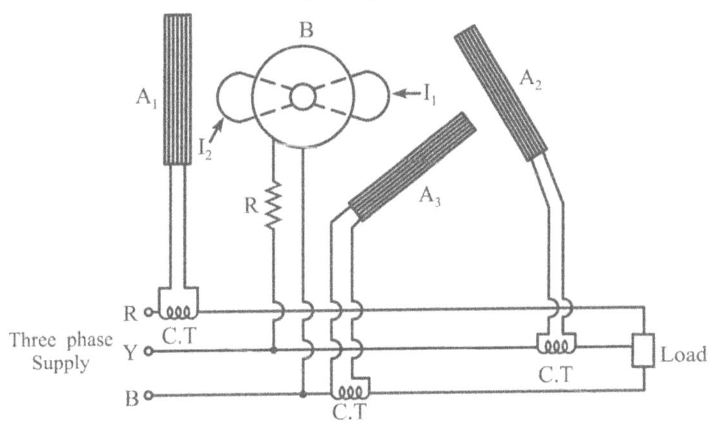

Fig. 2.13(a) Rotating field type moving iron power factor meter

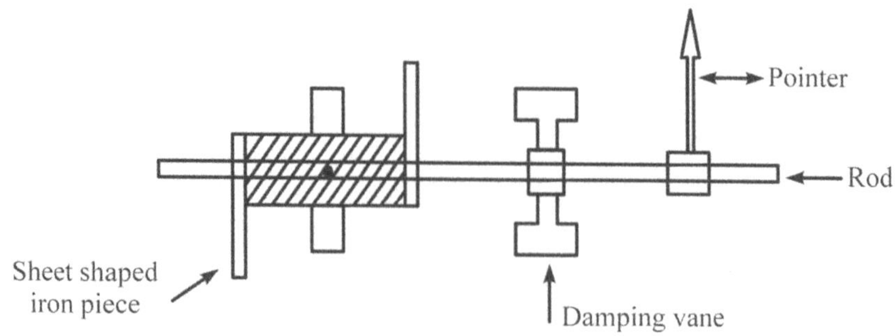

Fig. 2.13(b) Moving system

2.15.2 Alternating Field Type Moving Iron Power Factor Meter

It consists of three moving iron elements mounted on a single spindle one over the other with iron magnetic separating piece and are spaced 120° in different parallel planes. Each iron piece is magnetised by a separate pressure coil which are star connected to the three phase supply. A pair of coil forming current coil are placed on each side of the moving coil and is connected to one of the three phase supply. The spindle of the moving system carries a pointer and damping vanes without any control system. With a slight modification, the system can be used for unbalanced or single or two phase circuit also (Fig. 2.14).

The non-magnetic pieces denoted as 'S'. The Fig. 2.14(a). shows the construction of this instrument where C_1, C_2, and C_3 are iron sectors, these iron sectors are magnetised by the coils B_1, B_2 and B_3 these are voltage coils.

The current coil is divided into two equal parts A_1 and A_2 parallel to each other, the current coil carries one of the three line currents. One that A_1 of the current coil is on one side of the moving system and other A_2 on other side.

When the instrument is connected to supply, the moving system of the instrument deflects to such a position that the mean torque upon one of the iron pieces is neutralised by the two other torques so that the resultant is zero. At this position, the deflection of the iron piece, which is magnetised phase as the current coil, is equal to the phase angle between the voltage and current of three phase supply. The effects of iron loss and inductance of pressure coils are negligible. No resultant rotating magnetic field is set up by the three pressure coils which are placed at different levels. Hence any tendency for the moving system to rotate continuously following the drag of such a field is eliminated. The readings are not affected by variation of supply frequency, voltage, wave forms as in other instruments.

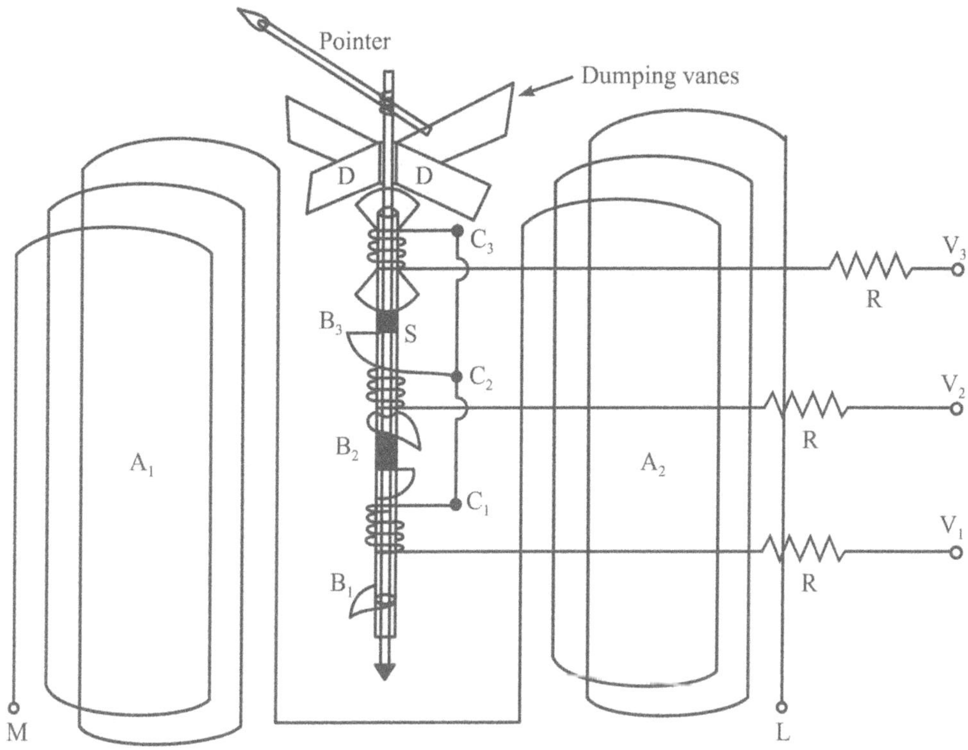

Fig. 2.14(a) Alternating field type moving iron power factor meter

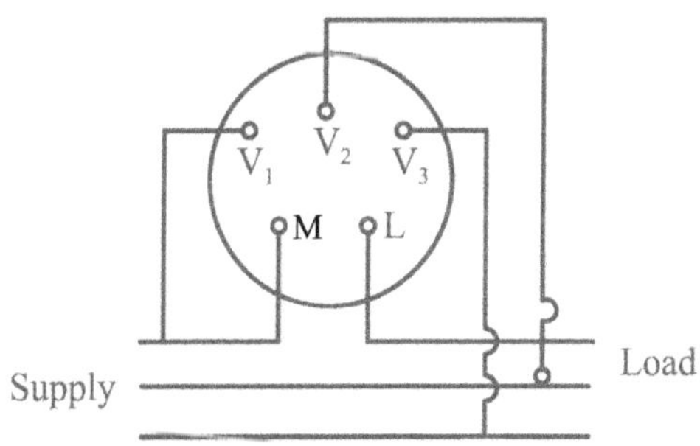

Fig. 2.14(b) Symbolic representation

2.16 Frequency Meters

There are several methods of measuring frequency such as measuring the speed of alternator, using 'AC' bridge and oscillators etc. But the general principle on which it works are:

(a) Phenomenon of mechanical resonance

(b) Phenomenon of electrical resonance

(c) Variations of impedance with changing frequency.

Under the frequency measurement only three distinct types are discussed in the succeeding paragraphs.

They are:

1. Vibrating Reed Type frequency meter
2. Electrical Resonance type frequency meter
3. Weston frequency meter.

2.16.1 Vibrating Reed Type Frequency Meter

It consists of a number of a thin steel strips called Reeds arranged along side and close to an electromagnet which is energised by the supply whose frequency is to be determined. The electromagnetic coil has high resistance and the meter, for all practical purposes, behaves as a voltmeter. The normal dimensions of the reeds are 4 mm wide and 0.05 mm thick and differ in size with one another in length. Each reed carries a load at its upper edge and the weights also differ from one another. The frequency to which it responds can be obtained from

$$f_n = \frac{1_x}{2\pi l^2}\sqrt{\frac{E_x}{\Delta k\,(1 + 4.1\,k)}}$$

where,

l = length of the reed

x = thin of the reed

E_x = Modulus of elasticity of reed material

Δk = Density of the material

k = Ratio of the mass of the reed

The electromagnet is wound with large number of turns with fine wire and is connected in series to a resistance across the supply such that the induction effect is least and negligible. When the instrument is connected across the supply, alternating flux is set up and due to a attractive force for every half cycle the reed whose natural frequency is double that of the supply frequency responds and vibrates with maximum amplitude. The corresponding frequency can be noted from the scale opposite to the vibrating reed. If the two adjacent reeds vibrate with equal amplitude, then supply frequency measuring. The usual range of this type of instrument is between 47 Hz to 53 Hz. By polarizing the reeds, the range of the instrument can be doubled.

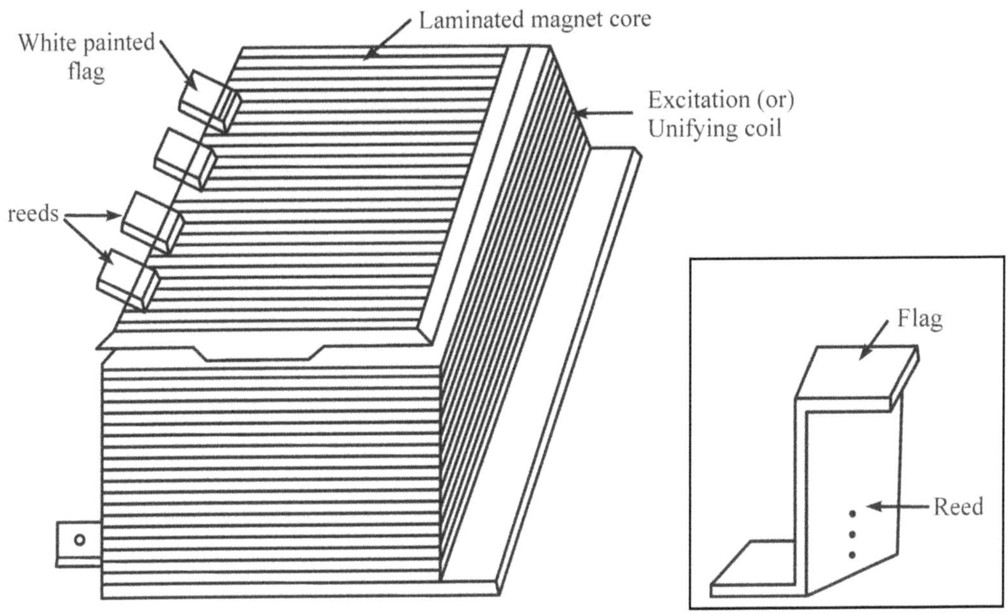

Fig. 2.15 Vibrating reed type frequency meter

2.16.2 Electrical Resonance Type Frequency Meter

It consists of a fixed magnetising coil and a moving coil as shown in Fig. 2.16. The fixed coil is mounted at one end of a laminated core, while the moving coil is pivoted in such a way as to allow the fixed coil inside it while in motion. The moving coil also carries a

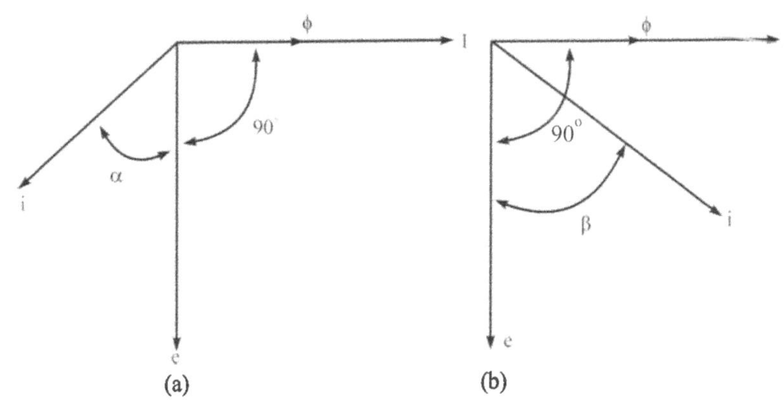

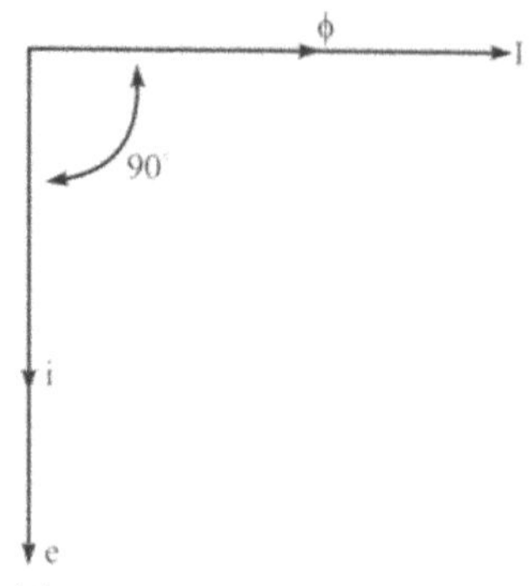

(c)

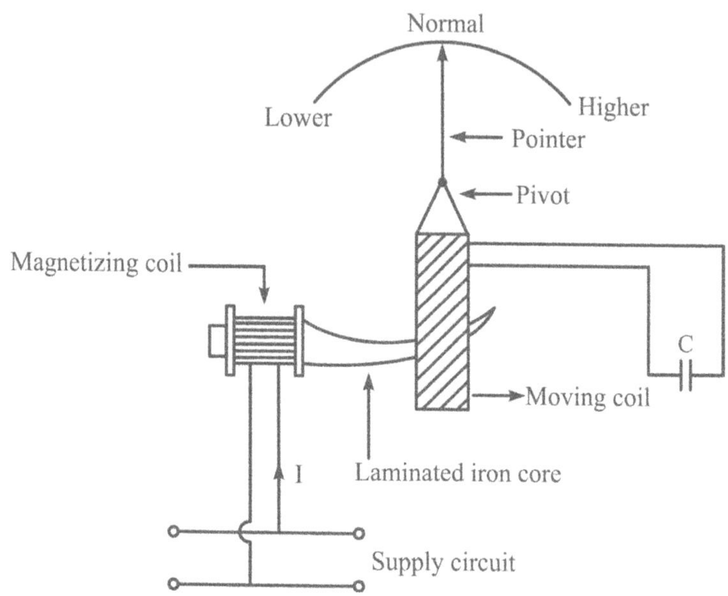

Fig. 2.16 Electrical resonance type frequency meter

pointer and is connected across a capacitor. When the fixed oil is energised by the supply, it sets up a flux ϕ in the core and the e.m.f is induced in moving coil which lags behind the flux by 90° and the P.F of the current in the moving coil depends upon the circuit constant and the coil. If the load is inductive, the current will lag behind e.m.f and vice versa. A suitable capacitance across the moving coil balances the inductive effect so that the current in the moving coil will be in phase with the induced e.m.f. The torque acting on the moving system will be proportional to I. i cos $(90 \pm \beta)$ where I is the current in the fixed oil as i is the current in the moving coil and β is the phase difference between current and induced e.m.f in the moving coil. If 'β' is zero, torque is zero.

An important advantage of the instrument is that the great sensitivity is achieved, as the inductance of the moving coil changes slowly with variation of its position on the core. This meter is also called "*Ferro-dynamic Frequency Meter*".

2.16.3 Weston Frequency Meter

It operates on the principle of variation of impedance with frequency and almost corresponds to moving iron instrument. It consist of two sets of coils $A_1 - A_2$, and coil $B_1 - B_2$ placed with their magnetic axis at right angles to each other. A soft iron is placed at the center of the coil group on pivoted needle which carries a pointer as shown in Fig. 2.17. The pointer moves over a graduated scale with the air damping provided to it and without a control spring. The coil A is connected is series with an inductance L_A, across a resistance R_A and in the similar way, coil B is connected in series with resistance R_B across an inductance L_B so that the voltage available to coil B is the impedance drop across L_B. The inductance 'L' in series is to damp any harmonic wave form of the supply.

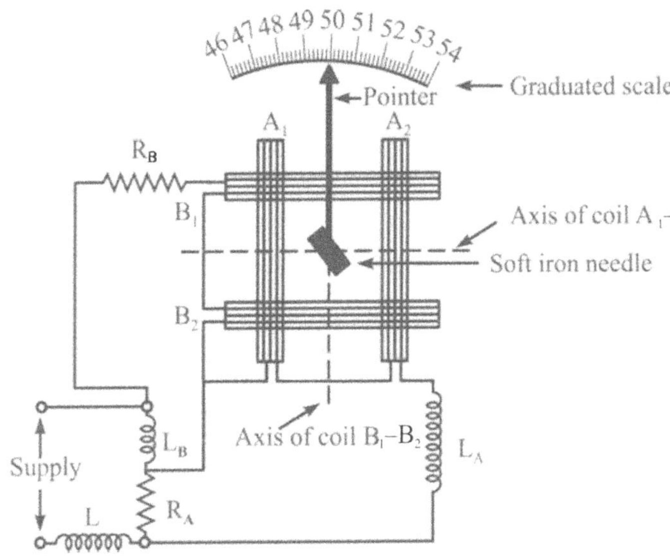

Fig. 2.17 Weston frequency meter

Depending upon the current in the coils A & B and the resultant field due to them, the soft iron takes a position and deflected position is from the centre zero depending upon the lagging and leading power factor of the load. If the frequency increases, the current in coil decreases because of inductance in coil A due to increase in L_A while the voltage across R_A remains constant. On the other hand, the current in coil B increases since voltage drop increases. In view of voltage drop across L_B, thus, coil B exerts more force on the needle and the needle deflects towards B axis and vice versa if the frequency decreases.

2.17 Synchroscopes

A synchroscopes is used to determine the correct instant for cling the switch which connects an alternator of the power station busbars. This process of connecting at the current instant or 'synchronizing' is necessary when an unloaded "Incoming" machine is to be connected to the busbar in order to share the load. The conditions which are to be fulfilled before paralleling an alternator are:

1. The frequency must be equal
2. The terminal voltage must be equal
3. The phase sequence should be equal
4. At the instant when paralleling takes place, the amplitude of voltage should be equal.

The simplest method for synchronizing alternator with busbar is through synchronising lamp method, but it is not suitable for large power stations, where alternator generates high voltage. Hence synchroscope which are basically indicating instruments of special form of power factor meter are used.

These are mainly two types:

(a) Weston synchroscope (or) Electrodynamometer:

(b) Moving iron synchroscope

2.17.1 Weston Synchroscope (or) Electrodynamometer Type Synchroscope

The construction is similar to the spring controlled dynamometer type of watt meter. In this type, fixed coils are designed to carry small current and is connected in series with a resistance and the unit is connected across a busbar as shown in Fig. 2.18. The moving coil has a resistance and capacitance in series with it and the unit is connected across the incoming machine terminals. The values of inductance and capacitance are so chosen that the currents in the fixed and moving coils are in quadrature when the two supplies to be parallel are in synchronism. At this position no torque is produced in the moving system and the pointer attached to it indicates synchronism. Even when the emf are 180° apart, the currents in the coil will be in quadrature and to avoid, lamp is also used. The lamp will not glow when the emfs are 180° apart. The lamp is connected in the secondary windings of a special transformer which has two primary windings – one of which is connected to the busbar and the other to the incoming machines as is evident from the Fig. 2.18. The instrument has an opal glass scale and the words slow and fast are written on both sides of center zero. The synchronising lamp is placed behind the scale. If two frequencies are not same, the needle oscillates over the centre zero position.

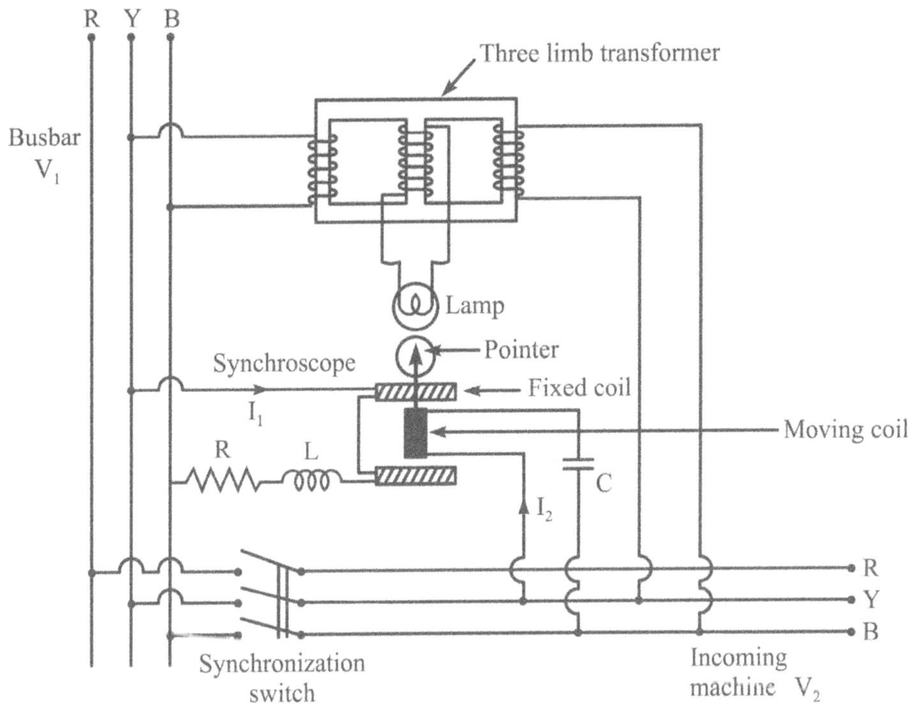

Fig. 2.18 Weston type synchroscope

2.17.2 Moving Iron Synchroscopes

The construction is similar to *single phase moving Iron P.f Meter*. It has two iron pieces common spindle as shown in Fig. 2.19. The fixed coil AA' is connected across the busbar while the pressure coils are connected across the terminals of the incoming alternator. In one of the pressure coils a non inductive resistance 'R' is connected in series while in the other an inductance 'L' coil is connected in series.

When the incoming frequency is equal to that of bus bar, the synchroscope works as a single phase p.f meter and so indicates the phase difference between the two emfs. When the incoming frequency is different to that of busbar, the spindle rotates continuously at a speed in revolutions per second equal to the difference of the frequencies of the two systems. The directions of rotation depends whether the incoming system speed is too fast or slow. The instrument has 360° scale and the displacement from the zero position gives the difference in phase between the two emfs and when the pointer is stationary at center zero, it shows the synchronising time.

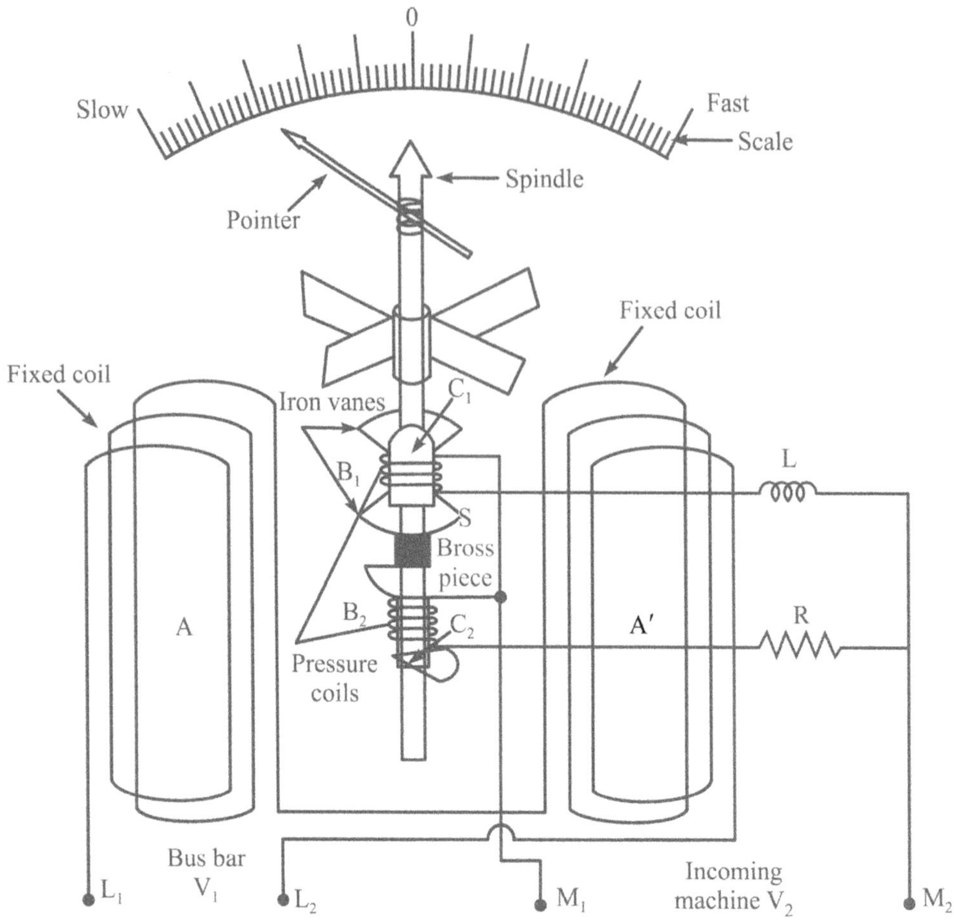

Fig. 2.19 Moving iron synchroscope

Problems with Solutions

1. A current transformer with 5 primary turns has a secondary burden consisting of a resistance of 0.16 Ω and an inductive reactance of 1.12 Ω. When the primary current is 200 A, the magnetising current is 1.5 A and the iron loss current is 0.4 A. Determine the expression used. The number of secondary turns needed to make the current ratio 100:1 and also the phase angle under these conditions

Solution:

Given data,

Number of primary winding turns, $N_p = 5$

Secondary circuit burden resistance $R_s = 0.16\ \Omega$

Inductive reactance, $X_s = 1.12\ \Omega$

Primary current, $I_p = 200$ A

Magnetizing current, $I_\mu = 1.5$ A

Iron loss current, $I_\omega = 0.4$ A

Transformation ratio, $R = 100:1$

To find

Number of secondary winding turns, $N_s = ?$

Phase angle $\theta = ?$

The transformation ratio is given by $R = \frac{I_p}{I_s}$

$$\therefore \qquad \frac{100}{1} = \frac{I_p}{I_s} \Rightarrow I_s = \frac{I_p}{100} = \frac{200}{100} = 2\text{ A}$$

The angle between the secondary winding induced e.m.f (E_s) and its current (I_s) is given by,

$$\delta = \tan^{-1}\left(\frac{X_s}{R_s}\right)$$

$$= \tan^{-1}\left(\frac{1.12}{0.16}\right) = 81.87^\circ \quad \therefore\ \delta = 81.87^\circ$$

Transformation ratio is also given by,

$$R = n + \frac{I_\mu \sin\delta + I_\omega \cos\delta}{I_s}$$

where, $$n = \text{turns ratio} = \frac{N_s}{N_p}$$

$$\therefore \qquad n = R - \frac{I_\mu \sin\delta + I_\omega \cos\delta}{I_s}$$

$$= 100 - \frac{1.5 \times \sin(81.87) + 0.4 \times \cos\ (81.87)}{2}$$

$$= 100 - \frac{1.54}{2} = 99.23$$

$$\Rightarrow \qquad \frac{N_s}{N_p} = 99.23 \Rightarrow N_s = N_p \times 99.23$$

$$\Rightarrow \qquad N_s = 5 \times 99.23 = 496.15 = 495 \text{ turns}$$

$\therefore$ $N_s = 496$ turns

$\therefore$ the number of secondary winding turns = 496.

Phase angle $$\theta = \frac{180}{\pi}\left[\frac{I_\mu \cos\delta - I_\omega \sin\delta}{nI_s}\right]$$

$$\Rightarrow \qquad \theta = \frac{180}{\pi}\left[\frac{1.5\times\cos(81.87) - 0.4\times\sin(81.87)}{99.23\times 2}\right]$$

$$= \frac{180}{\pi}\left[\frac{-0.184}{99.23\times 2}\right] = -0.0531^\circ$$

$\therefore$ phase angle $\theta = -0.0531^\circ$ ***Ans.***

2. In a deflection frequency meter, working on the principle of electrical resonance, there are two parallel circuits each consisting of an inductance and capacitance in series. In circuit has $C_1 = 1\ \mu F$ and is tuned to a frequency $f_1 = 60$ Hz; the other has $C_2 = 1.5\ \mu F$ and is tuned to a frequency f_2 below 50 Hz. The resistance of each circuit is $R_1 = R_2 = 100\ \Omega$. What must be the inductance of the second circuit and to what frequency must it be tuned in order that the current in both the circuits shall be same at a frequency of 50 Hz?

Solution:

Given data,

Components of circuit –1,

Resistance, $R_1 = 100\ \Omega$

To find inductance $L_1 = ?$

Capacitance, $C_1 = 1\ \mu F$

Resonance frequency, $f_1 = 60$ Hz

Component of circuit –2

Resistance, $R_2 = 100\ \Omega$

Capacitance, $C_2 = 1.5\ \mu F$

Resonance frequency, $f_2 < 50$ Hz

To find inductance $L_2 = ?$

For $I_1 = I_2$ at 50 Hz

Then $L_2 = ?$

$f_2 = ?$

The resonance frequency of a series circuit is, we know that L – C circuit is

$$f_r = \frac{1}{2\pi\sqrt{LC}}$$

$$f_1 \Rightarrow \frac{1}{2\pi\sqrt{L_1C_1}} = 60 = \frac{1}{2\pi\sqrt{L_1 \times 1 \times 10^{-6}}}$$

$$= 10^{-6} \times 4 = \left(\frac{1}{120\pi}\right)^2$$

$$L_1 = 7.036 \text{ H}$$

At 50 Hz, impedance of circuit – 1

$$Z_1 = R_1 + j\left(\omega_L - \frac{1}{\omega C_1}\right)$$

$$= 100 + j\left(2\pi \times 50 \times 7.036 - \frac{1}{2\pi \times 50 \times 1 \times 10^{-6}}\right)$$

$$Z_1 = 100 - j\,972.674\ \Omega$$

Impedance of circuit –2,

$$Z_2 = R_2 + j\left(\omega L_2 - \frac{1}{\omega C_2}\right)$$

$$Z_2 = 100 + j\left(2\pi \times 50L_2 - \frac{1}{2\pi \times 50 \times 1.5 \times 10^{-6}}\right)$$

$$Z_2 = 100 + j\,(100\,\pi L_2 - 2122.066)$$

It is given that the current through both the circuits is equal. This is possible only when impedance of both the circuits are equal.

i.e., $Z_1 = Z_2$

$\Rightarrow$ $100 - j\,972.674 = 100 + j\,(100\,\pi L_2 - 2122.066)$

$\Rightarrow \quad 100\pi L_2 = 1149.392$

$\Rightarrow \quad L_2 = 3.659$ H

$\therefore$ inductance of second circuit, $L_2 = 3.659$ H

$\therefore$ Resonant frequency of circuit –2.

$$f_2 = \frac{1}{2\pi\sqrt{L_2C_2}} = \frac{1}{2\pi\sqrt{3.659 \times 1.5 \times 10^{-6}}}$$

$$= 67.93 \text{ Hz}$$

$\therefore$ frequency = 67.93 Hz

In this problem, it has been mentioned that circuit – 2 is tuned to frequency of less then 50 Hz. But actually it is tuned to 67.93 Hz i.e., above 50 Hz.

3. An alternating current of 50 Hz has a maximum values of 10 Amps. What will be the value of current 1/600 second after it has reached the maximum.

***Solution*:** 8.66 amps

Ex: Maximum at $\frac{\pi}{2}$ = 5 ma (50 Hz) = 20 ms)

$$\frac{1}{600} = 1.67 \text{ ms}$$

$\theta = wt = 314 \times 6.67 = 2094 \times 10^{-3}$ rad

θ in degree = 110. 99

$\sin \theta = 0.866$

$I_m \sin \theta = 8.66$ A

Review Questions

1. What are the limitations of Electro Static Instruments
2. Explain advantages and disadvantages of Static Instruments
3. Explain comparison of controlling systems
4. Explain Torque equation of electro static instruments
5. Draw a neat sketch and explain operation of Quadrant – types electro static voltmeter
6. Explain briefly Heterostatic connection and derive Torque equation
7. Explain briefly Idiostatic connection

8. Explain – Kelvin multi cellular voltmeter with a neat diagram
9. Explain – theory of attracted disc type voltmeter
10. Explain –construction and operation of Kelvin absolute electrometer
11. State and explain the extension of range of electrostatic voltmeters
12. Derive multiplying procedure equation for capacitance multipliers?
13. Explain construction details and operations of C.T
14. Explain the construction features of C.T with the help of Neat sketches.
15. Derive the actual ratio of C.T equation
16. Derive the phase angle of C.T
17. What are the major causes of Errors in C.T? Explain them in detail
18. Draw the equivalent circuit of C.T?
19. What is the effect of change in burden on secondary?
20. "Never open the secondary circuit of a C.T while its primary is energized". Justify.
21. Explain-constructional details and operation of P.T
22. Derive the actual ratio of P.T equation?
23. Derive the phase angle of P.T?
24. What are the characteristic of P.T. and explain in detail?
25. What are the errors in P.T?
26. Explain the advantages and disadvantages of Instruments Tranformers.
27. Obtain the difference between current transformer and Potential transformer.
28. Discuss about the difference between instruments and power transformers
29. With a neat sketch explain the working principle of a single-phase dynamometer type p.f meter?
30. What are the advantages and disadvantages of moving Iron p.f meter over the dynamometer type?
31. With a neat sketch explain about rotating field type moving Iron p.f meter
32. With a neat sketch explain alternating field type moving iron p.f meter?
33. Writer short notes on frequency meters
34. Explain vibrating read type frequency meter
35. With a neat sketch explain electrical resonance type frequency meter.
36. With a neat sketch explain Weston frequency meter

37. Write short notes on synchroscopes?
38. With a neat sketch explain Weston type synchroscope?
39. With a neat explain moving Iron synchroscope?

Quiz Questions

1. In the dynamometer type of P.F meter resistor R and Inductor 'L' is connected
   (a) in series to moving coils (b) in series to fixed coils
   (c) in parallel to fixed coils (d) in parallel to moving coils
2. Frequency is measured with the help of
   (a) merz-price MD indicator (b) dynamometer type pf meter
   (c) vibrating reed type meter (d) trivector meter
3. Synchroscope is an instrument to
   (a) measure speed of alternator
   (b) be used parallelising alternator to busbar
   (c) measure frequency of supply
   (d) measure p.f of load
4. The normal range of frequency measurement in Reed type frequency meter is
   (a) 0 to 50 Hz (b) 45 Hz to 100 Hz
   (c) 47 to 53 Hz (d) any frequency
5. In the electrical resonance frequency meter, the moving coil is connected across
   (a) a capacitor (b) the supply
   (c) an inductor (d) none
6. In the Weston frequency meter to damp out the harmonics of the wave forms
   (a) a resistor is connected is series
   (b) an inductor is connected in series
   (c) a resistor and inductor is connected in series
   (d) a capacitor is connected in series
7. The controlling torque in single phase p.f meter is provided by:
   (a) spring (b) gravity
   (c) shiffness of suspension (d) none of the above

8. In a reed type frequency meter, all the reeds:
   (a) have the same natural frequency
   (b) have different natural frequency
   (c) have different natural frequency but the difference in natural frequency of adjacent is $\pm$ 0.5 Hz
   (d) none of the above

9. The ratio and phase angle errors in a well designed current transformer are kept specified limits by using.
   (a) ferrite core
   (b) strip wound core
   (c) some fractional turns
   (d) in-buit compensating capacitors

10. Precautions are essential for ensuring that the secondary of a current transformer is not open circuited when the primary circuit carries a current because
   (a) the ferromagnetic core may develop residual magnetism
   (b) dangerously high voltages might develop across the secondary
   (c) the reflected impedance may prevent the flow of current in the primary circuit

11. Instrument transformers are known to introduce magnitude and phase errors in measurement. These are primarily due to
   (a) improper connections on the primary side
   (b) measurement errors inherent in the meter connected to the transformer secondary
   (c) open and short circuit parameters of the instrument transformers
   (d) none of the above

12. Ratio error in a C.T is reduced by
   (a) increasing leakage reactance
   (b) using core with high resistance
   (c) turns compensation
   (d) reducing burden

13. The secondary of a C.T always short circuited through a low resistance or an ammeter
   (a) to get accurate measurement
   (b) to reduce the error
   (c) to establish current in secondary
   (d) none

14. The secondary voltage of P.T is always
   (a) 230 V
   (b) 11 KV
   (c) 110 V
   (d) 132 KV

15. The sensitivity of a instrument having a full scale deflection current of 50 μA in Ω/volts
    (a) 200 (b) 2,000
    (c) 20 (d) 20,000
16. In a circuit, the p.f meter is connected similar to that of
    (a) watt meter (b) voltmeter
    (c) amper-hour meter (d) galvanometer
17. The secondary circuit of C.T should be .... while its primary is energised
    (a) open circuited (b) short circuited
    (c) removed (d) none
18. A 0-300 V voltmeter has a error of + 20% of FSD. If true voltage is 30 V, the range of readings will be
    (a) 24 V–36 V (b) 29.4 V – 30.6 V
    (c) 20 V–40 V (d) none
19. Instrument transformers are used in AC system for help in the measurement of
    (a) I & V (b) power and energy
    (c) p.f (d) all of the above
20. A p.f meter has
    (a) 4 control springs (b) 2-control springs
    (c) 1-control springs (d) no control springs
21. VA ratings of instrument transformer is nearly:
    (a) 0.1 VA (b) 10 VA
    (c) 1000 A (d) 1 VA
22. PT's are used in
    (a) A.C current measurement (b) A.C voltage measurement
    (c) D.C current measurement (d) D.C volt measurement
23. Nominal ratio of C.T is defined as the ratio of
    (a) no. of primary turns/no. of secondary turns
    (b) rated primary current/no. of secondary turns
    (c) rated secondary current/rated primary current
    (d) no. of secondary turns/No. of primary turns

24. PT's are used in
    (a) A.C current measurement
    (b) A.C voltage measurement
    (c) D.C current measurement
    (d) D.C volt measurement

25. Nominal ratio of C.T is defined as the ratio of
    (a) no. of primary turns/no. of secondary turns
    (b) rated primary winding current/rated secondary winding current
    (c) rated secondary current/rated primary current
    (d) no. of secondary turns/no. of primary turns

Questions and Answers

1. A.C.T is overloaded when

Ans. When its secondary terminals are open-circuited

2. The errors in C.T are mainly due to

Ans. Core loss

3. The VA rating of instrument transformers is normally near

Ans. 10 VA

4. The standard secondary voltage for a P.T is

Ans. 110 V

6. Usually a C.T has

Ans. A higher overload capacity than a P.T

7. The problem of electro-static coupling in a transformer is acute at

Ans. High frequencies

9. The thermocouple instrument will measure

Ans. Current

10. If the secondary winding of a current transforms is open when current is flowing in the primary

Ans. There will be very high secondary voltage

12. Which one of the following quantities has the same dimension in both electromagnetic and electrostatic systems.

1. current 2. electric energy 3. electric power.

Select the correct answer using the codes given

Ans. 2 and 3

13. Absolute null method is used for testing of ________

Ans. Potentiometer

14. A.C.T is overloaded when

Ans. When its terminals are open-circuited

15. The errors in C.T are mainly due to

Ans. Core loss

16. The VA rating of instrument transformers is normally near

Ans. 10 VA

17. The standard secondary voltage for a P.T is

Ans. 110 V

18. To measure flux, devices used are based on

Ans. Lorenz method

19. Usually a C.T has

Ans. A higher overload capacity than a P.T

20. The problem of electro-static coupling in a transformer is acute at

Ans. High frequencies

21. An alternating current of 50 Hz has a maximum values of 10 Amps. What will be the value of current 1/600 second after it has reached the maximum.

Ans. 8.66 amps

Ex: Maximum at $\frac{\pi}{2} = 5$ ma (50 Hz) = 20 ms)

$$\frac{1}{600} = 1.67 \text{ ms}$$

$$\theta = wt = 314 \times 6.67 = 2094 \times 10^{-3} \text{ rad}$$

$$\theta \text{ in degree} = 110.\ 99$$

$$\sin\theta = 0.866$$

$$I_m \sin\theta = 8.66 \text{ A}$$

22. Which instrument has identical calibration for a.c as well as d.c values?

Ans. Hot-wire type

23. In the circuit shown, through oversight, the meters have been interchanged. This is likely to result in

Ans. Damage to ammeter

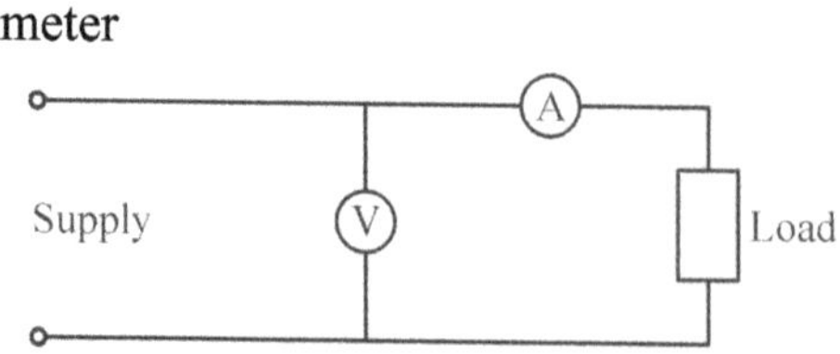

24. The thermocouple instrument will measure

Ans. Current

25. The secondary winding of a current transforms is open when current is flowing in the primary

Ans. There will be very high secondary voltage

26. Which instrument is used on a transfer instruments?

Ans. Electrodynamameter

27. If a d.c voltmeter is made from an ammeter having a load of 100 μ A then its sensitivity (in K Ω / V) will be

Ans. 10

Quiz Questions Key

1. (a)	2. (c)	3. (b)	4. (c)	5. (a)
6. (b)	7. (d)	8. (c)	9. (a & b)	10. (b)
11. (c)	12. (c)	13. (c)	14. (c)	15. (d)
16. (a)	17. (b)	18. (a)	19. (d)	20. (d)
21. (b)	22. (b)	23 (b)	24. (b)	25. (b)

CHAPTER 3

Measurement of Power

3.1 Introduction

In the case of a D.C circuit, the measurement of power can be done by using voltmeter and ammeter since Power = VI. Where as, in the case of A.C, the power measurement by simple voltmeter and ammeter is not possible since P = VI cos ϕ where cos ϕ is the power factor of the circuit. In such cases, a meter with combination of ammeter and voltmeter and which can also taken into account the P.F necessary. Such an instrument is called wattmeter. The torque produced is due to the interaction of current and voltage. A wattmeter gives the reading proportional not only to the current and voltage but also cosine of the angle between them.

Power in D.C. Circuits

The power taken by a load from a D.C. supply is given by the product of readings of an ammeter and a voltmeter when connected in the circuit as shown in Fig. 3.1.

$$\text{Power, } P = VI \text{ watt}$$

It should be remembered that the ammeter and voltmeter require power for their operation and allowance must be made for this. Either of the two connection shown in Fig. 3.1 (a) and (b) may be used. In Fig. 3.1(a) the ammeter is connected between load and voltmeter. Therefore the voltmeter not only indicates the voltage 'V_L' across the load but in addition voltage drop V_a across the ammeter. If R_a is the resistance of the ammeter, voltage drop across ammeter

$$V_a = I R_a$$

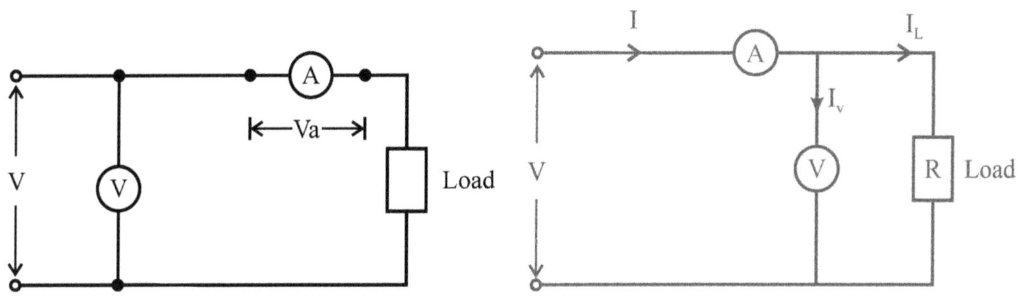

(a) (b)

Fig. 3.1 Power measurement in D.C circuit

Power consumed by load $= V_L I = (V - V_a)\, I$

$$= VI - V_a I$$

$$= VI - I^2 Ra$$

= 'P' indicated by instruments – power loss in ammeter.

$\therefore$ power indicated by instruments = power consumed in load + power loss in ammeter.

In Fig. 3.1 (b), voltmeter is connected between load ammeter. Therefore Ammeter not only indicates the current through the load but in addition current in the voltmeter also.

Current through the voltmeter $I_V = V/R_V$ where R_V = resistance of voltmeter.

$$\therefore \text{ Power consumed by load} = VI_L = V\,(I - I_V) = V\left(\frac{I}{R_V}\right) = VI - \frac{V^2}{R_V}$$

= power indicated by instruments – power loss in voltmeter.

$\therefore$ Power indicated by instruments = Power consumed in load + power loss in voltmeter.

Thus in both the cases, the power indicated by the instruments is equal to the power consumed by the load plus the power consumed by the instrument nearer to the load terminals.

In order to obtain the true power, corrections must be applied for power loss in instruments. Under normal conditions the value of power loss in instruments is quite small as compared with the load power and therefore, the error introduced on this account is small. However, when the output of the serious errors, for a permanently wired installation, when power measurements are require, it is a distinct advantage to install a wattmeter in place of voltmeter and ammeter. Wattmeter and ammeter are used. The labour involved is reduced and the accuracy is also increased.

Power in A.C Circuit

In an A.C circuit, the instantaneous power with time and is the product of the instantaneous values of voltage and current in the circuit.

Thus $V = V_m \sin \omega t,$ $i = i_m \sin (\omega t \pm \phi)$

$$P = Vi = V_m I_m \sin wt \sin (\omega t + \phi) = \text{instantaneous power}$$

Then the average power consumption is given by

$$P_{ac} = \frac{1}{2\pi} \int_0^{2\pi} P d(\omega t) = \frac{1}{2\pi} \int_0^{2\pi} V_m I_m \sin(\omega t) \sin(\omega t \pm \phi) d(\omega t)$$

This gives the average power consumption in A.C circuit as,

$$P_{AC} = VI \cos \phi \text{ watts.}$$

where,

$$V = \text{r.m.s value of the voltage} = \frac{V_m}{\sqrt{2}}$$

$$I = \text{r.m.s value of the current} = \frac{I_m}{\sqrt{2}}$$

$\cos \phi$ = power factor of circuit

ϕ = power factor angle = VI

The above expression shows that ammeter and voltmeter is not sufficient to measure power. The measurement of $\cos \phi$ is also required. Therefore in a.c. circuit, the "WATTMETER" is used which senses the angle between voltage and current and directly gives the power consumption of the circuit in watts considering the effect of $\cos \phi$ in to account.

3.2 Electrodynamometer Type Instruments

The electrodynamometer type instrument is a "transfer instrument". A transfer is one which is calibrated with a D.C source and used without any modifications for A.C measurements. Such a transfer instrument has same accuracy for A.C and D.C measurements. The electrodynamometer type instruments are often used in accurate A.C voltmeter and ammeters, not only at the power line frequency but also in the lower audio frequency range. With some little modifications, it can be used as a wattmeter for the power measurement.

The necessity for A.C caliberation of moving iron instruments as well as other types of instruments which cannot be correctly caliberated requires the use of a transfer type of instrument. A transfer instrument is one that may be caliberated with D.C source and then used without modification to measure A.C. This requires the transfer type instrument to have same accuracy for both D.C and A.C, which the electrodynameter instruments have.

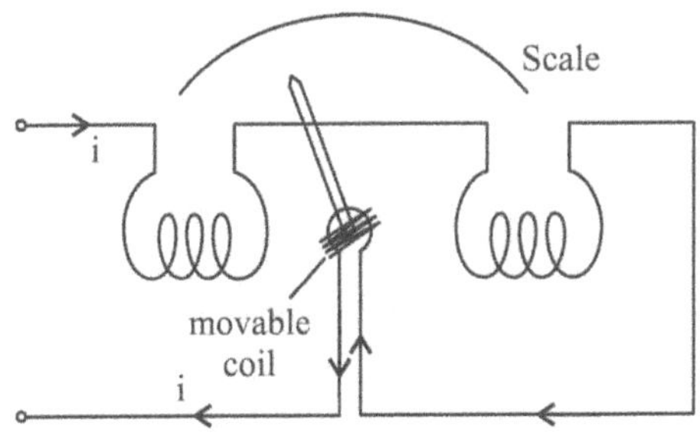

The caliberation of an A.C instrument may be performed as. The transfer instrument is first caliberated on D.C. The caliberation is then transferred to A.C instrument on alternating current, using operating conditions under which the later operates properly. Electrodynamic instruments are capable of service as transfer instruments. These are used A.C voltmeters and ammeters both in the range of power frequencies and lower part of the audio frequency range. They are used as watt-meters, varmeters and with some modification as power factor meters and frequency meters.

Operating principle: The working principle of this instrument taking up a permanent magnet moving coil instrument and considering how it would behave on A.C. It would have torque in are direction during one half of cycle and an equal effect in opposite direction during other half of cycle. If the frequency were very low, the pointer would swing back and forth around zero point. However for an ordinary meter, the interia is so great that on power frequency the pointer does not go very far in either direction but merely stays around zero. If, however, we were to reverse the direction of the flux each time the current through movable coil reverse, a unidirectional torque would be produced for both positive and negative halves of the cycle. In this instrument the field can be made coil of the field coil is connected in series with movable coil.

Construction

1. **Fixed coils:** The field is produced by a fixed coil which is divided into two sections to give more uniform field.
2. Moving coils: It is wound either as a self sustaining coil or else on a non metallic former.

Control: The controlling toque is provided by two control springs.

Moving system: It is mounted on aluminium spindle.

Damping: Air friction damping is employed for these instruments.

Shielding: As field produced by fixed coils is weak, shielding is necessary to protect instrument from stray magnetic fields. This is made by a high permeability alloy.

Cases and Scales: Cases are so constructed so as to remain dimensionally stable over longer periods of time where scales are hand drawn, using machine sub-divising equipment.

Torque Eq:

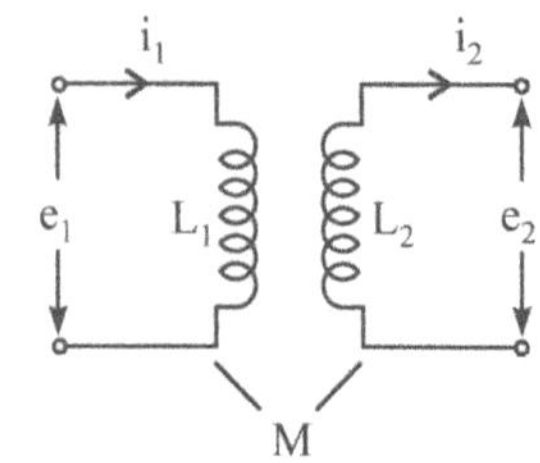

Let, i_1 = instantaneous current in fixed coils; A

i_2 = instantaneous current in moving coil

L_1 = Self inductance of fixed coils, H

L_2 = Self inductance of moving coils, H

M = Mutual inductance between fixed and moving coils.

Flux linkages of coil, $\psi_1 = L_1i_1 + Mi_2$ & in coil 2, $\phi_2 = L_2i_2 + Mi_1$

Electrical i/p energy $= e_1\, i_1\, dt + e_2\, i_2\, dt = i_1\, d\phi_1 + i_2\, d\phi_2$

as $\quad e_1 = \dfrac{d\phi_1}{dt}$ and $e_2 = \dfrac{d\phi_2}{dt}$

$$= i_1 d(L_1i_1 + Mi_2) + i_2\, d(L_2i_2 + Mi_1)$$

$$= i_1L_1di_1 + i^2{}_2dt_1 + i1i_2dM + i_1Mdi_2 +$$

$$i_2L_2di_2 + i^2{}_2dL_2 + i_1i_2dM + i_2Mdi_1 \quad .....(3.1)$$

Energy in magnetic field $= \dfrac{1}{2}i_1^2L_1 + \dfrac{1}{2}i_2^2L_2 + i_1i_2M$

Change in energy stord $= d\left[\dfrac{1}{2}i_1^2L_1 + \dfrac{1}{2}L_2i_2^2 + i_1i_2M\right]$

$$= i_1L_1di_2 + (L_1i_1 + Mi_2) + i_2\, d(L_2i_2 + Mi_1)$$

$$= i_1L_1di_1 + (i_1^2/2)dL_1 + i_2L_2di_2 + \frac{i_2^2}{2}dL_2 + i_1Mdi_2 + i_2Mdi_1 + i_1i_2dM \quad .....(3.2)$$

From principle of energy conservation,

$$\text{Total i / p energy} = \text{Change in energy} + \text{Mechanical energy}$$

$\therefore$ Mechanical energy $= \dfrac{1}{2}i_1^2dL_1 + \dfrac{1}{2}i_2dL_2 + i_1i_2dM$ (i.e., by (3.1), (3.2))

$$= i_1i_2dM \;(\because\; L_1, L_2 \text{ are constant and } dL_1 = dL_2 = 0)$$

If T_i is instantaneous deflecting torque, $d\phi$ is change in deflection, then mechanical energy = work done = $T_id\theta$

Thus, $T_id\theta = i_1i_2dM$ and $T_i = i_1i_2dM/d\theta$ .....(3.3)

Op^n with DC: Let I_1 = Current in fixed coils; A

I_2 = Current in moving coil, A

From 3.3 $T_d = I_1 I_2 dM / d\theta$ .....(3.4)

Eq. 3.4 shows that T_d depends on product of I_1, & I_2 and rate of change of mutual inductance. This T_d deflects moving coil to such a position where T_c of spring is equal to T_d. Let θ = final ready deflection then, T_c (controlling torque) = $k\theta$.

At steady state, $T_d = T_c \Rightarrow I_1 I_2 \frac{dM}{d\theta} = k\theta$ or $\theta = \frac{I_1 I_2}{K} \cdot \frac{dM}{d\theta}$ .....(3.5)

Op^n with AC: Let i_1 and i_2 be instantaneous values of current carried by coils

$\therefore$ $T_i = i_1 i_2 dM / d\theta$ from (3.3); T_i = instantaneous deflecting torque

The average deflection torque over complete cycle,

$$T_d = \frac{1}{1}\int_0^T T_i dt = \frac{dM}{d\theta}\frac{1}{T}\int_0^T i_1 i_2 dt;$$

T = time for one complete cycle.

3.2.1 Advantages and Disadvantages of Electrodynamic Instruments

Advantages

1. These voltmeters are very useful where accurate rms values of voltage, irrespective of waveforms, are required
2. These are free from hysteresis errors
3. These instruments can be used on both A.C and D.C. They are also used as a transfer instruments
4. These instruments are free from hysteresis and eddy current losses
5. They are in light weight
6. They have a precision grade accuracy
7. Low power consumption

Disadvantages

1. They are more expensive than other type of instruments
2. These instruments have a low sensitivity due to a low torque to weight ratio
3. These instruments are sensitive to overloads and mechanical impacts. Therefore care must be taken while handling them
4. The operating current of these instruments is large due to the fact that they have weak magnetic field
5. They have a non-uniform scale

3.2.2 Errors in Electrodynamometer Instruments

1. **Stray magnetic field error**

   As similar to moving iron instrument, the operating field in electrodynamometer instrument is very weak. Hence external magnetic field can interact with the operating field to cause change in the deflection, causing the error. To reduce the effect of stray magnetic field, the shields must be used for the instruments.

2. **Eddy current errors**

   In the metal parts of the instrument the eddy currents get produced. The eddy currents interact with the instrument current, to cause change in the deflecting torque; to cause error. Hence metal parts should be kept as minimum as possible. Also the resistivity of the metal parts used must be high, to reduce the eddy currents.

3. **Torque to weight ratio**

   The reasonable deflecting torque, m.m.f of the moving coil must be large enough. Thus m.m.f = NI hence current through moving coil should be high or number of turns should be large. The current through coil not be made very high because it may cause excessive heating of springs. Large number of turns hence is the only option but it increases weight of the coil. This makes the system heavy reducing torque to weight ratio. This can cause frictional errors in the reading.

4. **Temperature error**

   The temperature errors are caused due to the self heating of the coil, which causes change in the resistance of the coil. Thus temperature compensating resistors can be used in the precise instrument to eliminate the temperature errors.

5. **Frequency errors**

   The change in the frequency causes to change self inductance of moving coil and fixed coil. This causes the error in the reading. The frequency error can be reduced by having equal time constants for both fixed and moving coil circuit.

3.2.3 Comparison of various Types of Instruments

Meter Type	Type of Control	Type of Damping	Suitability	Accuracy of which the instrument is capable	Applications
PMMC	Spring	Eddy current	D.C	Precision (or) industrial	Used for d.c current and voltage measurements in low and medium impedance circuit

Contd...

Meter Type	Type of Control	Type of Damping	Suitability	Accuracy of which the instrument is capable	Applications
Electrodyna-mometer	spring	Air-friction	D.C and A.C	Precision d.c and A.c 25 to 150 Hz industrial	Used mainly as wattmeter. Widely used as a calibration instrument and as a transfer instrument
Moving iron	Spring (or) Gravity	Air-friction	D.C and A.C	Precision d.c A.c up to 100 Hz in (laboratory)	up to 100 A on d.c and A.c up to 5×10^6 Hz 10 mA to 1 A
Thermocouple	Spring	Air-friction	D.C and A.C	Industrial	Voltmeter ranges up to 750 V.

3.3 Single Phase Dynamometer Wattmeter

An electrodynamometer type wattmeter is used to measure power. It has two coils, fixed coil which is current coil and moving coil is pressure coil or voltage coil. The current coil carries the current of the circuit while pressure coil carries current proportional to the voltage in the circuit. This is achieved by connecting a series resistance in voltage circuit. Circuit diagram shown in Fig. 3.2.

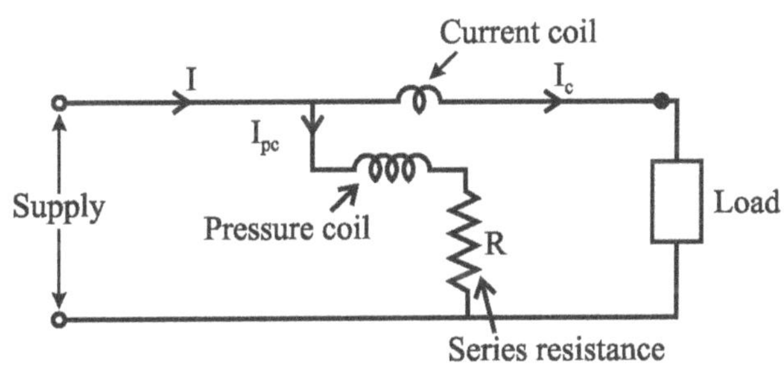

Fig. 3.2 Electrodynamometer wattmeter

I = R.M.S value of current

V = R.M.S value of supply voltage

R = series resistance

I_c = current through current coil

I_{pc} = current through pressure coil

Torque equation:

From the theory of electrodynamic instruments,

$$T_i = i_1 \, i_2 \frac{dM}{d\theta} \qquad \text{.....(3.6)}$$

V = instantaneous voltage

$= V_m \sin \omega t$

$$= \sqrt{2}\, V \sin \omega t \qquad \text{.....(3.7)}$$

Due to high series resistance, pressure coil is treated to be purely resistive.

$$i_{pc} = \text{instaneous value} = \frac{V}{R_p} \text{ where } R_p = r_{pc} + R$$

$$i_{pc} = \frac{\sqrt{2}V}{R_p} \sin \omega t = \sqrt{2}\, I_{pc} \sin \omega t$$

If current coil current lags the voltage by angle ϕ then its instaneous value is

$$i_c = \sqrt{2}\, I_c \sin (\omega t - \phi) \qquad \text{.....(3.8)}$$

$i_1 = i_c$ and $i_2 = i_{pc}$

Hence, $\omega t - \phi$

$$Ti = \left[\sqrt{2}\, I_{pc} \sin \omega t\right] \left[\sqrt{2}\, I_c \sin\ (\omega t - \phi)\right] \frac{dM}{d\theta}$$

$$= 2I_c\, I_{pc} \sin (\omega t) \sin (\omega t - \phi) \frac{dM}{d\theta}$$

$$T_i = I_c\, I_{pc} [\cos \phi - \cos (2\omega t - \phi)] \frac{dM}{d\theta} \qquad \text{.....(3.9)}$$

Note: Thus instantaneous torque has a component of power which varies as twice the frequency of current and voltage.

T_d = Average deflecting torque

$$\boxed{T_d = \frac{1}{T} \int_0^T T_i \, d(\omega t)}$$

$$= \frac{1}{T} \int_0^T I_c\, I_{pc}\ [\cos\phi - \cos\ (2\omega t - \phi)] \frac{dM}{d\theta}\, d(\omega t)$$

$$T_d = I_c\, I_{pc} \cos\phi \,\frac{dM}{d\theta},$$

where, $$I_{pc} = \frac{V}{R_p} \qquad .....(3.10)$$

For a spring controlled wattmeter

$$T_c = K\theta \qquad .....(3.11)$$

but $$T_d = T_c$$

$\therefore$ $$I_c\, I_{pc} \cos\phi \,\frac{dM}{d\theta} = K\theta$$

$$\theta = \frac{1}{K} I_c I_{pc} \cos\phi \,\frac{dM}{d\theta} = K_1\, I_c\, I_{pc} \cos\phi \qquad .....(3.12)$$

where $$K_1 = \frac{1}{K}\,\frac{dM}{d\theta}$$

$\therefore$ $$\theta = K_1\, I_c \,\frac{V}{R_p} \cos\phi = K_2 P \qquad .....(3.13)$$

where $K_2 = \dfrac{K_1}{K_P}$ and $P = VI_c \cos\phi$ = power

$\therefore$ $$\boxed{\theta \propto P}$$

Note: Thus the wattmeter deflection when calibrated gives the power consumption of the circuit.

3.3.1 Reading on Wattmeter

The Fig. 3.3 shows symbolic representation of wattmeter.

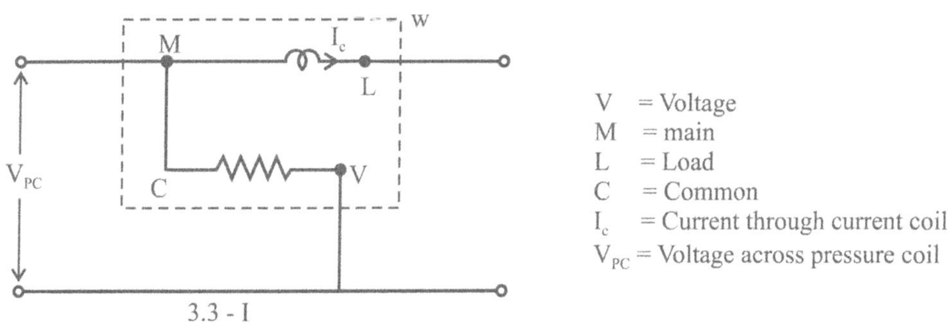

Fig. 3.3 Symbolic representation of wattmeter

The wattmeter reading is

$$W = V_{pc} I_c \cos \boxed{V_{pc} \wedge I_{pc}}$$

[**Note:** the angle between V_{pc} and I_{pc} may or may not be power factor angle ϕ. It depends on the wattmeter connection in the circuit]

If wattmeter is connected in a three phase circuit, such that $I_c = I_{ph}$ and $V_{pc} = V_{ph}$, then only $\boxed{V_{pc} \wedge I_{pc}} = \phi$.

Note: In a three phase circuit, angle between V_{pc} and I_{pc} is to be obtained from the corresponding phasor diagram.

3.3.2 Shape of Scale of Electrodynamometer Wattmeter (Theory)

The deflection is given by $\theta = \frac{V}{R_p} I_c \cos\phi \frac{1}{K} \frac{dM}{d\theta}$

$$= K_2 P \frac{dM}{d\theta}$$

Where $K_2 = \frac{1}{K\,R_P}$ and $P = VI_c \cos\phi$ = power measured thus the deflection is directly proportional to power being measured and the scale is uniform over the range in which $\left(\frac{dM}{d\theta}\right)$ remains constant.

Practically the waltmeters are designed such that $\frac{dM}{d\theta}$ remains almost constant over a range of 40° to 50° on either side of zero mutual inductance position.

The 'M' varies linearily in this zone with respect to θ. Thus if zero mutual inductance position is kept in the middle of the scale then 'M' varies linearily for the flections upto 80° to 100° and thus scale is uniform over the range of 80° to 100°. Practically this covers the entire scale range the shape of scale and variation in mutual induction is shown in the Fig.

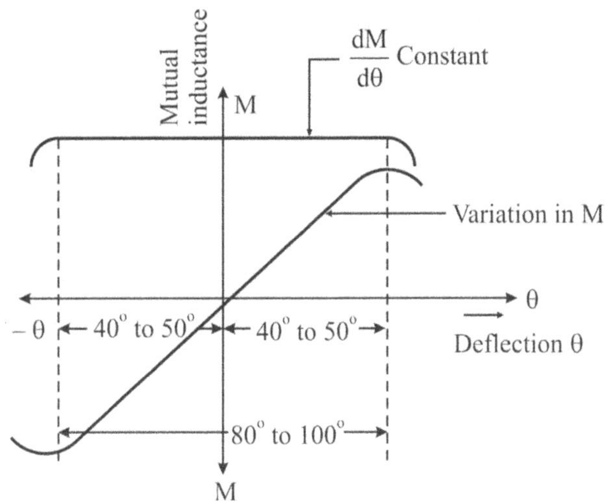

Fig. 3.4 Shape of scale of dynamometer wattmeter

3.4 Error due to Pressure Coil Inductance

Let

r_p = resistance of pressure coil

L = Inductance of pressure coil

R = resistance in series with pressure coil

R_p = total resistance of pressure coil circuit = $r_p + R$

V = voltage applied to pressure coil circuit

I = current in the current coil circuit

I_p = current in the pressure coil circuit

Z_p = inductance of pressure coil circuit

$$= \sqrt{(R + r_p)^2 + (\omega L)^2}$$

It was implied in the discussion of the idealized wattmeter that the current in the pressure coil in phase with the applied voltage. If the pressure coil of the wattmeter has an inductance, the current in it will lag the voltage by an angle 'β' where

$$\therefore \quad \beta = \tan^{-1} \frac{\omega L}{R_p} = \tan^{-1} \frac{\omega L}{r_p + R}$$

It will be seen from Fig. 3.5, that for lagging power factor, the angle between current in the current coil circuit and current in the pressure coil circuit is less than ϕ, by which the load current lags the applied voltage. The angle between the pressure coil current and the current in the current coil is

$$\phi' = \phi - \beta$$

$\therefore$ The actual wattmeter reading is

$$= \frac{I_1 I_2}{K} \cos\phi \frac{dM}{d\theta} \quad \text{here} \quad I_1 = I,\ I_2 = I_p = \frac{V}{2p},\ \phi = \phi'$$

$$\text{Actual wattmeter reading} = \left(\frac{V}{z_p K}\right) I \cos(\phi - \beta) \frac{dM}{d\theta}$$

Now $$z_p = \frac{R_p}{\cos\beta};\ R_p = Zp \cos\beta$$

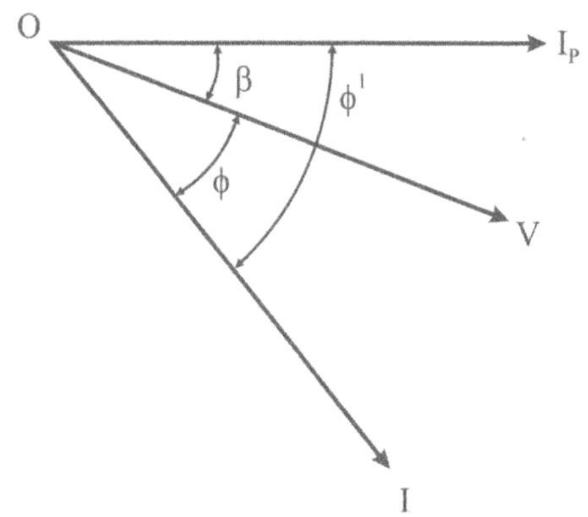

Fig. 3.5

$$\therefore \text{actual wattmeter reading} = \frac{VI\cos(\phi - \beta)}{K\,(R_p/\cos\beta)} \cdot \frac{dM}{d\theta}$$

$$= \frac{VI}{KR_p} \cos\beta . \cos(\phi - \beta) \frac{dM}{d\theta}$$

In the absence of inductance $z_p = R_p$ and $\beta = 0$ and therefore the wattmeter will read true power under these conditions.

$$\therefore \text{True power} = \frac{I_p I}{K} \cos\phi \frac{dM}{d\theta} = \frac{VI\cos\phi}{K\,R_p} \cdot \frac{dM}{d\theta}$$

$$\text{Ratio} = \frac{\text{true power}}{\text{actual wattmeter reading}} = \frac{\left[\frac{VI\cos\phi}{K\,R_p}\right]\cdot\frac{dM}{d\phi}}{\left[\frac{VI\cos(\phi-\beta)}{K\,R_p}\right]\frac{dM}{d\theta}}$$

$$= \frac{\cos\phi}{\cos\beta\cos(\phi-\beta)}$$

$$\therefore\ \text{True power} = \frac{\cos\phi}{\cos\beta\cos(\phi-\beta)} \times \text{Actual wattmeter reading}$$

$\therefore$ The correction factor to be applied to get the true power from the actual wattmeter reading is

$$\text{Correction factor} = \frac{\cos\phi}{\cos\beta\cos(\phi-\beta)} \qquad \text{...for lagging p.f}$$

Similarly for leading p.f

$$\text{the correction factor} = \frac{\cos\phi}{\cos\beta\cos(\phi+\beta)}$$

$\therefore$ Error = Actual wattmeter reading – True power.

$$\text{Percentage error} = \frac{\text{wattmeter reading} - \text{true power}}{\text{true power}} \times 100$$

3.5 Compensation for Inductance of Pressure Coil

Many wattmeters are compensated for errors caused by inductance of pressure coil by means of a capacitor connected in parallel with a portion of multiplier (series resistance) as shown in Fig. 3.6.

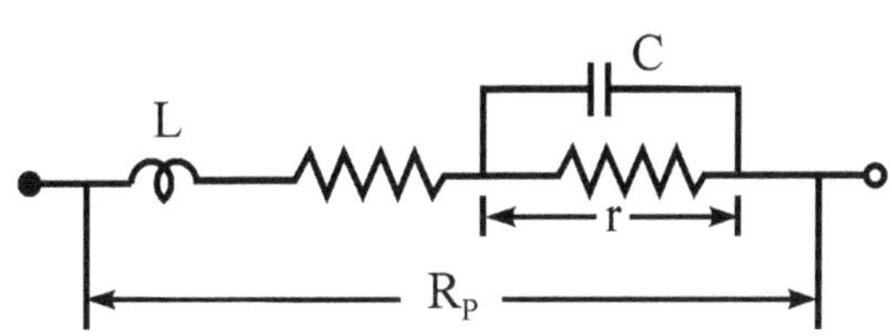

Fig. 3.6

The total impedance of the circuit may be written as:

$$Z_p = (R_p - r) + j\omega L + \frac{(r)(-j/\omega C)}{r - \dfrac{j}{\omega C}}$$

$$\therefore \qquad Z_p = (R_p - r) + jwL + \frac{r - j\omega\, Cr^2}{1 + \omega^2 C^2 r^2}$$

If the value of circuit constants are so chosen that for power frequencies $\omega^2 c^2 r^2 << 1$

We have approximately, $Z_p \approx (R_p - r) + j\omega L + r - j\omega Cr^2$

$$\approx R_p + j\omega\,(L - Cr^2)$$

If we make $\quad L = Cr^2$

Then $\quad Z_p \approx Rp$ and $\beta = 0^\circ$

Thus the error caused by pressure coil inductances almost completely eliminated.

This type of compensation is very slightly affected by change in frequency and can be used for frequencies at which $\omega^2 c^2 r^2 << 1$.

The frequency range over which the above compensation holds good is up to 10 kHz.

3.6 Error due to Pressure Coil Capacitance

The pressure coil circuit may posses capacitance in addition to inductance. This capacitance is mainly due to inter turn capacitance of the series resistance. The effect of capacitance is opposite to that produced by inductance. Therefore the wattmeter read high on lagging P.F of the load.

The phase angle between pressure coil current and the applied voltage depends upon the reactance of the pressure coil circuit. The inductance resistance is normally greater than the capacitance reactance and therefore the phase angle varies with frequency increasing with increase in frequency.

3.7 Error due to Mutual Inductance Effect

Errors are caused due to mutual inductance between current and pressure coil of the wattmeter. These errors are quite low at power frequency but they increase and become more important as the frequency increases.

The effect of mutual inductance is to increase the phase angle for the connection in which the pressure coil is connected on the load side and to decreases the phase angle when the current coil is connected on the load side. Thus change in phase angle is

$$\tan^{-1}\left(\frac{\omega M}{R_p}\right)$$

If the self capacitance are neglected, the following expression may be used for computing power.

$$P_w = P + I^2R_c + \left(\frac{L+M}{R_p}\right)\omega\ VI \sin\phi$$

where P_w is the wattmeter induction and P is the power in the load. This expression is used when the current coil is on the load side.

The mutual inductance between the two coils varies as the moving coil (pressure coil) deflects thus the error caused by mutual inductance varies with relative position of the two coils.

3.8 Error caused because of Connections

There are two alternate methods of connecting a wattmeter in a circuit. These are shown in Fig. 3.7 (a) and (b).

Errors are introduced in the measurements owing to power loss in the current and pressure coils.

In the connection of Fig. 3.7(a) the pressure coils is connected on the supply side and therefore the voltage applied to the pressure coil is the voltage across the load plus the voltage drop across the current coil. Thus the wattmeter measures the power loss in the current coil in addition to the power consumed by load.

Power indicated by wattmeter = Power consumed by load + power loss in current coil.

$$= \text{Power consumed by load} + I^2R_c$$

where R_c resistance of current coil.

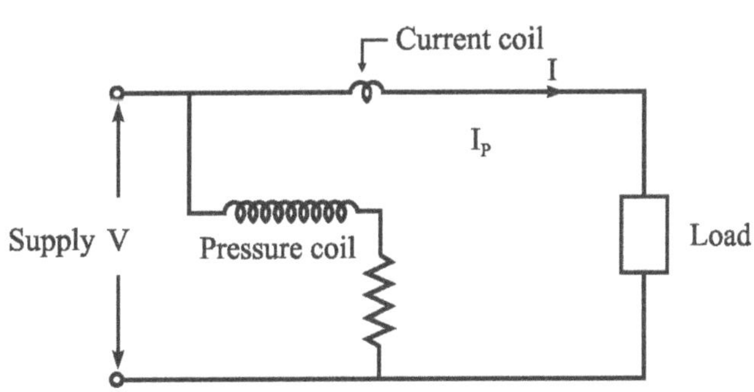

Fig. 3.7(a)

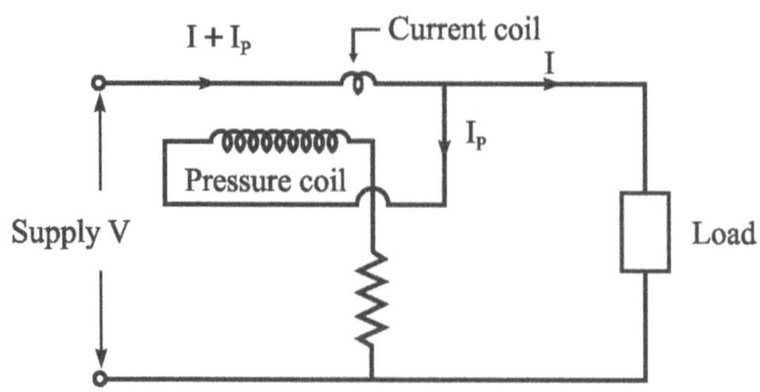

Fig. 3.7(b)

In connection of Fig. 3.7(b), the current coil is on supply side and therefore it carries the pressure coil current phase the load current. Hence the wattmeter reads the power consumed in the load plus the power loss in the pressure coil.

∴ Power indicated by wattmeter = power consumed by load + power loss in the pressure coil circuit

$$= \text{power consumed by load} + \frac{v^2}{R_p}$$

If the load current is small, the voltage drop in the current coil is small; so that connection of Fig. 3.7(a) introduces a vary small error. On the other hand, if the load current is large, the value of pressure coil current is very small when compared with the load current. Hence power loss in the pressure coil will be small as compared with the load power and therefore connection of shown in Fig. 3.7(b) is preferable.

It should be noted that in an uncompensated wattmeter, the reading of the wattmeter including the power loss in the coil connected on the load side. The use of above connections for different applications. [connection (a) for small current and (b) for large current] holds good only if the power loss in the instrument is to be neglected. However, if accuracy requires that the power loss in the instruments be taken into account, connection (b) is preferable. This is because for a constant voltage, the power loss v^2 / R_p in the pressure coil current is constant and therefore, the true power consumed by the load can be obtained by subtracting v^2 / R_p from the wattmeter reading. The value of v^2 / R_p can be calculated or observed directly by simply reading the wattmeter with load disconnected.

3.9 Compensation

The power taken by the pressure coil circuit is constant if the voltage is constant and increases a smaller percentage of total as a larger amount of power is measured. However, in cases where the load current is large and the power factor is small, connection Fig. 3.7 (b) may result in large error. Since the total power measured small hence in wattmeters which are designed for low power factor measurements, a compensating coil may be used in the instrument to compensate for the error caused by power losses in the pressure coil circuit. The arrangement using a compensating coil is shown in Fig. 3.8.

In connection 3.7(b) the current coil carries a current of I + I_p and produces a field corresponding to this current. The compensating coil is connected in series with the pressure coil circuit.

The pressure coil circuit in Fig. (3.8) and is made as nearly as possible identical and coincident with the current coil.

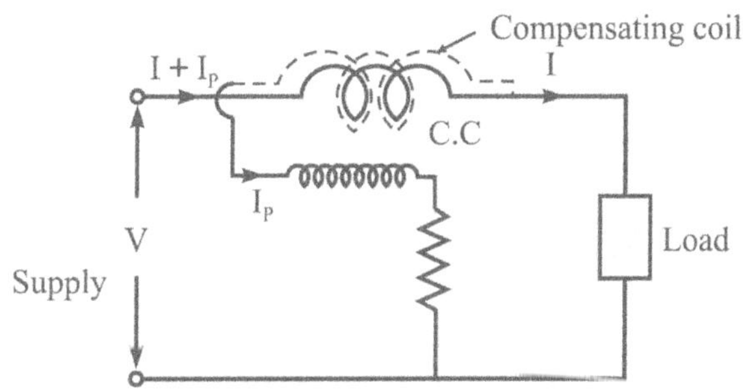

Fig. 3.8

It is so connected that it opposes the field of the current coil.

The compensating coil carries a current I_p, and produces a field corresponding to this current. This field acts in opposition to the current coil field. Thus the resultant field is due to current I only, hence the error caused by the pressure coil current flowing in the current coil is neutralized.

Other errors namely, eddy current errors, stray magnetic field error, error caused by vibration at moving system are less important and some are not easily calculated.

3.10 Low Power Factor Electrodynamic Type Wattmeter

If in case of any circuit is operating at low P.F, then power in that circuit is difficult to measure with ordinary electrodynamometer wattmeter, the reading of the wattmeter is inaccurate on account of following reasons.

1. The inductance of pressure coil introduces considerable error at low power factors.
2. The deflecting torque on the moving system is small as the power factor is low, even though the current and pressure coils are fully excited.

In order to get accurate reading from the wattmeter when it is measuring low power, extra adjustments are required to be made so that there will be compensation of the errors.

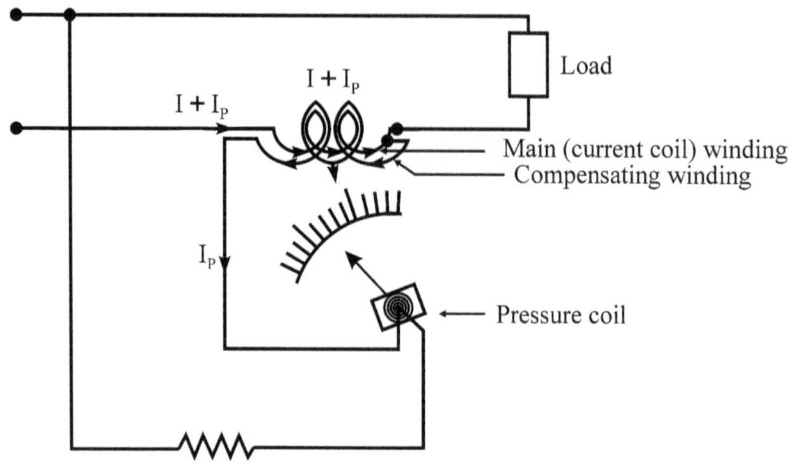

Fig. 3.9

when power to be measured is low the current in the circuit is high as the P.F is low. Thus in this case, pressure coil can not be connected to supply side as otherwise large error will be produced because of large current flowing in current coil and corresponding power loss in current coil circuit is measured by wattmeter.

If pressure coil is connected to load side, power consumed by pressure coil is measured by wattmeter which is a appreciable in comparison with power to be measured, which is small. Hence it is necessary to compensate for pressure coil current in low p.f wattmeter. The compensated wattmeter is shown in Fig. 3.9. As shown in the Fig. 3.9, the compensating coil is connected in series with the potential coil and is made as identical and coincident with current coil as possible.

The current coil carries current $I + I_p$ and produces its own field proportional to this current. The compensating coil carries current I_p and produces field proportional to this current. This field acts in opposite direction to the field produced by current coil.

Thus the resultant field is due to current 'I' only, hence error due to pressure coil current is neutralized. Thus at no load condition, the wattmeter should not deflect as the resultant current coil field is zero.

In case of low p.f wattmeter, the pressure coil circuit is designed for low resistance to increase the current flowing through it so as to have increased torque. In low p.f wattmeter, the value of pressure coil current is 10 times the current in case of high p.f wattmeter.

We have already seen in previous section that the pressure coil inductance introduces error whose magnitude is given by VI sin ϕ tan β. If power factor is low then Ø is large and hence sinØ is large. Thus the error introduced in the measurement is appreciable which must be compensated. It is compensated by connecting a capacitor across a part of series resistance in the pressure coil circuit which is shown in Fig 3.10.

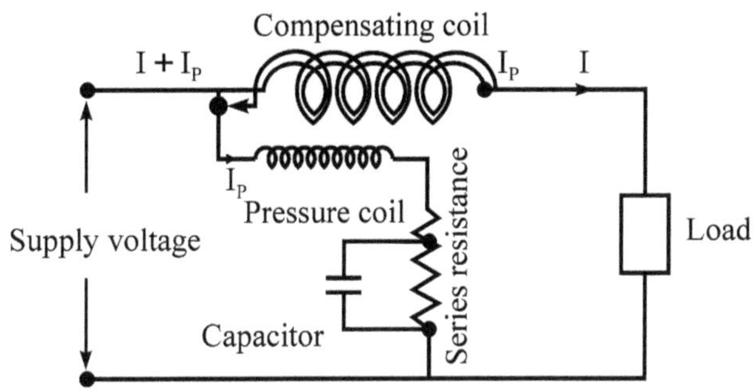

Fig. 3.10 Low power factor wattmeter

The average deflecting torque equation of an electrodynamometer type wattmeter is given by

$$T_d = I_1 I_2 \cos\phi \frac{dM}{d\theta}$$

$$T_d \propto \cos\phi$$

From above equation, we can say that the deflecting torque is directly proportional to the p.f. Hence, at low p.f, the deflecting torque is low. Also there are some inductance in the result due to the pronounced effect of induction of pressure coil at load p.f. Hence in order to overcome the above difficulties, additional features are provided in the electrodynamometer type wattmeter to make a low p.f wattmeter.

3.11 Power in Three Phase System

In a three phase system, the load can be star connected or delta connected having impedance per phase as,

$$Z_{ph} = R_{ph} + jX_{ph} \ \Omega/ph = |Z|\angle\phi \ \Omega/ph$$

the reactance X_{ph} may be positive or negative depending on inductive or capacitive load. It is positive for inductive and negative for capacitive load.

ϕ = Angle between V_{ph} and I_{ph}

	For star connected load	For deltaconnected load
1.	$V_L = \sqrt{3}\, V_{ph}$	$I_L = \sqrt{3}\, I_{ph}$
2.	$I_L = I_{ph}$	$V_L = V_{ph}$
3.	$V_R, V_Y, V_B = V_{ph}$	$I_{ph} = I_{RY}, I_{YB}, I_{BR}$
4.	$V_{RY}, V_{YB}, V_{BR} = V_L$	$I_L = I_R, I_Y, I_B$
5.	$V_{BR} = \overline{V}_B - \overline{V}_R$	$\bar{I}_B = \bar{I}_{BR} - \bar{I}_{YB}$
6.	$\overline{V}_{RY} = \overline{V}_R - \overline{V}_Y$	$\bar{I}_R = \bar{I}_{RY} - \bar{I}_{BR}$
7.	$\overline{V}_{YB} = \overline{V}_Y - \overline{V}_B$	$\bar{I}_Y = \bar{I}_{YB} - \bar{I}_{RY}$

For any load, stor or delta connected, the 3-ϕ total power is given by,

$$P = \sqrt{3}\ V_L I_L \cos\phi \qquad = 3\,[V_{ph} I_{ph} \cos\phi]$$

The wattmeters must be properly connected in a three phase system to measure the total power.

3.12 The Main Difference between LPF and UPF Wattmeters

	LPF Wattmeters	UPF Wattmeters
1.	In case of LPF wattmeter the deflection torque is small even when the current and pressure coil are fully energized ($\because T_d \propto \cos\phi$).	In case of UPF wattmeters, deflecting torque is large.
2.	The pressure coil current is very high.	The pressure coil current is small.
3.	The error caused by pressure coil inductance is large.	The error caused by the pressure coil inductance is zero.
4.	The resistance of pressure coil circuit is small.	The resistance of pressure coil circuit is large.
5.	In LPF wattmeters, the results are inaccurate.	In UPF wattmeters, the results are accurate.
6.	The control torque is small.	The control torque is large.

3.13 Examples of Wattmeter Connections and Corresponding Readings

Case I

Consider delta connected inductive load and wattmeter connected as shown to measure power.

Find the reading on it. Will measure phase power? Let us find it out.

For wattmeter: $I_C = I_R = I_L$ = line current

while $V_{pc} = V_{RY} = V_L = V_{ph}$ as delta load

$$W = V_{pc}\, I_c \cos (V_{pc} \wedge I_c)$$

$$= V_{RY}.\, I_R.\, \cos (V_{Ry} \wedge I_R)$$

Now $\quad I_R = I_L$ and $V_{RY} = V_L = V_{ph}$

Hence angle between V_{RY} and I_R is not 'ϕ'.

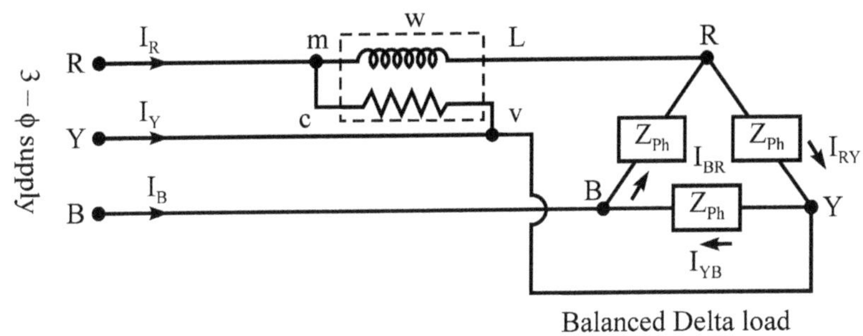

Fig. 3.11

Assume load having cos ϕ lagging power factor

$$I_{ph} = I_{RY} = I_{YB} = I_{BR}$$

$$\phi = V_{ph}\, I_{ph}$$

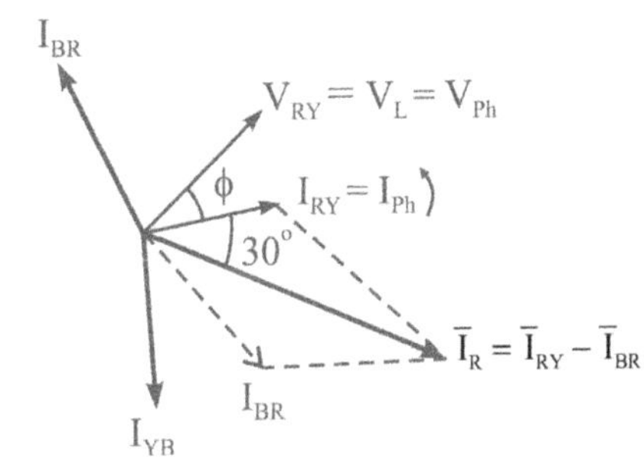

Fig. 3.11(a)

To find $V_{RY} \wedge I_R$, let us draw phasor diagram as shown in the Fig. 3.11(a)

For delta connected load

$$\bar{I}_R = \bar{I}_{RY} - \bar{I}_{BR}$$

Phase current I_{RY} lags phase voltage V_{RY} assuming that load p.f. is cos ϕ lagging.

From phasor diagram it is clear that

$$V_{RY} \wedge I_R = (30 + \phi)^{\circ}$$

$$\therefore \quad W = V_{RY} \times I_R \times \cos(30 + \phi)$$

$$\therefore \quad W = V_L I_L \cos(30 + \phi) \text{ W},$$

this is not a phase power reading.

Case II

Now let us shift the same wattmeter in such a way that it has to read phase power V_{ph} I_{ph} $\cos\phi$. For this $I_c = I_{ph} = I_{RY}$ or I_{YB} and $V_{pc} = V_{ph} = V_L = V_{RY}$ or V_{YB} or V_{BR}. Accordingly wattmeter coils must be connected such that $I_c = I_{ph}$ and $V_{pc} = V_{ph}$, as $I_{ph} \wedge V_{ph} = \phi$ and then it will read V_{ph} I_{ph} $\cos\phi$ which is phase power. The connections can be as shown in Fig. 3.12.

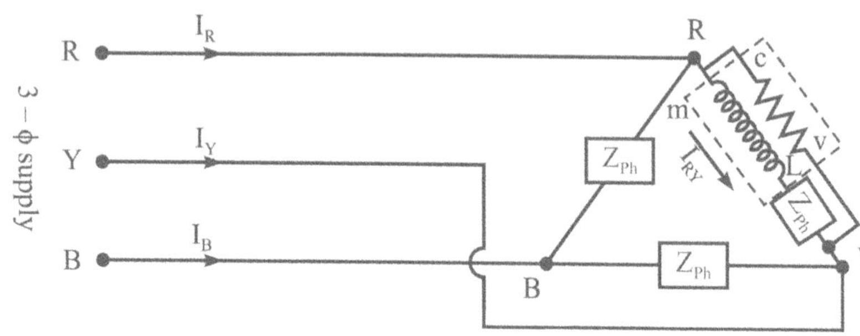

Fig. 3.12

For wattmeter: $\quad I_c = I_{ph} = I_{RY}$ and $V_{pc} = V_{RY} = Vph$

$\therefore$ wattmeter will read, $\quad W = I_c V_{pc} \cos(I_C \wedge V_{pc}) = I_{RY} V_{RY} \cos(I_{RY} \wedge V_{RY})$

$$W = I_{ph} V_{ph} \cos\phi = \text{phase power}$$

Note: wattmeter will not always measure phase power. Its reading depends on the angle between current sensed by its current coil and voltage sensed by its pressure coil. This depends on its connections in the circuit.

Three wattmeter measuring power in three phase may be connected to measure power. But connecting wattmeter to measure phase power is not always possible, if neutral point of star connected load is not available outside. Similarly in delta connected load to measure phase current it is necessary to open delta to insert current coil of the wattmeter as discussed above, which is not practicable. The best method of measuring power whether load is star or delta connected, balanced or unbalanced, neutral is available or not is, using only two wattmeters. The method is called **two wattmeter method.**

Let us discuss first the single wattmeter method and three wattmeter method along with the corresponding limitations before discussing the two wattmeter method.

3.13.1 Single (or) One Wattmeter Method

This can be only used for balanced three phase load. When the load is BALANCED, total power can be calculated as,

$$P = 3\ V_{ph}\ I_{ph} \cos \phi = 3 \times [\text{wattmeter reading}]$$

hence one wattmeter is to be used to measure single phase power and then reading is to be multiplied by three.

Note: the wattmeter must be connected in such a way that its current coil must carry I_{ph} and its voltage coil must be across V_{ph}.

Case I: Star Connected Load

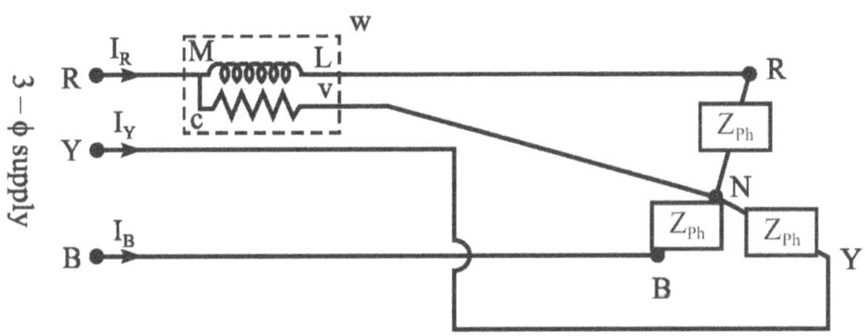

Fig. 3.13 Wattmeter measuring single phase power

$$I_{ph} = I_L = I_R = I_C \text{ as load is star connected.}$$

But voltage coil must be connected as to measure V_{ph} i.e., as shown in the Fig. 3.13. The voltage coil across one line terminal and other to a neutral point.

$$V_{ph} = V_{pc} = V_{RN}$$

$$W = V_{pc}\ I_c \cos (V_{ph} \wedge I_c)$$

$$\therefore\ W = V_{RN}\ I_R \cos (V_{RN} \wedge I_R)$$

$$\therefore\ W = V_{ph}\ I_{ph} \cos \phi$$

Total power $P = 3V_{ph}\ I_{ph} \cos \phi = 3W$ watts.

Let Z_{ph} is having power factor as cos ϕ lagging. Then phasor diagram can be drawn as in Fig. 3.14. For Z_{ph} having cos ϕ leading P.F only change is I_{ph} will read V_{ph} by ϕ degrees while for Z_{ph} having cos ϕ = 1 i.e., unity p.f (resistive load), V_{ph} and I_{ph} will be in phase with each other ($\phi = 0^\circ$).

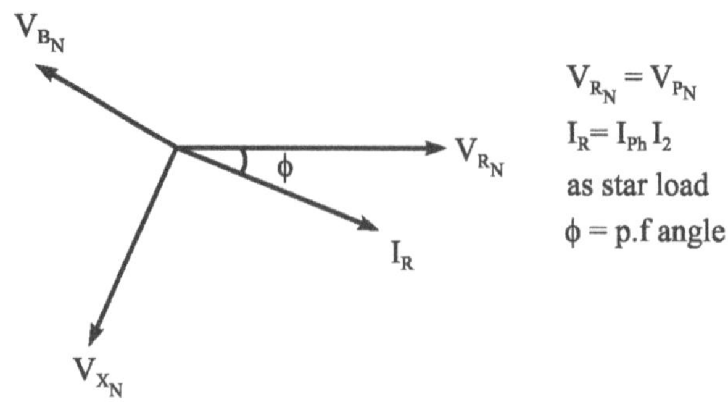

Fig. 3.14

Case II: Delta Connected Load

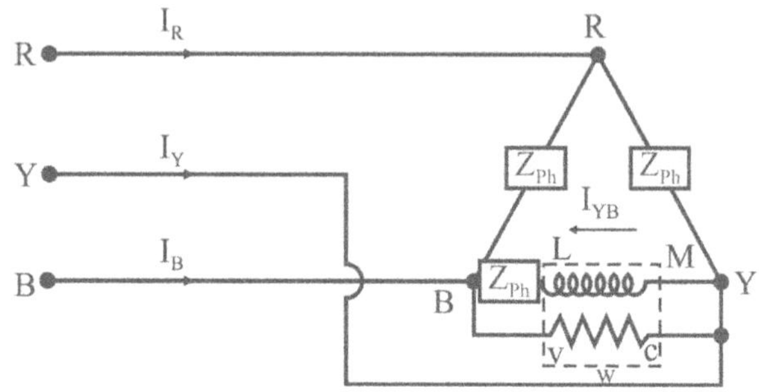

Fig. 3.15 Wattmeter measuring single phase power

In this connection, $I_c = I_{YB} = I_{ph}$

while $V_{pc} = V_{YB} = V_{ph}$

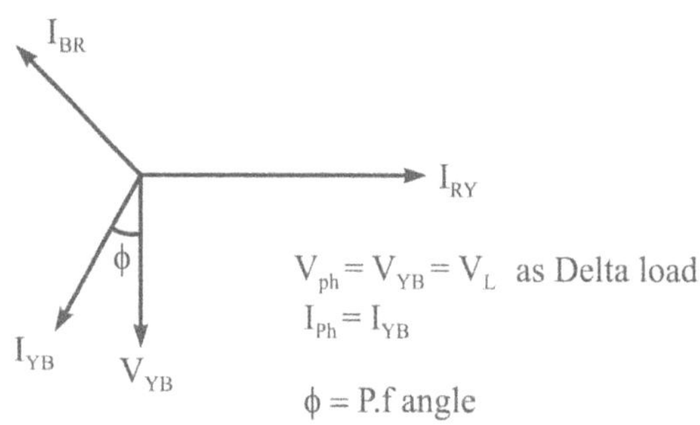

Fig. 3.16

here as current coil must be read I_{ph}, it is necessary to open delta and insert current coil in it. $W = V_{pc}\, I_c \cos (V_{pc} \wedge I_c)$

$\therefore \qquad W = V_{YB} I_{YB} \cos (V_{YB} \wedge I_{YB}) = V_{ph} I_{ph} \cos \phi$

Total power = 3W watts.

If load is having cos ϕ lagging power factor, phasor diagram can be shown as in Fig. 3.16, but this method has certain disadvantages.

3.13.2 Disadvantages

1. Applicable only for balanced loads
2. Using this method for star connected loads, neutral point must be available other wise voltage coil can not be connected so as to measure phase voltage. Similarly for delta connected load, it must be possible to open the closed delta so as to measure phase current. This may not be possible in practise

3.14 3-Wattmeter Method

If the load is unbalanced then we can use three wattmeters which will measure power consumed in each phase and then all the three readings can be added to get the total power. If load is unbalanced, power consumed in each phase will be different and hence three wattmeters are necessary. The connection of the wattmeter are exactly similar to the connection of one wattmeter method discussed above.

The connections for star connected load can be shown as in Fig. 3.17.

If load is balanced $w_1 = w_2 = w_3$

If load is unbalanced $w_1 \neq w_2 \neq w_3$

Total power $= w_1 + w_2 + w_3$ for unbalanced load

$= 3\,w_1$ or $3w_2$ or $3w_3$ for balanced load.

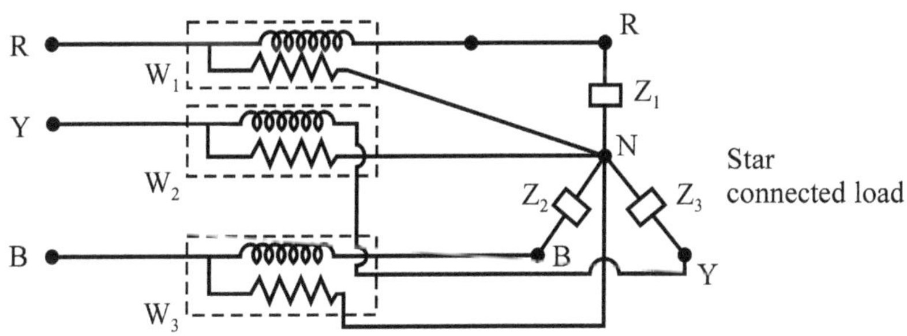

Fig. 3.17 Star connected load

The connection for delta connected load can be as shown in Fig. 3.18.

If load is balanced, $w_1 = w_2 = w_3$

And total power is, $P = w_1 + w_2 + w_3 = 3w_1$ (or) $3w_2$ (or) $3w_3$.

In this method, the power can be measured for unbalanced load but the disadvantages of requirement of neutral point for star load and arrangement for insertion of current coil in closed delta, still continues.

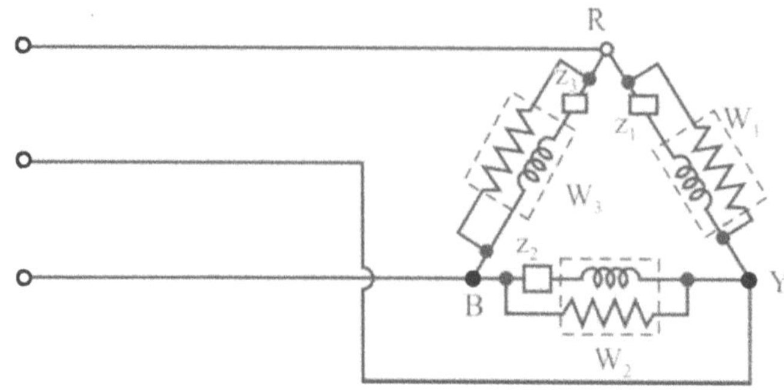

Fig. 3.18 Delta connected load

3.15 Blondel's Theorem

The Blondel's theorem is about number of wattmeter required in a three phase system for the measurement of total power. It states that,

If a Network is supplied using 'n' conductors, the total power is the sum of the readings of 'n' wattmeters so arranged that a current coil of each wattmeter is in each line and the corresponding pressure coil is connected between that line and a common point.

If the common point is located on one of the lines then the power may be measured by n – 1 wattmeters. Thus in a three phase system, if common point for connecting pressure coils is located on one of the lines then only two wattmeter are sufficient for measuring the power. This is called two wattmeter method.

3.16 Two Wattmeter Method

Methods of Connection: The current coils and the two wattmeter are connected in any two lines which the voltage coil of each wattmeter is connected between its own current coil terminal and the line without a current coil. For example, the current coils are inserted in the lines R and Y then the pressure coils are connected between R-B for one wattmeter and Y-B for other wattmeter, as shown in the Fig. 3.19.

The connections are same for star or delta connected load. It can be shown that when two wattmeter are connected in this way, the algebraic sum of the two wattmeter reading gives the total power dissipated in the three phase circuit.

If W_1 and W_2 are the two wattmeter reading then the total power is

$$W = W_1 + W_2 = \text{three phase power}$$

3.16.1 Proof of 2-Wattmeter Method

Consider star connected load and two wattmeter connected as shown in Fig. 3.19.

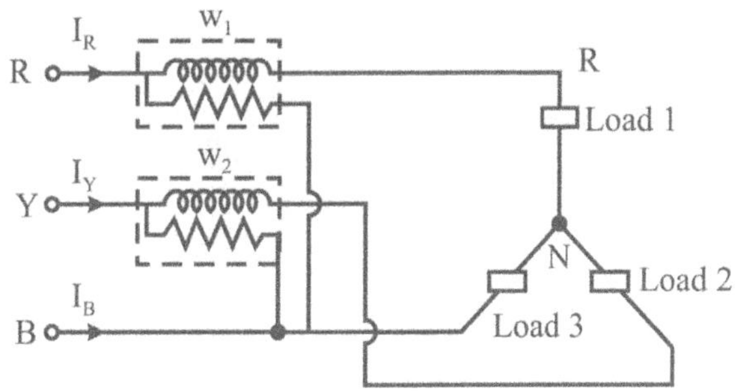

Fig. 3.19 Two wattmeter method for star connected load

For Balanced Load

Let us consider the r.m.s value of the current and voltages to prove that sum of two wattmeter gives total power consumed by three phase load.

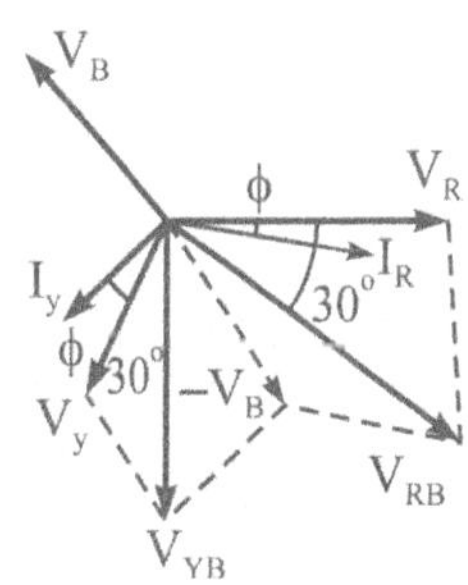

Fig. 3.19(a)

$$W_1 = I_R \times V_{RB} \times \cos (I_R \wedge V_{RB})$$

$$W_2 = I_Y \times V_{YB} \times \cos (I_Y \wedge V_{YB})$$

To find angle between (I_R and V_{RB}) and (I_y and V_{yB}) let us draw phasor diagram. Assuming load p.f. be cos ϕ lagging.

$$\overline{V_{RB}} = \overline{V_R} - \overline{V_B} \quad \text{and} \quad \overline{V_{YB}} = \overline{V_Y} - \overline{V_B}$$

$$V_R \wedge I_R = \phi \quad \text{and} \quad \phi = V_Y \wedge I_Y$$

$$V_{ph} = V_R = V_Y = V_B \quad \text{and} \quad V_L = V_{RB} = V_{YB}$$

$$I_{ph} = I_R = I_Y = I_L$$

from Fig. 3.19 (a)

$$I_R \wedge V_{RB} = 30 - \phi$$

$$I_Y \wedge V_{YB} = 30 + \phi$$

$$W_1 = I_R\, V_{RB} \cos (30 - \phi) \qquad \therefore\ W_1 = V_L\, I_L \cos (30 - \phi)$$

and

$$W_2 = I_Y\, V_{YB} \cos (30 + \phi) \qquad \therefore\ W_2 = V_L I_L \cos (30 + \phi)$$

$$\therefore \quad W_1 + W_2 = V_L I_L [\cos(30 - \phi) + \cos(30 + \phi)]$$

$$= V_L I_L [\cos 30 \cos\phi + \sin 30 \sin\phi + \cos 30 \cos\phi - \sin 30 \sin\phi]$$

$$= 2 V_L I_L \cos 30 \cos\phi = 2 V_L I_L \frac{\sqrt{3}}{2} \cos\phi$$

$$\therefore \quad W_1 + W_2 = \sqrt{3}\, V_L I_L \cos\phi = \text{Total 3 phase power}$$

Consider delta connected balanced load (Fig. 3.20).

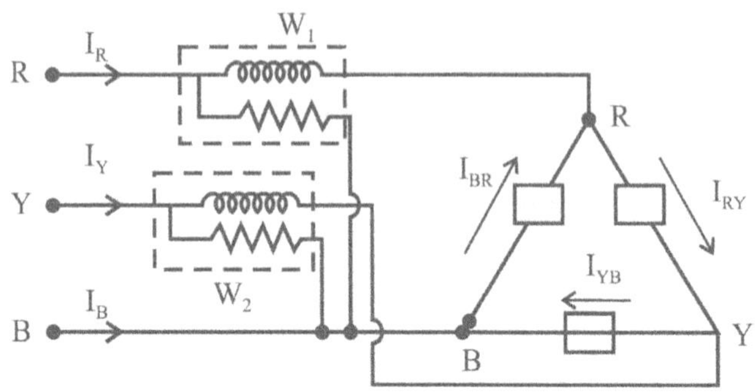

Fig. 3.20 Two wattmeter method for delta connected load

for w_1, $\quad I_c = I_R$ and $V_{pc} = V_{RB}$

for w_2, $\quad I_c = I_Y$ and $V_{pc} = V_{YB}$

$$W_1 = I_R V_{RB} \cos(I_R \wedge V_{RB})$$

$$W_2 = I_Y V_{YB} \cos(I_Y \wedge V_{YB})$$

To find $I_R \wedge V_{RB}$ and $I_Y \wedge V_{YB}$, let us draw phasor diagram (Fig. 3.21). Assume load having cos ϕ lagging p.f.

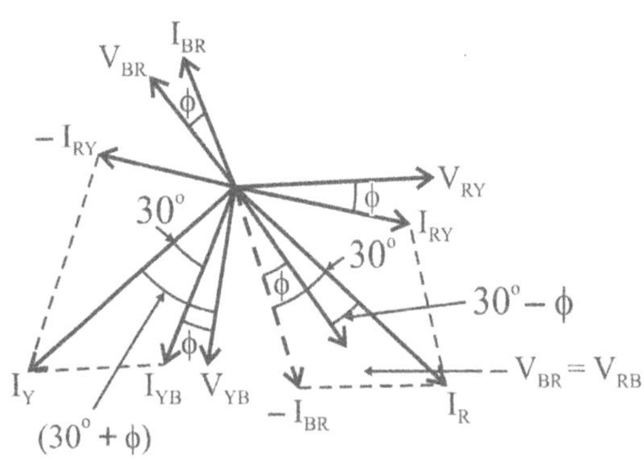

Fig. 3.21 Delta connected load, lagging p.f

$$V_{RB} = V_{YB} = V_{phase} = V_{line}$$

and $$I_R = I_Y = I_{line}$$

$$\bar{I}_R = I_{RY} - I_{BR} \qquad \text{and} \qquad \bar{I}_Y = \bar{I}_{YB} - \bar{I}_{RY}$$

$$W_1 = I_R\, V_{RB} \cos(30 - \phi) = I_L V_L \cos(30 - \phi)$$

$$W_2 = I_Y\, V_{YB} \cos(30 + \phi) = I_L V_L \cos(30 + \phi)$$

$\therefore$ $$W_1 + W_2 = V_L I_L\, [\cos(30 - \phi) + \cos(30 + \phi)]$$

$$= \sqrt{3}\, V_L I_L \cos\phi = \text{Total power consumed by three phase load}$$

For load having leading power factor, I_{ph} will lead V_{ph} by angle ϕ and it can be observed from phasor diagram that W_1 reading and W_2 reading will get interchanged as $30 - \phi$ will become $30 + \phi$ and vice-versa.

Refer Fig. 3.21 (a) for reading Power factor phasor diagram.

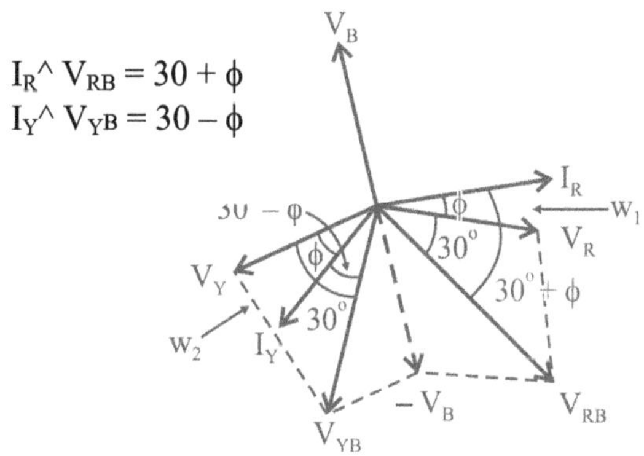

Fig. 3.21(a) Star connected leading p.f

I_R leads V_R by ϕ, I_Y loads V_Y by ϕ,

$$W_1 = I_R\, V_{RB} \cos(I_R \wedge V_{RB})$$

$\therefore$ $$W_1 = V_L I_L \cos(30 + \phi)$$

$$W_2 = I_Y\, V_{YB} \cos(I_Y \wedge V_{YB})$$

$\therefore$ $$W_2 = V_L I_L \cos(30 - \phi)$$

Reading of w_1 and w_2 are interchanged compared to lagging power factor load.

but $$W_1 + W_2 = \sqrt{3}\, V_L I_L \cos\phi$$

$$W_1 + W_2 = \text{Total 3 phase power}$$

Note: As observed that whether load is STAR of DELTA, the expression for 'W_1' and 'W_2' remain same.

- For star or delta leading p.f load, $W_1 = V_L I_L \cos(30 + \phi)$
  $W_2 = V_L I_L \cos(30 - \phi)$
- For star or delta lagging p.f load, $W_1 = V_L I_L \cos(30 - \phi)$
  $W_2 = V_L I_L \cos(30 + \phi)$
- For star or delta unity p.f load, $\cos\phi = 1$, and $\phi = 0^\circ$
  $W_1 = W_2 = V_L I_L \cos 30^\circ$

3.16.2 Power Factor Calculation by Two-wattmeter Method

In case of balanced load, the p.f. can be calculated from W_1 and W_2 readings. For balanced, lagging p.f. load,

$$W_1 = V_L I_L \cos(30 - \phi)$$

$$W_2 = V_L I_L \cos(30 + \phi)$$

$$W_1 + W_2 = \sqrt{3}\, V_L\, I_L \cos\phi \qquad .....(3.14)$$

$$W_1 - W_2 = V_L I_L [\cos(30 - \phi) - \cos(30 + \phi)]$$

$$= V_L I_L [\cos 30 \cos\phi + \sin 30 \sin\phi - \cos 30 \cos\phi + \sin 30 \sin\phi]$$

$$= V_L I_L [2 \sin 30 \sin\phi] = V_L I_L \left[2 \times \frac{1}{2} \times \sin\phi\right]$$

$$\therefore \qquad W_1 - W_2 = V_L I_L \sin\phi \qquad .....(3.15)$$

From eq. (3.14) and (3.15)

$$\frac{W_1 - W_2}{W_1 + W_2} = \frac{V_L I_L \sin\phi}{\sqrt{3}\, V_L I_L \cos\phi} = \frac{\tan\phi}{\sqrt{3}}$$

$$\tan\phi = \frac{\sqrt{3}\,(W_1 - W_2)}{(W_1 + W_2)}; \qquad \phi = \tan^{-1}\left[\frac{\sqrt{3}\,(W_1 - W_2)}{(W_1 + W_2)}\right]$$

$$\text{p.f} \cos\phi = \cos\left\{\tan^{-1}\left[\frac{\sqrt{3}\,(W_1 - W_2)}{(W_1 + W_2)}\right]\right\}$$

Note: For leading p.f we get $\tan\phi$ negative, but cosine of negative angle is positive.

So $\cos\phi$ is always positive, but its nature must be determined by observing sign of $\tan\phi$.

$$\cos\phi = \cos\left\{\tan^{-1}\left(\frac{\sqrt{3}\,(w_1 - w_2)}{w_1 + w_2}\right)\right\}$$

Hence phase angle ϕ and p.f cos ϕ can be determined from readings of two wattmeter

1. **When both wattmeter read same,** $w_1 = w_2$, $\tan\theta = 0$, $\cos\theta = 1$, the load is purely resistasistive.
2. **When only one wattmeter reads,** $w_1 = 0$, $\tan\theta = \sqrt{3}$, $\cos\theta = 0.5$, $\theta = 60^\circ$ lagging.
3. $60^\circ < \theta < 90^\circ$, $0.5 < \cos -$ (P.F) < 0, **one wattmeter reading is positive and another negative as wattmeter can not indicate negative readings**; coil has to be reversed, preferably the pressure coil. $W_t = W_2 - W_1$.
4. **When the p.f is leading by θ,** $W_1 = VI\cos(30 + \theta)$ and $W_2 = VI\cos(30 - \theta)$ i.e., the readings of the two wattmeters are only interchanged.
5. **If the wattmeter readings are equal and opposite** $W_1 = -W_2$, power factor $0 < \cos\phi < 0.5$, the load is purely reactive. So $\tan\theta$ depends on phase sequence and when wattmeter reading w_1 and w_2.

3.17 Effect of Power Factor on Wattmeter Readings

Range of 'ϕ'	Range of power factor	w_1 sign	w_2 sign	Remarks
1. $\phi = 90^\circ$	$\text{Cos}\,\phi = 0$	positive	negative	$\lvert w_1\rvert = \lvert w_2\rvert$
2. $90^\circ < \phi < 60^\circ$	$0 < \cos\phi < 0.5$	positive	negative	
3. $\phi = 60^\circ$	$\cos\phi = 0.5$	positive	0	
4. $60^\circ < \phi < 0^\circ$	$0.5 < \cos\phi < 1$	positive	positive	
5. $\phi = 0^\circ$	$\cos\phi = 1$	positive	positive	$w_1 = w_2$

3.18 Reactive Volt-Amperes by Two Wattmeter Method

We have observed that

$$W_1 - W_2 = V_L I_L \sin\phi$$

the total reactive volt ampere for a three phase circuit is given by

$$Q = \sqrt{3}\ V_L I_L \sin\phi = \sqrt{3}\ (W_1 - W_2)\ \text{VAR}$$

thus reactive volt-amperes of a three phase circuit can be obtained by multiplying the difference of two wattmeter reading by $\sqrt{3}$.

so,

$$P = \text{Active power} = \sqrt{3}\ V_L I_L \cos\phi = W_1 + W_2\ \text{W or kW}$$

$$Q = \text{Reactive power} = \sqrt{3}\ V_L I_L \sin\phi = \sqrt{3}\ (w_1 - w_2)\ \text{VAR (or) KVAR}$$

$$S = \text{Apparent power}\ \sqrt{3}\ V_L I_L,\ \text{VA (or) KVA}$$

3.19 Disadvantages and Advantages of Two Wattmeter Method

Disadvantages

1. These are not applicable for three phase 4 wire system.
2. The signs of w_1 and w_2 must be identified and noted down correctly otherwise it may lead to the wrong results.

Advantages

1. Neutral point for star connected load is not necessary to connect the wattmeters.
2. This method is applicable for balanced as well as unbalanced loads.
3. Only two wattmeters are sufficient to measure total three phase power.
4. Total reactive volt amperes can be obtained using two wattmeter readings for balanced loads.
5. If the load is balanced, not only the power but power factor also can be determined.
6. The delta connected load, need not be opened for connecting the wattmeters.

3.20 Modified Version of 2-Wattmeter as One Wattmeter Method

This type of method can be used only for balanced loads. In this method, only one wattmeter is used. Its current coil is introduced in any one line while its voltage coil is connected to other two lines, one after the other in sequence, with the help of two way switch.

The sum of the two readings gives total active power in three phase circuit. Its connection can be shown as in Fig. 3.22.

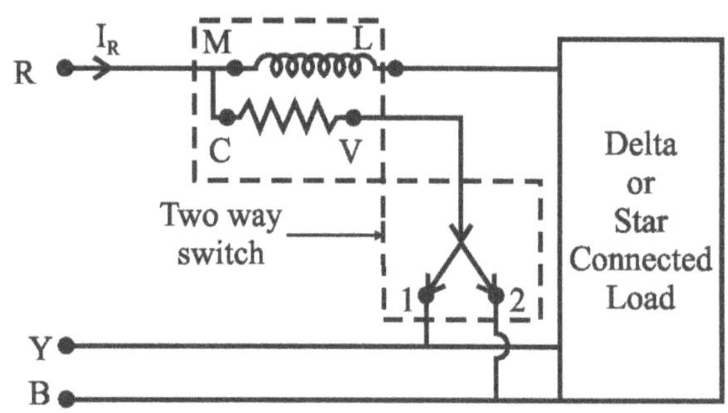

Fig. 3.22 Modified two wattmeter method as one wattmeter method

In this Fig. 3.22 current coil introduced in line R and hence it carries current I_R. While one end of pressure coil is connected to 'R' only and second end gets connected to 'Y' when switch is in position 1 and gets connected to 'B' when switch is position 2.

Assume star connected load

Let us find the two readings on that wattmeter. Assume load having p.f. cos ϕ lagging.

In switch position 1:

$$I_c = I_R, \qquad V_{pc} = V_{RY}$$

$$W_1 = I_R V_{RY} \cos (I_R \wedge V_{RY})$$

In switch position 2:

$$I_c = I_R, \qquad V_{pc} = V_{RB}$$

$$\therefore \qquad W_2 = I_R V_{RB} \cos (I_R \wedge V_{RB})$$

Let us find I_R, $\wedge V_{RY}$ and $I_R \wedge V_{RB}$ from phasor diagram as shown in the Fig. 3.23.

$$\overline{V}_{RB} = \overline{V}_R - \overline{V}_B$$

$$\overline{V}_{RY} = \overline{V}_R - \overline{V}_Y$$

I_R lags V_R by angle 'ϕ'

$$\therefore \qquad I_R \wedge V_{RY} = (30 \quad \phi)$$

$$I_R \wedge V_{RB} = (30 + \phi)$$

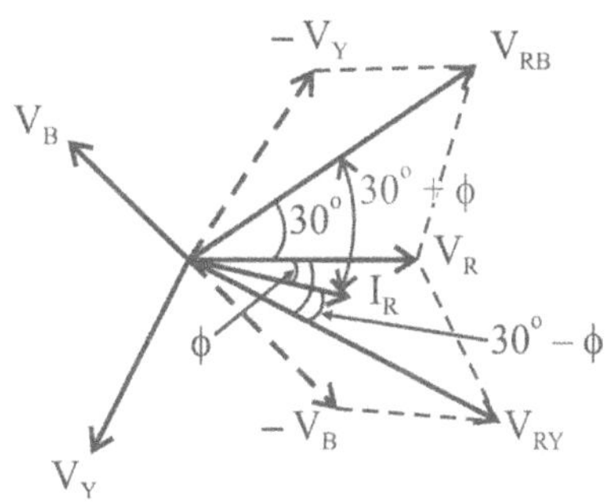

Fig. 3.23

$\therefore$ in position 1, $W_1 = I_R V_{RY} \cos (30 - \phi) = I_L V_L \cos (30 - \phi)$

and in position 2, $W_2 = I_R V_{RB} \cos (30 + \phi) = I_L V_L \cos (30 + \phi)$

$$\therefore \qquad W_1 + W_2 = \sqrt{3}\, V_L I_L \cos \phi = \text{Total power}$$

Note: any unbalanced condition of the load may create large errors in this method hence not preferred over two wattmeter method.

3.21 One Wattmeter Method for Reactive Volt Amperes Measurement

This can be used for balanced load. In this method, current coil of wattmeter is connected in any one line and pressure coil is connected across remaining two lines. The connection is as shown in Fig.3.24.

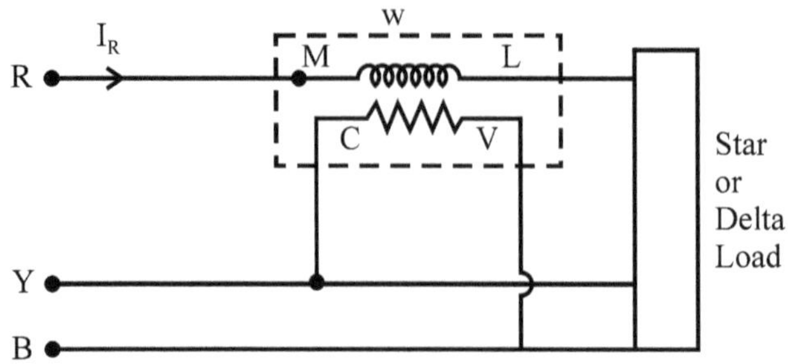

Fig. 3.24

$$W = I_c\, V_{pc} \cos (I_c\, V_{pc}) = I_R\, V_{YB} \cos (I_R \wedge V_{YB})$$

To find $I_R \wedge V_{YB}$, assume load to be star connected having cos ϕ. Lagging p.f. Phasor diagram is as in the Fig. 3.24 (a).

$$\overline{V}_{YB} = \overline{V}_Y - \overline{V}_B$$

$$I_R \wedge V_{YB} = 90 - \phi$$

$$w = I_R\, V_{YB} \cos (90 - \phi)$$

$$= I_R\, V_{YB} \sin \phi = I_L\, V_L \sin \phi$$

Thus in this method the wattmeter reading is,

$$W = V_L I_L \sin \phi$$

but total reactive volt amperes are $\sqrt{3}\ V_L I_L \sin \phi$.

Note: this reading ω must be multiplied by $\sqrt{3}$ to get total reactive volt amperes.

$$\sqrt{3}\, W = \text{Total reactive volt amperes}$$

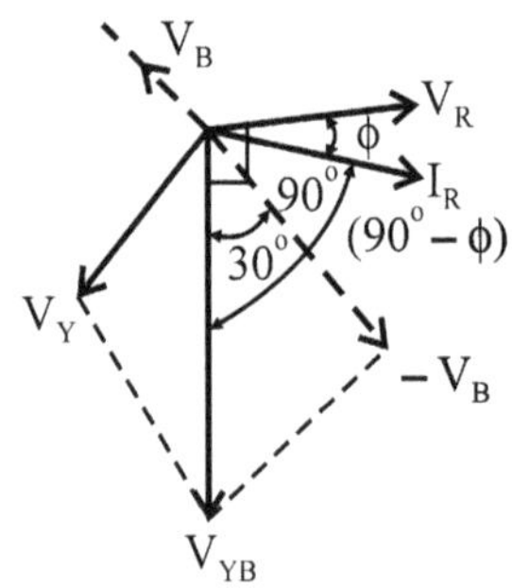

Fig. 3.24(a)

3.22 Extension of Range of Wattmeter using Instrument Transformer

For very high voltage circuit, the high rating wattmeters are not available to measure the POWER. The range of wattmeter can be extended using instrument transformer in such high voltage circuit. The connections are shown in the Fig. 3.24 (b).

The primary winding of current transformer is connected in series with the load; secondary is connected in series with an ammeter and the current coil of wattmeter. The primary winding of power transformer is connected across the supply and secondary is connected across voltmeter and the pressure coil of the wattmeter. One secondary terminal of each transformer and the casings are grounded.

Now both current transformer and power transformer have error like ratio error and phase angle error. For precise measurements, these errors must be considered. If not considered, these errors may cause inaccurate measurements. The correction must be applied to such errors to get the accurate results.

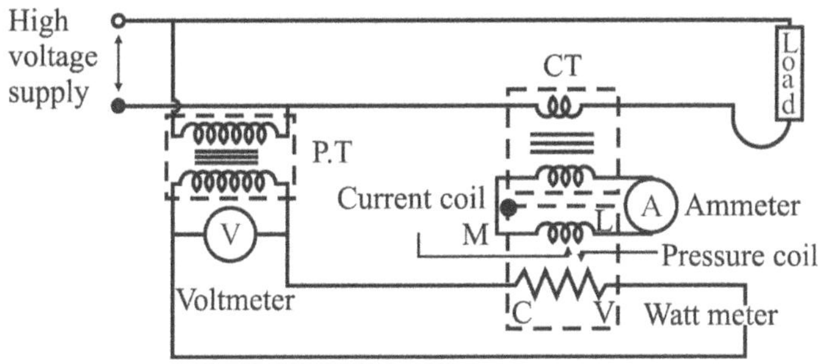

Fig. 3.24(b) Power measurement using C.T and P.T

3.22.1 Phasor Diagrams and Correction Factors

Consider the various parameters as,

I = load current

ϕ = phase angle between V and I

V_s = wattmeter pressure coil voltage (or) voltage across secondary of P.T

I_s = wattmeter current coil current (or) current in secondary of C.T

I_p = current in pressure coil of wattmeter

α = phase angle between current coil and pressure coil of wattmeter

δ – phase angle of P.T

θ = phase angle of C.T

β = angle by which I_p lags V_s due to inductance of pressure coil.

The phasor diagrams for lagging and leading p.f loads are shown in the Fig. 3.25(a) and (b) respectively.

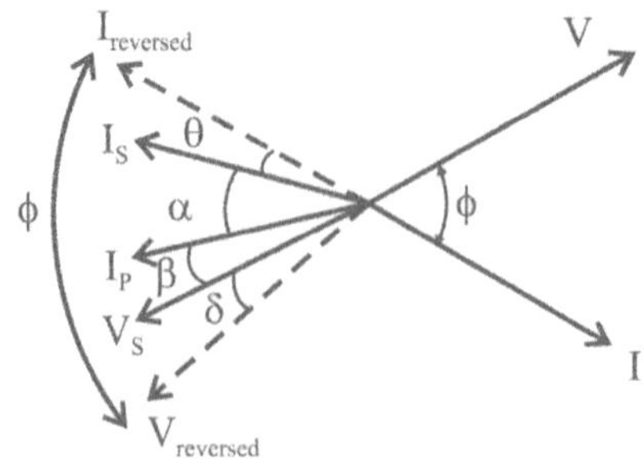

Fig. 3.25(a) Lagging p.f

$$\phi = \alpha + \beta \pm \delta + \theta$$

Lagging Power Factor

For lagging P.F, θ is positive i.e., I_S leads reversed I, while phase of P.T may be positive or negative i.e., δ can be positive or negative. For phasor diagram shown in Fig. 3.25 (a) δ is negative i.e., V_s lags reversed 'V' hence,

$$\left.\begin{array}{lll} \phi = \alpha + \beta + \delta + \theta & \text{--------} & \delta \text{ negative} \\ \phi = \alpha + \beta - \delta + \theta & \text{--------} & \delta \text{ positive} \end{array}\right\} \text{Lagging p.f}$$

Leading Power Factor

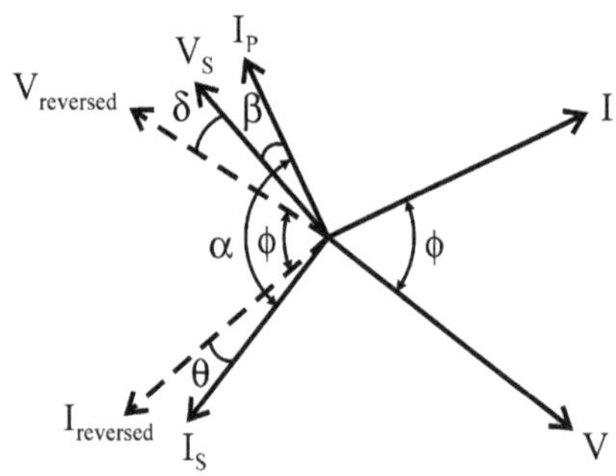

Fig. 3.25(b) Leading p.f

$$\phi = \alpha - \theta \pm \delta - \beta$$

For leading P.F., the φ is given by

$$\left.\begin{array}{lll} \phi = \alpha - \beta - \delta - \theta & \text{--------} & \delta \text{ negative} \\ \phi = \alpha - \beta + \delta - \theta & \text{--------} & \delta \text{ positive} \end{array}\right\} \text{Leading p.f}$$

3.22.2 Correction Factor

The correction factor, neglecting transformation ratio errors is,

$$K = \frac{\cos\phi}{\cos\beta \cos\alpha}$$

where $\alpha = \phi - \beta - \delta - \theta$ -------- lagging p.f (with δ negative)

and $\alpha = \phi + \beta + \delta + \theta$ -------- lagging p.f (with δ negative)

[**Note:** change the sign of δ in the expression of α if δ is positive.]

True power = K × Actual ratio of C.T × Actual ratio of P.T × Wattmeter reading

∴ True power = K × [R.C.F. of C.T] × [R.C.F. of P.T] × [Nominal ratio of C.T] × [Nominal ratio P.T] × [Wattmeter reading]

3.23 3-ϕ Power Measurement using Current Transformer and Power Potential Transformer

This concept of using C.T and P.T for single ϕ power measurement can be extended for three phase power measurement. The connections are basically similar to the two wattmeter method as shown in the Fig. 3.26.

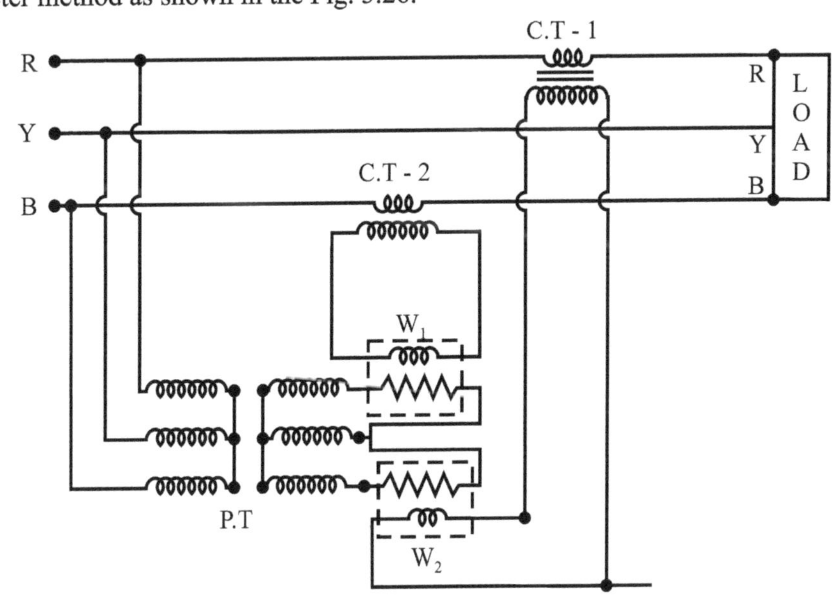

Fig. 3.26 Three phase power measurement using C.T and P.T

3.24 The Advantages and Disadvantages of Instrument Transformer for Extension of Range of A.C. Instrument

Advantages

1. It provides electrical isolation between the high tension lines and the meter which provides safety for the observer or operating personnel.
2. In absence of it, the instruments suffer from various problems like insulation, stray fields etc.
3. Its power consumption is very low.
4. In case of clamp on ammeter, there is no need to break the bus-bar also. The core hang on the bus-bar and the bus-bar itself acts as a single Tern primary.
5. The readings are independent of 'R', 'L' and 'C'; constant due to which a linear relationship exists for the entire range.
6. Heavy current and voltage can be measured by using moderate rating, cheap and accurate measuring instrument. So overall cost is reduced.
7. From a single measuring instrument, may other instruments are operated.
8. Handling and replacing is quite simple.

Disadvantages

1. In instrument transformer, transformation ratio is not equal to the turns ratio and also the value is not constant. But depends upon the magnetizing and loss components of exiting current, secondary winding, load current and power factor. This introduces errors in current and voltage measurement.
2. The basic requirement in power measurement is that the secondary winding current must be displaced by 180° exactly from that of primary winding current. But in case of instrument transformer used in power measurement, secondary winding current will never be 180° out of phase with primary winding current.

3.25 3-ϕ Wattmeter

Similar to single phase dynamometer wattmeter a three phase dynamometer wattmeter is available. It consist of two sets of fixed and moving coils, mounted together in one case. The moving coils are placed on the spindle. The Fig. 3.27 shows the construction of a three phase wattmeter.

Due to two sets of current and pressure coils, its connection is similar to the connection of two single phase wattmeter to measure three phase power. Each element experience a torque which is proportional to the power measured by that element. The net torque on the moving system is the sum of the deflecting torques produced on each of the two elements.

$$\therefore \quad T_{d1} \propto W_1 \quad \text{and} \quad T_{d2} \propto W_2$$

$$\therefore \quad T_d \propto (T_{d1} + T_{d2}) \quad \propto \quad (W_1 + W_2) \propto W$$

where T_d = Total deflecting torque

W_1 = Power measured by element - 1

W_2 = Power measured by element - 2

W = total power

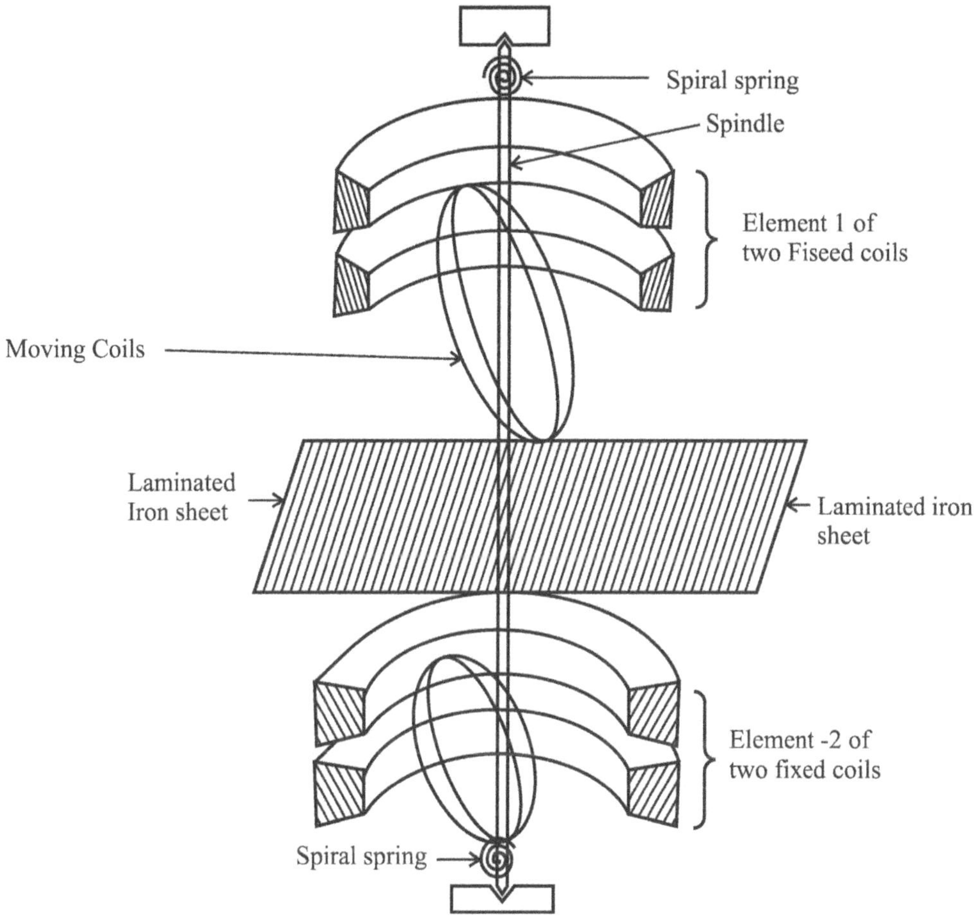

Fig. 3.27 Three phase two element wattmeter

Thus the total deflecting torque is proportional to the total power. As the coils are mounted very near to each other, errors due to mutual interference are possible. To eliminate such errors, the laminated iron shield is placed in between the two elements.

The compensation for mutual interference can be obtained by using the resistance as shown in Fig. 3.28.

The value of 'R' can be adjusted using R' to compensate the errors due to mutual effects between the two elements.

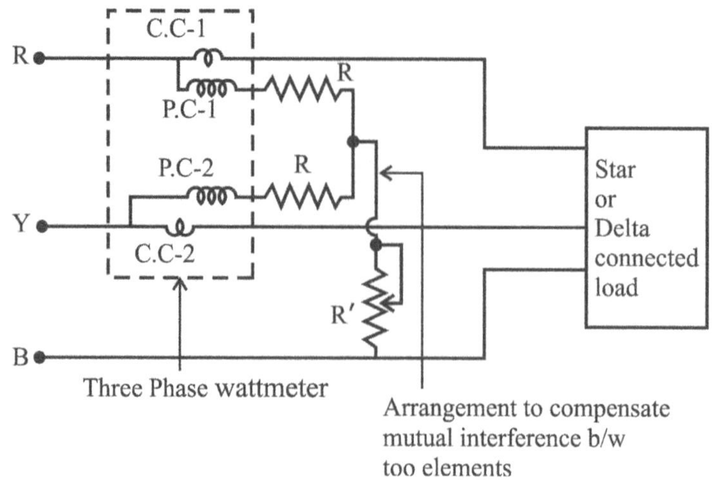

Fig. 3.28 Connections of three phase wattmeter

3.25.1 Elimination of Mutual Effects between the Two Elements of the Wattmeter

The power measured by the meter will be correct only when there is no interaction between the field of both the elements. i.e., only when the mutual effect between the two elements should be absent. This can be achieved by placing an iron shield in between the elements.

Further, the presence of any mutual effect will be compensated by connecting a variable resistor called as compensating resistor R^1 in series with the pressure coils as shown in Fig. 3.28. This method of compensating the mutual effect is called as Weston's method.

By varying the resistor R' the currents through the pressure coils are varied which in tern varies the fields, so as to compensate the mutual effects.

Problems with Solutions

1. A voltage 100 sin wt + 40 cos (3wt – 30°) + 50 sin (5 wt + 50°) v is applied to the pressure circuit of a wattmeter and through the current coil is passes a current of 8 sin wt + 6 cos (5 wt – 120°) A. What will be the reading of the wattmeter?

***Solution*:**

Given data

Supply voltage, $V = 100 \sin wt + 40 \cos(3wt - 30^\circ) + 50 \sin(5wt + 50)$ V

Current, $I = 8 \sin wt + 6 \cos(5wt - 120)$ A

To determine the wattmeter reading, the voltage and current expression must be written in the same trigonometric function, so that the phase relation is obtained between the various harmonics and fundamental.

Thus, re writing the voltage and current expression

$$V = 100 \cos (wt - 90) + 40 \cos (3wt - 30) + 50 \cos (5wt - 40)$$

$$I = 8 \cos (wt - 90) + 6 \cos (5wt - 120)$$

$\therefore$ wattmeter reading will be P = V1

$$= \frac{100 \times 8}{2} \cos (-90 - (-90) + 0 + \frac{50 \times 6}{2} \cos [-40 - (-120)]$$

$$= 400 \times 1 + 150 (0.1736)$$

$$= 426 \text{ watts.}$$

2. A V = 100 sin wt + 40 cos (3wt – 30°) + 50 sin wt + 6 cos (5wt – 120°)A. What will be the reading on the wattmeter? What percentage of this power is due to fundamental frequency?

***Solution*:**

Given data:

The given data is incomplete, assuming voltage applied across the pressure coil circuit to be $Y = 100 \sin wt + 40 \cos (3wt + 30^\circ) + 50 \sin (5\, wt + 45^\circ)$V.

Current passing through the current coil to be

$$I = 8 \sin wt + 6 \cos (5wt - 120^\circ) \text{ A}$$

reading on wattmeter, P = ?

percentage of P due to fundamental frequency = ?

from wt = θ, we know that

Instantaneous power $P = V \times I$

$$= [100 \sin \theta + 40 \cos (3\theta - 30^\circ) + 50 \sin (5\theta + 45^\circ)] \times$$

$$[8 \sin \theta + 6 \cos (5\theta - 120^\circ)]$$

$$= 800 \sin^2 \theta + 600 \sin \theta \cos (50 - 120^\circ) + 320 \sin \theta \cos$$

$$(3\theta - 30^\circ) + 240 \cos (3\theta - 30^\circ) \cos (5\theta - 120^\circ) + 400 \sin \theta$$

$$\sin (5\theta + 45^\circ) + 300 \sin (5\theta + 45^\circ) \cos (5\theta - 120^\circ)$$

Average power over cycle.

$$P = \frac{1}{2\pi}\int_0^{2\pi} p d\theta = \frac{1}{2\pi}\int_0^{2\pi} VI\, d\theta$$

$$= \frac{1}{2\pi}\int_0^{2\pi}$$

3. Two wattmeters used to measure the power input in a 3-phase circuit indicate 1000 w and 5000 w respectively. Find p.f of the circuit

1. When both wattmeter readings are positive and
2. When the latter is obtained by reversing the current coil connections.

***Solution*:**

Given data:

$P_1 = 1000$ W

$P_2 = 500$ W

to find, $\cos\phi = ?$

Case (1):

When both wattmeter readings are + ve

$$\phi = \tan^{-1}\left[\sqrt{3}\left(\frac{P_1 - P_2}{P_1 + P_2}\right)\right]$$

$$\phi = \tan^{-1}\left[\sqrt{3}\left(\frac{1000 - 500}{1000 + 500}\right)\right]$$

$$\phi = 30^o$$

$\therefore$ Power factor $\cos f = \cos(30^o) = 0.866$

Case (2):

When the latter is obtained by reversing the current coil connections

$P_1 = 1000$ w, $P_2 = -500$ w

$$\phi = \tan^{-1}\left[\sqrt{3}\left(\frac{1000 - (-500)}{1000 + (-500)}\right)\right]$$

$$\phi = \tan^{-1}(5.196) = \phi = 79.106$$

$\therefore$ power factor, $\cos\phi = \cos(79.106) = 0.18899$

4. The power input to a 3 phase balanced load is measured by two wattmeter method. The ranges and accuracies of the instrument used and the readings obtained are as under

Wattmeter	:	w_1	w_2
(FSR) Full scale range	:	1500	1500
Max. error as % of FSR	:	0.5	0.5
Reading obtained	:	$\pm$ 500	1000

Estimate the maximum uncertainty meter load p.f compared from the two wattmeter readings.

***Solution*:**

FSR w_1 and w_2 = 1500 w
max error = 0.5%

$$\text{Limiting error at full scale} = 1500 \times \frac{0.5}{100} = 7.5 \text{ W}$$

Percent error at meter reading of $\pm$ 500 w

$$= \frac{7.5}{500} \times 100 = \pm\, 1.5\%$$

$$\text{max. possible error reading } w_1 = 500 \times \frac{1.5}{100} = \pm\, 7.5 \text{ W}$$

i.e., $= 507.5\text{w}$

$$\text{max possible error reading } w_2 = 1000 \times \frac{0.75}{100}$$

$$= \pm\, 7.5 \text{ w}$$

i.e., 1007.5 W

max uncertainly in p.f.

$$\therefore \quad \cos\phi = \cos\left[\tan^{-1} \times \sqrt{3}\left(\frac{w_2 - w_1}{w_2 - w_1}\right)\right]$$

$$= \cos\left[\tan^{-1} \times \sqrt{3} \times \frac{1007.5 - (-\,507.5)}{1007.5 - 507.5}\right]$$

$$= \cos\left[\tan^{-1} \frac{1515\sqrt{3}}{500}\right]$$

$$= \cos\tan^{-1}(5.248)$$

$$= 0.1872$$

5. A 3-ϕ, 400 V meter load has a P.F. of 0.75 lagging. Two wattmeters connected to measure the power, show the input to be 26 kW. Find the reading on each instrument.

***Solution*:**

Power drawn $\quad P = 26 \text{ kW} = 26,000 \text{ W}$

Power factor, $\quad \cos\phi = 0.75$

$$\phi = \cos^{-1}(0.75) = 41.4^\circ$$

Since power drawn, $P = \sqrt{3}\, V_L I_L \cos\phi$

$$\therefore \qquad V_L I_L = \frac{P}{\sqrt{3}\cos\phi} = \frac{26,000}{\sqrt{3}\times 0.75} = 20,015$$

Reading on wattmeter $w_1 = V_L I_L \cos(30^\circ - \phi)$

$$= 20,015 \cos(30^\circ - 41.4^\circ)$$

$$= 19,620 \text{ w or } 19.62 \text{ kW}$$

Reading on wattmeter $w_2 = V_L I_L \cos(30^\circ + \phi)$

$$= 20,015 \cos(30^\circ + 41.4^\circ)$$

$$= 6,384 \text{ w or } 6.384 \text{ kW}$$

$$\therefore \quad w_1 = 19.62 \text{ kW} \quad w_2 = 6.38 \text{ kW.}$$

6. The reading on two wattmeters connected for the measurement of 3-phase power are 6 kW and 0.8 kW, the latter reading being obtained after reversal of the current coil connection. Compute p.f ?

***Solution*:**

wattmeter $w_1 = 6 \text{ kW} = 6000 \text{ W}$

wattmeter $w_2 = -0.8 \text{ kW} = -800 \text{ W}$

Total power $= w_1 + w_2 = 6000 + (-800) = 5200 \text{ W}$

$$\text{Power factor } \cos\phi = \cos\left[\tan^{-1}\frac{w_1 - w_2}{w_1 + w_2}\sqrt{3}\right]$$

$$= \cos\left[\tan^{-1}\frac{6800}{5200}\sqrt{3}\right] = \cos[\tan^{-1} 2.265]$$

$\cos 66.1784 = 0.4039$

Review Questions

1. Explain an operation of electrodynamometer type instrument?
2. Explain the advantages and disadvantages of Electro dynamic instruments?
3. What are the errors in electro dynamometer instruments? Briefly explain in each?
4. Explain the comparison of various types of instruments?
5. What is power? State the importance of wattmeter while measuring power?
6. Explain theory of single phase dynamometer watt meter?
7. Describe the error due to pressure coil inductance?
8. Describe the compensation for Inductance of pressure coil?
9. What are the errors due to 1. Pressure coil capacitance 2. Mutual inductance
10. Write about the shape of scale of dynamometer type wattmeter
11. Which errors are possible to occur in wattmeter and how are they compensated?
12. Explain the working principle and operation of low power factor electro dynamic type waltmeter?
13. Write short notes on power in three phase systcm
14. Explain the main difference between LPF and UPF wattmeters.
15. Describe the wattmeter connection and corresponding rending with examples?
16. Explain how power can be measured by using one watt meter method single phase power?
17. Explain how the power in a 3-phase system in measured by the use of 1. only one wattmeter 2. two watt meters
18. What is Blondel's Theorem?
19. How the power factor is determined by the use of 1. One wattmeter 2. Two watt meters method?
20. Explain an effect of power factor in wattmeter
21. What are the advantages and disadvantages of two wattmeters method?
22. Explain with a neat circuit diagram how reactive power can be measured in balanced 3-phase system using single wattmeter?
23. Explain how 3-phase power can be measured with the help of two wattmeters with suitable connections.
24. Explain how 3- phase power measurement using C.T and P.T?

25. What are the advantages and disadvantages of Instrument transformer for extension of range of A.C Instruments
26. Explain with neat sketch 3-phase two element wattmeter?
27. Explain how do you measure the total power in a 3-phase circuit with the help of two wattmeters only.
28. Write short notes on 1. Low power factor wattmeter 2. Three phase wattmeter
29. Explain the error in wattmeter reading due to method of connections?
30. Explain the features of low power factor wattmeter. Prove that the error due inductance of pressure coil in the wattmeter is [tanθ tanβ] times actual power
31. How reactive volt amperes are measured by 2-wattmeter method?

Quiz Questions

1. In two-wattmeter method of power measurement, one of the wattmeters will show negative reading when the load of angle is strictly
   (a) less than 30° (b) less than 60°
   (c) greater than 30° (d) greater than 60°
2. Electro dynamic instruments can be used for the power measurement of
   (a) A.C voltage (b) D.C voltage
   (c) A.C power (d) all
3. The type of wattmeter widely used is
   (a) electro static (b) electro dynamic
   (c) induction type (d) hot-wire type
4. In two-wattmeter method of power measurement, the difference in two wattmeter readings gives
   (a) active power (b) reactive power
   (c) apparent power (d) none
5. In electro dynamometer type wattmeters, current coils designed for carrying heavy currents use standard wire or laminated conductors.
   (a) to reduce iron losses
   (b) to reduce hysterisis
   (c) to reduce eddy current losses in conductor
   (d) all of the above

6. One wattmeter method is used to measure power in three phase circuit when
   (a) the power factor is unity
   (b) the supply frequency is 50 Hz
   (c) the load is balanced in all three phases
   (d) the supply voltage is 220 V
7. Three phase four-wire energy meter is used to measure
   (a) single phase energy (b) two phase energy
   (c) three phase balanced energy (d) three phase unbalanced energy
8. Minimum number of voltmeters required to measure power in an unbalanced 4-wire system is
   (a) one (b) two
   (c) three (d) four
9. For handling currents larger than the rating of wattmeter current coil
   (a) A resistance is connected in series with C.C
   (b) A resistance is connected in parallel with C.C
   (c) A suitable capacitor is connected across the C.C
   (d) Current coil is connected in conjunction with C.T
10. In the measurement of 3-ϕ power by 2-wattmeter method if the two wattmeter readings are equal, the Pf of the circuit is
    (a) 0.08 lagging (b) 0.8 leading
    (c) 0 (d) Unity
11. The attraction or repulsion between two electrically charged objects is called
    (a) electromagnetic deflection (b) electrostatic force
    (c) electroscopic force (d) magnetic force
12. A wattmeter is reading backward upscale reading can be had by reversing
    (a) connections of CC only (b) connections of PC only
    (c) either (a) or (b) (d) both (a) and (b)

Questions and Answers

1. The main error in a dynamometer type wattmeter is due to

Ans. Pressure coil inductance

2. Induction type wattmeter operation is based on

Ans. Eddy current flow

3. In three phase power measurement the power factor of load will be

Ans. $\sqrt{3}\frac{(\omega_1 - \omega_2)}{\omega_1 + \omega_2}$

4. In two wattmeter method of measuring 3-phase power, power factor is 0.5, then one of the wattmeters will read

Ans. Zero

5. In the circuit shown, through oversight, the meters have been interchanged. This is likely to result in

Ans. Damage to ammeter

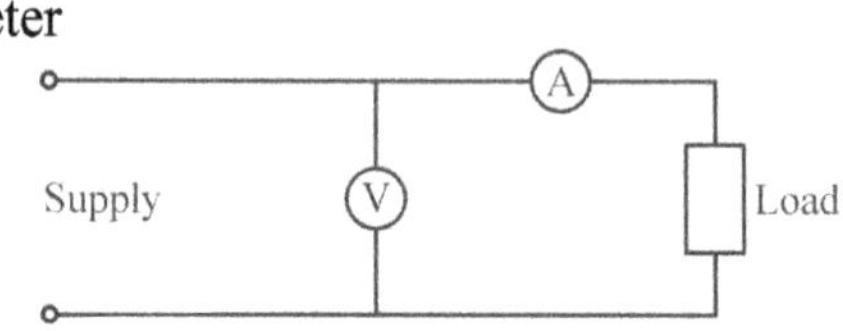

6. Electrodynamometer can be used also to measure

Ans: Power, power factor, reactive power, frequency

7. The two-wattmeter method of measurement of 3-phase power can be used for

Ans. Balanced source with three-phase, three-wire unbalanced load

8. Two wattmeters connected to measure the power in a 3 ϕ, balanced data connected load read W_1 = 12 kW, W_2 = 15 kW respectively. If the same load will be connected in star the wattmeters would read.

Ans. W_1 = 5 kW, W_2 = 1 kW

9. A dynameter type wattmeter has scale

Ans.

10. The minimum number of wattmeters required to measure power in an unbalanced 3-phase 3-wire system is ________

Ans. 2

11. The type of wattmeter widely used is ……

Ans. Electro dynamic

12. Induction type energy meter measures electric energy in ________

Ans. kwhr

13. Induction type energy meters are free from ________ errors

Ans. Frictional

14. A dynamometer type wattmeter has ________ scale

Ans. Uniform

15. The minimum number of wattmeters required to measure power in an unbalanced 3-ϕ 3-wire system is ________

Ans. 2

16. The type of wattmeter widely used is ________

Ans. Electro dynamic

17. In three phase power measurement the power factor of load will be

Ans. $\sqrt{3}\dfrac{(\omega_1 - \omega_2)}{\omega_1 + \omega_2}$

18. In two wattmeter method of measuring 3-phase power, power factor is 0.5, then one of the wattmeters will read

Ans. Zero

19. Electrodynamometer can be used also to measure

Ans. Power, power factor, reactive power, frequency

20. The two-wattmeter method of measurement of 3-ϕ power can be used for

Ans. Balanced source with three-phase, three-wire unbalanced load

21. Two wattmeters connected to measure the power in a 3 ϕ, balanced data connected load read W_1 = 12 kW, W_2 = 15 kW respectively. If the same load will be connected in star the wattmeters would read.

Ans. $W_1 = 5$ kW, $W_2 = 1$ kW

Match the Following

1. Type of wattmeter Conditions of power measurement

(a) Electrostatic	(r) power at UPF and at high values
(b) Induction	(p) power at 50 Hz only
(c) Dynamometer	(s) power at frequency ranging from d.c to 150 Hz
(d) Thermal	(q) power at frequency ranging from d.c to RF

Quiz Questions Key

1. (d)	2. (c)	3. (b)	4. (b)
5. (c)	6. (c)	7. (d)	8. (b)
9. (c)	10. (d)	11. (b)	12. (c)

Match the Following Key

1. a-r,b-p,c-s,d-q

CHAPTER 4

Measurement of Energy

4.1 Introduction

It may be stated generally that the measurement of energy is essentially the same process as the measurement of power, except that the instrument used must not merely indicate the power, or rate of supply of energy, but must take into account also the length of time for which this rate of supply is continued. Actually, energy or "supply" meters do not indicate power directly. For a given amount of energy supplied to a circuit their registrations should always be the same, no matter what the instantaneous values of the power during the time in which the energy is supplied.

The energy is defined as the power delivered over a time interval

$$\text{Energy} = \text{Power} \times \text{Time}$$

The electrical energy is defined as the work done over a time interval 't' and mathematically expressed as

$$E = \int_0^t \text{Power}(dt) = \int_0^t vi.dt$$

where, v = voltage in volts

i = current in amperes

The energy is measured in Joules (J) or Watt-Sec (W-S). Thus energy of one joule means the power of 1 watt over a time interval of 1 second. An electrical energy can also be expressed in the unit watt-hour (Wh) or kilowatt-hour (kWh). Thus one kilo watt-hour energy means the expenditure of 1 kW power over a time internal of 1 hour. The domestic electric energy expenditure is measured in kWh and 1 kWh is called 1 unit of energy.

4.2 Single Phase Energy Meter (1-ϕ)

Induction type instruments are most commonly used as 'Energy meters'. Energy meter is an integrating instrument which measures quantity of electricity. Induction type of energy meters are universally used for domestic and industrial applications. These meters record the energy in kilowatt-hours (kWh).

The fig. 4.1 shows the induction type single phase energy meter. It works on the principle of induction i.e., on the production of eddy currents in the moving system by the alternating fluxes. These eddy currents induced in the moving system interact with each other to produce a driving torque due to which disc rotates to record the energy.

In the energy meter there is no controlling torque and thus to driving torque only, a continuous rotation of the disc is produced. To have constant speed of rotation braking magnet is provided.

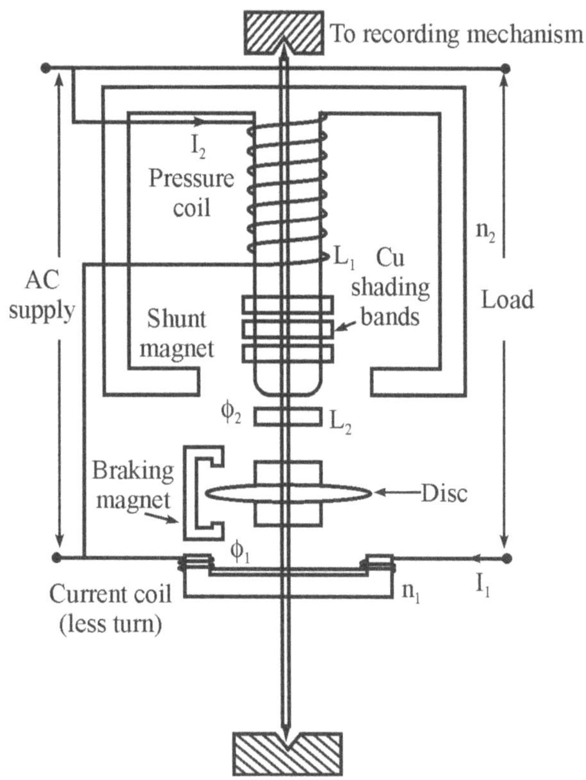

Fig. 4.1 Induction type single phase energy meter

4.3 Construction of Induction Type Energy Meters

The construction varies in details from one manufacturer's product to the next. The differences are very minor in nature.

There are four main parts of the operating mechanism:

1. Driving system
2. Moving system
3. Braking system
4. Registering system

4.3.1 Driving System

It consists of two electro magnets whose core is made up of silicon steel laminations. The coil of the electromagnets, called "current coil", is excited by load current which produces flux further. The coil of another electromagnet is connected across the supply and it carries current proportional to supply voltage, this coil is called "pressure coil". These two electro magnets are called "series" and "shunt" magnets respectively.

The flux produced by shunt magnet is brought in exact quadrature with supply voltage with the help of copper shading bands whose position is adjustable.

4.3.2 Moving System

Light aluminium disc mounted in a light alloy shaft is the main part of moving system. This disc is positioned in between series and shunt magnets. It is supported between jewel bearings. The moving system runs on hardened steel pivot. A pinion engages the shaft with the counting mechanism. There are no springs and no controlling torque.

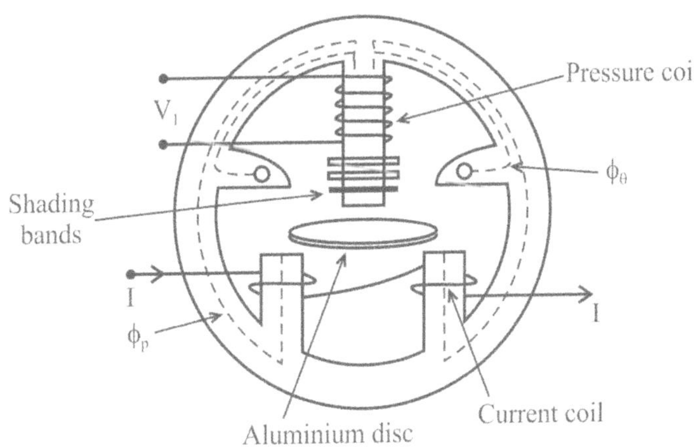

Fig. 4.2 Single phase energy meter

4.3.3 Braking System

A permanent magnet is placed near the aluminium disc for braking mechanism. This magnet reproduce it's own field. The disc moves in the field of this magnet and a braking torque is obtained. The position of this magnet is adjustable and hence braking torque is adjusted by shifting this magnet to different radial positions. This magnet is called "braking magnet."

4.3.4 Registering (Counting) Mechanism System

The function of a registering or counting mechanism is to record continuously a number which is proportional to the revolutions made by the moving system. By a suitable system, a train of reduction gear the pinion on the rotor shaft drives a series of five or six pointers. These rotate on round dials which are marked with ten equal divisions. The pointer type of register is shown in Fig. 4.3(a). Cyclo-meter register as shown in Fig. 4.3(b) can also be used.

Practically the pointer type registering mechanism is used. The pointer indicates one kwh when the disc completes certain number of revolutions. The second dial represents 10 kwh, third 100 kwh while on the other sides, dials measuring 1/100 and 1/10 kwh are also provided. The Fig.4.3(a) shows the pointer type register while the Fig. 4.3(b) shows the cyclometer type register. In some meters the cyclometer type registering mechanism is used.

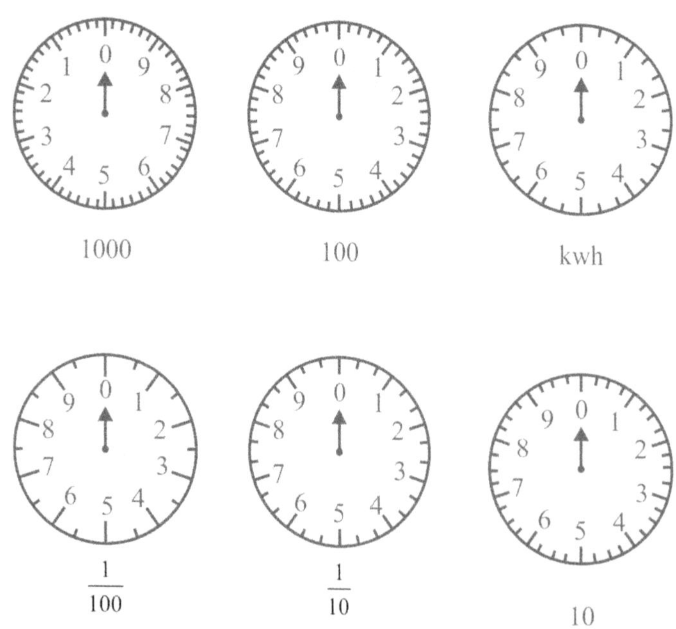

Fig. 4.3(a) Pointer type of register

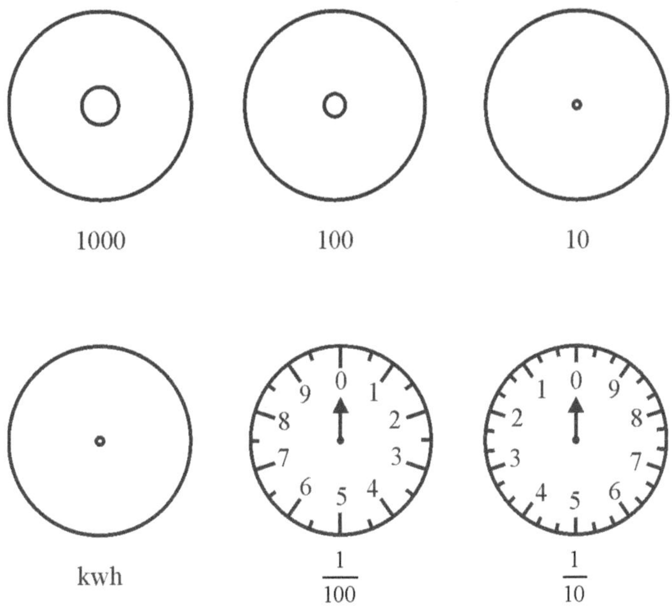

(b) Cyclometer register

Fig. 4.3(b) Registering mechanisms used in induction energy meter

4.4 Theory and Operation of Single Phase Energy Meters

A simple functional diagram of the driving system of the meter is shown Fig. 4.4. Since the pressure coil is carried by shunt magnet M_2 which is connected across the supply. It carries current proportional to the voltage. Series magnet M_1 carries current coil which carries the load current. Both these coils produce alternating fluxes ϕ_{sh} and ϕ_{se} respectively. These fluxes are proportional to current in their coils. Parts of each of these fluxes link with the disc and induces e.m.f in it. Due to these e.m.fs eddy currents are induced in the disc. The eddy current induced by the electromagnet M_2 react with magnetic field produced by M_1. Also eddy currents induced by electromagnet M_1 react with magnetic field produced by M_2. Thus each portion of the speed of disc is controlled by the C shaped magnet called braking magnet. When disc rotates in the air gap, eddy currents are induced in disc which oppose the cause producing them i.e. relative motion of disc with respect to magnet hence braking torque T_b is generated. This is proportional to speed 'N' of disc. By adjusting position of this magnet, desired speed of disc is obtained. Spindle is connected to recording mechanism through gears which record the energy supplied.

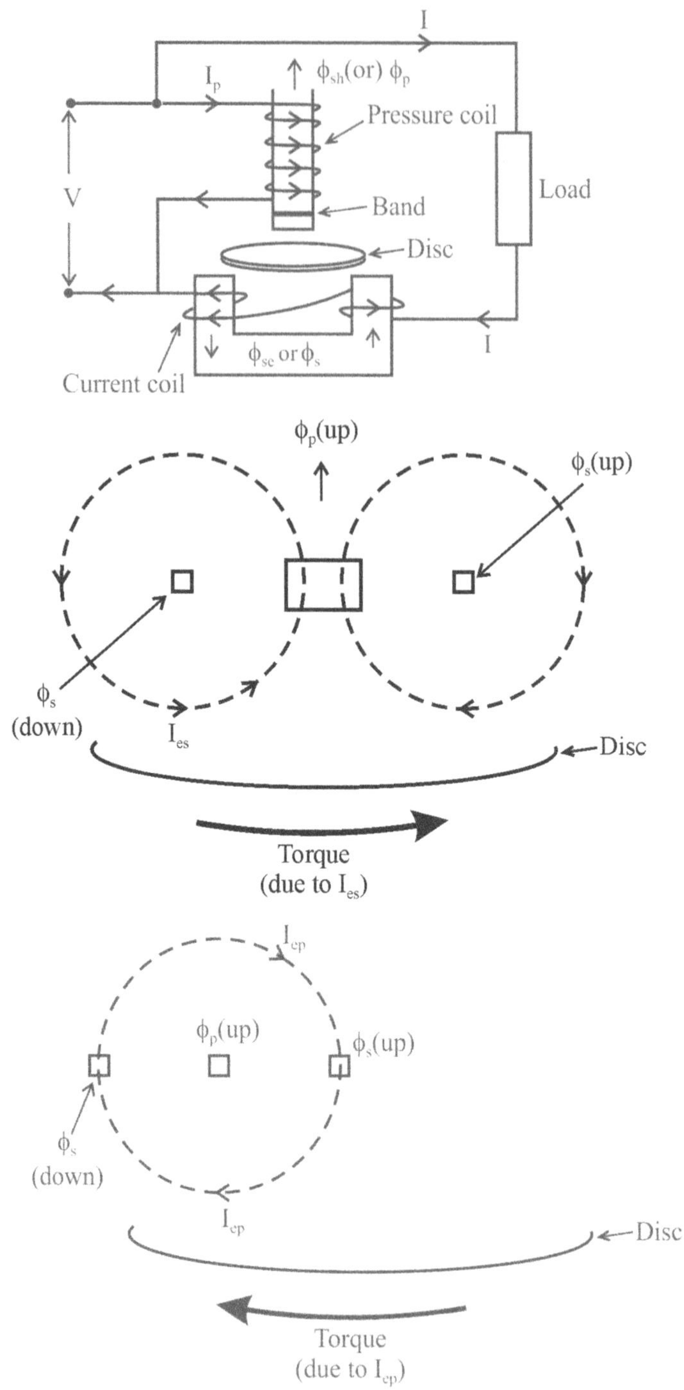

Fig. 4.4 Functional diagram of induction type energy meter

The current I_p produces the total flux ϕ_{pt} which has two components ϕ_g and ϕ_p. The major portion is ϕ_g, which flows through the side gaps as the reluctance of this part is very small. While ϕ_p flows across the air gap and across the disc and is responsible to produce the eddy e.m.f E_{ep} in the disc which produces the eddy current I_{ep} in the disc.

The ϕ_p is small and is in phase with I_p. It is proportional to I_p and hence to supply voltage 'V' as it produces I_p through pressure coil ϕ_p lags the supply voltage V by an angle slightly less than 90°. The current coil comes the load current I and produces the flux ϕ_s. This is proportional to I and in phase with it. This flux is responsible to induce eddy e.m.f E_{es} in the disc which produces the eddy current I_{es} in the disc.

This interacts with the flux ϕ_p to produce the torque while the eddy current I_{ep} interacts ϕ_s to produce the torque. These two torques are opposite in direction and the net torque produced is the difference between these two torques.

4.5 Torque Equation

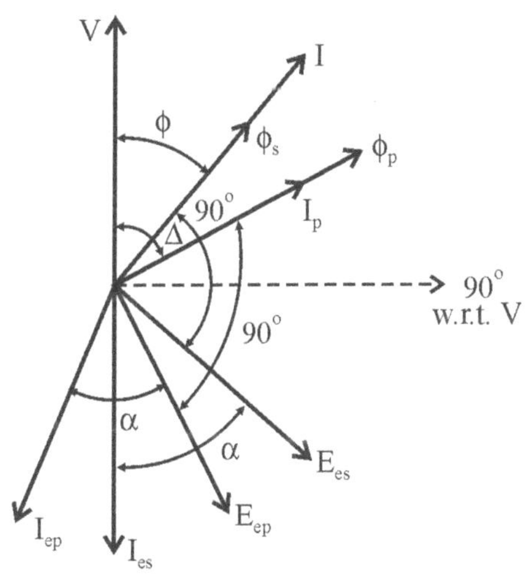

Fig. 4.5 Phasor diagram of 1-ϕ induction type energy meter

V = supply voltage

I = load current

= current coil current

Δ = Phase angle between V and I_p

Δ ≈ 90°

α = Phase angle of eddy currents.

I_{ep} = eddy current due to E_{cp}

I_{es} = eddy current due to E_{es}

I_p = pressure coil current

E_{ep} = eddy e.m.f induced due to ϕ_p

E_{es} = eddy e.m.f induced due to ϕ_s

The current I_p lags V by Δ and Δ is made 90° using copper shading bands. The current I lags V by ϕ which depends on the load. The flux ϕ_s and I are in phase. The E_{cp} lags ϕ_p by 90° while E_{es} lags ϕ_s by 90°. The eddy currents I_{es} and I_{ep} lags E_{es} and E_{ep} respectively by angle α.

The interaction between ϕ_p and I_{es} produced torque T_1. While the interaction between ϕ_s and I_{ep} produces torque T_2.

$$\therefore\ T_1 \propto \phi_p I_{es} \cos(\phi_p \wedge I_{es}) \text{ and } T_2 \propto \phi_s I_{ep} \cos(\phi_p \wedge I_{ep})$$

$$\phi_p \wedge I_{es} = \alpha + \phi \text{ and } \phi_s \wedge I_{ep} = 180 - \phi + \propto \ \ldots\ldots\ \Delta = 90^\circ$$

$$\therefore\ T_d \propto T_1 - T_2 \propto \{[\phi_p I_{es} \cos(\alpha + \phi)] - [\phi_s I_{ep} \cos(180 - \phi + \alpha)]\}$$

Now $\phi_p \propto V$, $\phi_s \propto I$, $I_{es} \propto \phi_s \propto I$, $I_{ep} \propto \phi_p \propto V$

$$\therefore\ T_d \propto VI[\cos(\alpha + \phi) - \cos(180 - \phi + \alpha)]L$$

$$\propto VI[\cos\alpha\cos\phi - \sin\alpha\sin\phi) - \phi\cos(180 - \phi)\cos\alpha - \sin(180 - \phi)\sin\alpha))]$$

$$\propto VI[\cos\alpha\cos\phi - \sin\alpha\ \sin\phi - \cos(180 - \phi)\cos\alpha + \sin(180 - \phi)\sin\alpha)]$$

$$\propto 2VI\cos\alpha\cos\phi \ \ldots\ldots\ldots\ \cos(180 - \phi) = -\cos\phi,\ \sin(180 - \phi) = \sin\phi$$

$$\therefore\quad T_d = k_3 VI\cos\phi \ (\alpha \text{ is constant}) \quad \ldots..(4.1)$$

Note: Thus the deflection torque (T_d) is proportional to the true power in the circuit.

If 'Δ' is considered.

$$T_d \propto VI\sin(\Delta - \phi) \quad \ldots..(4.2)$$

But practically Δ is achieved to be exactly 90° with the help of copper shading bands so that T_d is proportional to power in the circuit.

The braking torque is due to eddy currents induced in the aluminium disc. The magnitude of eddy current is proportional to the speed N of the disc. Hence the braking torque T_b is also proportional to the speed 'N'.

$$T_b \propto N \quad \text{i.e.} \quad T_b = K_4 N \quad \ldots..(4.3)$$

For the steady speed of rotation, $T_d = T_b$

$$\therefore\quad K_3 VI\cos\phi = K_4 N$$

$$\therefore\quad N = KVI\cos\phi = K\,[\text{power}] \quad \ldots..(4.4)$$

$$\text{Total number of revolutions} = \int_0^t N dt = \int_0^t K(\text{power})dt$$

$$\therefore\ \text{Total number of revolutions} = K\int_0^t P dt = K \times \text{energy} \quad \ldots..(4.5)$$

Thus the number of revolutions of the disc in a given time is the energy consumption by the circuit in that time.

$$\therefore \qquad K = \text{Meter constant} = \frac{N}{\text{energy}} = \frac{\text{Number of revolution}}{\text{kWh}} \qquad .....(4.6)$$

Thus the number of revolutions of the disc per kWh of energy consumption is called the "meter constant".

4.6 Errors and Compensations

There are various errors present in the 1-ϕ induction type energy meter. The driving system can cause the errors disc to inaccurate phase angles, abnormal frequencies, effect of temperature on the resistance and unsymmetrical magnetic circuit. The braking system also can causes error due to change in the strength of the braking magnet, change in resistance of the disc, abnormal friction of moving disc etc. To get accurate reading, these errors are required to be compensated. Hence some adjustments are provided in the energy meter to minimize these errors.

4.6.1 Temperature Effect and Compensation

An increase in temperature results in increased resistance to the eddy current path reducing the driving torque and braking torque. It also increases the resistance of the shunt coil causing smaller current to flow due to which the driving torque is reduced. At higher temperatures, the flux from braking magnet is also less. The net result is to neutralise each others effect and error caused due to temperature is not appreciable.

This has following effects:

1. Increase in eddy current resistance path
2. Small reduction in shunt magnet flux
3. Reduction in torque produced by all shading bands
4. Decrease in angle of log α of the eddy currents
5. Reduction in angle of lag between V and ϕ_p.

These effects are compensated by providing a temperature shunt on the brake magnet. Special magnetic materials such as mutemp is used for the shunt whose permeability decreases considerebly as temperature increases. Thus provides temperature compensation and does not allow the disc to rotate faster as temperature increases.

4.6.2 Lag Adjustment or Power Factor Adjustment

It is absolutely necessary that meter should measure correctly for all P.F. conditions of the loads. This is possible when the flux produced due to current in the pressure coil lags the applied voltage by 90°. But the iron loss and resistance of winding do not allow the flux to lag by exact 90° with respect to the voltage.

The arrangements are used to adjust this angle to 90° is called "lag arrangement". A magnetic shunt circuit is introduced in the device which allows the main portion of the shunt magnet flux to bypass the gap in which the disc rotates. It is possible to produce an m.m.f. in the proper phase relation to the shunt magnet flux to bring ϕ_{sh} i.e., ϕ_p in exact quadrature with the voltage.

This is shown in the Fig. 4.6. The lag coil is used in addition to shunt coil. The lag coil is few turns of fairly thick wire placed around the central limb of the shunt magnet. The part of shunt flux i.e., ϕ_p links with the lag coil to reduce an e.m.f in it. This produces the lag coil current I_L. This current produces a m.m.f. A T_L in phase with I_L. Thus now the phase of ϕ_p is decided by the combined m.m.f of lag coil and shunt coil. This can be adjusted by adjusting the resistance connected across the lag coil. When resistance increases, current and m.m.f of lag coil decreases which decreases the value of the lag angle of coil hence ϕ_{sh} lags behind the voltage by exactly 90°.

Instead of lag coil and resistance, many times copper shading bands are placed on the central limb of the shunt magnet. These bands are adjustable. By moving these bands along the axis of the central limb, the lag adjustment can be achieved.

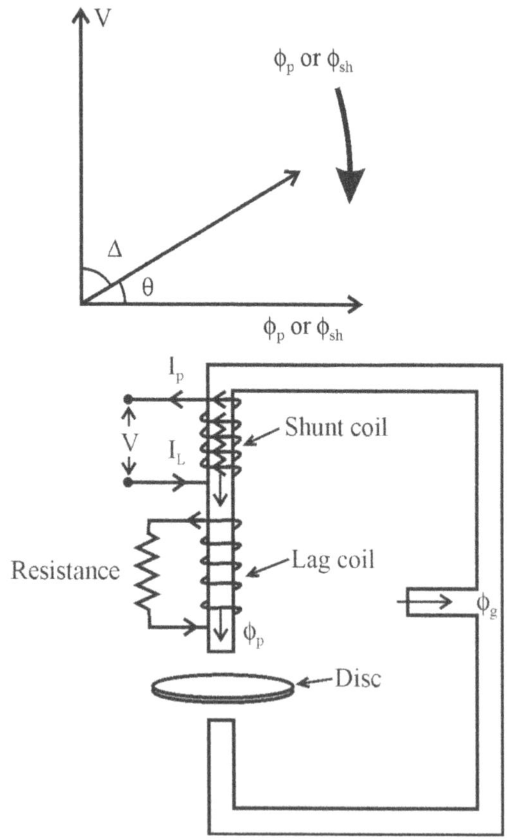

Fig. 4.6 Lag adjustment

When bands are moved upwards the e.m.f induced in them increases increasing the m.m.f produced, hence lag angle increases. When bands are moved down, the m.m.f produced by the bands decreases which decreases the lag angle. Thus the ϕ_{sh} can be brought in exact quadrature with the voltage V. Thus adjustment is also called "power factor adjustment, quadrature adjustment or inductive load adjustment".

4.6.3 Light Load Adjustment or Friction Compensation

In case of proper design of the bearings and registering mechanism, there is bound to exist some friction. Due to this, speed of the meter gets affected which cause the error in the measurement of the energy.

To compensate for this, a metallic loop or strip is provided between central limb of shunt magnet and the disc. Due to this strip an additional torque independent of load is produced which acts on the disc in the direction of rotation. This compensates for the friction and meter can be made to read accurately. This is shown as L_2 in the fig. 4.7.

The shading loop L_2 is also called "light load plate" the interaction between the portions of the flux which are shaded and unshaded by this loop and the currents they induce in the disc generates a small driving torque whose value can be adjusted by lateral movement of the loop L_2. This additional driving torque overcomes the frictional error. This torque is practically independent of the load and depends on line voltage hence remains constant. The frictional error is dominant at rated voltage and very low current i.e., at light loads. The shading loop can be moved laterally to adjust the speed to provide necessary compensation.

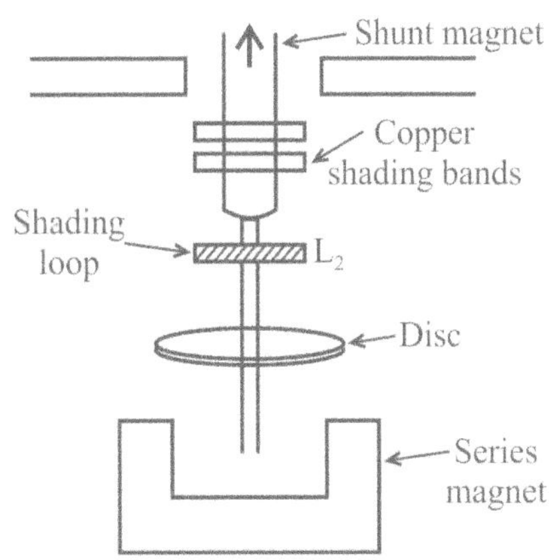

Fig. 4.7 Shading loop for friction compensation

4.6.4 Creeping or Creeping Compensation

In some meters a slow, but continuous, rotation is obtained when the voltage coils are excited but with no-load current flowing. This may be due to incorrect friction compensation, to vibration, to stray magnetic fields, or to the fact that the voltage of the supply circuit is in excess of the normal.

To prevent such "creeping" two holes, or slots, are cut in the disc on opposite sides of the spindle. The disc tends to remain stationary when one of the holes comes to the edge of the field of the shunt magnet.

In some cases, a small piece of iron wire is attached to the edge of the disc. The force of attraction of the brake magnet upon this iron wire is sufficient to prevent continuous rotation of the disc under no-load conditions.

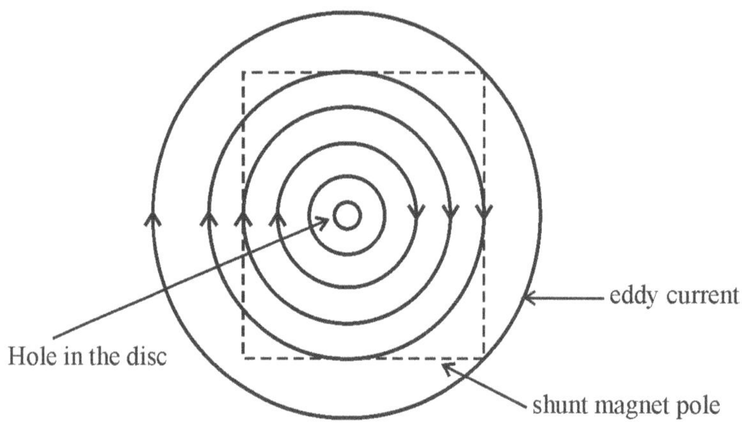

Fig. 4.8 Creeping adjustment

4.6.5 Main Speed Adjustment

The measurement of energy is dependent on the speed of the rotating disc. For accurate measurement, speed of the disc must be also proportionate. The speed of the meter can be adjusted by means of changing the effective radius of the braking magnet. Moving the braking magnet in the direction of the spindle, decreases the value of the effective radius, decreasing the braking torque. This increases the speed of the meter. While the movement of the braking magnet in the outward direction i.e. away from the centre of the disc, increases the radius, decreasing the speed of the disc. The fine adjustments of the speed can be achieved by providing an additional flux divertor.

4.6.6 Over-load Compensation

When the disc rotates in the field of series magnetic field under load conditions, it cuts the series flux and dynamically, e.m.f is induced in the disc. This produces eddy currents in the disc which interacts with series magnet flux to produce braking torque. This is proportional to square of the current. This is called self braking torque and at large loads its value is very high, causing serious errors in the measurement. To minimize this braking torque, the full load speed of the disc is kept very low about 40 r.p.m. The current coil series flux is kept minimum compared to shunt magnet flux. Practically an overload compensating device in the form of a saturable magnetic shunt for the series magnet core is used. This is shown in the Fig. 4.9. At high loads, magnetic shunt saturates and diverts some of the series magnetic flux. This compensates for the self braking torque.

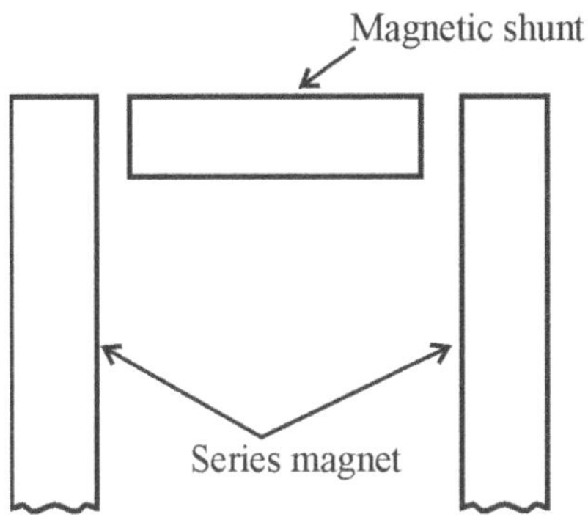

Fig. 4.9 Overload compensation

4.6.7 Voltage Compensation

A certain amount of variation is permitted in the declared voltage of supply. Therefore, energy meters must be compensated for this variation. Voltage variations cause errors owing to two reasons.

1. The relationship between shunt magnet flux and the supply voltage is not linear owing to saturation in iron parts
2. The shunt magnet flux ϕ_p produces a dynamically induced e.m.f in the disc which in turn results in a self-braking torque which is proportional to square of the supply voltage

The voltage compensation is provided by the saturable magnetic shunt which diverts a large proportion of the flux into the active path when the supply voltage increases. The compensation can be provided by increasing the side limb reluctance, by providing holes in the side limbs.

4.7 Advantages and Disadvantages of Induction Type Energy Meter

Advantages

The various advantages of induction type energy meter are:

1. It is cheap in cost
2. It has more accuracy

3. Its range can be extended with the help of instrument transformers.
4. It requires less maintenance
5. Its construction is simple and strong
6. It has high torque to weight ratio, so frictional errors are less and we can get accurate reading.

Disadvantages

1. The creeping can cause errors.
2. Lack of symmetry in magnetic circuit may cause errors.
3. The main disadvantage is that it can be used only for a.c. circuits.

4.8 Three Phase Energy Meter

In a three phase 4-wire system, the measurement of energy is to be carried out by a three phase energy meter. For three phase 3-wire system, the energy measurement can be carried out by two element meter, the connections of which are similar to the connections of two watt meter for power measurement in a three phase 3-wire system. So these meters are classified as:

1. Three element energy meter and
2. Two element energy meter

4.9 Three Element Energy Meter

This type of meter consists of three elements. The construction of an individual element is similar to that of a single phase energy meter. The pressure coils are denoted as P_1, P_2 and P_3. The current coils are denoted as C_1, C_2 and C_3. All the elements are mounted in a vertical line in common case and have a common spindle, gearing and registering mechanism. The coils are connected in such a manner that the net torque produced is sum of the torques due to all the three elements. These are employed for three phase, four wire system where fourth wire is a neutral wire.

The current coils are connected in series with the lines while pressure coils are connected across a line and a neutral. Fig. 4.10 shows a three phase energy meter.

One unit of three element, three phase element is always cheaper than three units of single phase energy meter. But due to interaction between eddy currents produced by one element with the flux produced by another element, there may be errors in the measurement by three phase energy meter. Such errors may be reduced by suitable adjustments.

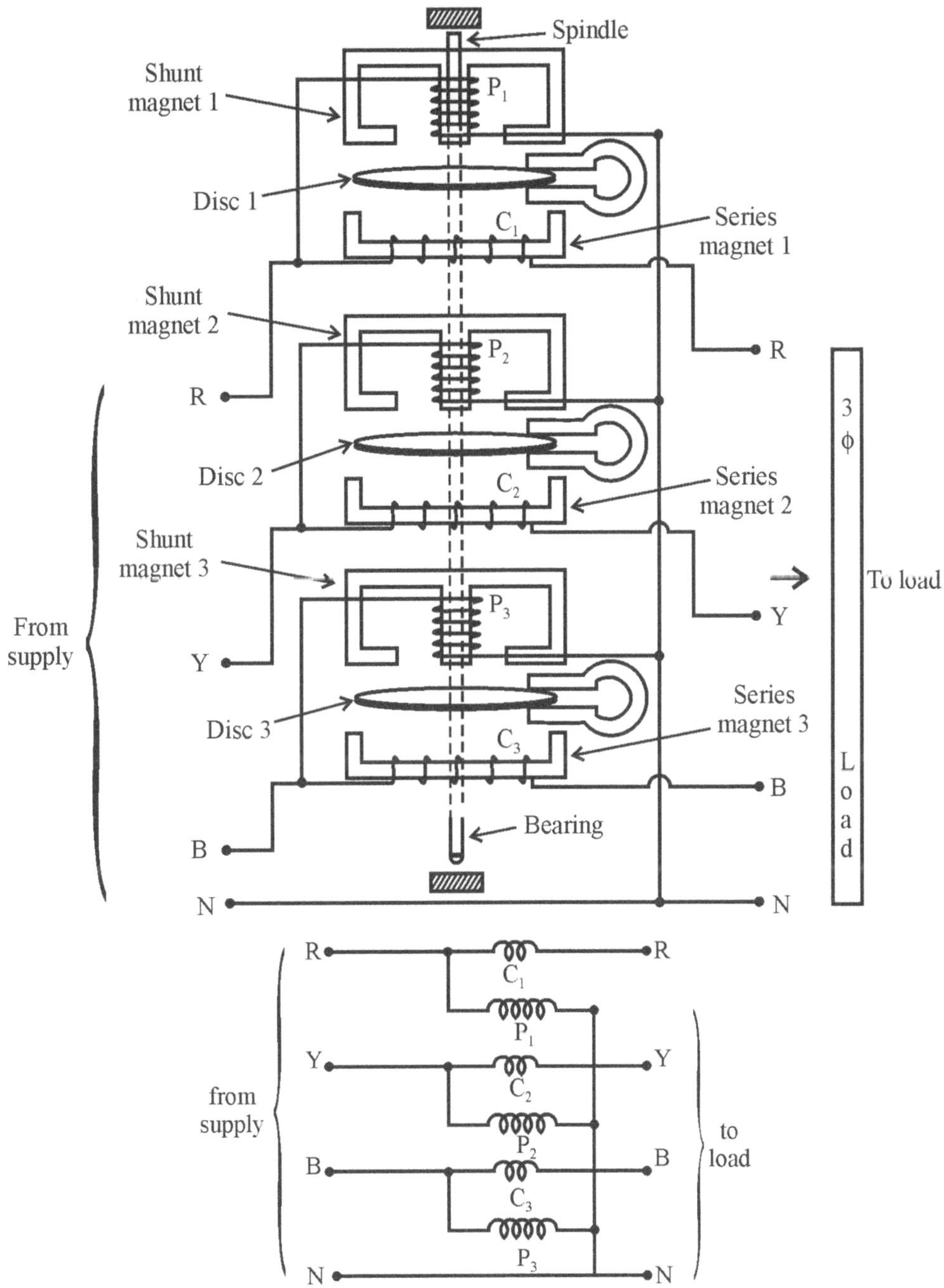

Fig. 4.10 Three element energy meter

4.9.1 Two Element Energy Meter

The construction of three phase energy meter appears as an assembly of two single phase energy meters in one case. The Fig. 4.11 is a two element energy meter and a simplified connection diagram.

This energy meter is used for three phase three-wire system. The meter is provided with two discs each for an element. The shunt magnet is carrying pressure coil while a series magnet carries a current coil. The pressure coils are connected in parallel and the current coils in series. The connections are similar to the connections of two watt meters for power measurement in three phase 3-wire system. Torque is produced in same manner as in a single phase energy meter, in each element. The total torque on the registering mechanism connected to moving system, is sum of the torques of the individual elements.

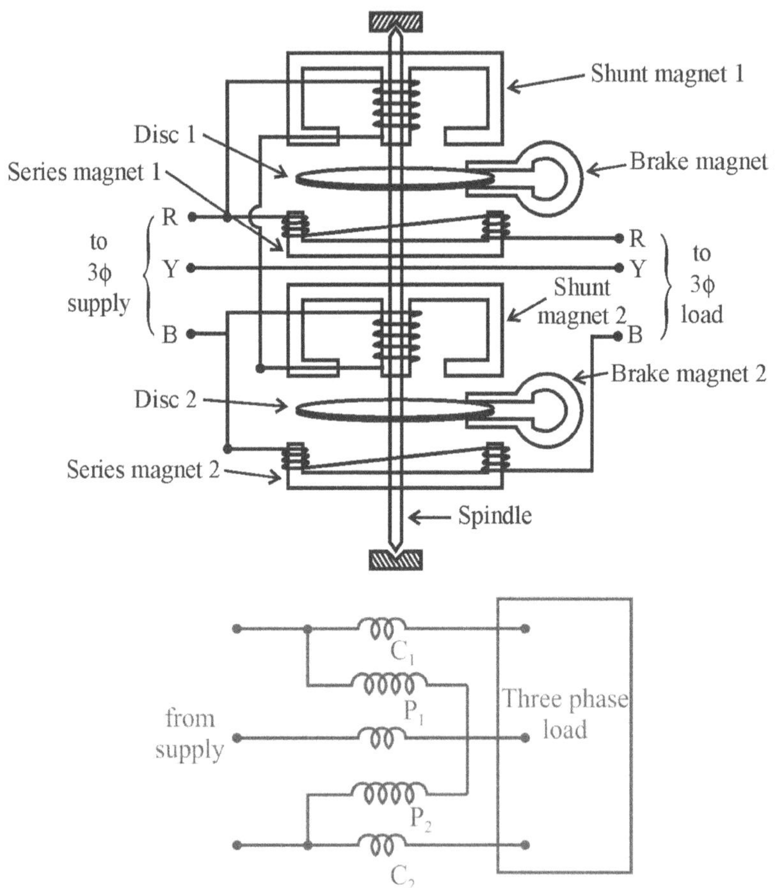

Fig. 4.11 Two element energy meter

4.10 Calibration of an Energy Meter

The calibration of the energy meter means to find out the error in the measurement of energy by energy meter.

Every energy meter has its own characteristic constant specified by the manufacturer which relates the energy measured in joules and the number of revolutions of the disc. For example, say 'X' revolutions corresponds to the measurement of 'Y' joules. But practically the value of 'X' is very large and can not be measured in the laboratory. Hence using this constant, energy recorded for certain less number of revolutions say 5, is calculated in the laboratory. For the calibration purpose this energy is denoted as E_r. Thus, E_r can be calculated from 'X' as,

$$E_r = \frac{5X}{Y} \text{ joules}$$

To have zero error, the actual energy consumed by the load for the time corresponding to the 5 revolutions must be same as E_r. This energy is called actual energy consumed or the true energy denoted as E_t. Experimental set up used in the laboratory to obtain the value of E_t is shown in the Fig. 4.12.

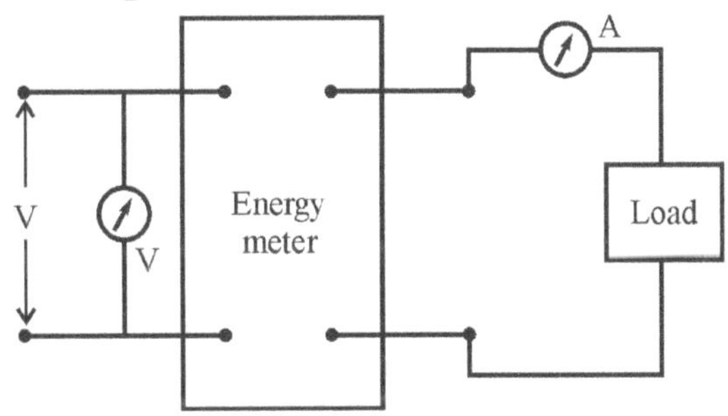

Fig. 4.12

For various loads, the time required to complete the 5 revolutions of disc is measured with the help of stop watch. The voltage and current readings are observed on the ammeter and volt meter connected in the circuit. The readings can be tabulated as:

Sr. No.	Voltage (V)	Current (A)	Time of revolutions	True energy $E_t = VIt$ J
1.				
2.				
3.				

Now E_r is fixed for the 5 revolutions, while E_t is obtained practically. Hence error for each load condition can be obtained as

$$\% \text{ error} = \frac{E_r - E_t}{E_t} \times 100$$

The graph of % error against the load current I can be obtained, which is called "calibration curve" for the energy meter. When there is no load, I = 0 and hence true energy E_t is also zero. Hence E_r is also zero; hence the error is also zero. Thus calibration curve passes through origin. The errors can be positive or negative. Such a curve is shown in the Fig. 4.13.

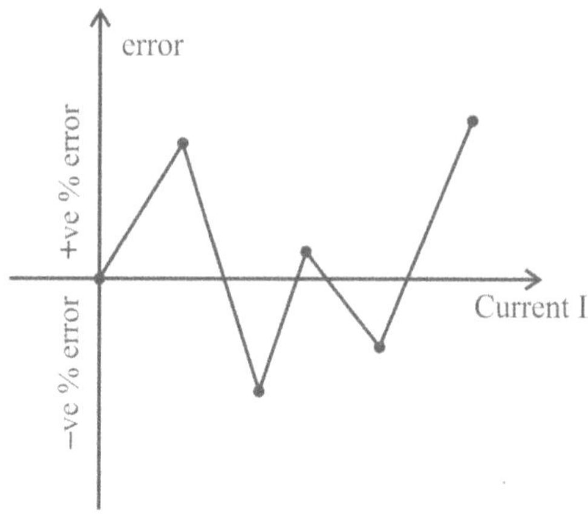

Fig. 4.13

Once the calibration curve is obtained by observation of the curve, in which range of the load current error is severe, can be easily predicted and if error is not within the permissible limits then by using the various adjustments discussed earlier, the error can be minimised.

4.11 Current Transformer and Potential Transformer uses in Energy Measurement

The single phase energy meter connections are exactly similar to the connections of a wattmeter along with C.T and P.T for power measurement shown in Fig. 4.14. The pressure coil of wattmeter is replaced by pressure coil of energy meter and current coil of wattmeter is replaced by current coil of energy meter.

The Fig. 4.14 shows a two element energy meter with C.T and P.T. The connections with three element, three phase energy meter are shown in the Fig. 4.14(a).

But three phase energy meter connections along with C.T and P.T are little bit different than single phase but the basic principle of extending the ranges by using C.T and P.T remains the same.

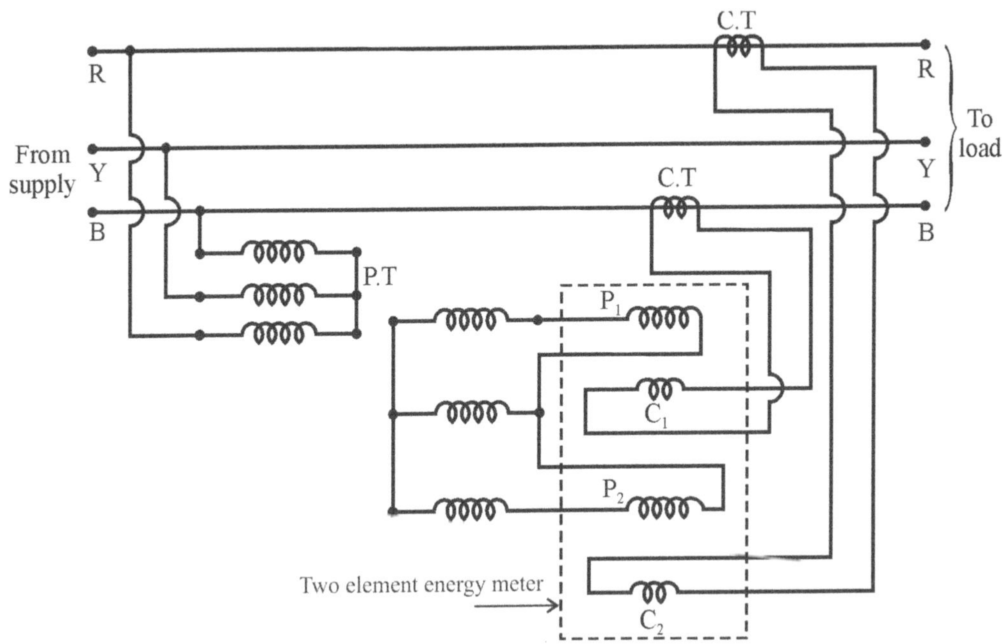

Fig. 4.14 Two element energy meter with C.T and P.T

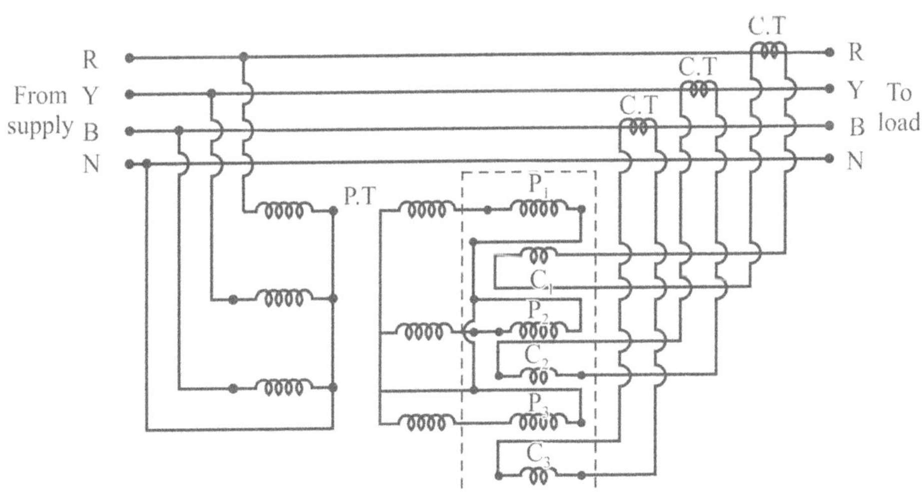

Fig. 4.14(a) Three element energy meter with C.T and P.T

4.12 Average Demand Indicator (or) (Merz Price Maximum Demand Indicator)

This type of demand indicator is not a separate instrument but is a fitting which can be attached as a unit with any type of energy meter. Thus the maximum demand indicator together with the energy meter registers the total energy consumption and the maximum value of the average power over equal intervals of time which may be of 15 or 30 minutes duration.

4.12.1 Construction and Operation

It consists of a special disc mechanism which drives the pointer through gearing arrangement, which is coupled to the energy meter spindle. The dial system is coupled to the energy meter spindle for a fixed interval of time. Generally this time interval is of 30 minutes duration. After the end of this interval, reset device resets the driving mechanism bringing to zero position. But the pointer is held by special friction device which indicates the energy consumed during that interval of time. This pointer position remains fixed unless and until in next intervals of time, the energy consumed exceeds the one indicated by pointer. Thus pointer then indicates new energy consumption, which is more than the previous. Thus in all, the pointer indicates the maximum demand expressed in energy consumed per half hour, for any given period of time.

The Fig. 4.15 the Merz price maximum demand indicator. The pin provided drives the pointer forword for the set period of time interval. When the period ends, the 'cam' controlled by a timing gear momentarily disengages the pinion with help of bell crank mechanism. Under the spring force, the driving mechanism and pin comes to zero position. The pointer remains at its position indicating the energy consumed during the past interval of time.

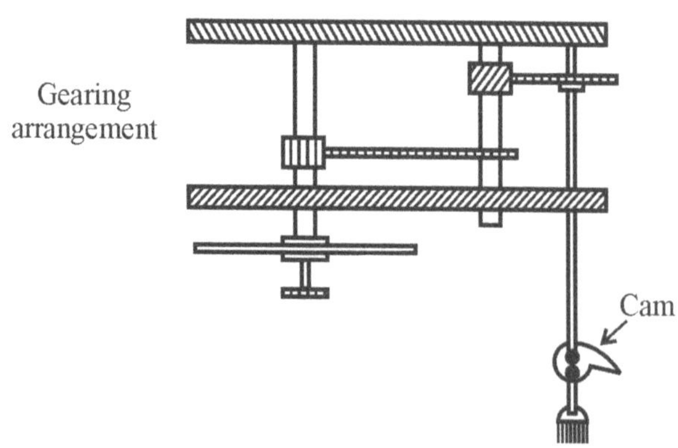

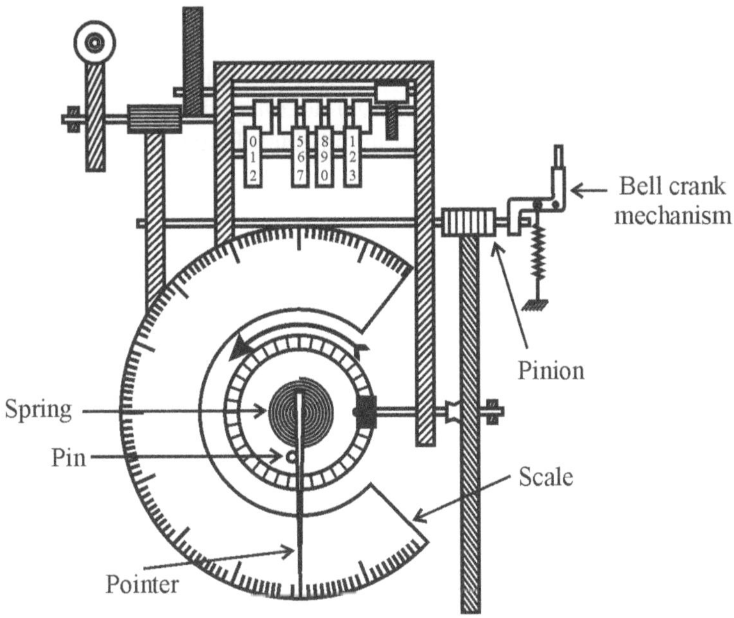

Fig. 4.15 Merz price maximum demand indicator

In some cases reset of the driving mechanism is achieved using a switch actuated by small synchronous motor of the electric clock type.

The average maximum demand can be obtained as:

$$\begin{bmatrix}\text{Average maximum} \\ \text{demand in kW}\end{bmatrix} = \frac{\text{Maximum energy recorded over a time interval in kwh}}{\text{time interval in hours}}$$

4.12.2 Advantages

The basic advantages of Merz price max demand indicator type of meter are, mainly:

1. The scale is uniform
2. Accurate maximum demand is measured

4.12.3 Disadvantages

1. This meter is costly since it involves mechanisms of great complexity.
2. The serious disadvantage is, if the maximum demand occurs some time after the start of one time interval and continues over only a part of the next interval. The real maximum is not indicated since it is split up into two different time intervals. The thermal lagged meters whose indication does not depend upon preselected times but whose integration is continuous, do not suffer from this disadvantage as they have a very long response time.

4.13 Phantom Loading

The phantom loading is also called "FICTICIOUS" loading. when the current rating of a meter under test is high. A test with actual loading arrangements would involve a considerable waste of power. In order to avoid this "Phantom" or "Fictitious" loading done.

Phantom loading consists of supplying the pressure circuit from a circuit of required normal voltage and the current circuit from a separate low voltage supply. It is possible to circulate the rated current through the current circuit which with a low voltage supply as the independence of this circuit is very low. With this arrangement, the total power supplied for the test is that due to the small pressure coil current at normal voltage, plus that due to the current circuit current supplied at low voltage. The total power, therefore required for testing the meter with phantom loading is comparatively very small.

The arrangement for the phantom loading is shown in the Fig. 4.16.

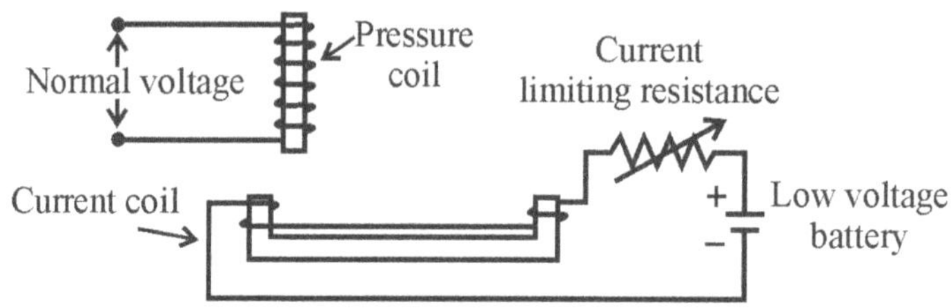

Fig. 4.16 Phantom loading of energy meter

4.14 Testing of Energy Meters

The term "testing" includes the checking of the actual registration of the meter as well as the adjustments done to bring the errors of the meter within the prescribed limits.

A.C energy meters

This type of meters should be tested for the following conditions:

1. At one intermediate load with unity power factor.
2. At 100% or 125% of marked current with unity power factor.
3. At 5% of marked current with power factor.
4. At marked current and 0.5 lagging power factor.

In addition meters should be subjected to the following tests.

1. *Starting test:* At 0.5% of marked value of current and with normal voltage, the meter should start and run. Accuracy is not checked at this current.
2. *Creep test:* With as applied voltage of 110% of its marked value and with the current circuit open, the meter should not revolve through more than one revolution i.e. it should not creep.

4.14.1 Testing of Meter using Rotating Sub Standard Meter

If the constants of meters are same then error can be directly obtained, but if meter constants are different then error is required to be observed.

Let

K_s = Meter constant in number of revolutions per kwh for sub standard meter

K_X = Meter constant in number of revolutions per kwh, for meter under test

N_s = number of revolutions made by substandard meter

N_X = number of revolutions made by meter under test

E_r = Energy recorded by Meter under test = $\frac{N_X}{K_X}$

E_t = True energy recorded by substandard meter = $\frac{N_S}{K_s}$

$$\therefore \qquad \% \text{ error} = \frac{\frac{N_X}{K_X} - \frac{N_S}{K_s}}{\frac{N_S}{K_s}} \times 100$$

Before conducting the test, the meters are allowed to run for 15 to 30 minutes with full load, to attain steady temperature.

4.14.2 Meter Testing Circuit using Phase Shifting Device

The meter testing circuit are meant for phantom loading. Hence separate supplies are used for the pressure coil circuits and current coil circuit. The phase shifting arrangement for either voltage or current is necessary so that test at any power factor can be conducted.

Method 1: In this method, the phase shifting device is in the pressure coil circuit. The pressure coil is supplied through an input transformer having output voltage of 250 V. The auto transformer on primary side helps to adjust the voltage accurately to 250 V. The current coil is supplied separately through a transformer the primary of which is connected to same supply as that of pressure coil circuit. To adjust the necessary current, the toppings are provided on the transformer secondary. Fine adjustment is possible using variable resistor.

The phase shifting transformer is employed in the circuit supplying the pressure coil circuit. Thus the phase of the pressure coil can be changed by changing the rotor of the phase shifting transfer. Hence any power factor condition, required for testing can be obtained. The arrangement is shown in Fig. 4.17.

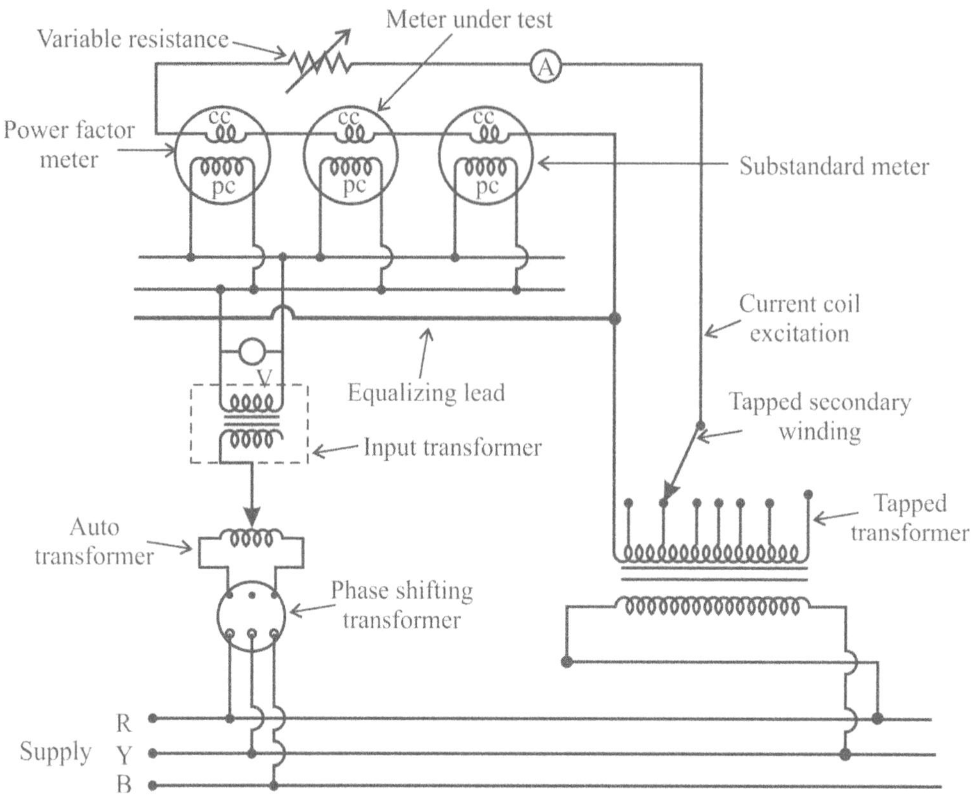

Fig. 4.17 Phantom loading of A.C meter using rotating substandard meter

Method 2: In this method, instead of transformers, two separate alternators mounted on the same shaft are used to excite current and pressure coil circuit. The stator of one alternator can be moved, i.e., the position of one stator can be changed with respect to other stator. Thus the phase shifting of one circuit with respect to other can be achieved. The alternator with movable stator is in current coil circuit. The angle through which the stator is moved gives the phase angle between voltage and current. The frequency of both supplies is same as alternators run at same speed. The arrangement is shown in the Fig. 4.18.

The equalizing lead is provided as to maintain the potential difference between current and pressure coil of the meters to be zero. The method is also called Phantom loading with R.S.S meter'.

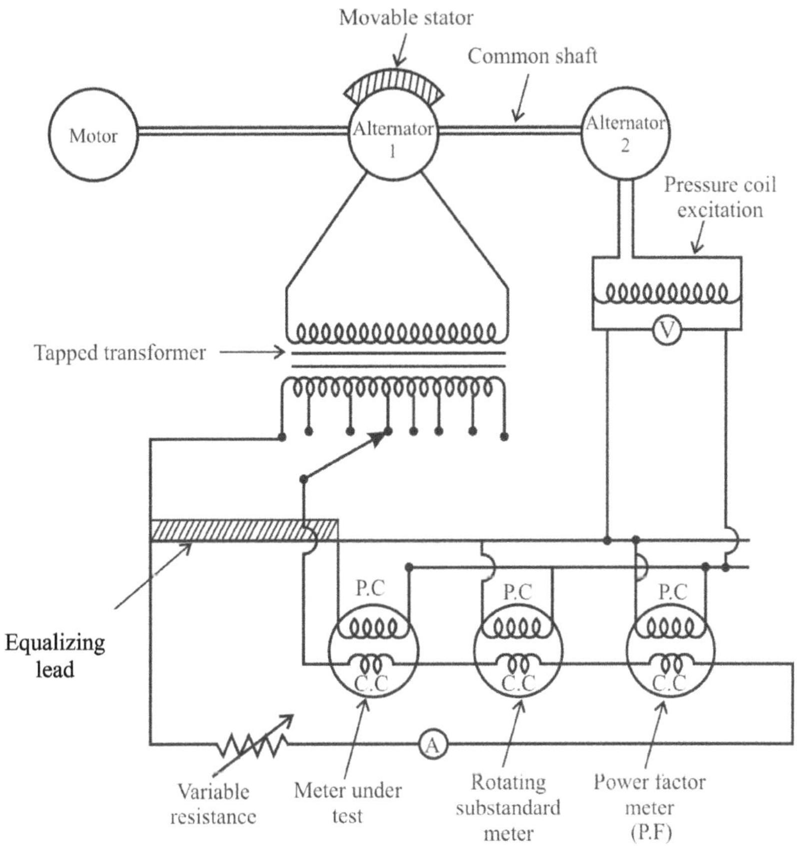

Fig. 4.18 Phantom loading and A.C meter using rotating substandard meter

4.15 Trivector Meter

The meter which measures "the KVAh and KVA of the maximum demand simultaneously" is called a "trivector" meter. It registers the true values of KVAh correct to within $\pm$ 1%. The meter consists of an energy meter and a reactive KVAh meter in the same case with a special summator between them. Both the meters drive the summator with the help of complicated gear train which arranges, the summator to register KVAh correctly at all power factor.

$$\text{KVAh} = \sqrt{(\text{kwh})^2 + (\text{KVARh})^2}$$

It uses five differing gearing systems,

1. Watt-hour meter driving alone at normal speed. This corresponds to unity power factor.

2. Watt-hour speed slightly reduces. Reactive meter speed considerably reduced corresponding to phase angle 22.5° and power factor 0.925.
3. Both speeds reduced by the same factor. Phase angle 45°, power factor 0.707.
4. Watt-hour speed considerably reduced. Reactive meter speed slightly reduced. Phase-angle 67.5°, power factor 0.38.
5. Reactive meter driving alone at normal speed zero power-factor.

The ratchet coupling is linked to run main common register shaft to which final drive from each gear system is connected. Thus the shaft is always driven by the direct drive which has the maximum speed. At that time all other four slower shafts are idle on ratchets.

As the power factor changes, other gear drive system drives the shaft at higher speed and drive shifts to different ratchet. For a given V-I product, the speed of the kwh meter varies as cos ϕ where ϕ varies between 0° to 90° lagging. The variation of percentage speed against the phase angle curves are provided for the trivector meter as shown in Fig. 4.19.

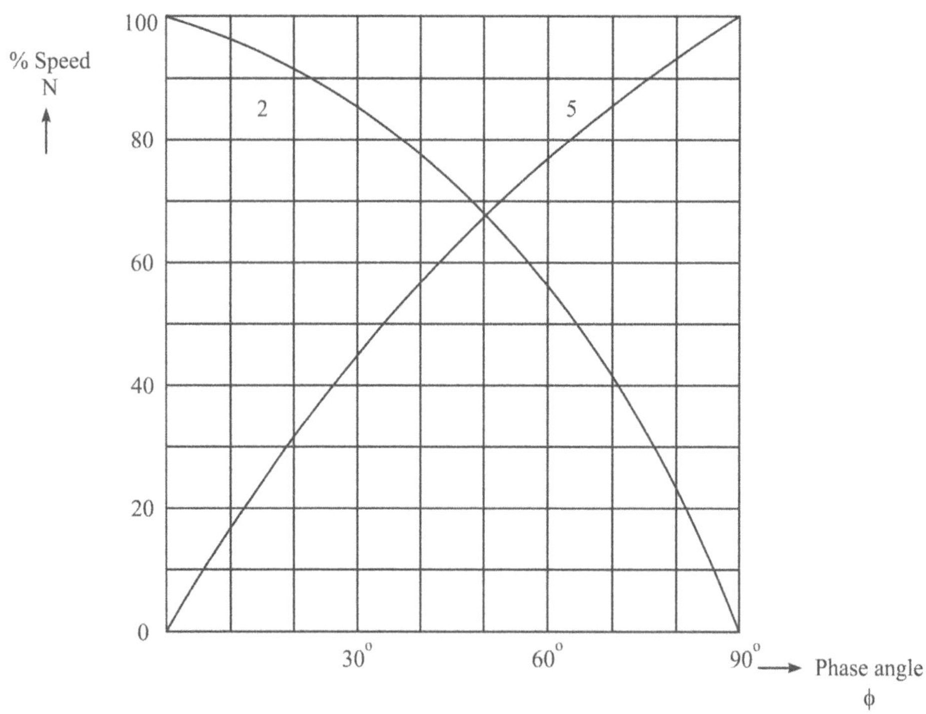

Fig. 4.19 Speed phase angle curves for trivector meter

4.16 Tariff

Introduction

The cost of generation of electrical energy consists of "fixed cost" and "running cost". Since the electricity generated is to be supplied to the consumers, the total cost of generation has to be recovered from the consumers. Tariffs or energy rates are the different methods of charging the consumers for the consumption of electricity. It is desirable to charge the consumer according to the maximum demand (kW) and energy consumed (kWh). The tariff chosen should recover the fixed cost, operating cost and profit etc., incurred in generating the electrical energy.

"The rate at which electrical energy is supplied to a consumer is known as tariff".

4.17 Objectives of Tariff

Like other commodities, electrical energy is also sold at such a rate so that it not only returns the cost but also earns reasonable profit. Therefore, a tariff should include the following items:

1. Recovery of cost of producing electrical energy at the power station.
2. Recovery of cost of operation and maintenance of supply of electrical energy e.g., metering equipment, billing etc.
3. Recovery of cost on the capital investment in transmission and distribution systems.
4. A suitable profit on the capital investment.

4.18 Important Terms and Factors

The following terms and factors are important while studying the economic aspects of energy and tariff.

1. *Connected load:* It is the sum of continuous ratings of all the equipments which are connected to the system of supply. The sum of connected load of all the consumers is the connected load to the power station.
2. *Average load:* It is defined as the average of loads occurring on the power station at a specified period which may be a day or a month or a year.
3. *Maximum demand:* It is the highest demand of the load on the power station for a given period. It is already stated that the load on the station goes on changing from time to time. The maximum of all the demands that have occurred during a given period is the maximum demand. The connected load is greater than the maximum demand as all the consumers do not switch on their connected loads simultaneously. The installed capacity of the station can be determined by knowing the maximum demand.

4. *Load factor:* It is defined as the ratio of average load to the maximum demand during a specified period.

$$\text{Load factor} = \frac{\text{Units generated in given period}}{\text{Maximum demand} \times \text{time of operation of plant}}$$

5. *Capacity factor:* It is the ratio of actual amount of energy that is produced to the maximum possible energy which could be produced during a specified period. It represents reserve capacity of the plant. There must be some reserve capacity for a plant to meet increased load demand.

6. *Plant use factor:* It is defined as the ratio of energy produced to the product of plant capacity and number of hours for which the plant was in operation.

7. *Demand factor:* The ratio of maximum demand to the connected load is called demand factor. Its value is less than 1. The capacity of plant equipment can be decided by using demand factor.

8. *Diversity factor:* It is defined as the ratio of the sum of individual maximum demands to the maximum demand on power station.

The load associated with various types of consumers will not have their maximum demands accuring at the same time. Hence sum of individual maximum demands is always greater than maximum demand on power station and diversity factor is always greater than 1. With great diversity factor, the cost of generation of power is less.

4.19 Types of Tariff

There are several types of tariff. However, the following are the commonly used types of tariff.

4.19.1 Simple Tariff

When there is a fixed rate per unit of energy consumed, it is called a "simple tariff" or "uniform rate tariff".

In this type of tariff, the price charged per unit is constant i.e., it does not vary with increase or decrease in number of units consumed. The consumption of electrical energy at the consumer's terminals is recorded by means of an energy meter. This is the simplest of all tariff and is readily understood by the consumers.

The main disadvantage is "the cost per unit delivered is high".

4.19.2 Flat Rate Tariff

When different type of consumers are charged at different uniform per unit rates, it is called a "flat rate tariff".

In this type of tariff, the consumers are grouped into different classes and each class of consumers is charged at a different uniform rate. For instance, the flat rate per kWh for lighting load may be 60 paise whereas it may be slightly less (say 55 paise / kwh) for power load. The different classes of consumers are made taking into account of their diversity and load factors. The advantage of such a tariff is that it is more fair to different types of consumers and is quite simple in calculations.

The main disadvantage is that "since the flat rate tariff varies according to the way the supply is used, separate meters are required for lighting load, power load etc. This makes the applications of such a tariff expensive and complicated.

4.19.3 Block Rate Tariff

When a given block of energy is charged at a specified rate and the succeeding blocks of energy are charged at progressively reduced rates, it is called a "block rate tariff".

In block rate tariff, the energy consumption is divided into blocks and the price per unit is fixed in each block. The price per unit in the first block is the highest, and it is progressively reduced for the succeeding blocks of energy. For example, the first 30 units may be charged at the rate of 60 paise/unit, the next 25 units may be charged at the rate of 30 paise/unit.

The main advantage of such a tariff is that the consumer gets an incentive to consume more electrical energy. This increases the load factor of the system, and hence the cost of generation is reduced. However, its principal defect is that it lacks a measure of the consumer's demand. This type of tariff is being used for majority of residential and small commercial consumers.

4.19.4 Two-Part Tariff

When the rate of electrical energy is charged on the basis of maximum demand of the consumer and the limits consumed it is called a "two-part tariff".

In two-part tariff, the total charge to be made from the consumer is split into two components viz. Fixed charges and running charges.

The fixed charges depend upon the maximum demand of the consumer while the running charges depend upon the number of units consumed by the consumer. Thus, the consumer is charged at a certain amount per kW of maximum demand plus a certain amount per kWh of energy consumed i.e.

$$\text{Total charges} = \text{Rs}\ (b \times \text{kW} + c \times \text{kWh})$$

where b = charge / kW of maximum demand

c = charge / kWh of energy consumed

This type of tariff is mostly applicable to industrial consumers who have appreciable maximum demand.

Advantages

1. It is easily understood by the consumers.
2. It recovers the fixed charges which depend upon the maximum demand of the consumer but are independent of the units consumed.

Disadvantages

1. There is always error in assessing the maximum demand of the consumer.
2. The consumer has to pay the fixed charge irrespective of the fact whether he has consumed or not the electrical energy.

4.19.5 Three Part Tariff

When the total charge to be made from the consumer is split into three parts viz. Fixed charge, semi-fixed charge, and running charge, it is known as a "three-part tariff".

i.e. Total charge = Rs (a + b × kW + c + kWh)

where a = fixed charge made during each billing period.

It includes interest and depreciation on the cost of secondary distribution and labour cost of collecting revenues.

b = charge / kW of maximum demand

c = charge / kWh of energy consumed.

It may be seen that by adding fixed charge or consumers charge (i.e., a) to two-part tariff, it becomes three-part tariff. The principal objection of this type of tariff is that the charges are split into three components. This type of tariff is generally applied to big consumers.

4.19.6 Power-Factor Tariff

The tariff in which power factor of the consumers load is taken into consideration is known as "power factor tariff".

In an A.C system, p.f plays an important role. A low power factor increases the rating of station equipment and line losses. Therefore, a consumer having low power factor must be penalised. The following are the important types of "P.F tariff".

4.19.6.1 KVA maximum demand tariff

It is a modified form of two part tariff. In this case, the fixed charges are made on the basis of maximum demand in kVA and not in kW. As kVA is inversely proportional to P.F, therefore a consumer having low power factor has to contribute more towards the fixed charges. This type of tariff has the advantage that it encourages the consumers to operate their appliance and machinery at improved power factor.

4.19.6.2 Sliding scale tariff

This is also known as average-power factor tariff. In this case, an average power factor, say 0.8 lagging, is taken as the reference. If the power factor of the consumer falls below this factor, suitable additional charges are made. On the other hand, if the power factor is above the reference a discount is allowed to the consumer.

4.19.6.3 kWh and kVAR tariff

In this type, both active power (kWh) and reactive power (kVAR) supplied are charged separately. A consumer having low power factor will draw more reactive power and hence shall have to pay more charges.

4.19.6.4 Maximum demand tariff

It is similar to two-part tariff with the only difference that the maximum demand is actually measured by installing maximum demand meter in the premises of the consumer. This removes the objection of two-part tariff where the maximum demand is assessed merely on the basis of the ratcable value. This type of tariff is mostly applied to big consumers. However, it is not suitable for a small consumer (e.g. residential consumer) as a separate maximum demand meter is required.

Review Questions

1. Draw a neat sketch showing the construction of a angle phase induction type energy meter. Give the theory and operation of the instrument.
2. Derive the Torque equation of single phase induction type energy meter?
3. Explain the various types of errors occurring in an energy meter and also the methods of compensation to overcome these errors.
4. Explain advantages and disadvantages of induction type energy meter
5. Describe the construction and working of three element energy meter
6. Describe the construction and working of two element energy meter?
7. Explain theory of calibration of an energy meter
8. What is creeping? How can it be prevented?
9. Explain the current T/F and potential T/F uses in energy indicator?
10. Explain advantages and disadvantages of merz price maximum demand indicator?
11. What is phantom loading? When it is used?
12. What is Tarrif? explain briefly the various types of tarrif used.
13. Write a short note on Testing of energy meters?

14. Explain briefly about trivector meter with neat sketch
15. What is Tarrif ? explain the various types of tariff used?
16. Writer a short notes on poly phase wattmeter?

Quiz Questions

1. Name the methods used for power measurement in three phase 4 wire circuits.
   (a) single wattmeter method (b) two wattmeter method
   (c) three wattmeter method (d) four wattmeter method
2. A 200 V, 1-Ø energy meter has constant load current of 4 A is passing through for 5 hours at a p.f of 0.5. If the energy meter disc makes 2000 revolutions during this period, what is the meter constant.
   (a) 1000 rev/kwh (b) 500 rev/kwh
   (c) 200 rev/kwh (d) 2000 rev/kwh

Solution: Meter constant K = no. of revolution/kwh

$$= \frac{2000}{\dfrac{200 \times 4 \times 0.5 \times 5}{1000}} = \frac{2000}{2} = 1000 \text{ rev/kwh}$$

3. A watt-hour meter measures
   (a) voltage (b) current
   (c) energy (d) power
4. An oscilloscope cannot be used to indicate
   (a) frequency (b) peak signal voltage
   (c) energy (d) wave shape
5. In two watt meter method of power measurement, if one of the watt meter shows zero reading, then it can be concluded that
   (a) power factor is unity (b) power factor is zero
   (c) power factor is 0.5 lagging (d) power factor is 0.5 leading
6. To measure the applied power supply voltage to the electrical circuit, the voltmeter should be placed
   (a) In series with the circuit that works from the supply
   (b) Between the positive pole of the supply and the circuit working from the supply
   (c) Between the negative pole of the supply and the circuit working from the supply
   (d) In parallel with the circuit that works from the supply

7. If the readings of the two watt meters in the 2 watt meter method of power measurement are 4.5 kW and 3.5 kW respectively and the latter reading has been obtained after reversing the current coil of the watt meter. What will be the total power in kW?
   (a) 1 (b) 3.5 (c) 4.5 (d) 8

8. A utility meter's reading indicates
   (a) current (b) power
   (c) voltage (d) energy

9. The most suitable material used as spring material for most of the applications, except in low resistance measurements
   (a) platinum silver (b) phosphor bronze
   (c) silicon – bronze (d) hard rolled silver

10. Which instrument has to be suitable to measure the single phase energy?
    (a) ammeter (b) voltmeter
    (c) watt meter (d) induction type energy meter

11. In the measurement of 3-phase power by 2-wattmeter method if the two watt meter readings are equal, the P.f of the circuit is
    (a) 0.08 lagging (b) 0.8 leading
    (c) 0 (d) unity

12. In an induction type energy meter
    (a) these is no brake magnet (b) there is a control spring
    (c) disc revolves continuously (d) these is no temperature error

13. Creep in an energy meter is prevented by
    (a) adjusting brake magnet (b) two holes on the disc
    (c) control spring (d) adjusting shading band

14. Which of the following types of instrument is an integrating instrument
    (a) P.f meter (b) energy meter
    (c) watt meter (d) frequency meter

Questions and Answers

1. What are the constructional parts of dynamometer type wattmeter?

Ans. Fixed coil
Moving Coil

Current limiting resister

Helical spring

Spindle attached with pointer

Graduated scale

2. Write down the deflecting torque equation in dynamometer type wattmeter.

Ans. $T_d \alpha VI \cos \phi$

3. State the disadvantages of Dynamometer type wattmeter.

Ans. Readings may be affected by stray magnetic fields.

At low power factor it causes error.

4. Name the errors caused in Dynamometer type wattmeter.

Ans. Error due to pressure coil inductance

Error due to pressure coil capacitance

Error due to methods of connection

Error due to stray magnetic fields

Error due to eddy current.

5. How the errors caused by pc inductance is compensated.

Ans. By connecting a capacitor in parallel to the resister.

6. How the errors caused by methods of connection is compensated.

Ans. By using compensating coil.

7. What are the special features to be incorporated for LPF wattmeter?

Ans. Pressure coil circuit

Compensation for Pressure coil current

Compensation for Pressure coil inductance.

8. Define Phantom loading.

Ans. Method by which energizing the pressure coil circuit and current coil circuits separately is called phantom loading.

9. State the use of phantom loading.

Ans. Power loss is minimized.

10. Name the methods used in Wattmeter calibration.

Ans. By comparing with std wattmeter.

By using voltmeter ammeter method

By using Potentiometer.

11. What are the types of energy meters?

Ans. Electrolytic meters

Motor meters.

Clock meters

12. Name the constructional parts of induction type energy meter

Ans. Current coil with series magnet

Voltage coil with shunt magnet

Aluminium disc

Braking magnet

Registering mechanism.

13. How voltage coil is connected in induction type energy meter

Ans. It is connected in parallel to supply and load.

14. How current coil is connected in induction type energy meter

Ans. It is connected in series to the load.

15. Why Al disc is used in induction type energy meter

Ans. Aluminum is a nonmagnetic metal.

16. What is the purpose of registering mechanism

Ans. It gives a valuable number proportional to the rotations.

17. What is the purpose of braking mechanism

Ans. It provides necessary braking torque.

18. Define creeping.

Ans. Slow but continuous rotation of disc when pc is energized and cc is not energized.

19. State the reason why holes are provided in Al disc

Ans. To avoid creeping, holes are provided on both sides of Al disc.

20. One single phase wattmeter operating on 230 V and 5 A for 5 hours makes 1940 revolution. Meter constant in revolution is 400. What is the power factor of the load?

Ans. 0.84

Ex: $$\text{Energy consumed} = \frac{\text{No of revolutions}}{\text{meter constant}} = \frac{1940}{400} = 4.85 \text{ KWH}$$

$$\text{Load in KW} = \frac{\text{Energy consumed in KWH}}{\text{Time in hrs}} = \frac{4.85}{5} = 0.97$$

$$\text{P.F.} = \frac{0.97 \times 10^3}{230 \times 5} = 0.84$$

21. Induction type energy meter measures electric energy in ________

Ans. kwh

22. Induction type energy meters are free from ________ errors

Ans. Hysteresis

23. An instrument that measures electrical energy is an ________ type instrument.

Ans. Integrating

24. The main error in a dynamometer type wattmeter is due to

Ans. Pressure coil inductance

25. Induction type wattmeter operation is based on

Ans. Eddy current flow

Quiz Questions Key

1. (c)	2. (a)	3. (c)	4. (c)	5. (c)
6. (d)	7. (a)	8. (d)	9. (b)	10. (d)
11. (d)	12. (b)	13. (b)	14. (b)	

CHAPTER 5

Potentiometers

5.1 Introduction

A potentiometer is essentially, a piece of apparatus by means of which e.m.f's are compared. If one of two e.m.f is known, the other may be determined by comparison with the known one, and thus the potentiometer is used for the measurement of e.m.f's by comparison with a standard e.m.f.

The unknown e.m.f is compared with a known e.m.f which is obtained from a standard cell or any reference voltage source. The main advantage of the comparison method of the e.m.f measurement is that the potentiometer is capable of providing high degree of accuracy as the measurement result is not dependent on the actual deflection of the pointer. Instead of that the accuracy is dependent on the accuracy with which the reference voltage is known.

Basically a potentiometer uses *balance or null condition* during the measurement of unknown e.m.f. Actually no current flows in the circuit of the unknown e.m.f during measurement. Thus no power is consumed in such circuit.

Since a potentiometer measures voltage, it can also be used to determine current simply by measuring the voltage drop produced by the unknown current passing through a known standard resistance.

The potentiometer is extensively used for a calibration of voltmeters and ammeters and had in fact became the standard for the calibrations of these instruments. For the above mentioned advantages, the potentiometer had became very important in the field of electrical measurements and calibration.

5.2 Principle of Potentiometer

The potentiometer works on the principle of opposing the unknown e.m.f by a known e.m.f with the negative terminals of both the e.m.fs connected together, while the positive terminals connected together through a galvanometer as shown in the Fig. 5.1.

When the e.m.fs are of same values, there is no deflection on galvanometer. Thus to measure the unknown e.m.f by using above method, the known e.m.f used must be variable. Another important requirement is that known e.m.f should be varied to give a larger number of known values but it is practically very difficult.

Hence alternatively, the unknown e.m.f is connected in parallel with and in opposition to a voltage drop measured across the resistor as is shown in the Fig. 5.2.

The potentiometers are classified as d.c potentiometers and a.c potentiometers. There are various form of the d.c potentiometers used widely practically. The main advantages of this method is that the current in the resistor can be varied easily to obtain any desired voltage with very fine adjustment. The voltage drop across resistor can be determined by calibrating the resistor with standard cell.

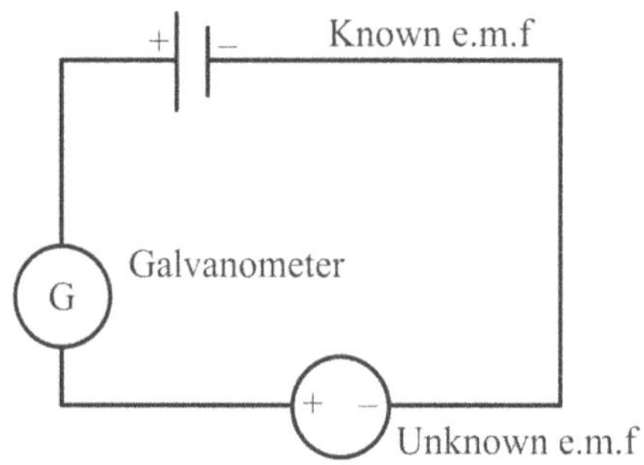

Fig. 5.1 Representation of method of balancing e.m.f

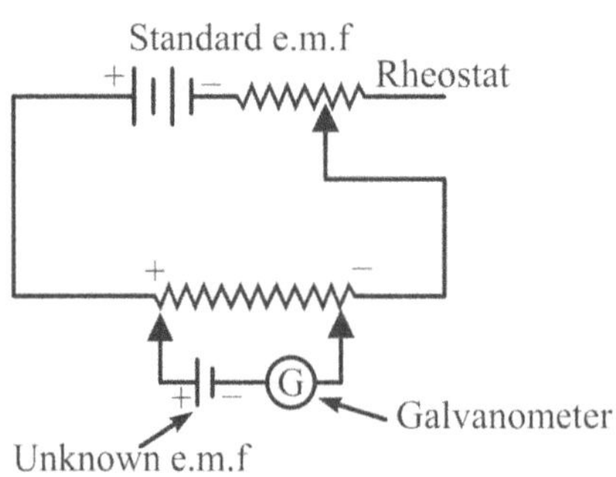

Fig. 5.2 Alternative method of balancing e.m.f

The basic simplest type of the d.c potentiometer is the slide wire potentiometer. Let us study the slide wire potentiometer in detail.

5.3 Potentiometers for use with Direct Current

(Or)

Slide Wire D.C. Potentiometer

The basic slide wire d.c potentiometer is as shown in the Fig. 5.3.

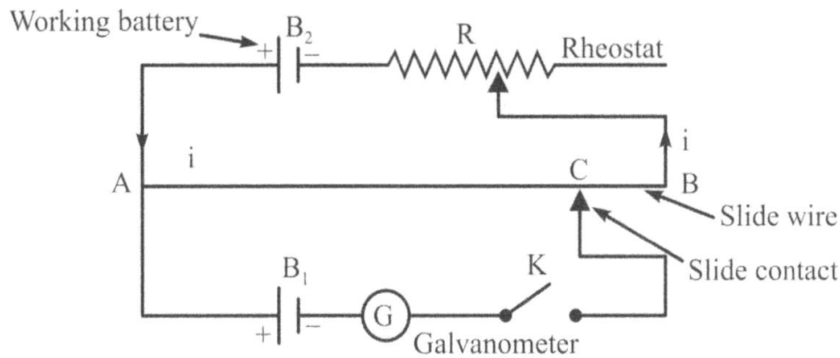

Fig. 5.3 Basic slide wire potentiometer

The circuit consists of a slide wire A, B of uniform cross section and unit length. Generally slide wire is made up of MANGANIN. Let 'r' be the resistance per unit length of slide wire. The battery B_2 supplies a current through the slide wire which is limited with the help of regulating resistance i.e. Rheostat. The battery B_1, whose e.m.f is to be measured is connected in series with a galvanometer G and switch 'K'. If the key 'K' is opened, the current through slide wire is 'i'. If the sliding contact is at position 'C', let the length 'AC' be 'l' units, then the voltage drop across AC is given by '*irl*'.

If the key 'K' is closed which puts the battery B_1 in the circuit. The battery B_1, whose e.m.f is to be measured is connected such that the voltage drop along the slide wire and e.m.f of B_1 oppose each other. The deflection in the galvanometer 'G' depends on the magnitudes of voltage drop across the slide wire portion 'AC' and e.m.f of 'B_1'. If the voltage drop across length '*l*' of the slide wire is greater than e.m.f of battery B_1, then the current will flow in the direction 'A' to 'C', through the galvanometer. Similarly if the battery B_1 is greater than the voltage drop across the length '*l*' of the slide wire, then the current will flow in the direction 'C' to 'A' through the galvanometer. The most important condition exhibiting the basic principle of the potentiometer is that no current flows through the galvanometer when the two e.m.fs are equal.

Generally a scale is provided along with the slide wire which enables to measure the length of portion 'AC'.

Hence to measure e.m.f of a battery, first adjust a current through slide wire with switch 'K' is open. Then insert battery whose e.m.f is to be measured by closing switch 'K'. Adjust sliding of the portion of the slide wire with the help of scale provided. Then the unknown e.m.f 'E' of battery is given by

E_1 = e.m.f of battery B_1.

E_2 = e.m.f of battery B_2.

$$E = i\,(rl) \qquad \text{.....(5.1)}$$

where 'r' is the resistance per unit length, 'i' is the working current adjusted using rheostat 'R'.

If e.m.fs of the two batteries 'B_1' and 'B_2' are to be compared then insert the first battery 'B_1' in series with the galvanometer. Measure the length of the slide wire portion say 'l_1'. Repeat the same procedure with battery B_2 in the circuit. Let the length measured be 'l_2'.

Let e.m.f of batteries B_1 and B_2. We can write

$$E_1 = i\,(rl_1) \qquad \text{.....(5.2)}$$

$$E_2 = i\,(rl_2) \qquad \text{.....(5.3)}$$

Then we can write

$$\frac{E_1}{E_2} = \frac{l_1}{l_2} \qquad \text{.....(5.4)}$$

From above equation, it is clear that the ratio of two lengths gives the ratio of the two e.m.f's.

If one of the batteries used is a standard cell, say battery B_2 of known voltage, the e.m.f of battery B_1 is given by

$$E_1 = E_2\left(\frac{l_1}{l_2}\right) \qquad \text{.....(5.5)}$$

5.3.1 Precautions

While using basic slide wire potentiometer following precautions must be taken.

1. A small resistance should be used in series with the galvanometer to protect it during the initial adjustments of contact 'C'. It also takes care that no appreciable current is taken from the standard cell.
2. The supply battery B_2 should be of high capacity so that a constant current flows through the slide wire throughout the measurement.
3. The accuracy of the measurement depends on how accurately ratio (l_1/l_2) is determined. Hence in such elementary potentiometer the length of the slide wire used should be large enough so that percentage error in measurement reduces.

Now-a-days the potentiometers used for precise measurement, by connecting a number of resistance coils in series with short slide wire, the effect of larger length can be obtained.

5.3.2 Standardization of Potentiometer

Standardization of a potentiometer is a process of adjusting the working current supplied by the supply battery such that the voltage drop across a portion of sliding wire matches with the standard reference source.

A battery of sufficient capacity i.e., 'B_2' is connected in series with a rheostat 'R_h' which regulates the working or standard current flowing the slide wire.

The practical set up for standardizing a d.c potentiometer is as shown in the Fig. 5.4.

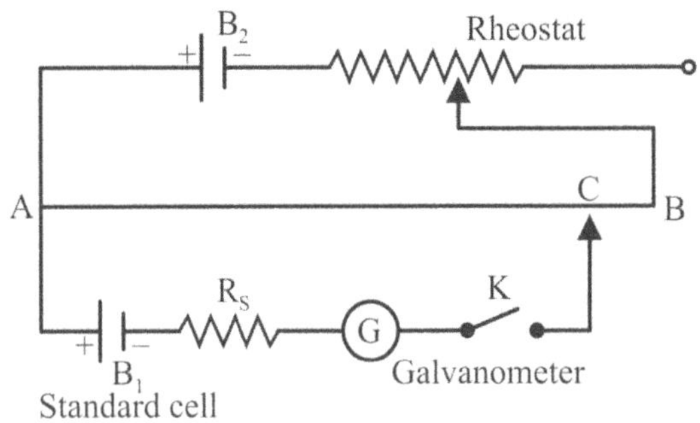

Fig. 5.4 Practical setup for d.c potentiometer standardisation

A standard cell B_1 usually a Weston standard cell of e.m.f 1.0186 volts is connected to galvanometer and a switch 'K' through a series resistance R_s. By properly adjusting R_s full sensitivity of the galvanometer can be obtained.

A slide wire with total length 200 cm and resistance of 200 Ω is connected which is indicated by point A and B. During standardization process, K is closed and the sliding contact is placed at the mark of 101.86 cm along slide wire as indicated by point C in Fig. 5.4. Then we can observe some deflection in the galvanometer. Now by adjusting the value of rheostat R_h, we can get null deflection in the galvanometer.

Under the condition of null deflection, the voltage drop along 101.86 cm portion of the slide wire equals the e.m.f of standard Weston cell. This is nothing but the standardization of a potentiometer and once the potentiometer is standardized, the Rheostat is not disturbed. In other words, the working or standard current is kept constant.

After standardizing a potentiometer, it is used as direct reading potentiometer as the voltage along the slide wire at any point is proportional to the length of the slide wire where the point is obtained by moving sliding contact along the wire to get null deflection in the galvanometer for any battery whose e.m.f is to be measured.

5.4 The Crompton's D.C. Potentiometer

R.E Crompton first modified the simple slide-wire form of potentiometer described above. The general arrangement of his form of the instrument being shown in Fig. 5.5.

The basic slide wire potentiometer considered in the previous section is not the practical form of the d.c potentiometer. The main draw back of the basic slide wire is that the length illustrated can not be read with great precision. Not only this, but the long slide wire is very awkward as the size of the instrument increase due to it. Hence in practical laboratory form of a d.c potentiometer, the size is reduced as instead of long slide wire calibrated dialed resistors or small circular wires of one or more turns are used.

This potentiometer consists of a graduated slide wire 'AC' which is connected in series with large number of coils. The coils are selected such that a resistance of each coil is equal to the resistance of the slide wire. Instead of one sliding contact used in the basic slide wire potentiometer, two sliding contacts P_1 and P_2 are used here. The first sliding contact slides over the slide wire while the second one P_2 over the studs of the resistance coils. The battery 'B' is of sufficient capacity to provide working current and it is connected in series with regulating resistance. The regulating resistance is realized by series connection of two variable resistances R_1 and R_2 where R_1 consists of number of resistance coils while R_2 in the form of slide wire.

A galvanometer G and switch 'K' are connected in series between point P_1 and P_2 through multiple circuit switch or change over switch. The main advantage of the multiple circuit switch is that we can connect a standard cell 'S' (such as Weston cell) between terminals S^+ and S^- or the batteries whose e.m.fs are to be measured between the terminals 1^+ and 1^- or 2^+ and 2^-. The terminals are marked with positive (+) and negative (–) signs so as to avoid range to the potentiometer due to the wrong polarity of battery connected.

First the potentiometer is standardized by using a standard cell 'S'. Say Weston cell with e.m.f of 1.0186 V. as shown in Fig. 5.5. The multiple circuit switch is thrown to the terminals S^+ and S^-. The sliding contact P_1 and P_2 are set at 0.0186 along the slide wire and 1.0 on the stud respectively. The switch 'K' is closed and the null deflection in the galvanometer is obtained by varying R_1 and R_2. Here R_1 is used for coarse adjustment while R_2 is used for fine adjustment. Once the potentiometer is standardised, resistance R_1 and R_2 are left undisturbed. Now the potentiometer is direct reading potentiometer.

Now connect a battery whose e.m.f is to be measured between terminals 1^+ and 1^- with the polarities indicated by the terminals and change over the multiple circuit switch to the terminals 1^+ and 1^-. By adjusting P_1 and P_2, the potentiometer can be again balanced which gives the reading of e.m.f of the battery under measurement directly. For example as shown in Fig. 5.5, if P_1 is at 0.05 and P_2 is at 1.1 then the e.m.f of the battery is 1.15 V.

The main draw back of this instrument is that it is necessary to re-standardise the potentiometer if it takes considerable time for the potentiometer balance adjustment during the measurement. It is advisable to allow the current to flow through the slide wire, be for the measurement to set stable balance of potentiometer in quick time.

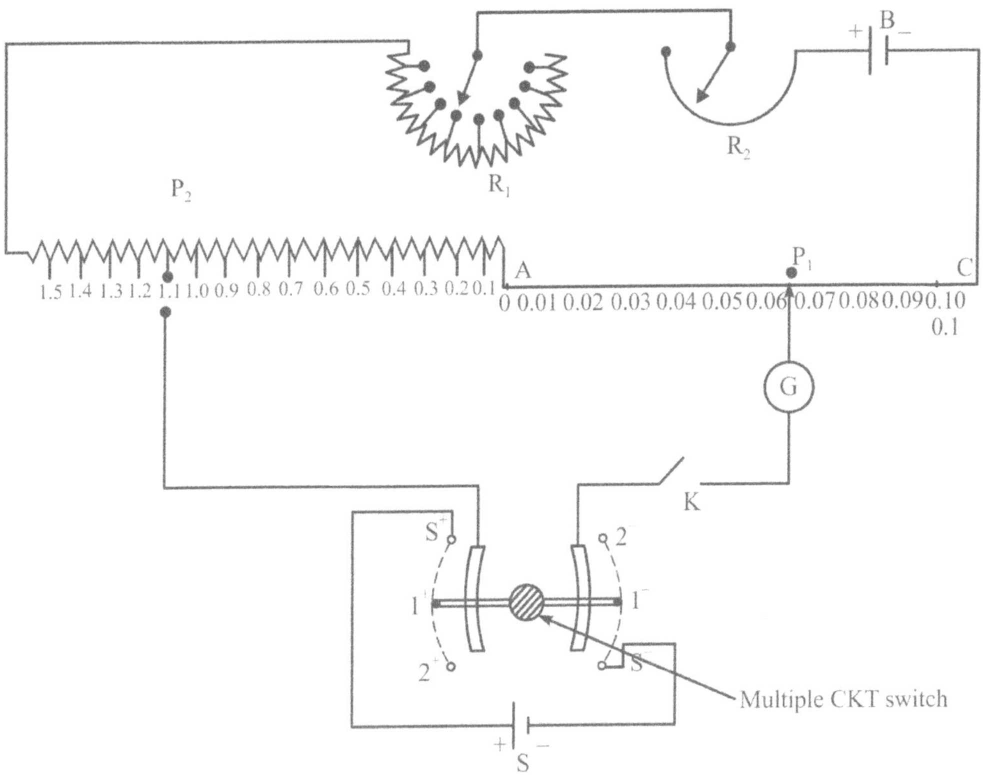

Fig. 5.5 Crompton's D.C potentiometer

5.5 Duo-range Potentiometer

By keeping direct reading feature, the basic potentiometer may be modified to add a second range with usually a second factor such as 0.01. Such a potentiometer is called *duo-range potentiometer.*

The duo-range potentiometer is the modified version of Crompton's D.C potentiometer in which additional range selector switch is used as shown in the Fig. 5.6. The working battery B of sufficient capacity is connected in series with two variable resistances R_1 and R_2 which are used for regulating the current through potentiometer during standardization. Let the total resistance of measuring circuit including the main dial and slide wire be R_m, where R_m is the addition of the main dial resistance and slide wire resistance. The slide wire resistance is 10 Ω. The resistance of each coil in main dial is same as that of the slide wire i.e. 10 Ω and there are 15 steps and so the total main dial resistance is 150 Ω. Let the current through measuring circuit be 10 mA. Hence the total drop across R_m is given by

$$V_{Rm} = I_m \times R_m = \left(10 \times 10^{-3}\right)(160) = 1.6 \text{ V}$$

The second range can be obtained by changing position of switch 'S' from 'M' to 'N' i.e., to × 0.1 range. When this range is selected, the current I_m must reduce to $\frac{1}{10}$th of it's original value.

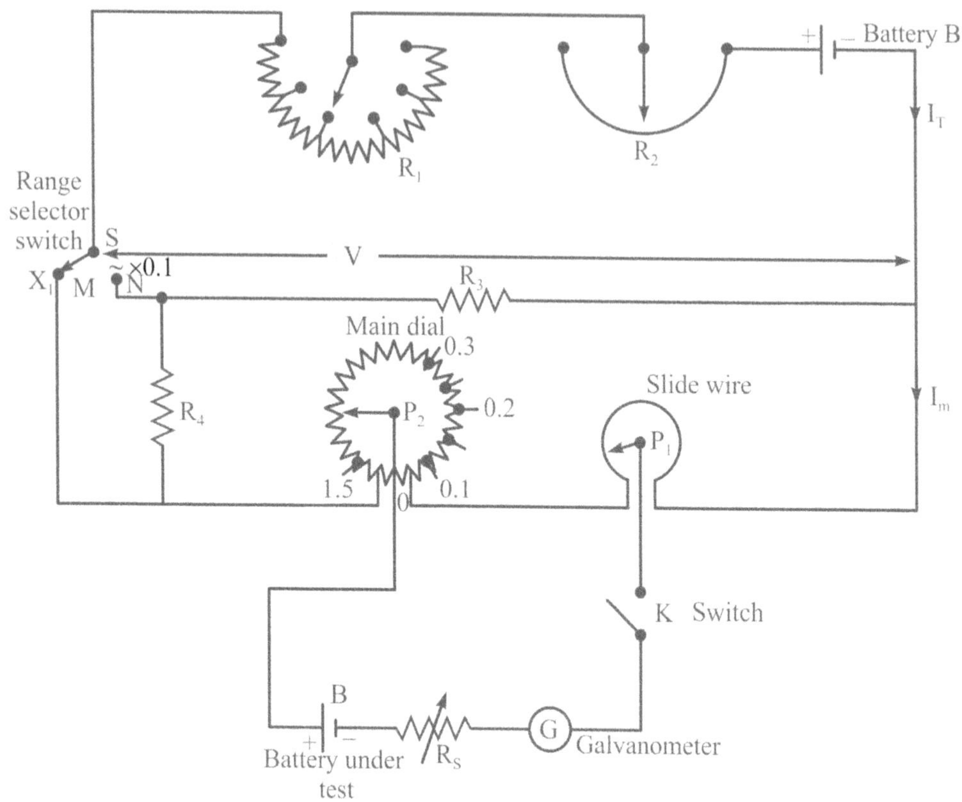

Fig. 5.6 Duo-range D.C. potentiometer

When the switch 'S' is at position m the current adjusted equals 10 mA. So when switch S is moved to position N, I_m must reduce to 1 mA in order to produce voltage drop of 0.16 v across R_m.

It is very important to design a circuit in such a way that the instrument should enable the user to change the measuring range without standardizing the instrument again and again either by adjusting R_1 and R_2 or by changing the voltage of working battery B. In other words, once the instrument is standardized for X_1 range. It should not be needed to standardize again for X 0.1 range. This condition can be fulfilled only if the voltage 'V' remain same for both the positive of range selector switch 'S' and this is possible only if the total battery current I_T is same for both the ranges.

The operation of duo-range potentiometer can be explained in simple form as below. First consider that the range selector switch 'S' is at position M i.e., on range of X_1. Then the total measuring resistance R_m gets shunted by series combination of range resistor R_3 and R_4 as shown in the Fig. 5.7(a). The current through R_m is I_m while total current is I_T. Now consider that the range selector switch 'S' is moved to position N i.e., on range of $\times 0.1$. Then the range resistor R_3 shunts the series combination of total measuring resistor R_m and range resistor R_4 as shown in the Fig. 5.7(b). Now the current flowing through the branch consisting R_m and R_4 is I_m' which is (0.1) I_m and still total current is I_T. In order to have the same current I_T for both ranges, the condition is,

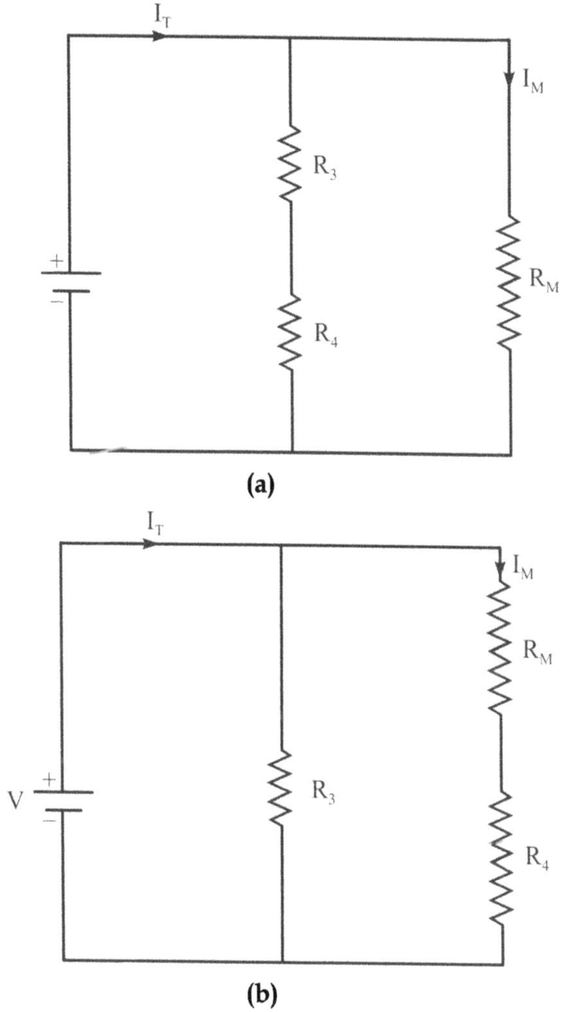

Fig. 5.7(a) & (b) Simplified circuit of duo-range potentiometer for different range selections

$$(R_3 + R_4) \| R_M = R_3 \| (R_M + R_4)$$

$$\frac{(R_3 + R_4)R_M}{R_3 + R_4 + R_M} = \frac{R_3(R_M + R_4)}{R_3 + R_4 + R_M}$$

$$\therefore R_3R_M + R_4R_M = R_3R_M + R_3R_4 \qquad \text{.....(5.6)}$$

$$R_M = R_3$$

Above equation (5.6) indicates that range resistance R_3 must be selected same as total measuring resistance R_M so as to keep total current, supplied by battery B, same for both the ranges.

Now the second condition states that the current $I_m^{'}$ when switch 'S' is at position N must be equal to 0.1 I_m where I_m, is the current through R_m when switch 'S' is at position M.

$$\therefore \quad I'_M = 0.1 I_M$$

$$\therefore \quad \frac{V}{R_M + R_4} = 0.1\left[\frac{V}{R_M}\right]$$

$$\therefore \quad R_M = 0.1(R_M + R_4)$$

$$\therefore \quad R_4 = 9R_M = 9R_3 \qquad \text{.....(5.7)}$$

Thus by properly designing values of R_3 and R_4 we can achieve high resolution in measurement using duo-range potentiometers.

5.5.1 Advantages of Duo-range Potentiometer

Following are the advantages of the duo-range potentiometer.

1. Due to the inherent accuracy of dial resistors as compared to that of slide wire the accuracy of reading is increased.
2. Due to the dial range, the precision of reading is increased by one decimal point.

5.5.2 Simple Potentiometer with Independent Calibrating Circuit

In general, the calibration of the potentiometer is done by using standard cell dial circuit. It provides a standard cell balance resistance to be used for a particular standard cell used. Many times it is necessary to check the standard cell balance during the measurement also. So under this condition, the operator has to check the standard cell balance using separate standard cell dial circuit without disturbing settings of the potentiometer. But practically the potentiometer settings has to be readjusted after every standard cell balance check. This disadvantage is overcome by using simple potentiometer with independent calibrating circuit incorporated in it. Now-a-days, all the modern potentiometers incorporate separate standard cell dial circuit as shown in the Fig. 5.8. The potentiometer shown in the Fig. 5.8 the calibration circuit A-D can be set for any range of standard cell from 1.016 V to 1.02 V. the resistance RS provides 1.016 V while the standard slide wire AC provides 0.04 V.

These changes in the e.m.f of standard cell due to the changes in temperature are also accounted for the operation using this calibration circuit. The slide wire of the standard cell dial measures the e.m.f of the standard cell which is also connected to the potentiometer through the switch. First the switch is put at CALIBRATE position and the rheostat is adjusted such that the current through galvanometer 'G' is zero. Thus with this adjustment the working current is fixed at proper value. Then the switch is put at OPERATE position. The unknown e.m.f is then measured by using measuring circuit consisting measuring circuit dial and slide wire. The main advantage of this type of the potentiometer is that during measurement also the calibration circuit is checked for the set value of the working current by simply throwing switch to CALIBRATE position, without disturbing the settings of potentiometers. These facilities to increase speed of measurement by reducing the time that is required for the calibration using separate standard cell dial.

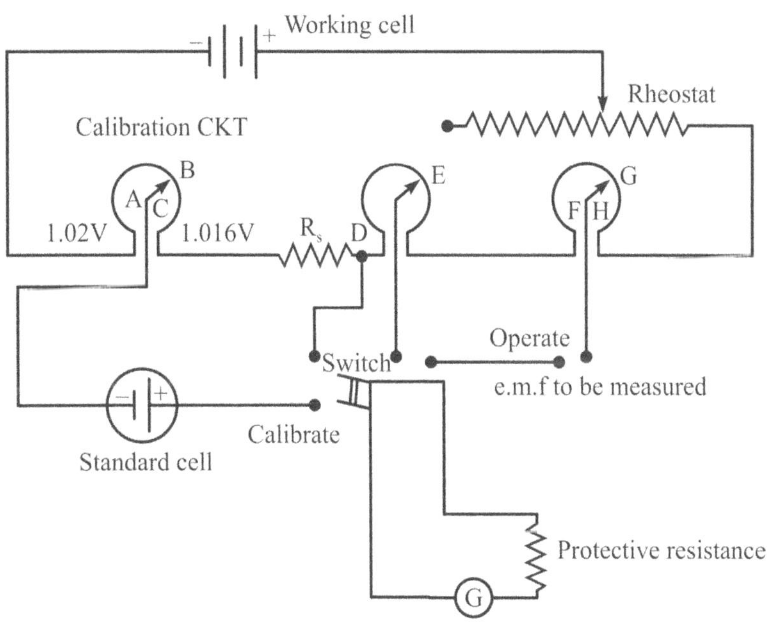

Fig. 5.8 Potentiometer with independent calibration circuit

5.6 Vernier Potentiometer

By using basic simple potentiometers, the precision of 100 μV for readings up to 1.6 V can be obtained. Using such instruments it is very difficult to get accurate readings mainly due to the uniformity of slide wire and maintaining good potential contacts. Some of the practical applications demand highly précised and accurate measurements. The limitations due to the slide wire in basic potentiometers are eliminated in the vernier potentiometer. The instrument with normal range of 1.6 V with 10 μV precision and lower range of 0.16 V with 100 μV precision is as shown in the Fig. 5.9.

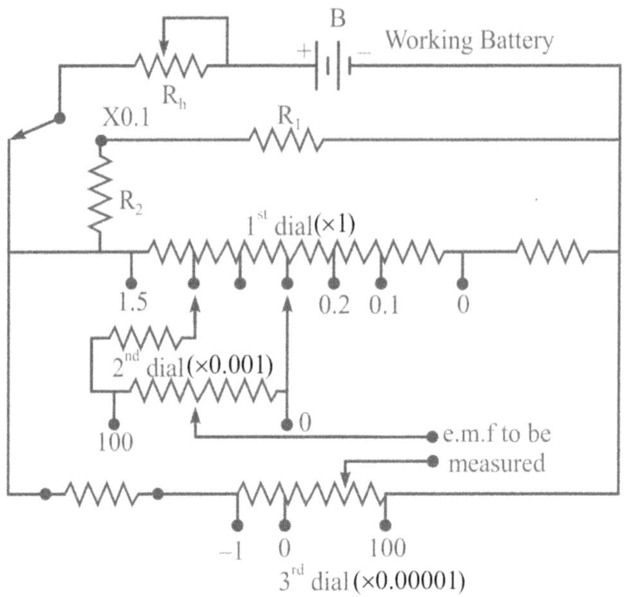

Fig. 5.9 Vernier potentiometer

The main difference in the simple potentiometer and vernier potentiometer is that it uses three measuring dials. The slide wire is not used in this type of the potentiometer. The main dial i.e., the first dial measure upto 1.5 V on (X_1) range in steps of 0.1 volts. The second dial reads upto 0.1 V in steps of 0.001 volts on (X0.001) range.

It consists of 102 studs. The third dial again with 102 studs measures from 0.00001 to + 0.001 volts on (X 0.00001) range. This third dial provides true zero and negative setting. The function of range selector switch S, and range resistance R_1 and R_2 is already explained in previous section. The resistance of the second dial shunt two of the coils of the main dial. The moving arm of the second dial carries two contacts which are placed two studs away from each other.

The vernier potentiometer reads voltages of 0.00001V (10 μV) on X_1 range while 1 μV on X 0.1 range to have voltage reading of 0.1 μV one more range of X 0.01 may be provided but it is not possible to read such small voltage as stray thermal and contact potentials in potentiometer, galvanometer and measuring circuit are uncontrollable. These potentials are of the order of one to several micro volts. These potentials can be reduced by properly selecting metals for resistor, terminals and connecting leads. Sometimes thermal shields are also used to enhance process of reduction of potentials.

5.7 Brook's Deflectional Potentiometer

This potentiometer is used for applications where the voltage to be measured is continuously changing. With conventional type of potentiometer it becomes impossible to obtain exact balance even when the changes are slow since it is difficult to follow the changes by manipulation of the dails.

The Fig. 5.10 shows a deflectional potentiometer. In this potentiometer only one or two main dials, consisting of decade resistance boxes are used. A center zero galvanometer circuit included in the circuit to indicate deflection. The galvanometer circuit includes a set of compensating resistors R_1, R_2. The values of the compensating resistors is such that the resistance of the potentiometer circuit as viewed from terminal where the unknown e.m.f is applied remains constant irrespective of the position of the sliding contacts. This means that current through the galvanometer will always be proportional to the out of balance current whatever may be the setting of the main dials. Thus the galvanometer scale can be calibrated to read the out of balance e.m.f directly. The value of the unknown em.f is obtained by adding the galvanomcter reading to the main dial setting. The main dial setting is kept nearly equal to the e.m.f being measured.

R_1:R_2:compensating resistors

The potentiometer is used to monitor the value of temperature where a thermocouple is employed. The output of the thermocouple is fed to the unknown e.m.f terminals.

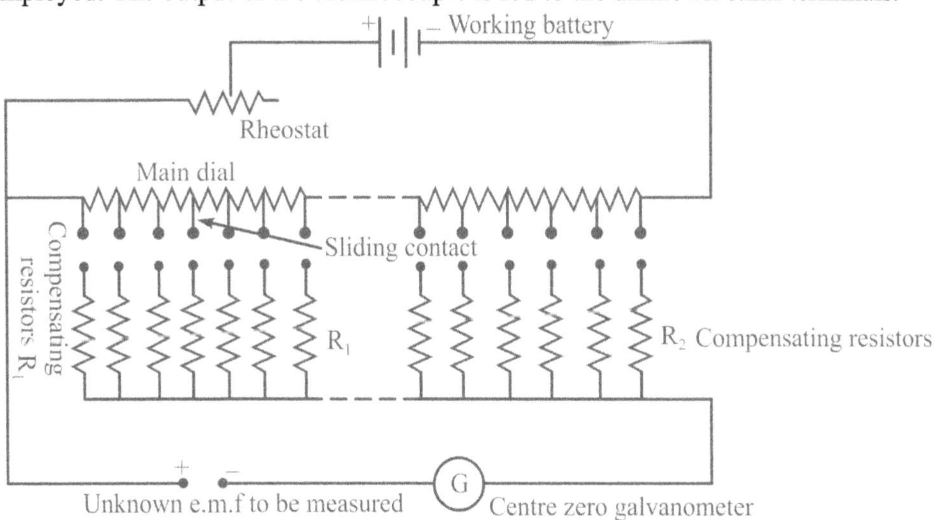

Fig. 5.10 Brook's deflectional potentiometer

5.8 Potentiometer with True Zero

In the basic potentiometer it is very difficult to obtain true zero as the two contacts can not coinside correctly.

To overcome this drawback the circuit is modified as shown in the Fig. 5.11.

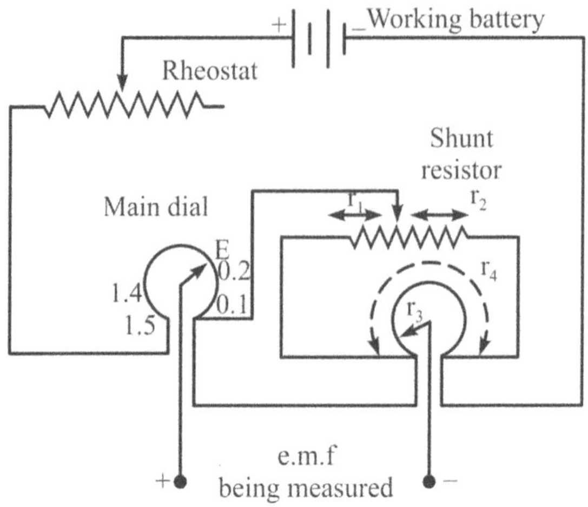

Fig. 5.11 Modified potentiometer with true zero.

The slide wire BC is shunted by shunt resistor which is tapped at D. This serves as 0, on main dial. When the contact is in a positions, then $\frac{R_1}{R_3} = \frac{R_2}{R_4}$ is zero. If the slider can travel lower than zero position providing negative readings. The slider movement above zero gives positive reading. The typical range of slider wire of from – 0.005 V to + 0.1.5 V.

5.9 Volt-Ratio Box

As per the discussion of D.C potentiometers in earlier sections, the maximum voltage measured is less than 2 V. In practical application if the voltage to be measured is greater than 2 V, then along with the potentiometer a volt-box (or) ratio-box (or) volt-ratio box is used. It is based on the concept of potential divider. It consist of a high resistance having number of tappings with properly divided its resistance between various pairs of tappings are as shown in Fig. 5.12.

Consider that the voltage of the order of 150 V to be measured. Then this voltage is connected between the terminals 150 V and common terminal. The leads to the potentiometer are taken from two tapping points say 50 Ω as shown in the Fig. 5.12. Now if the voltage on the potentiometer is 1.3 V then the actual voltage to be measured is given by

$$V_{unknown} = 1.3\left(\frac{5000}{50}\right) = 130 \text{ V}$$

Thus by using simple concept of potential divider, we can measure unknown voltage.

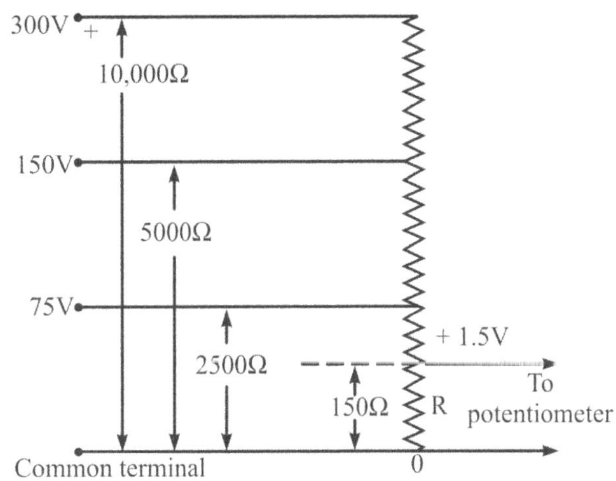

Fig. 5.12 Volt-ratio box

5.10 Applications of D.C Potentiometer

In addition to measurement of voltage, the potentiometer is the usual basis for calibration of all voltmeters and watt meters. The potentiometer may also be used for measurement of current, power and resistance. Since the potentiometer is a d.c device, the instrument to be calibrated must be d.c moving iron or electrodynamometer types.

Let us discuss briefly.

5.10.1 Calibration of Voltmeter

The practical set up for calibration of voltmeter is as shown in the Fig. 5.13(a) & (b). In both the circuits, the concept used is of potential divider. The main important requirement of these circuits is the use of very stabilised D.C supply. The only difference is that in the second circuit two rheostats used which are useful in coarse and fine adjustment. With these adjustments it is possible to adjust a voltage such that the pointer of voltmeter exactly coincides the major division. The voltage across voltmeter is stepped down using volt-ratio box. For high accuracy it is advisable to measure voltages near the maximum range of potentiometer.

Let the reading of volt meter indicated be V_{ind}. The actual voltage measured using potentiometer is given by

$$V_{act} = \frac{\text{(Potentiometer reading)}}{\text{(V.R. box ratio)}}$$

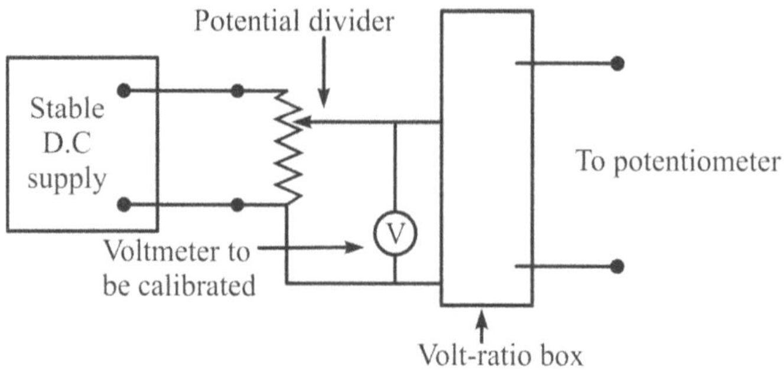

Fig. 5.13(a)

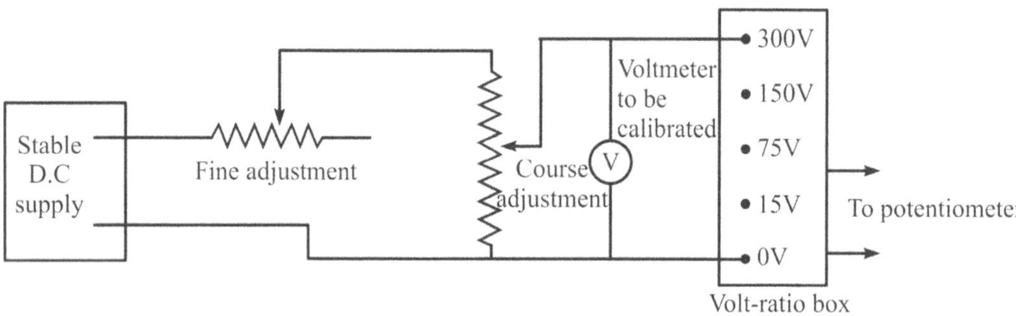

Fig. 5.13(b) Calibration of voltmeter

If these two readings are not matching, a positive or negative error is indicated. The percentage error can be calculated as,

$$\% \text{ error} = \frac{V_{ind} - V_{act}}{V_{ind}} \times 100$$

A calibration curve can be drawn by plotting percentage error against the indicated voltage.

5.10.2 Calibration of Ammeter

An ammeter to be calibrated is connected in series with a standard resistance R_s of suitable value (Fig. 5.14). The current supplied by a D.C supply passes through ammeter as well as standard resistance R_s using d.c potentiometer, a voltage across standard resistance can be measured. Thus, current flowing through R_s (and ammeter also) is given by

$$I = \frac{V_s}{R_s}$$

where, V_s = voltage across R_s measured using potentiometer.

R_s = resistance of standard resistor.

since the resistance of standard resistance is known accurately and also the voltage across R_s is measured using standardised potentiometer, the method of calibrating ammeter is very accurate. If the actual current 'I_{act}' and the current indicated by ammeter 'I_{ind}' are not matching, error is indicated. The percentage error is given by.

$$\%\ \text{error} = \frac{I_{ind} - I_{act}}{I_{ind}} \times 100$$

where I_{ind} = current indicated by ammeter and

$$I_{act} = I_s = \frac{V_s}{R_s}$$

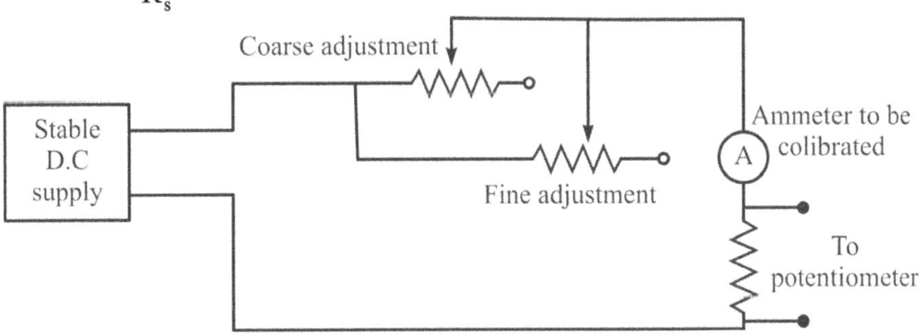

Fig. 5.14 Calibration of ammeter

The calibration curve of ammeter can be obtained by plotting % error against the reading of ammeter i.e., I_{ind}.

5.10.3 Measurement of Power

The circuit diagram for measurement of power is as shown in the Fig. 5.15. The power across load can be calculated as,

$$P = IV$$

In above circuit, voltage across standard resistance R_s is measured using potentiometer, let it be V_{Rs}. Then current flowing through it is given by

$$I = \frac{V_{Rs}}{R_s} \qquad \text{.....(5.8)}$$

For the measurement of voltage V across load, volt-ratio box is connected across load. The output of volt-ratio box is then connected to potentiometer. Let it be V. Then the voltage across load is given by

$$V = KV_R \qquad .....(5.9)$$

where K is the multiplying factor of volt-ratio box and V_R is actual reading of potentiometer when it is connected across volt-ratio box. Thus from eq. (5.8) and (5.9), the power is given by

$$P = VI = (KV_R)\left(\frac{V_{Rs}}{R_s}\right)$$

$$P = K\left(\frac{V_R V_{Rs}}{R_s}\right) \qquad .....(5.10)$$

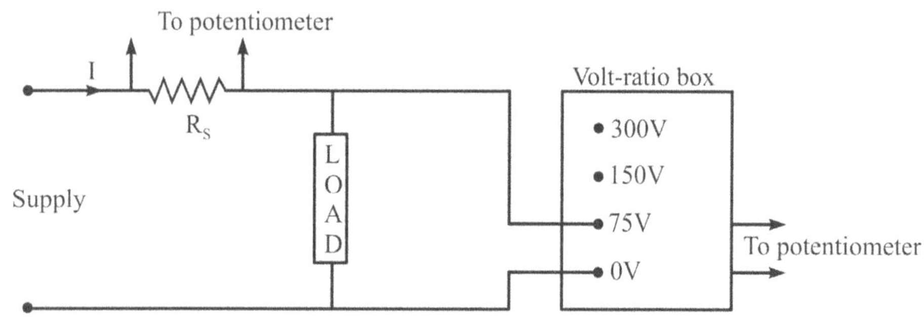

Fig. 5.15 Measurement of power

5.10.4 Calibration of Wattmeter

By using d.c potentiometer it is possible to calibrate wattmeter too. Practical set up is shown in the Fig. 5.16.

A low voltage supply supplies a current to current coil (CC) of watt meter. This current is adjusted by using a rheostat R_h in series with low voltage supply. The potential divider circuit is supplied by a high voltage supply. The voltage is stepped down by volt-ratio box and the tappings are adjusted accordingly. A voltmeter measures voltage V and ammeter measure current I which gives true power as,

$$W_{ind} = VI$$

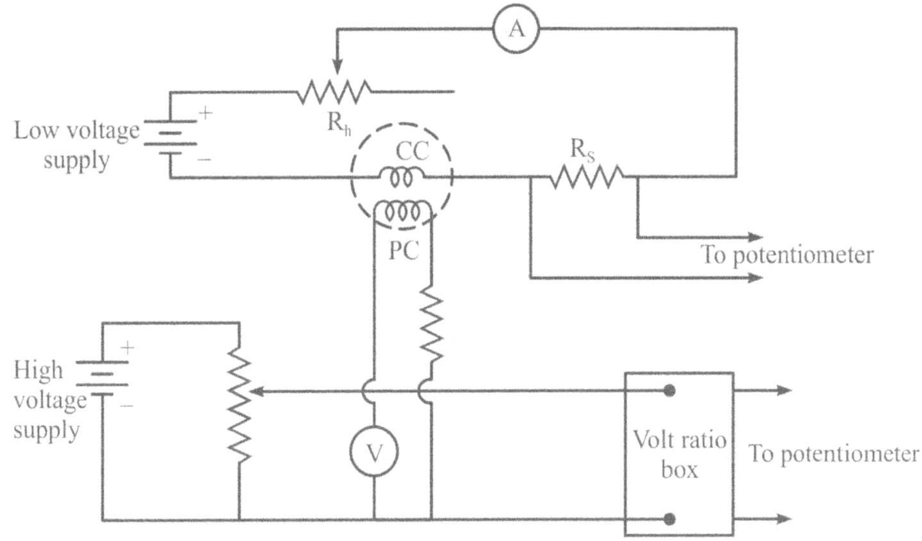

Fig. 5.16 Calibration of wattmeter

This value can be compared with a value indicated by watt meter. If two values are not matching, a positive or negative error is indicated which is given by

$$\% \text{ error} = \frac{W_{ind} - W_{act}}{W_{ind}} \times 100$$

5.10.5 Measurement of Resistance

The set up of measurement of resistance is as shown in Fig. 5.17.

A resistor whose resistance is to be measured is connected in series with a standard resistor of resistance R_s. The current through the circuit is supplied by a stable D.C supply and it is controlled by rheostat R_h. The current is adjusted such that the drop across each resistor is of the order of 1 V. Due to the current I, voltages are developed across R_s and R. Both are then measured by using a d.c potentiometer.

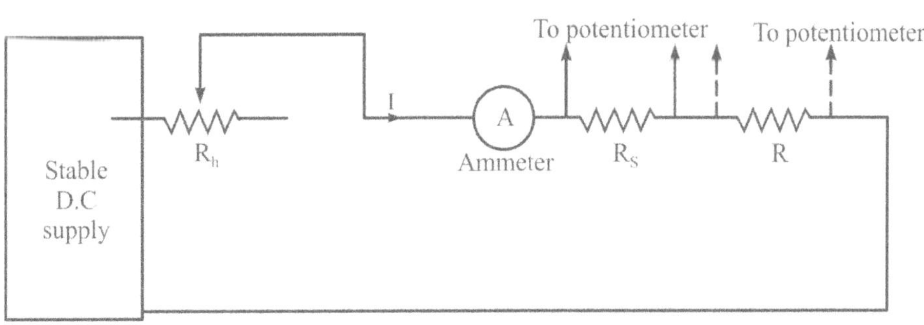

Fig. 5.17 Measurement of unknown resistance

Let the voltage across standard resistance be V_{Rs}. Then, we can write,

$$V_{Rs} = IR_s \qquad .....(5.11)$$

Let the voltage across unknown resistance be V_R. Then, we can write

$$V_R = IR \qquad .....(5.12)$$

Dividing eq. (5.11) by (5.12), $\dfrac{V_R}{V_{Rs}} = \dfrac{R}{R_s}$

Hence the unknown resistance is given by

$$R_s = R\left(\frac{V_R}{V_{Rs}}\right)$$

The basic requirement of above measurement method is that the current flowing through the circuit should remain same during measurement of voltages across R and R_s. This need can be fulfilled by using a stablilized D.C supply at the input.

5.11 A.C Potentiometer

The basic principle of operation of d.c potentiometer and an A.C potentiometer is exactly same, but in a D.C potentiometer the balance between voltage drop across the slide wire and two magnitude of unknown voltage is obtained, while in an a.c potentiometer, the two voltages should be balanced in magnitude as well as phase.

Broadly, A.C potentiometer are classified on the basis of methods of measurement of unknown voltages. There are two types of a.c potentiometers:

1. "Polar type A.C potentiometer" in which the magnitude and phase angle of unknown voltage are measured on different scales directly. The phase angle measured with respect to some reference phaser. As the voltage measured is represented in polar from as $V\underline{|\theta}$; the A.C potentiometer is called polar type A.C potentiometer.
2. "Co-ordinate type A.C potentiometer" in which the two components of an unknown voltage are measured on two different scales. One of the components measured is in phase component while remaining is "quadrature" component. Both the compounds are 90° out of phase with each other. If the in phase components and quadrate components are represented by V_A and V_B respectively then the magnitude and phase angle of an unknown voltage can be represented as given below.

$$V = \sqrt{V_A^2 + V_B^2} \text{ and}$$

$$\theta = \tan^{-1}\left(\frac{V_B}{V_A}\right)$$

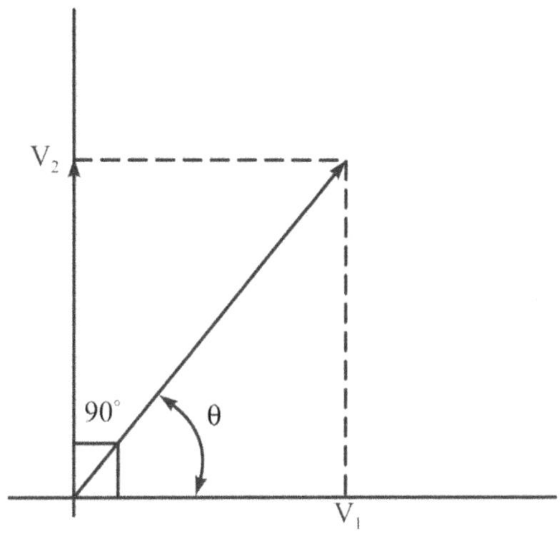

Fig. 5.18 Phasor diagram of polar type potentiometer

For the measurement of all angles up to 360°, the provision is made in such potentiometer to read positive as well negative values of V_A and V_B. As the two components of the unknown voltage represent rectangular form of voltage, the potentiometer is called co-ordinate type of A.C potentiometer.

5.11.1 The Basic Requirement of A.C Potentiometers

1. While comparing two voltages in A.C potentiometer, it is necessary to measure the potentiometer voltage accurately as A.C reference is not available in the circuit.
2. In case of A.C potentiometer, the basic requirement is that, at all the instants of time, both the voltages being compared must be equal with respect to magnitude and phase. Hence the current in the potentiometer circuit must have same phase and frequency as compared to the voltage being measured. Hence in A.C potentiometer, the current in the potentiometer circuit is derived from the voltage being measured.
3. In A.C potentiometer, another requirement in the control of stray e.m.f in fields and coupling circuit. As for as possible such e.m.fs must be eliminated as they may get added vectorically with the a.c voltage being measured.

4. The A.C source must supply exactly sinusoidal signal. It is necessary in A.C potentiometer as the detectors used in it are vibration galvanometers, being timed circuits. These galvanometers may respond harmonics present in one or both the voltages. Due to this, achieving balance point is a very difficult task.

5.12 Drysdale-Tinsely Polar Type A.C Potentiometer

Being a polar type, Drysdale-Tinsley a.c potentiometer measure, unknown e.m.f in terms of its magnitude and phase angle. It consists of basic d.c potentiometer along with some auxiliary components such as, Drysdale phase shifter and electrodynamometer type ammeter. Let us study Drysdale phase shifter construction first.

Drysdale phase shifter is also called "Phase shifting transformer". The drysdale phase shifting transformer consists of stator and rotor. The stator is made up of silicon steel and is provided with two (or) three phase windings. The rotor of the transformer is also provided with winding in slots. A small air gap is provided in between the stator and rotor. When the supply is given, the current flow through the stator windings and generate the flux in the air gap. Whenever the rotor winding cuts the flux, an e.m.f is induced in the rotor windings. The rotor position can be adjusted by adjusting the rotor angle and the phase displacement of induced e.m.f is adjusted by adjusting the rotor position with respect to null-pointer. In order to measure the deflected angle of the rotor, a scale and pointer is provided at the top of the instrument.

Instead of using single phase supply, the phase shifter can be operated by using a phase splitting device. This device consists of two windings displaced at an angle of 90°. Among the two windings, one is connected to the supply and the other is connected in series with a resistor and capacitor, which are adjusted till the currents in two windings become equal. Now, the induced e.m.f in rotor windings due to first stator winding is given by,

$$E_1 = KI \sin \omega t \cos \theta$$

It is so, because of non inductive wound coils in potentiometer, a switch is employed to enable the potentiometer either D.C side (or) A.C side.

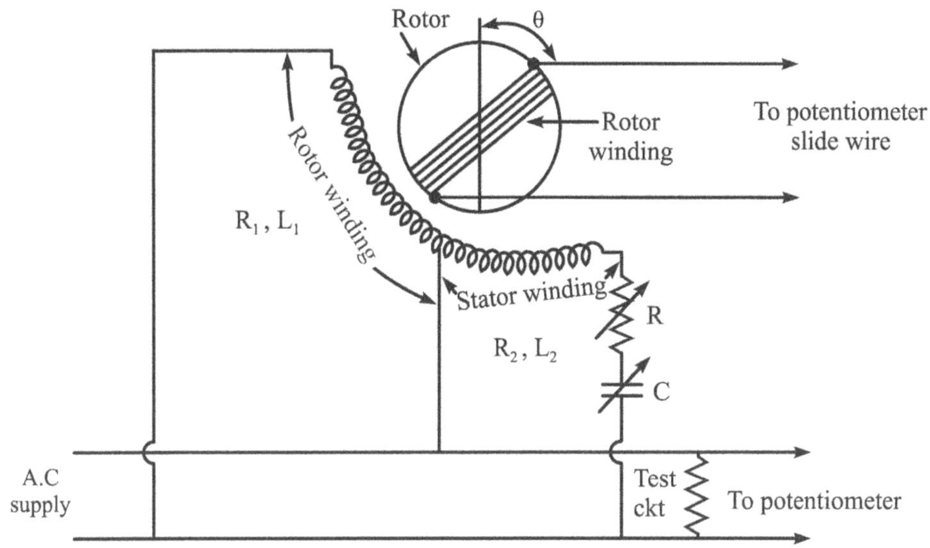

Fig. 5.19 Drysdale phase shifting transformer connection diagram

5.13 Standardisation of Drysdale-Tinsley A.C Potentiometer

In standardisation of a.c potentiometer, both D.C as well as a.c standardisations are done. The d.c standardisation is done first by replacing vibration galvanometer by D'Arsonal galvanometer. A standard cell such as Weston cell is used for d.c standardisation. Then by adjusting sliding contacts null deflection in galvanometer is achieved. The reading of a precision ammeter included in battery supply is noted. During a.c standardisation again vibration galvanometer is used. The ammeter is still included in the supply circuit but now this circuit is without standard cell. By properly adjusting resistance in the circuit, the r.m.s value of current in slide wire is made same as that of d.c current noted in d.c standardisation.

5.13.1 Measurement of Unknown e.m.f

The circuit diagram for measurement of unknown e.m.f using an A.C potentiometer is shown in the Fig. 5.20.

An e.m.f to be measured is connected across terminals A– A'. The sliding contacts P_1 and P_2 and the position of rotor in phase shifter are adjusted simultaneously till the balance is obtained as indicated by the null deflection of vibration galvanometer.

At balance, the magnitude of the unknown e.m.f is obtained from P_1 and P_2, and the phase angle is obtained from the scale reading which is mounted on the top of the instrument. Thus the unknown e.m.f can be expressed in polar form or $E\underline{|\theta^{\circ}}$.

Similarly, the e.m.f induced in rotor due to second stator winding is given by

$$E_2 = KI \sin(\omega t + 90) \cos(\theta + 90)$$

Therefore, the resultant e.m.f is given by,

$$E = E_1 + E_2$$

$$E = KI \cos\theta \sin\omega t + KI \sin(\omega t + 90) \cos(\theta + 90)$$

$$E = KI [\cos\theta \sin\omega t + \sin(\omega t + 90) \cos(\theta + 90)$$

But $\sin(\omega t + 90) = \cos\omega t$ and $\cos(\omega t + 90) = -\sin\theta$

$\therefore$ $E = KI [\sin\omega t \;\cos\theta + \cos\omega t\,(-\sin\theta)$

$$= KI [\sin\omega t \;\cos\theta - \cos\omega t \sin\theta$$

$$E = KI \sin[\omega t - \theta]$$

From the above expression it is clear that the rotor e.m.f has constant amplitude and the phase angle is given by the rotor deflection θ.

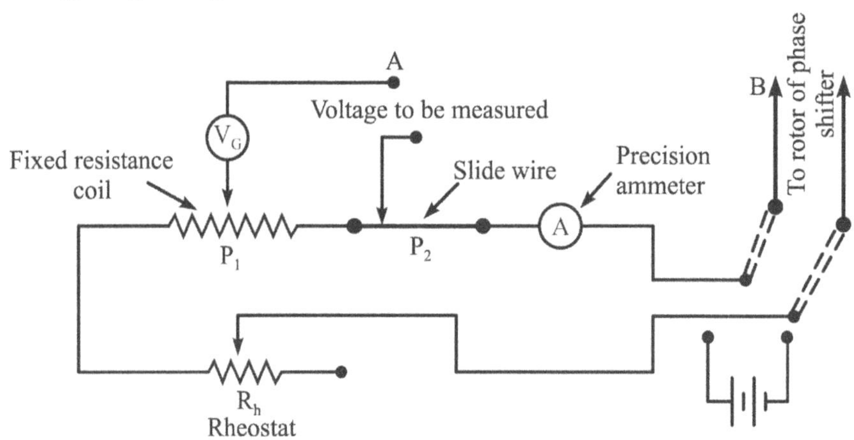

Fig. 5.20 Measurement of unknown e.m.f.

5.13.2 Transfer Instrument

The transfer instrument uses an electrodynamometer type ammeter to standarize an A.C potentiometer. The process of standardisation of a.c potentiometer can be obtained by switching the slide wire circuit on to D.C supply by using a standard cell (or) battery. The standard current is known. This standard current is measured by using electrodynamometer type ammter. When a transfer instrument is operated under A.C supply, the r.m.s value of the slide wire current is remained at the same value as on D.C supply.

5.14 Self Balancing Potentiometer

The self-balancing potentiometers are used in the field of instrumentation as an automatic device or meter. The given potentiometer measures the thermo-electric e.m.f induced in the thermocouple (so it indirectly measures the temperature different between its junctions).

Initially the meter is standardized by throwing the SPDT switch to calibrate position then, the switch is thrown to operate position and so. The potential different between the reference source voltage and the thermo electric e.m.f cause an unbalance in the circuit. In case of ordinary potentiometers, this unbalance cause a deflection in the galvanometer. But here, this potential difference is filtered and converted into A.C and it is amplified to feed the control winding of two phase A.C motor. The A.C converter consists of a reed excited from an A.C supply. As the supply is A.C, the reed starts vibrating and hence operates the SPDT switch connected to the transformer terminals. For each cycle of vibration, the current through the transformer reverses. In this way D.C is converted into A.C. An e.m.f is induced in the secondary winding of the transformer, which is amplified with an A.C amplifier and is given to the control winding of the two phase induction motor.

This amplified alternating voltage will be 90° out-of-phase with that of A.C supply connected to the motor. The direction of rotation of the motor shaft depends on whether this 90° (out-of-phase) is a leading or a lagging angle. This in turn depends upon the polarity of the unbalanced D.C e.m.f because if the polarity of amplifier input voltage reverses, the output voltage is shifted by 180°. Finally the motor drives the slider so that the circuit becomes balanced. At balance there will be no input supply to the transformer and hence no excitation is given to the control winding and the motor stops. In this way, the self balancing potentiometer automatically balance the circuit.

The circuit diagram of a self balancing potentiometer is shown in the Fig. 5.21.

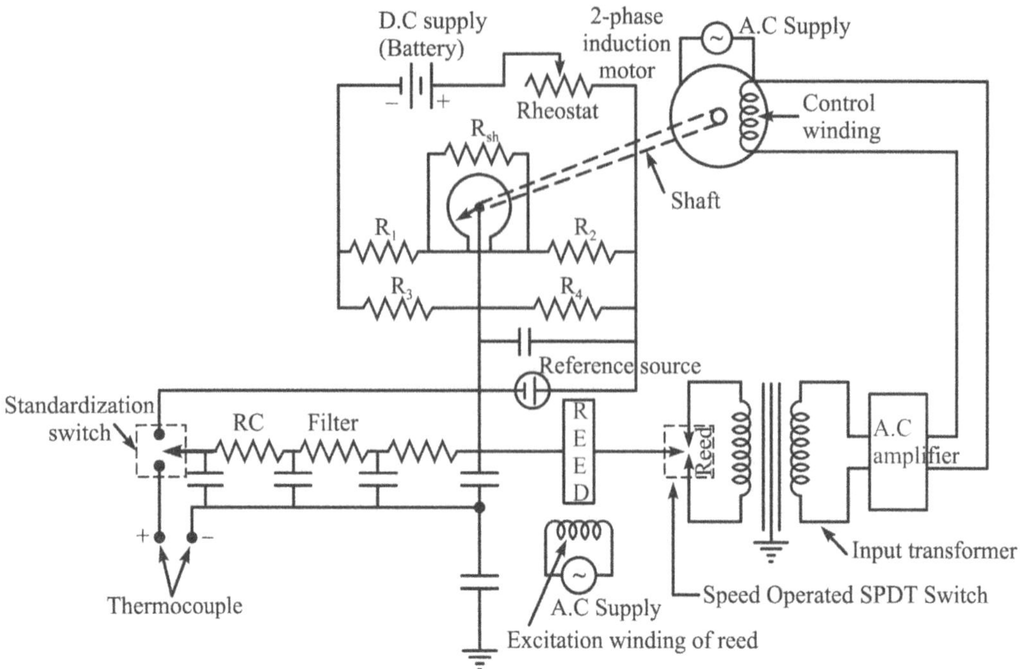

Fig. 5.21 Self balancing potentiometer

5.15 Gall-Tinsley Co-ordinate Type A.C Potentiometer

This potentiometers consists of two separate potentiometers enclosed in same common case. The circuit diagram of Gall-tinsley co-ordinate potentiometer is as shown in the Fig. 5.22.

The in-phase and quadrature potentiometers consists of sliding contact BB' and CC' respectively. The rheostats R and R' are also provided in the respective potentiometers for the adjustments of current. By using different arrangement, the supplies for the potentiometers are obtained.

A vibration galvanometer 'V_G' is tuned to the supply frequency and it is connected in series with a switch 'K' and electrodynamometer type ammeter.

S_1 and S_2 are the sign changing switches which are necessary for reversing direction of unknown e.m.f. The unknown voltage is introduced using selector switch 'S_3' which is having 4 pair of terminals.

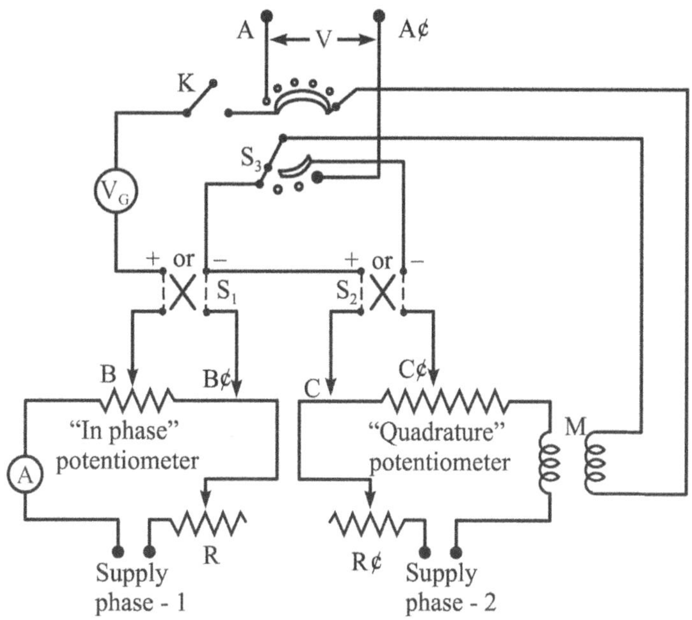

Fig. 5.22 Connections of Gall-Tinsley potentiometers

Phase Spilling Circuit: Consider Fig. 5.23, the transformer T_1 and T_2 are step down transformers and they supply about 6 to 8 volts to potentiometer. By using variable resistor and capacitor the supply for T_2 is obtained. By adjusting 'R' and 'C' qudrature is obtained.

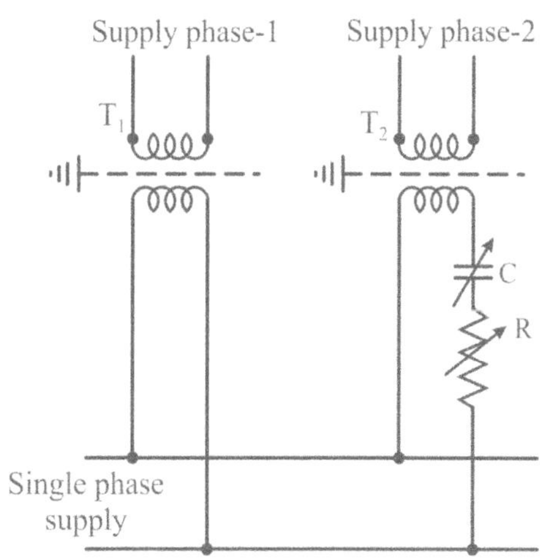

Fig. 5.23 Phase splitting circuit

Standardisation of Gall-Tinsley A.C potentiometers

First of all d.c standardisation of potentiometer is done by using standard cell and D Arsonal type galvanometer. Then without disturbing this setting, a.c standardisation is done by adjusting slide wire current to give zero deflection. Then previous galvanometer is replaced by vibration galvanometer and also direct current supply is replaced by a.c supply. Then the rheostat is adjusted till the current in the quadrature potentiometer wire is same as that in the in-phase potentiometer magnitude wise. These two currents must be exactly in quadrature.

5.15.1 Measurement of Unknown e.m.f

The e.m.f to be measured is connected across the terminals $A - A'$ using selector switch S_3 (Fig. 5.22). The sliding contacts of both the potentiometers are adjusted till the null deflection is obtained in the vibration galvanometer.

Under the balance condition, the in-phase components of the unknown e.m.f is obtained from in-phase potentiometer while the quadrature component of the unknown em.f is obtained from quadrature potentiometer. If needed the polarity of the test voltage may be reversed by using sign changing switches S_1 and S_2 to balance the potentiometers.

If V_A and V_B are the two potentiometer readings, the magnitude and phase angle of unknown e.m.f are given by

$$V = \sqrt{V_A^2 + V_B^2} \text{ and } \theta = \tan^{-1}\left(\frac{V_B}{V_A}\right)$$

Thus the e.m.f can be expressed in complex form as

$$V = \pm V_A \pm jV_B$$

5.15.2 Measurement of Unknown Inductance using A.C Potentiometers

The circuit diagram for the measurement of unknown inductance of a inductor (or coil) is shown in the Fig. 5.24.

In the figure,

R_s = known standard, non-inductive resistance

R_c = internal resistance of the coil

L = inductance of the coil

The circuit is supplied from a single phase a.c supply of known frequency.

In this method, two voltage measurements are carried out by switching the 'DPDT' switch to both the positions.

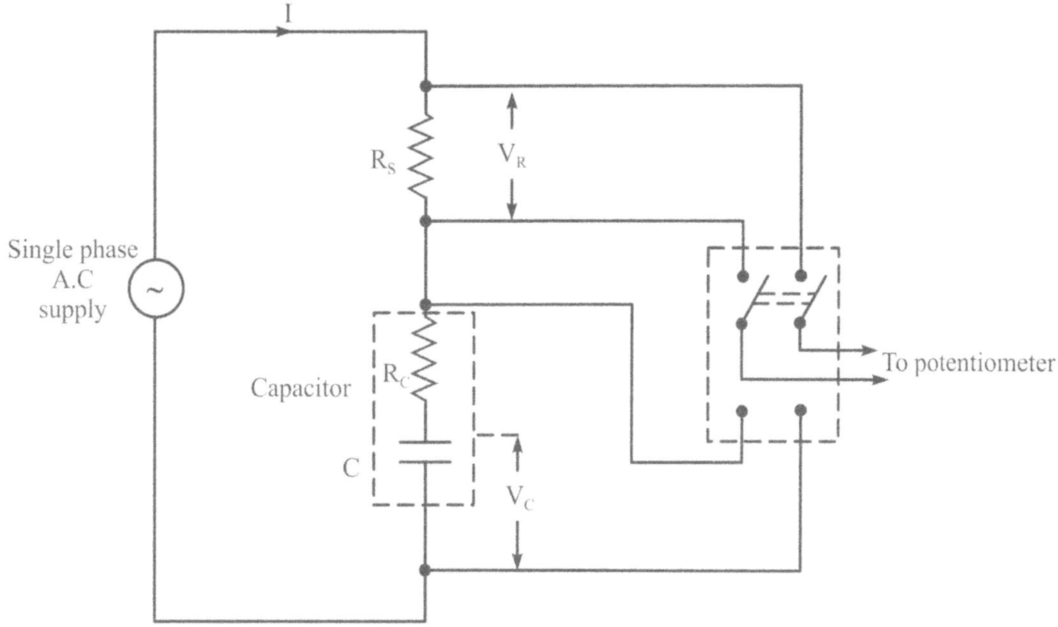

Fig. 5.24

Let the readings be

(i) Voltage across R_s $\qquad \overline{V}_R = V_R \angle \theta_R$

(ii) Voltage across the coil $\quad \overline{V}_C = V_C \angle \theta_C$

∴ the current through the circuit

$$\overline{I} = \frac{\overline{V}_R}{R_s} = \frac{V_R}{R_s} \angle \theta_R$$

$$Z_C = \frac{V_C}{I} = \frac{V_C \angle \theta_C}{\frac{V_R}{R_s} \angle \theta_R} = \frac{R_s V_C}{V_R} \angle \theta_C - \theta_R$$

∴ reactance of the coil

$$R_C = Z_C \cos(\theta_C - \theta_R) = \frac{R_S V_C}{V_R} \cos(\theta_c - \theta_R)$$

∴ reactance of the coil

$$X_C = Z_C \sin(\theta_C - \theta_R) = \frac{R_S V_C}{V_R} \sin(\theta_C - \theta_R)$$

$$\therefore \quad L_C = \frac{X_C}{2\pi f} = \frac{R_S V_C}{2\pi f V_R} \sin(\theta_C - \theta_R)$$

Hence, the value of unknown inductance can be obtained.

5.15.3 Measurement of Unknown Capacitance Using A.C Potentiometer

The circuit diagram for the measurement of unknown capacitance of a capacitor is shown in Fig. 5.25.

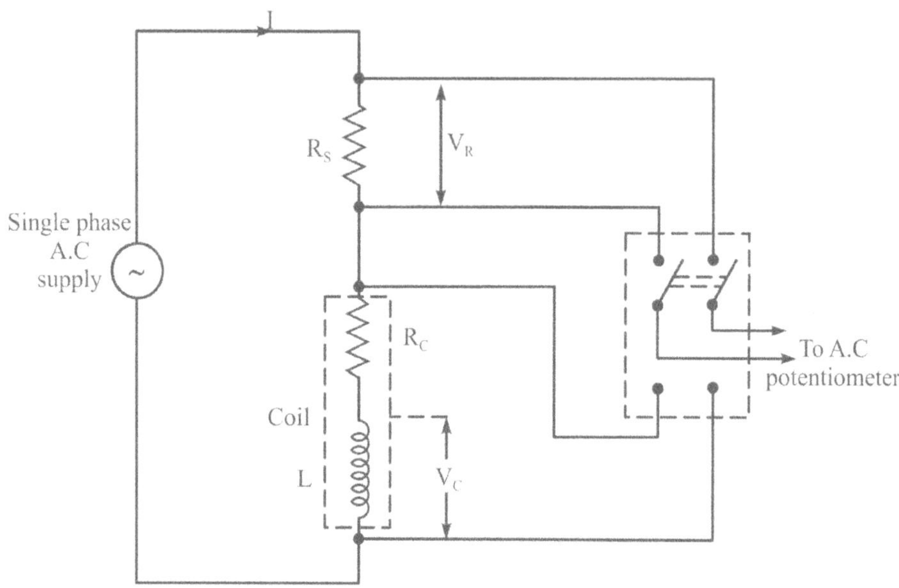

Fig. 5.25

where, R_s = known standard non-inductive resistance

R_c = series resistor representing the loss in the capacitor

C = Capacitance of the capacitor

The circuit is supplied from a single phase a.c supply of known frequency.

In this method, two voltage measurements are carried out by switching the DPDT switch to both the positions. Let the readings be

Voltage across $R_s = \overline{V_R} = V_R \angle \theta_R$

Voltage across the capacitor $\overline{V}_C = V_C\angle\theta_C$

$\therefore$ the current through circuit $I = \frac{V_R}{R_s} = \frac{V_R}{R_s}\angle\theta_R$

$\therefore$ impedance of the capacitor

$$Z_C = \frac{V_C}{I} = \frac{V_C\angle\theta_C}{\frac{V_R}{R_S}\angle\theta_R} = \frac{R_S V_C}{V_R}\angle\theta_C - \theta_R$$

$\therefore$ resistor of capacitor

$$R_C = Z_C\cos(\theta_C - \theta_R) = \frac{R_s V_C}{V_R}\cos(\theta_C - \theta_R)$$

$\therefore$ reactance of the capacitor

$$X_C = Z_C \sin(\theta_C - \theta_R)$$

$$= \frac{R_s V_C}{V_R}\sin(\theta_C - \theta_R)$$

$\therefore$

$$C = \frac{1}{2\pi f X_c} = \frac{V_R}{2\pi f R_s V_C}\sin(\theta_C - \theta_K)$$

Hence,

the value of unknown capacitance can be obtained.

5.16 The Application of Potentiometers

The applications of D.C potentiometer are:

1. Measurement of voltage
2. Measurement of resistance
3. Measurement of power
4. Calibration of volt meters
5. Calibration of ammeters
6. Calibration of watt meters

The application of A.C potentiometers are:

1. Calibration of volt meters

2. Calibration of ammeters
3. Calibration of watt meters
4. Calibration of energy meters etc.
5. Testing of magnetic material (i.e., A.C magnetic testing) at low flux densities
6. Measurement of voltage
7. Measurement of self reactance or inductance of a coil

5.17 Errors in a Potentiometers

The errors developed in a potentiometer are due to the following reasons.

1. Inaccurate measurement of frequency will produce error.
2. As we know that different parts of the instrument will have mutual inductance. This will effect the nominal value of mutual inductance, which in turn produces errors.
3. Due to the presence of harmonics in the supply, error are introduced.
4. Even a very minute difference in A.C and D.C reading of the reflecting dynamometer instrument may introduce errors.
5. Inter-capacitance, earth capacitance and mutual inductance of the slide wire coils will have effect on potential gradient, which may produce an error.

5.18 Apply a Correction for Thermo-emf in D.C Potentiometer Measurement

In order to ensure the accuracy of the measurement, care must be taken to eliminate the thermo-emf produced in the potentiometer. The following things can be done to reduce the effect of thermo-emf in the D.C potentiometer.

1. Manganin should be used for all the resistors in the potentiometer as it has a low temperature coefficient.
2. Casing should be used for all the part of potentiometer such that, same temperature in maintained across them
3. In order to further reduce the effect, copper terminals should be used
4. Also the slide wire should be made of platinum silver alloy and copper should be used for sliding contact so as to reduce the effect of thermo-emf.

5.19 Comparison between A.C and D.C Potentiometers

The Main comparison between A.C Potentiometer and a D.C Potentiometers are:

	D.C Potentiometer		A.C Potentiometer
1.	In case of D.C potentiometer, only magnitudes of the two voltages must be same to obtain the balance	1.	In case of A.C potentiometer, both magnitude and phase of the two voltages should be same to obtain the balance
2.	D.C potentiometers are more accurate	2.	A.C potentiometers are less accurate
3.	D.C potentiometers are less complex	3.	A.C potentiometers are more complex
4.	Galvanometer is used as null detector	4.	Vibrational galvanometer is used as null detector
5.	Reference source is present in D.C potentiometer	5.	There is no references source present in A.C potentiometer
6.	Balance point can be obtained easily	6.	Obtaining balance point is very difficult

5.20 The Modifications are needed in a D.C Potentiometer to be used for A.C Application are

In D.C potentiometers, the unknown D.C e.m.f is measured by taking a D.C voltage reference and the only condition for the balance is the magnitude of unknown e.m.f should be same as that of the drop across the part of slide wire. Whereas, in case of A.C measurement, the conditions that must be satisfied are:

1. both the voltage should have same frequency
2. their waveforms must be same
3. also, their magnitude must be same at all the instants.

By taking a separate source as reference (as in case of D.C potentiometers), the balance can never be achieved as we cannot estimate the above mentioned quantities of unknown e.m.f.

Hence, D.C potentiometer cannot be used for A.C measurement directly.

Precautions that are needed in a D.C potentiometer to be used for A.C applications

1. The basic D' Arsonval galvanometer is replaced with a vibration galvanometer as it is a tunable device. It can even compare the harmonics present in the wave forms.
2. Since, the magnitude, phase, frequency and waveform must be same, instead of some external source, the unknown e.m.f itself is taken as reference.
3. As the stray magnetic fields are likely to affect the measurement due to weak fields, they are to be eliminated or compensated.
4. The voltage across the potentiometer should be known with high accuracy. In case of D.C either a standard cell or zener reference are considered which is not the case with A.C.

5. All the resistors and the slide wire are to be replaced by a standard non-inductive resistors so that their impedance does not vary with frequency and wave form.

5.21 Applications of A.C Potentiometers

Similar to the d.c potentiometers, a.c potentiometers can be used for the calibration of voltmeter, ammeter, also they are used for testing energy meters and watt meters. Measurement of self reactance of a coil is also possible with the help of a.c potentiometers. Following are the few application of a.c potentiometers.

Voltmeter calibration: Low voltages up to 1.5 V can be measured directly. Higher voltages can be measured by using a volt-box (for medium voltages) or two capacitors in series (for high voltages) in conjunction with the potentiometers.

Ammeter calibration: The measurements of various alternating currents required: for such calibration may be made by the use of non-inductive standard resistors with the potentiometers. The method being similar to that adopted when the calibration is to be carried out with direct current.

Watt-meter and Energy-meter testing: The testing circuit for watt meters and energy meters is the same as that used in the case of d.c measurements. A phase sifting transformer is included in the potential circuit to vary the phase of voltage with respect to current so that the watt meters and energy meter are tested at various power factors.

Measurement of Self reactance coil: The practical set up for measurement of self reactance of a coil is as shown in the Fig. 5.26.

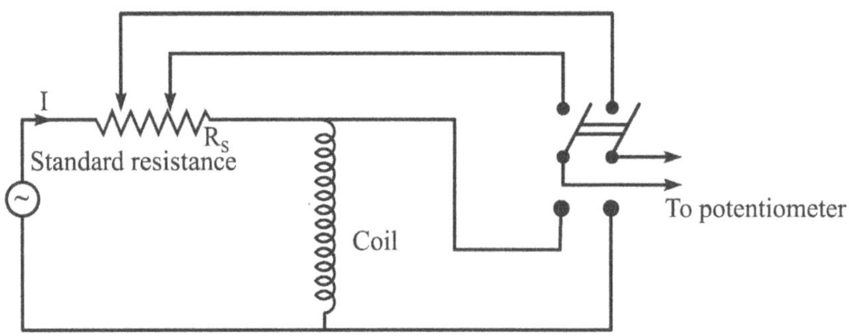

Fig. 5.26 Measurement of self reactance of coil

A coil whose self reactance is to be measured is connected in series with a standard resistance R_s. The current I in the circuit is supplied by a.c voltage source. The voltage drop across R_s is first measures using A.C potentiometer.

Let voltage across R_s be V_1. Similarly let the voltage measured across coil be V_c. For these measurements polar a.c potentiometer is used. Then we can write

Voltage across coil = $V_c = V_c \angle \theta_C$ in polar form

Voltage across $R_s = V_s = V_s \angle \theta_s$

The current through coil can be calculated as

$$i = \frac{V_s}{R_s} = \frac{V_s \angle \theta_s}{R_s}$$

The impedance of coil can be calculated as

$$Z = \frac{V_C}{i} = \frac{V_s \angle \theta_C}{\left[\frac{V_s}{R_s} \angle \theta_s\right]} = \frac{R_s V_C}{V_S} \angle \theta_C - \theta_s$$

We can write this impedance Z in rectangular form in real part and imaginary part as resistance and reactance.

The resistive part of impedance is given by

$$R = Z \cos(\theta_c - \theta_s) = \frac{R_s V_C}{V_s} \cos(\theta_c - \theta_s)$$

The reactive part of impedance is given by

$$X = Z \sin(\theta_c - \theta_s) = \frac{R_s V_C}{V_s} \sin(\theta_c - \theta_s) \qquad .....(5.13)$$

The eq. (5.13) represents reactance of the coil.

Problems with Solutions

1. If the slide wire has divisions marked in mm and each division can interpolated to one fifth, calculate the resolution of the instrument.

***Solution*:**

The maximum voltage that can be measured by the potentiometer is V_{max}

i.e., $V_{max} = IR_s$

$\therefore$ Voltage drop per mm $= \frac{V_{max}}{L} = \frac{IR_s}{L}$

$$= \frac{V_s}{L} \qquad \left[\because I = \frac{V_s}{R_s} \times \frac{L}{L}\right]$$

Since, the meter can read up to $1/5^{th}$ of a mm, the resolution of the meter is,

$$= \frac{V_s}{L} \times \frac{1}{5} = \frac{V_s}{5L}$$

2. A single range potentiometers has a 18 step dial switch where each step represents 0.1 V, the dial resistance are 10 Ω. The slide wire of the potentiometers is circular and has 11 turns and a resistance of 11 Ω each. The slide wire has 100 divisions and interpolation can be done to one fourth of a division. The working battery has a voltage of 6 volts and negligible internal resistance.

Calculate

(i) The measuring range of potentiometer

(ii) The resolution

(iii) Working current and

(iv) Setting of rheostat

***Solution*:**

Given that

Single range potentiometer

Battery voltage, V = 6 V

Number of steps n = 18

Voltage step, V = 0.1 V

Resistance of each dial, $R_d = 10\ \Omega$

Resistance of slide wire, Rs = 11 Ω

(i) Measuring range of potentiometer

Total resistance of the measuring circuits

$$R_m = \text{Resistance of dial} + \text{Resistance of slide wire}$$

$$= (n \times R_d) + R_s$$

$$= (18 \times 10) + 11$$

$$= 191\ \Omega$$

Range of potentiometers

$$= \frac{191 \times 0.1}{10}$$

$$= 1.91\ \text{V}$$ **Ans.**

(ii) Resolution

Voltage drop across the slide wire

$$= \frac{11 \times 0.1}{10} = 0.11 \text{ V}$$

Voltage drop across each turn

$$= \frac{0.11}{\text{Number of turns}}$$

$$= \frac{0.11}{11}$$

$$= 0.01 \text{ V}$$

Since, each turn is divided into 100 divisions, each division represents a voltage drop of

$$\frac{0.01}{100} = 0.0001 \text{ V}$$

Since, the interpolation is done for one fourth division

$$\therefore \text{ Resolution} = \frac{1}{4} \times 0.001$$

$$= 25 \ \mu\text{V}$$

(iii) Working current

The voltage drop across the circuit of the potentiometer is 1.91 V with a resistance of 191 V

$$\therefore \qquad I_m = \frac{V}{R_m}$$

$$= \frac{1.91}{191} = 0.01$$

$$= 10 \text{ mA}$$

(iv) Setting of Rheostat

Total resistance of the battery circuit,

$$R_T = \frac{V}{I_m} = \frac{6}{10 \times 10^{-3}}$$

$$= 600 \ \Omega$$

Resistance of potentiometer working circuit is 191 Ω

∴ setting of series rheostat

$$= 600 - 191$$

$$= 409\ \Omega$$

3. A basic slide wire potentiometer has a working battery voltage of 3 volts with negligible internal resistance. The resistance of slide wire is 400 Ω and its length is 200 cm. A 200 cm scale is placed along the slide wire. The slide wire has 1 mm scale divisions and it is possible to read up to of a division.

The instrument is standardized with 1.018 V standard cell with sliding contact at the 101.8 cm mark on scale. Calculate

(i) Working current

(ii) The measurement range

(iii) The resistance of series rheostat and

(iv) The resolution of the instrument

***Solution*:**

Assume the scale division to be 1/5

Given data

Battery voltage, V = 3 V

Resistance of slide wire R_s = 400 W

Length of slide wire L = 200 cm

Standard cell voltage, V_s = 1.018 V

Since, the instrument is standardized with sliding contact position at 101.8 cm mark on scale,

Resistance for the length of 101.8 cm of slide wire

$$= \frac{101.8}{L} \times R_s$$

$$= \frac{101.8}{200} \times 400$$

$$= 203.6\ \Omega$$

(i) Working current

$$I = \frac{\text{Voltage drop across 101.8 cm length}}{\text{Resistance of 101.8 cm length}}$$

$$= \frac{1.018}{203.6}$$

$$= 5 \times 10^{-3}$$

$$= 5 \text{ mA}$$

(ii) Resistance of series Rheostat

$$R_h = \text{Total resistance } (R_T) - \text{Resistance of slide wire } (R_s)$$

Total Resistance,

$$R_T = \frac{V_T}{L} = \frac{3}{5 \times 10^{-3}} = 600\ \Omega$$

$$\therefore \quad R_h = 600 - 400\ \Omega$$

$$= 200\ \Omega$$

(iii) Measurement range

$$V = IR_s$$

$$= 5 \times 10^{-3} \times 400$$

$$= 2 \text{ V}$$

(iv) Resolution of the instrument

The slide wire length of 200 cm represents 2 V.

$\therefore$ Voltage drop per 1 mm of length

$$= \frac{2}{200 \times 10} = 1 \text{ mV}$$

$$\therefore \text{resolution} = 1 \times 10^{-3} \times \frac{1}{5} = 0.2 \text{ mV}$$

4. Measurement for the determination of the impedance coil are made on a coordinate type of potentiometer. The results are: voltage across 1 Ω standard resistance in series with the coil +0.92 V on in phase dial and – 0.340 V on quadrate dial, voltage across 10:1 potential divides connected to the terminals of the coil, + 1.35 V on inphase dial and +1.28 V on quadrate dial. Calculate the resistance and reactance of the coil.

***Solution*:**

Given data

Ratio of potential divider = 10:1

Standard resistance, R = 1 Ω

Voltage across standard resistance,

on in-phase dial = +0.952 V

on quadrature dial = – 0.340 V

Voltage across 10:1 potential divider

on in-phase dial = +1.35V

on quadrature dial = + 1.28 V

Resistance of the coil R_C = ?

Reactance of the coil X_C = ?

The current through the coil is given as

$$I_C = \frac{(+\,0.952 - j\,0.340)}{1} = \frac{1.01\angle -19.65}{1}\,A$$

$$= 1.01\angle -19.65\ A$$

Voltage across the coil is

$$V_C = 10\,(1.35 + j\,1.28)$$

$$= (13.5 + j\,12.8)$$

$$= 18.6\ \angle\ 43.47\ V$$

Impedance of the coil

$$Z_C = \frac{V_C}{I_C}$$

$$= \frac{18.6\angle 43.47}{1.01\angle -19.65}$$

$$= 18.41\ \angle\ 63.12$$

$$Z_C = 8.32 + j\,16.42$$

The impedance of the coil is given by

$$Z_C = R_C + j\,X_C$$

∴ Resistance of the coil, $R_C = 8.32\ \Omega$

Reactance of the coil, $X_C = 16.42\ \Omega$

5. Calculate the inductance of a coil from the following measurements of an AC potentiometer.

Voltage drop across a 0.1 Ω standard resistor connected in series with the coil is 0.613 $\angle 12^\circ\ 6'$ volts.

Voltage across the test coil through a 100/1 volt ratio box is 0.781 $\angle 50^\circ\ 48'$ volts, frequency is 50 Hz.

Solution:

Given data

Standard resistance, $R = 0.1\ \Omega$

Voltage across standard resistor,

$$V_1 = 0.613\ \angle 12^\circ\ 6'\ V$$

Voltage across the test coil through a 100/1 volt ratio box,

$$V' = 0.781\ \angle 50^\circ\ 48'$$

Frequency, $f = 50$ Hz

Inductance of coil $L = ?$

Current through standard resistor is given by

$$I = \frac{V_1}{R}$$

$$= \frac{0.613\ \angle 12^0\ 6'}{0.1} = 6.13\ \angle 12^\circ\ 6'\ A$$

The coil is connected in series with standard resistor

∴ current passing through coil = current passing through standard resistor

$$= 6.13\ \angle 12^\circ\ 6'\ A.$$

Voltage drop across the coil,

$$V = 100 \times V'$$

$$= 100 \times 0.781\ \angle 50^\circ\ 48'$$

$$= 78.1\ \angle 50^\circ\ 48'\ V$$

Impedance of the coil,

$$Z = \frac{V}{I}$$

$$= \frac{78.1 \angle 50^\circ\ 48'}{6.13 \angle 12^\circ 6'}$$

$$= 12.74 \angle 38^\circ\ 42'\ \Omega$$

Reactance of the coil in terms of impedance is given by

$$X_L = Z \sin \phi$$

$$= 12.74 \sin 38^\circ\ 42'$$

$$= 12.74 \times 0.625$$

$$= 7.965\ \Omega$$

Inductance of the coil is,

$$L = \frac{X_L}{2\pi f}$$

$$= \frac{7.965}{2 \times \pi \times 50}$$

$$= 25.35 \text{ mH}$$

$\therefore$ inductance of the coil, $L = 25.35$ mH

6. In the measurement of power by a polar potentiometer, the following readings were obtained:

Voltage across a 0.2 Ω standard resistance in series with the load = $1.46 \angle 32^\circ$ V, voltage across a 200:1 potential divider across the line = $1.37 \angle 56^\circ$ V.

Estimate the current, voltage power and power factor of the load.

***Solution*:**

Standard resistance, $R_s = 0.2\ \Omega$

Voltage drop across R_s, $V_s = 1.46 \angle 32^\circ$ V

Voltage across 200:1 potential divider across line, $V_p = 1.37 \angle 56^\circ$ V

Current $I_L = ?$

Voltage $V_L = ?$

Power consumption $P_L = ?$

Load power factor $\cos \phi = ?$

Since, the standard resistance R_s and load are connected in series, current through load I_L is same as the current through 'R_s'.

$$\therefore \qquad I_L = \frac{V_s}{R_s} = \frac{1.46 \angle 32^\circ}{0.2} = 7.3 \angle 32^\circ \text{ A}$$

∴ magnitude of load current

$$|I_L| = 7.3 \text{ A}$$

Voltage across the load,

$$V_L = \frac{200}{1} \times 1.37 \angle 56^\circ$$

$$= 274 \angle 56^\circ \text{ V}$$

Magnitude of load voltage

$$|V_L| = 274 \text{ V}$$

The angle between voltage and current of the load,

$$\phi = 56^\circ - 32^\circ = 24^\circ$$

∴ power factor,

$$\cos\phi = \cos 24^\circ$$

$$= 0.9135 \text{ (lag)}$$

∴ power consumed by the load,

$$P_L = V_L I_L \cos\phi$$

$$= 274 \times 7.3 \times 0.9135$$

$$= 1827.18 \text{ W}$$

7. A co-ordinate type potentiometer is used for determination of impedance of a coil and the result obtained are:

Voltage across a 1.0 Ω resistor in series with the coil + 0.238 V on in-phase dial and – 0.085 V on quadrate dial voltage across 10 : 1 potential divider used with the coil + 0.337 V, on in-phase dial and + 0.232 V on quadrate dial.

Calculate the resistance and reactance of the oil.

***Solution*:**

Current through the coil $I = \dfrac{(+\,0.238 - j\,0.085)}{1.0} = 0.238 - j\,0.085 \text{ A}$

Voltage across the coil $V = 10\,(0.3375 + j\,0.232) = 3.375 + j\,2.32 \text{ V}$

∴ inductance of coil $z = \dfrac{V}{I} = \dfrac{3.375 + j\,2.32}{0.238 - j\,0.085} = 9.49 + j\,13.13\ \Omega$

Resistance of coil R = 9.49 Ω

Reactance of coil x = 13.13 Ω

Review Questions

1. Explain the principle of basic potentiometer with a neat diagram
2. Explain the operation of slide wire D.C potentiometer
3. What is standardisation of potentiometer? Why it is necessary?
4. With neat circuit diagram explain crompton's D.C. potentiometer.
5. Explain operation of Duo-range D.C. potentiometer with neat sketch also explain its advantages.
6. Explain about simple potentiometer with independent calibrating circuit.
7. Write short notes on

   1. Vernier potentiometer 2. Brook's deflectional potentiometer
8. Explain the operation of

   1. Potentiometer with true zero, 2. Volt-Ratio box
9. Write the application of D.C potentiometer
10. Describe the steps when D.C crompton potentiometer is used to measure as unknown resistance
11. What is A.C potentiometer? Give classification of a.c potentiometer
12. What are the basic requirement of A.C potentiometer?
13. With neat sketch explain Drysdale-Tinsely polar type AC potentiometer.
14. Explain how Drysdale-Tinsely A.C potentiometer is standardised.
15. How is the unknown e.m.f measured. Using Drysdale-Tinsley A.C potentiometer?
16. Write short notes on co-ordinate type A.C potentiometer.
17. How is the measure of unknown inductance using A.C potentiometers?
18. What are the applications of potentiometers?
19. Explain errors in a potentiometers.
20. Explain true zero in potentiometer.
21. Explain difference between A.C and D.C potentiometers.
22. What are the modifications needed for a D.C potentiometer to be used for A.C. applications?
23. Explain A.C potentiometers applications.

Quiz Questions

1. Which will not normally cause a large error in an ohmmeter reading?
   (a) a slight change in switchable internal resistance
   (b) a small voltage between points under test
   (c) a small change in the resistance to be measured
   (d) a slight error in the range switch position
2. Power frequency measurements the frequency meter used in
   (a) mechanical resonance type (b) weston frequency meter
   (c) ratio meter type (d) electrodynanometer type
3. A potentiometer is basically as:
   (a) deflection type instrument
   (b) null type instrument
   (c) deflection as well as null hyper instrument
   (d) a digital instrument
4. Standardization of potentiometer is done in order that, they become
   (a) precise (b) accurate
   (c) accurate and precise (d) accurate and direct reading
5. Brooks deflectional potentiometer is used when the unknown voltage:
   (a) is constant (b) is varying at a slow rate
   (c) is varying very rapidly (d) none
6. The standardization of a.c potentiometer is done by
   (a) using D.C standard sources and transfer instruments
   (b) directly using A.C standard voltage source
   (c) using A.C standard source and transfer instruments
   (d) using d.c standard source and d'Arsonval galvanometer
7. The e.m.f of Weston standard cell is measured using
   (a) M-I meter (b) M-C meter
   (c) digital voltmeter (d) potentiometer methods
8. Arnold method of testing is
   (a) comparison method of testing PT's
   (b) absolute method of testing PT's
   (c) comparison method of testing CT's
   (d) absolute methods of testing CT's

Questions and Answers

1. What is the basic principle used in potentiometer.

Ans. In potentiometer the unknown emf is measured by comparing it with a standard known emf.

2. Name the potentiometer material used.

Ans. German silver

Manganin wire

3. Define standardization.

Ans. It is the process by which adjusting the current flows through the potentiometer coil to make the voltage across the standard cell is equal.

4. State the applications of potentiometer.

Ans. Used for measurement of unknown emf

Used for ammeter calibration

Used for Voltmeter calibration

Used for wattmeter calibration

5. State the advantages of crompton potentiometer.

Ans. More accurate

Easy to adjust

6. What are the practical difficulties in ac potentiometers.

Ans. More complicated

Accuracy is seriously affected

Difficulty is experienced in standardization.

7. Classify ac potentiometers.

Ans. Polar potentiometer

Coordinate potentiometer.

8. How the phase angle is measured in polar type potentiometers.

Ans. It is measured from the position of phase shifter.

9. Name some ac potentiometers.

Ans. Drysdale Tinsley potentiometer

Gall Tinsley potentiometer

10. State the advantages of ac potentiometers.

Ans. Can be used for measurement of both magnitude and phase angle

Can be used for measurement of inductance of the coil.

It is used in measurement of errors in CTs

11. State the applications of ac potentiometers.

Ans. Measurement of self inductance.

Ammeter calibration

Voltmeter calibration

Wattmeter calibration.

12. State the advantages of instrument transformers.

Ans. Used for extension of range

Power loss is minimum

High voltage and currents can be measured.

13. State the disadvantage of instrument transformers.

Ans. Cannot be used for dc measurements.

14. What are the constructional parts of current transformer?

Ans. Primary winding

Secondary winding

Magnetic core.

15. Name the errors caused in current transformer.

Ans. Ratio error

Phase angle error

16. Define ratio error.

Ans. The ratio of energy component current and secondary current is known as the ratio error.

17. How the phase angle error is created.

Ans. It is mainly due to magnetizing component of excitation current.

18. State the use of potential transformer.

Ans. Used for measurement of high voltage

Used for energizing relays and protective circuits.

19. Name the errors caused in potential transformer.

Ans. Ratio error

Phase angle error.

20. How the CT and PT are connected in the circuits.

Ans. CT is connected in series and PT is connected in parallel.

21. The e.m.f of western standard cell is measured using

Ans. Potentiometer method

22. The e.m.f of western standard cell is measured using

Ans. Potentiometer method

23. A.P.T is a device which is

Ans. Electromagnetically coupled

24. Two helical springs are used in a D. Arsonval meter movement because

Ans. It compensates for temperature changes

Quiz Questions Key

1. (c)	2. (a)	3. (c)	4. (d)	5. (b)
6. (d)	7. (a)	8. (d)		

CHAPTER 6

Resistance Measurements

6.1 Introduction

The resistance measurements is an important as the measurement of any other electrical parameter. From the point of view of measurement, the basic knowledge of resistance measurement is necessary to understand the working of other instruments used for the measurement of other electrical quantities. Basically resistances are classified as low resistances, medium and high resistances. This classification is based on the value of the resistance, but practically this classification truly indicates different techniques or methods applied for the measurement of resistance such as low resistance measurement, medium resistance measurements and so on.

From the point of view of measurement, resistance can be classified generally as follows.

Low Resistance: (< 1 Ω)

All the resistance of the order of 1 Ω or less are classified as low resistance.

Medium Resistance: (1 Ω to 0.1 MΩ)

From 1 Ω onwords up to 0.1 MΩ the resistances are classified as medium resistance.

High Resistance: (0.1 MΩ >)

Resistance of the order of 0.1 MΩ and higher are classified as high resistance. The classification of resistance given above is not rigid but forms the basis for the methods used for the measurement of the resistance of different classes.

6.2 Classification of Low Resistance

Low Resistance

The various methods that can be employed for the measurement of low resistance are:

1. ***Voltmeter-Ammeter Method***

   This method is the simplest of all. It only requires an ammeter and a voltmeter which is easily available in the laboratory. However, it is not so accurate because of the voltage drop across ammeter and shunting effect of voltmeter.

2. ***Kelvin's Double Bridge Method***

   This method is most accurate compared to other methods. It eliminates the error, due to the contact and lead resistances.

3. ***Potentiometer Method***

   It is also one of the accurate methods of obtaining the value of resistance, but the accuracy of this method depends on the availability of stable D.C supply.

6.2.1 Voltmeter-Ammeter Method

This method is the simplest of all, as it requires an ammeter and a voltmeter which will be easily available in the laboratory. Also the connections are very simple and there is no lengthy procedure for the measurement of resistance. The ammeter as usual, will be connected in series and the voltmeter in parallel. The ratio of voltmeter to ammeter reading will give the value of resistance under measurement. Based on the relative connection of ammeter and voltmeter w.r.t. the unknown resistance, there are two methods. Mathematically the resistance value is given by

$$R_x = \frac{V}{I}$$

V = voltmeter reading

I = Ammeter reading

R_x = resistance value to be measured.

(a) Voltmeter connected to the side of unknown Resistance

The voltmeter reading is the true voltage across the resistance to be measured but the ammeter carries the total current which is sum of the current through resistance and the current through voltmeter (Fig. 6.1).

$$\therefore \qquad I_t = I \times I_v = \frac{V}{R_x} + \frac{V}{R_v}$$

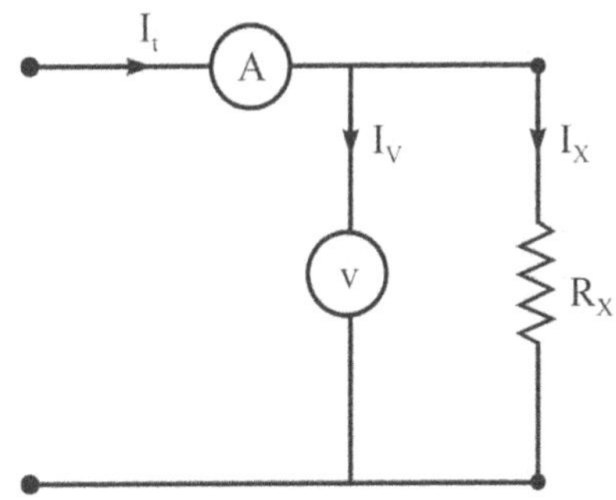

Fig. 6.1

where $\quad R_v$ = voltmeter resistance

$$\therefore \quad R_M = \frac{V}{I} = \frac{V}{\left[\frac{V}{R_x} + \frac{V}{R_v}\right]}$$

$$\therefore \quad R_M = \frac{R_x R_V}{R_x + R_v}$$

$$\therefore \quad R_x + R_v = \frac{R \times R_v}{R_M}$$

$$\therefore \quad R_x \left[1 - \frac{R_v}{R_M}\right] = -R_v$$

$$\therefore \quad Rx = \frac{-R_v R_M}{R_M - R_v} = \frac{R_M}{1 - \frac{R_M}{R_v}}$$

[adjusting negative sign]

Thus the actual value of resistance R_x is same as the measured value R_M, if $\frac{R_M}{R_v}$ is zero i.e., the voltmeter resistance is infinite.

Thus the method gives the measured value which is smaller than the actual value.

$$\text{error} = \frac{\text{Measured value} - \text{True value}}{\text{True value}}$$

But practically if the value of resistance to be measured is very small as compared to the voltmeter resistance ($R_X << R_v$) then the effect of $\frac{R_M}{R_v}$ becomes negligibly small and method gives true value of resistance. Hence this circuit is used for the LOW RESISTANCE measurement.

(b) Ammeter connected to the side of unknown resistance side

The ammeter carries the current which is the true current passing through the resistance i.e., I_X but the voltmeter reading is not the true voltage across the resistance. It is the sum of the voltages across the resistance and across the ammeter.

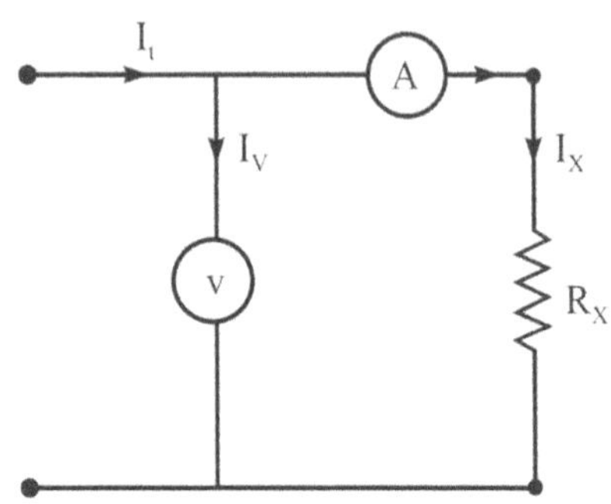

Fig. 6.2

$$\therefore \quad V = V_R + V_a = I_X R_X + I_X R_a$$

where R_a = ammeter resistance,

$$\therefore \quad R_m = \frac{V}{I} = \frac{I_X[R_X + R_a]}{I_X}$$

where R_m = measured value of resistance

$$\therefore \quad R_m = R_X + R_a$$

$$\therefore \quad R_X = R_m - R_a$$

thus the actual value of resistance is same as the measured value if the ammeter resistance is zero.

Actually the measured value is higher than the actual value of the resistance.

$$\text{error} = \frac{\text{Measured value} - \text{True value}}{\text{True value}}$$

But practically if the value of resistance is large compared to ammeter resistance ($R_x >> R_a$) then the effect of ammeter resistance becomes negligible. Hence, the circuit is used to measure the resistance in medium resistance range and is not suitable for the low resistance measurement.

The accuracy of voltmeter-ammeter method depends on the accuracy of the two meters. Similarly the reading is not available directly and required to be calculated from the readings. The division point between the two circuit is at the resistance value for which the relative errors due to both the methods are equal. This is possible for the resistance value given by

$$R_x = \sqrt{R_a \, R_v}$$

For the resistance values greater than this value, the circuit in Fig. 6.2(b) should be used while for the resistance values less than this, the circuit in the Fig. 6.1(a) should be used.

To avoid this difficulty, a practical circuit shown in the Fig. 6.3 can be used.

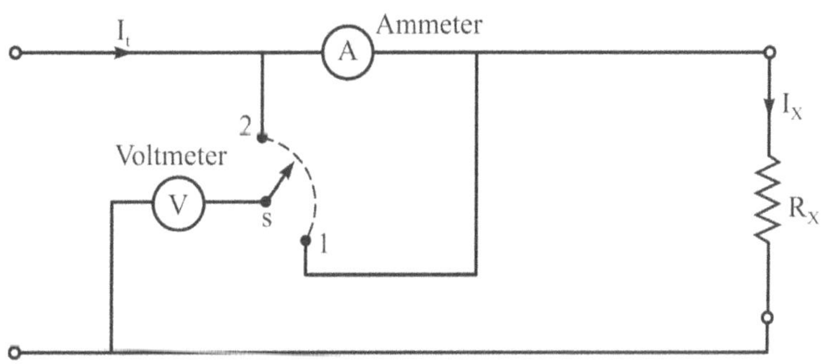

Fig. 6.3

while using this circuit, the switch S is used which is to be kept in position 1 and observe the ammeter reading. Change the switch to position 2, if the ammeter reading does not change, restore the switch in position 1. This indicates the value of resistance to be measured is low.

If the ammeter reading decreases after throwing the switch to position 2, keep it in the position 2. Itself to take the readings. This indicates that the resistance to be measured is a high value resistance, thus the accurate value of resistance can be obtained.

6.2.2 Kelvin Bridge - Measurement of Low Resistance

In the Wheatstone bridge, the bridge contact and lead resistance causes significant error, while measuring low resistance. Thus for measuring the values of resistance below 1 Ω, the modified form of Wheatstone bridge is used, known as Kelvin bridge.

The consideration of the effect of contact and lead resistances is the basic aim of the Kelvin bridge. The Fig.(6.4) shown below is the basic circuit of the Kelvin bridge.

The resistance R_x represents the resistance of the connecting leads from R_3 to R_x. The resistance R_x is the unknown resistance to be measured.

The galvanometer can be connected to either terminal a, b or terminal c. When it is connected to a, the lead resistance R_y gets added to R_x hence the value measured by the bridge, indicates much higher value of R_x.

If the galvanometer is connected to terminal c, then R_x gets added to R_3. This results in the measurements of R_x much lower then the actual value.

The point 'b' is in between the points a and c, in such a way that the ratio of the resistance from c to b and that from a to b is equal to the ratio of R_1 and R_2.

$$\therefore \qquad \frac{R_{cb}}{R_{ab}} = \frac{R_1}{R_2}$$

Now the bridge balance equation in its standard form is

$$R_1 R_3 = R_2 R_x \qquad .....(6.1)$$

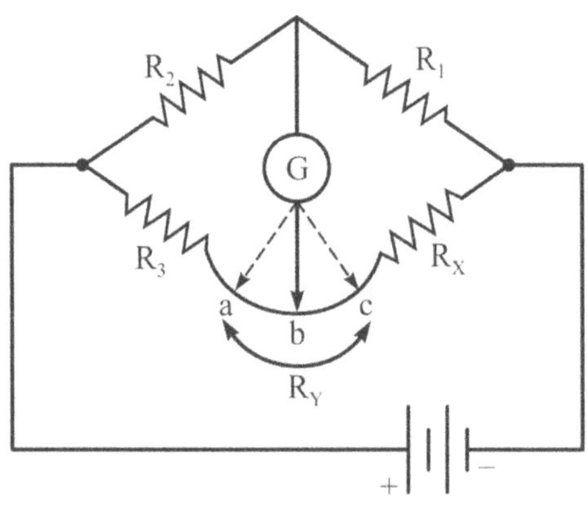

Fig. 6.4 Kelvin bridge

But R_3 and R_x now are changed to $R_3 + R_{ab}$ and $R_x + R_{cb}$ respectively due to lead resistance.

$$\therefore \qquad R_1 (R_3 + R_{ab}) = R_2 (R_x + R_{cb}) \qquad .....(6.2)$$

$$(R_x + R_{cb}) = \frac{R_1}{R_2} (R_3 + R_{ab}) \qquad .....(6.3)$$

Now we have, $$\frac{R_{cb}}{R_{ab}} = \frac{R_1}{R_2}$$

$$\therefore \quad \frac{R_{cb}}{R_{ab}} + 1 = \frac{R_1}{R_2} + 1 \qquad \text{[adding 1 to both sides]}$$

$$\frac{R_{cb} + R_{ab}}{R_{ab}} = \frac{R_1 + R_2}{R_2} \qquad \text{.....(6.4)}$$

But $$R_{cb} + R_{ab} = R_y$$

Total lead resistance

substituting in (6.4) we get

$$\frac{R_y}{R_{ab}} = \frac{R_1 + R_2}{R_2} \qquad \text{.....(6.5)}$$

$$R_{ab} = \frac{R_2 R_y}{R_1 + R_2} \qquad \text{.....(6.6)}$$

Now $$R_{cb} + R_{ab} = R_y$$

$$\therefore \quad R_{cb} = R_y - R_{ab} \qquad \text{.....(6.7)}$$

Substituting into eq. (6.6) and (6.7),

$$R_{cb} = R_y - \frac{R_2\, R_y}{R_1 + R_2} = R_y \left[1 - \frac{R_2}{R_1 + R_2}\right]$$

$$\therefore \quad R_{cb} = \frac{R_1\, R_y}{R_1 + R_2} \qquad \text{.....(6.8)}$$

Substituting these values of R_{cb} and R_{ab} in the eq. (6.3), we get,

$$R_x + \frac{R_1\, R_y}{R_1 + R_2} = \frac{R_1}{R_2}\left[R_3 + \frac{R_2\, R_y}{R_1 + R_2}\right]$$

$$\therefore \quad R_x + \frac{R_1\, R_y}{R_1 + R_2} = \frac{R_1\, R_3}{R_2} + \frac{R_1\, R_y}{R_1 + R_2}$$

But $$R_x = \frac{R_1\, R_3}{R_2} \qquad \text{.....(6.9)}$$

Thus equation (6.9) represents standard bridge balance equation for the Wheatstone bridge. Thus the effect of connecting lead resistance is completely eliminated by connecting the galvanometer to an intermediate position 'b'.

This principle forms the basis of the construction of kelvin's double bridge which is popularly called Kelvin bridge.

6.2.3 Kelvin's Double Bridge Method for Low Resistance Measurement

This bridge consists of another set of ratio arms hence called "double bridge". The Fig. 6.5 shows the circuit diagram of Kelvin's double bridge.

The second set of ratio arms is the resistance 'a' and 'b'. With the help of these resistances the galvanometer is connected to point '3'. The 'galvanometer gives null indication" when the potential of the terminal '3' is same as the potential of the terminal '4'.

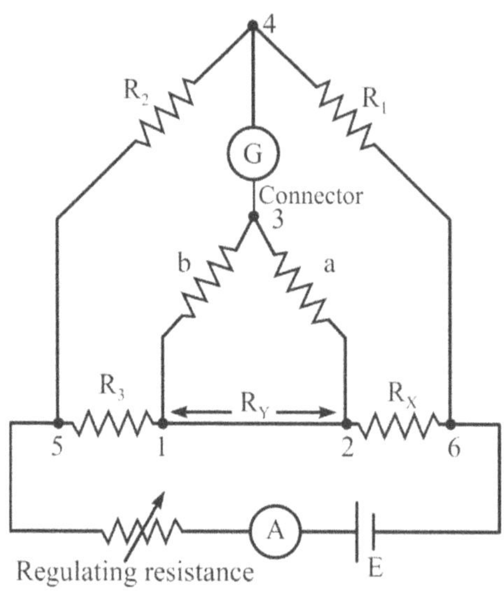

Fig. 6.5 Kelvin's double bridge

Thus, $E_{45} = E_{513}$ .....(6.10)

Where, E_{45} = potential across R_2

E_{513} = potential across R_3 and b.

The ratio of the resistance a and b is same as the ratio of R_1 and R_2

$$\frac{a}{b} = \frac{R_1}{R_2} \qquad .....(6.11)$$

Now $$E_{45} = R_2 \cdot \frac{E}{R_1 + R_2} \qquad .....(6.12)$$

Consider the path from 5 – 1 – 2 – 6, back to 5 through the battery 'E'. The resistance between the terminals 1-2 is the parallel combination of R_y and (a + b).

$$\therefore \quad E = I \times \left[R_3 + R_y\right]\left[(a + b) + R_x\right] \qquad .....(6.13)$$

$$E = I\left[R_3 + R_x + \frac{(a + b)\, R_y}{a + b + R_y}\right]$$

Substituting in

$$E_{45} = \frac{R_2}{R_1 + R_2} \times I\left[R_3 + R_X + \frac{(a + b) R_y}{a + b + R_y}\right] \qquad .....(6.14)$$

For E_{513}, consider the path from the terminal 5 to 2 as shown in the Fig. 6.6.

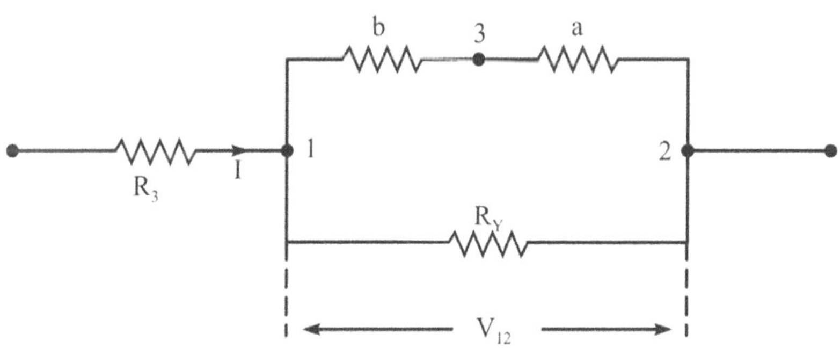

Fig. 6.6 Galvanometer carries zero current

Now,

from the Fig. 6.6, we can write

$$v_{12} = I \times \left[\frac{R_y\, (a + b)}{R_y + a + b}\right]$$

and $$v_{13} = \frac{b}{a+b} \cdot v_{12}$$

$$v_{13} = \frac{b}{a+b} \times I \left[\frac{R_y(a+b)}{R_y+a+b}\right] \quad .....(6.15)$$

$$\therefore \quad E_{513} = I\,R_3 + v_{13}$$

$$E_{513} = IR_3 + I\,\frac{b}{a+b}\left[\frac{R_y(a+b)}{R_y+a+b}\right]$$

$$E_{513} = I\left[R_3 + \frac{b}{a+b}\left[\frac{R_y(a+b)}{R_y+a+b}\right]\right] \quad .....(6.16)$$

Now (For balancing):

$$E45 = E513$$

$$\therefore \quad \frac{IR_2}{R_1+R_2}\left[R_3+R_x+\frac{(a+b)\,R_y}{a+b+R_y}\right] = I\left[R_3+\frac{b}{a+b}\left(\frac{R_y(a+b)}{a+b+R_y}\right)\right]$$

$$\therefore \quad R_3+R_x+\frac{(a+b)\,R_y}{a+b+R_y} = \left[1+\frac{R_1}{R_2}\right]\left[R_3+\frac{bR_y}{R_y+a+b}\right]$$

$$\therefore \quad R_3+R_x+\frac{(a+b)\,R_y}{a+b+R_y} = R_3+\frac{R_1\,R_3}{R_2}+\frac{b\,R_y}{R_y+a+b}+\frac{R_1\,b\,R_y}{R_2(R_y+a+b)}$$

$$\therefore \quad R_x = \frac{R_1\,R_3}{R_2}+\frac{b\,R_y}{R_y+a+b}+\frac{R_1 b\,R_y}{R_2(R_y+a+b)}-\frac{(a+b)\,R_y}{(R_y+a+b)}$$

$$\therefore \quad R_x = \frac{R_1\,R_3}{R_2}+\frac{b\,R_y}{(R_y+a+b)}\left[\frac{R_1}{R_2}-\frac{a}{b}\right]$$

But $$\frac{a}{b}=\frac{R_1}{R_2} \quad \text{thus} \quad \frac{R_1}{R_2}-\frac{a}{b}=0 \quad .....(6.17)$$

$$\therefore \quad R_x = \frac{R_1\,R_3}{R_2} \quad .....(6.18)$$

This is the standard equation of the bridge balance. The resistance a, b and R_y are not present in this equation. Thus "the effect of lead and contact resistance is completely eliminated".

Note: The important condition for this bridge balance condition is that the ratio of the resistance of ratio arms must be same as the ratio of the resistance of the second ratio arms

In a typical Kelvin's double bridge, the range of a resistance covered is 1 Ω to 10 μΩ with an accuracy of $\pm$ 0.05% to $\pm$ 0.2%.

6.2.4 Practical Kelvin's Double Bridge

The Fig. 6.7 shows a commercial Kelvin's double bridge, capable of measuring the resistance of very low range from 10 Ω to 0.00001 Ω.

The resistance R_3 is replaced by the standard resistance consisting nine steps of 0.001 Ω each, plus a calibrated manganin bar of 0.0011 Ω with a sliding contact. The required resistance can be selected by the switch S_3. The total resistance of the R_3 arm amount to 0.0101 Ω and in variable in steps of 0.001 Ω plus fractions of 0.0011 Ω by the sliding contact.

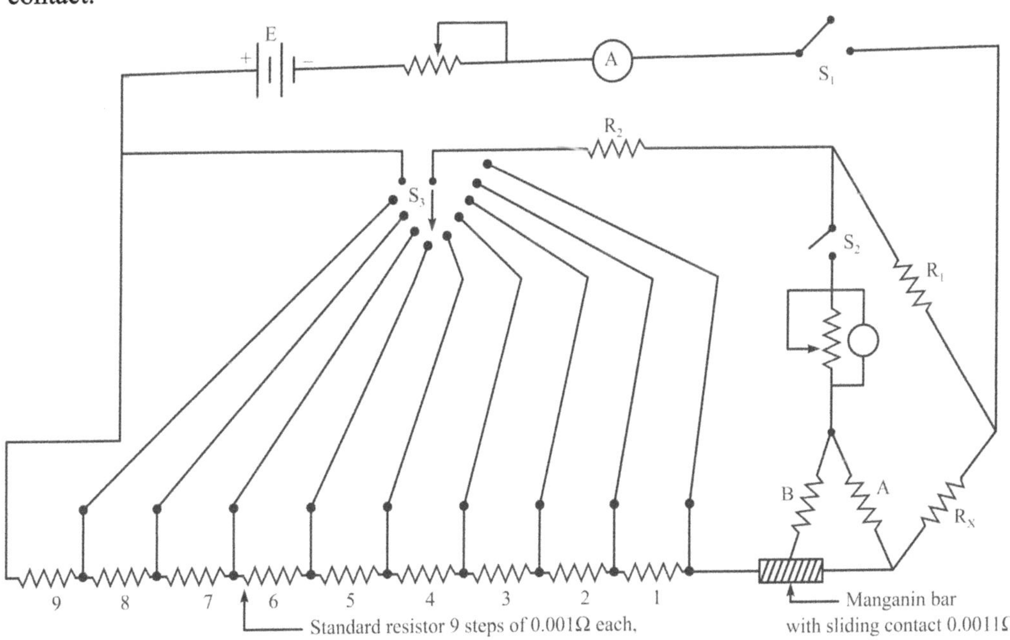

Fig. 6.7 Practical kelwin's double bridge

When both the contacts are switched to select the proper value of standard resistance, the voltage drop between the ratio arm connection points is changed but the total resistance around the battery circuit is unchanged. With this arrangement any contact resistance can be placed in series with the relatively high resistance values of the ratio arms. Due to this, the effect of contact resistance becomes negligibly small.

The ratio of R_1 and R_2 is selected in such a way that the larger part of the variable standard resistance is used and hence R_x is determined to the largest possible number of significant figures, this increases the measurement accuracy.

6.2.5 Potentiometer Method

This is explained in chapter 5.

6.3 Medium Resistance

Resistance ranging from 1 Ω to 0.1 MΩ are classified as medium resistance.

The various methods available for the measurement of medium value resistance are:

Ammeter–Volt Meter Method

Ammeter-voltmeter method being simple, can also be used for the measurement of medium resistance but the accuracy is again less owing to the same reason as that of low resistance measurement.

Substitution Method

Compared to Ammeter-voltmeter method, the substitution method is quite accurate. The error due to ammeter voltage drop and voltmeter shunts are not present in this method. Even though, the above errors are eliminated, there are many other causes of errors are this method i.e., error due to variations in battery e.m.f., error due to sensitivity of the instrument etc. Hence this method is not used frequently.

Wheatstone Bridge Method

This is one of the most common and popular method used for the measurement of medium resistance. It is most accurate and reliable method of measuring the resistance. However errors due to thermo e.m.f and heating effect are present in it. Also the cost of Wheatstone bridge in quite high.

Ohmmeter Method

The value of resistance can be directly measured by using an ohmmeter. The accuracy of an ohmmeter is moderate, but it is very simple and easy to use.

6.3.1 Ammeter-Voltmeter Method

It is the simplest among all the available methods. There are two types of connections available in this method. They are voltmeter across resistance and voltmeter across source as shown in the Figs. 6.7(a) and 6.7(b) respectively.

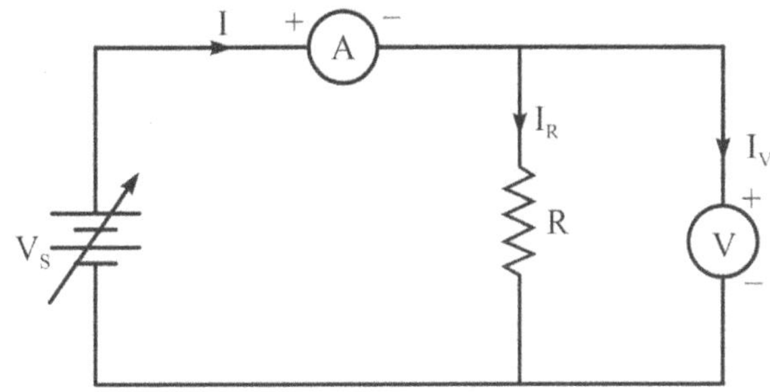

Fig. 6.7 (a)

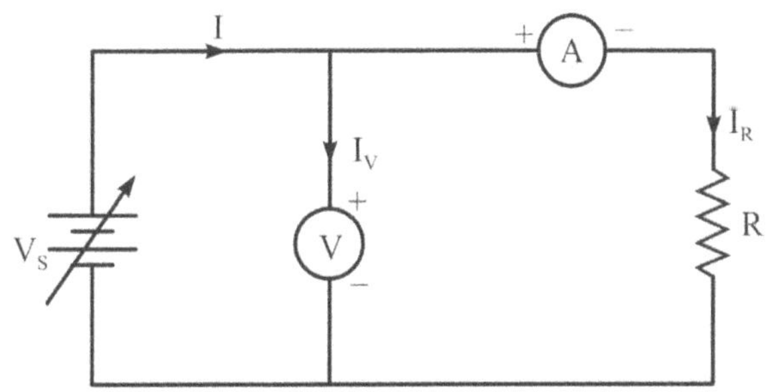

Fig. 6.7 (b)

In both the connections, the value of resistance can be obtained as

$$R = \frac{\text{Voltmeter reading (V)}}{\text{Ammeter reading (I)}}$$

but the obtained value of resistance in both the cases is not its actual value, due to the voltage drop across ammeter and shunting effect of voltmeter. Moreover, measurement accuracy is limited by the accuracy of the voltmeter and ammeter.

Hence this method is generally employed for the laboratory purpose where high accuracies are not needed.

6.3.2 Substitution Method

The connection diagram of the substitution method of measurement of medium resistance is shown in the Fig. 6.8.

V = voltage source

R = The rheostat to limit or adjust the ammeter (A) current

S = Standard variable resistance and

X = is the unknown resistance

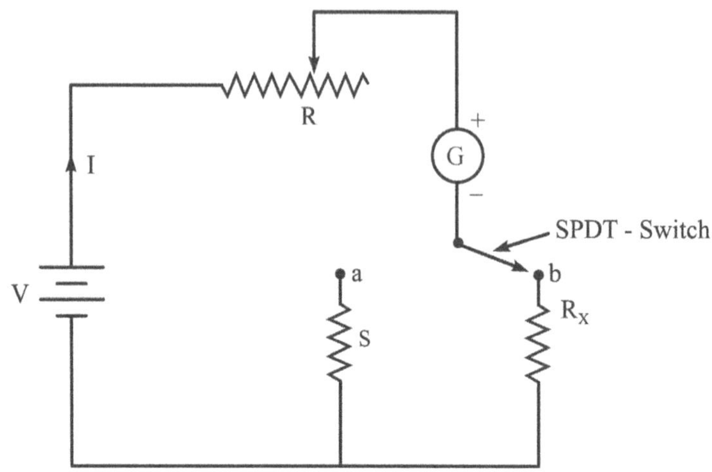

Fig. 6.8

The single pole double throw (SPDT) switch is thrown to position ‘b’ which connects the unknown resistance (X) into the circuit. The current through the ammeter is adjusted to some value (I) and it is noted. Now the SPDT switch is thrown to position ‘a’ which connects standard variable resistance (S) on to the circuit. The rheostat is kept unaltered and ‘S’ is varied until the ammeter reads the previously noted value (I).

The value of X is taken as this particular resistance of S, because when the currents through two resistors are equal. In similar conditions (i.e., applied voltage are equal) then their resistances will be equal.

The following are the factors on which the accuracy of this method depends upon:

1. Stability in battery voltage.
2. The accuracy with which the standard resistance (S) in known.
3. Ammeter sensitivity.
4. Stability in rheostat resistance (R).

6.3.3 Wheatstone Bridge Method

The bridge consists of four resistive arms together with a source of e.m.f. and a null detector. The galvanometer is used as a null detector.

The Fig. 6.9 shows the basic wheatstone bridge circuit.

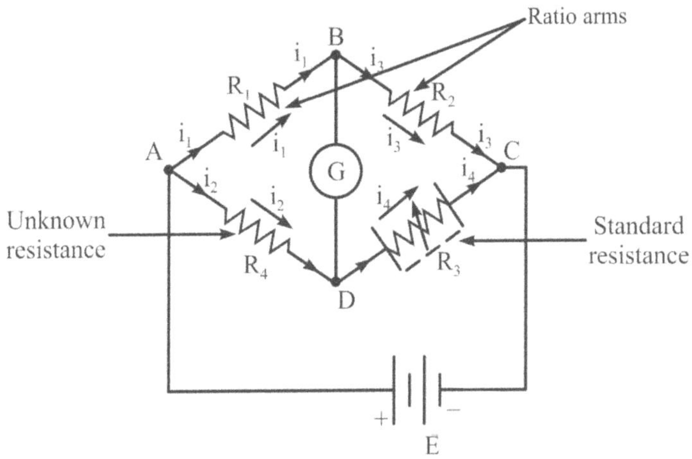

Fig. 6.9 Wheatstone bridge

The arms consisting the resistance R_1 and R_2 are called ratio "arms". The arm consisting the standard known resistance R_3 is called standard arm. The resistance R_4 is the unknown to be measured. The battery is connected between A and C while galvanometer is connected between B and D.

Balance Condition

When the bridge is balanced, the galvanometer carries zero current. It does not show any deflection. This bridge works on the "Principle of null deflection or null indication". To have zero current through galvanometer, the points B and D must be at the same potential. Thus potential across arm AB must be same as the potential across arm AD.

Thus, $$I_1R_1 = I_2R_4 \qquad .....(6.19)$$

As galvanometer current is zero

$$I_1 = I_3 \text{ and } I_2 = I_4 \qquad .....(6.20)$$

Considering the battery potential balanced condition

$$I_1 = I_3 = \frac{E}{R_1 + R_2} \quad \text{.....(6.21)}$$

and $$I_2 = I_4 = \frac{E}{R_3 + R_4} \quad \text{.....(6.22)}$$

using eq. (6.21) and (6.22) in eq. (6.19)

$$\frac{E}{R_1 + R_2} \times R1 = \frac{E}{R_3 + R_4} \times R_4$$

$$\therefore \quad R_1 (R_3 + R_4) = R_4 (R_1 + R_2) \Rightarrow R_1R_3 + R_1R_4 = R_1R_4 + R_2R_4$$

$$\therefore \quad R_4 = R_3 \times \frac{R_1}{R_2} \quad \text{.....(6.23)}$$

This is the required balance condition of Wheatstone bridge.

The main points can be observed that:

1. It depends on the ratio of R_1 and R_2 hence these arms are called "ratio arms".
2. The standard resistance R_3 can be varied to obtain the required balance.
3. As it works on null indication, the results are not dependent on the calibration and characteristics of galvanometer.

6.3.4 Industrial Form of Wheatstone Bridge

In an industrial or laboratory form of Wheatstone bridge, the resistance R_1, R_2 and R_3 are mounted in a box. The values are selected by the dial switches. The ratio R_1/R_2 is adjusted as per the requirement using "ratio selector switch". The resistance R_1 and R_2 generally consist of four resistor each of 10, 100, 1000 and 10,000 Ω respectively. The resistance R_3 can be adjusted using 4 dial or 5 dial decade arrangement and thus its value can be adjusted from 0 Ω to 10,000 Ω or 1,00,000 Ω depending upon 4 or 5 dial decades available. The battery is connected across the battery terminals while the unknown resistance is connected across X_1–X_2 terminals as shown in the Fig. 6.10. The galvanometer is connected across the terminals marked G. The Fig. 6.10 shows this commercial form of the Wheatstone bridge.

While using the bridge connect R_4, battery and the galvanometer. Select the ratio $\frac{R_1}{R_2}$ using ratio selection switch. Adjust R_3 till the galvanometer deflection is zero. Then R_4 is given by $\left(\frac{R_1}{R_2}\right)$ R_3. Repeat the set for various values of the ratio $\frac{R_1}{R_2}$ and find the average to minimise the errors.

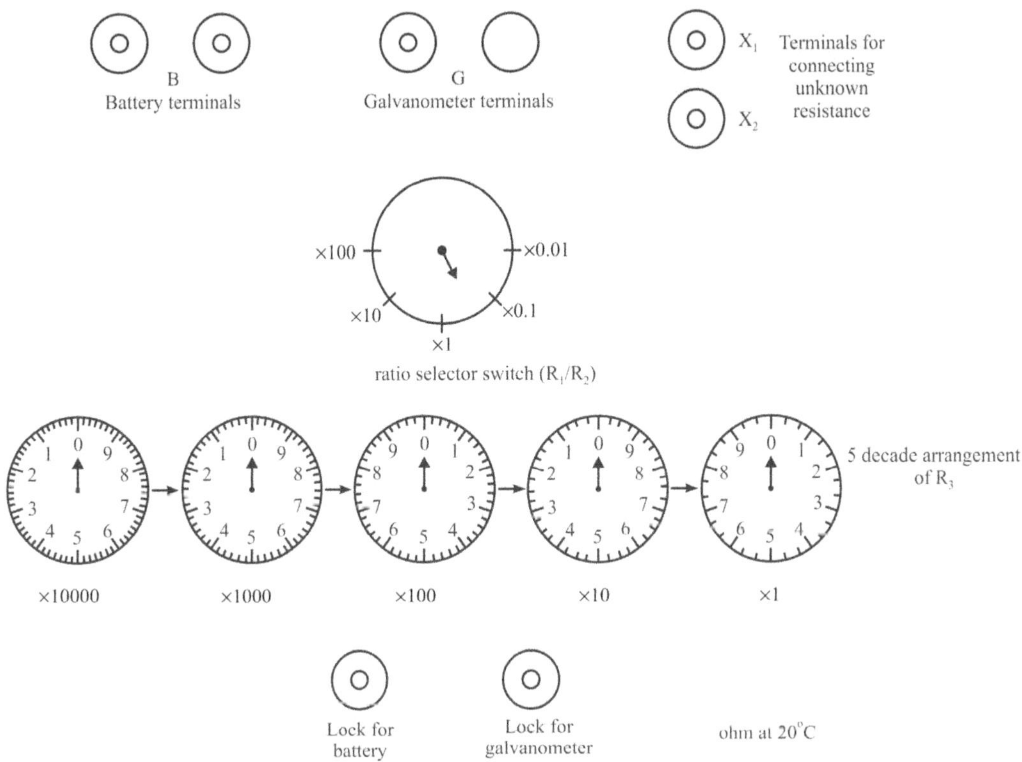

Fig. 6.10 Commercial form of Wheatstone bridge

6.3.5 Sensitivity of Wheatstone Bridge

When the bridge is balanced, the current through galvanometer is zero, but when bridge is not balanced, current flows through the galvanometer causing the deflection. The amount of deflection depends on the "sensitivity" of the galvanometer. This sensitivity can be expressed as amount of deflection per unit current

$$\text{Sensitivity S} = \frac{\text{deflection D}}{\text{current I}}$$

As the current is in microampere and deflection can be measured in mm, radians or degrees. The sensitivity is expressed as mm/μA, radians/μA or degrees/μA. More is the sensitivity of a galvanometer, more is its deflection for the same amount of current.

Another way of representing the galvanometer sensitivity is the amount of deflection per unit voltage across the galvanometer. This is called "voltage sensitivity" of the galvanometer. Mathematically it is denoted as

$$S_V = \frac{\theta}{e}$$

where,

e = voltage across galvanometer

θ = deflection of galvanometer

It is measured in degree per volts or radians per volts.

While the bridge sensitivity is defined as the deflection of the galvanometer per unit fractional change in the unknown resistance, it is denoted as S_B.

$$S_B = \frac{\theta}{\frac{\Delta R}{R}}$$

where $\frac{\Delta R}{R}$ = unit fractional change in unknown resistance.

6.4 Wheatstone Bridge under Small Unbalance

The bridge sensitivity can be calculated by solving the bridge for small unbalance.

At balance condition,

$$R_4 = R_3 \frac{R_1}{R_2}$$

i.e.,

$$\frac{R_4}{R_3} = \frac{R_1}{R_2}$$

Let the resistance R_4 is changed by ΔR creating the unbalance. Due to this, the e.m.f., appears across the galvanometer (Fig. 6.10 (a). To obtain this e.m.f., let us use thevenin's method. Remove the branch of galvanometer and obtain the voltage across the open circuit terminals.

$$E_{AB} = I_1 R_1 \quad \text{.....(6.24)}$$

$\therefore$

$$I_1 = \frac{E}{R_1 + R_2} \quad \text{.....(6.25)}$$

$$E_{AD} = I_2 (R_4 + \Delta R) ; \quad \text{.....(6.26)}$$

$$I_2 = \frac{E}{R_3 + R_4 + \Delta R} \qquad .....(6.27)$$

$$V_{BD} = V_{th} = E_{AD} - E_{AB} \qquad .....(6.28)$$

$$V^{th} = \frac{E(R_y + \Delta R)}{R_3 + R_y + \Delta R} - \frac{E}{R_1 + R_2} R_1$$

$$V_{th} = E \left\{ \frac{R_4 + \Delta R}{R_3 + R_4 + \Delta R} - \frac{R_1}{R_1 + R_2} \right\} \qquad .....(6.29)$$

As $\frac{R_4}{R_3} = \frac{R_1}{R_2}$ then $\frac{R_1}{R_1 + R_2} = \frac{R_4}{R_4 + R_3}$

using above relation in eqn (6.29)

$$V_{th} = E \left\{ \frac{R_4 + \Delta R}{R_3 + R_4 + \Delta R} - \frac{R_4}{R_3 + R_4} \right\}$$

$$= E \left\{ \frac{R_3R_4 + R_3\Delta R + R_4^2 + R_4\Delta R - R_3R_4 - R_4^2 - R_4\Delta R}{(R_3 + R_4)\,(R_3 + R_4 + \Delta R)} \right\}$$

$$= \frac{ER_3\Delta R}{(R_3 + R_4)^2 + (R_3 + R_4)\Delta R}$$

but as ΔR is very small, $(R_3 + R_4)\,\Delta R << (R_3 + R_4)^2$

$$V_{th} = V_g = \frac{E\,R_3\Delta R}{(R_3 + R_4)^2}$$

Now $S_B = \dfrac{\theta}{\dfrac{\Delta R}{R}}$ = bridge sensitivity

and $\frac{\Delta R}{R} = \frac{\Delta R}{R_4}$ as there is change in R_4

From the galvanometer sensitivity S_V,

Using θ in the expression of S_B.

$$S_B = \frac{S_V V_g}{\dfrac{\Delta R}{R_4}} = \frac{S_V\, ER_3\, \Delta R\, R_4}{(R_3 + R_4)^2} = \frac{S_v\, ER_3\, R_4}{R_3^2 + 2R_3R_4 + R_4^2}$$

$$\therefore \qquad S_B = \frac{S_V\, E}{\dfrac{R_3}{R_4} + 2 + \dfrac{R_4}{R_3}}$$

Thus the bridge sensitivity depends on the bridge parameters, the supply voltage and the voltage sensitivity of the galvanometer.

For higher or lower values of $\frac{R_3}{R_4}$, the sensitivity decreases considerably.

6.4.1 Thevenin's Equivalent and Galvanometer Current

The thevenin's voltage V_{th} across the galvanometer is already obtained. Let us obtain equivalent resistance as viewed across the terminals BD, when battery E is replaced by short circuit. Thus circuit becomes,

$$\therefore \qquad R_{eq} = (R_1 \| R_2) + (R_3 \| R_4)$$

while $\quad V_{TH} = E_{AD} - E_{AB}$ with R_4 not changed by ΔR

$$= I_2R_4 - I_1R_1 = \frac{E}{R_3 + R_4} R_4 - \frac{E}{R_1 + R_2} R_1$$

$$V_{TH} = E\left[\frac{R_4}{R_3 + R_4} - \frac{R_1}{R_1 + R_2}\right]$$

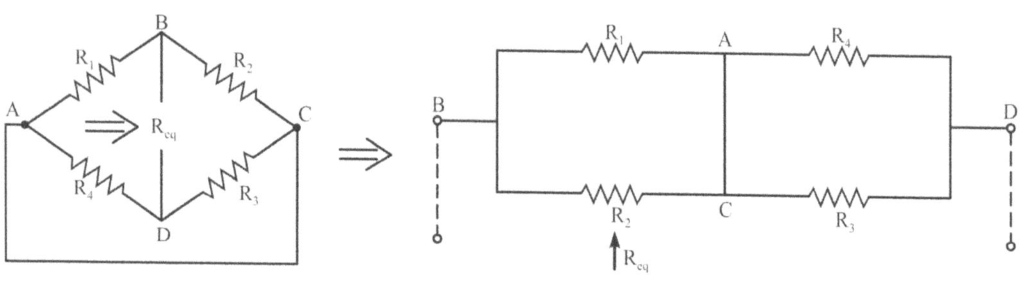

Fig. 6.11

Thus thevenin's equivalent is as shown in Fig. 6.11.

Let $\quad R_g$ = Galvanometer resistance

I_g = Galvanometer current

$$I_g = \frac{V_{TH}}{R_{eq} + R_g}$$

where $\quad V_{TH}$ = Thevenin's voltage

6.4.2 Galvanometer Current under Unbalanced Condition

Let the resistance R_4 is changed by ΔR which has caused the unbalanced in the bridge.

As derived earlier $$V_{TH} = \frac{E\,R_3\,\Delta R}{(R_3+R_4)^2}$$

and now $$R_{eq} = (R_1 \| R_2) + (R_3 \| R_4 + \Delta R)$$

$$= \frac{R_3\,(R_4 + \Delta R)}{R_3 + R_4 + \Delta R} + \frac{R_1\,R_2}{R_1 + R_2}$$

Neglecting ΔR compared to R_3 and R_4,

$$R_{eq} = \frac{R_3\,R_4}{R_3 + R_4} + \frac{R_1\,R_2}{R_1 + R_2}$$

$$\therefore \quad I_g = \frac{V_{TH}}{R_{eq} + R_g}$$

For bridge with equal "arms" $R_1 = R_2 = R_3 = R_4 = R$ then

$$V_{TH} = \frac{ER\,\Delta R}{4R^2} = \frac{E\,\Delta\,R}{4R}$$

and $$R_{eq} = \frac{R^2}{2R} + \frac{R^2}{2R} = R$$

$$\therefore \quad I_g = \frac{\frac{E \wedge R}{4R}}{R + R_g} = \frac{E\left(\frac{\Delta\,R}{4R}\right)}{R + R_g}$$

6.4.3 Sensitivity of Bridge in terms of Current Sensitivity of Galvanometer

The deflection of galvanometer for a small change in unknown resistance R_4 is

$$0 = S_v, \quad e = S_v, \quad V_g = \frac{S_v\,ER_3\,\Delta R}{(R_3 + R_4)^2}$$

which $$S_v = \frac{S_i}{R_{eq} + R_g}$$

where S_i = current sensitivity of galvanometer

$$\theta = \frac{S_i\,E\,R_3\,\Delta R}{\left(R_{eq} + R_g\right)(R_3 + R_4)^2}$$

and $$S_B = \frac{\theta}{\frac{\Delta R}{R_4}} = \frac{S_i\, E\, R_3\, R_4}{(R_{eq}+R_g)\,(R_3+R_4)^2}$$

where $$R_{eq} = (R_1 \| R_2) + (R_3 \| R_4 + \Delta R)$$

This is the bridge sensitivity in terms of current sensitivity of the galvanometer.

Note: practically R_{eq} can be assumed to be R_{TH} as ΔR is small compared to actual value of the resistances

$$\therefore \quad R_{eq} \approx R_{TH} = (R_1 \| R_2) + (R_3 \| R_4)$$

6.5 Measurement Errors or the Precision Measurement of Wheatstone Bridge

The Wheatstone bridge is used to measure the resistance in the range 1 Ω to few mega ohms. But certain errors occur during the measurement using the Wheatstone bridge.

The errors as follows:

1. **Resistance of Connecting Leads**

   The main error is because of limiting errors of the three known resistances hence very precise resistances are required having tolerance of 1% or even 0.1%. The insufficient sensitivity of the null detector may cause the error.

2. **Thermoelectric Effects**

   The galvanometer deflection is affected by the thermoelectric e.m.f. and which are present on measuring circuit due to the unbalance of bridge. The thermoelectric effect can be minimised (or) eliminated by reversing the battery connections through a quick acting switch and adjusting the galvanometer until no change in the deflection is observed. The results are obtained by taking an average of the two readings. In this way the thermoelectric effect may be eliminated.

   To prevent this, more sensitive galvanometer having copper coils and copper suspension system are used.

3. **Temperature Effects**

   An increase in temperature is accompanied by a rise in resistance of all copper and aluminium parts. The errors caused by change of resistance due to change of temperature produces serious errors in measurements. In case of copper which has a temperature co-efficient of 0.004% C, a change in temperature of 33.8° F will cause an error of 0.4%.

4. Contact Resistance

Serious errors may be caused by contact resistances of switches. A dial may have a contact resistance of about 0.003 Ω and thus a four dial resistance box has a contact resistance of about 0.012 Ω. This value is high, especially in measurement of low resistances.

6.6 Advantages of Wheatstone Bridge

The various advantages of Wheatstone bridge are:

1. The source e.m.f. and inaccuracies due to the source fluctuations do not affect the balance of the bridge. Hence the corresponding errors are completely avoided.
2. Due to null deflection method used, the accuracy and sensitivity is higher than direct deflection meters.

6.7 Limitations of Wheatstone Bridge

Wheatstone bridge can be used for the measurement of medium resistances only, due to the various limitation for the measurement of high and low resistance.

Limitations for Measurement of High Resistance

1. If the bridge potential is increased to maintain the bridge current then overheating in any arm may occur for which a special care has to be taken.
2. The leakages through the insulation of the bridge arm's may introduce errors in the reading.
3. Due to high resistances, the bridge current will be very low and thus the sensitivity reduces.

Limitations for Measurement of Low Resistances

1. The load resistance can be reduced, but the contact resistance produces uncertainity in the measurement which is very difficult to overcome.
2. In this measured value includes the resistance of leads and contacts.

6.8 Application of Wheatstone Bridge

The Wheatstone bridge is basically a D.C. bridge and used to measure the resistance in the range "1 Ω to low Mega ohm".

It is used to measure the d.c. resistance of various types of wires for the purpose of quality control of wire.

1. It is used to measure the resistance of motor winding, relay coils etc.
2. It is used by the telephone companies to locate the cable faults. The faults may be of the type line to line short or line to ground short.

6.9 Carey-Foster Slide Wire Bridge

A Carey-Foster slide wire bridge is the elaborated form of the Wheatstone bridge. This type of the bridge is most extensively used for the comparison of the two nearly equal resistances. The circuit arrangement for the Carey-Foster bridge is as shown in Fig. 6.12.

The arms consisting resistance P and Q are nominal equal ratio arms. The resistance R is the resistance under test while 'S' is the standard resistance. A slide wire of length 'L' is introduced between the resistance R and S as shown in the Fig. 6.12.

Initially a balanced condition is obtained by adjusting the sliding contact on the slide wire of a distance l_1 from the left hand side of the slide wire. After this, the positions of resistances R and S are interchanged and a new balance point is obtained. Let the distance from the left hand of the slide wire be l_2.

The first balance condition is given by,

$$\frac{P}{Q} = \frac{R + l_1 r}{S + (l + l_1)r} \quad \text{.....(6.30)}$$

where r = resistance of slide wire per unit length of the slide wire.

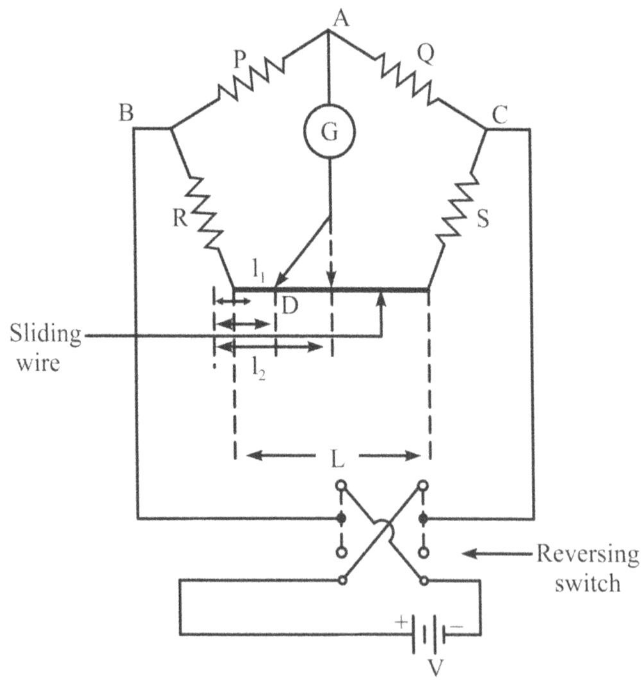

Fig. 6.12 Carey-Foster slide wire bridge

Similarly, the second balance condition is given by,

$$\frac{P}{Q} = \frac{S + l_2 r}{R + (l - l_2)r} \qquad .....(6.31)$$

From eqn (6.30) and (6.31), we can write

$$\frac{R + l_1 r}{S + (l - l_1)r} = \frac{S + l_2 r}{R + (l - l_2)r}$$

adding 1 on both the sides of above equation, we get,

$$\frac{R + l_1 r}{S + (l - l_1)r} + 1 = \frac{S + l_2 r}{R + (l - l_2)r} + 1$$

$$\frac{R + l_1 r + S + lr + l_1 r}{S + (l - l_1)r} = \frac{S + l_2 r + R + lr - l_2 r}{R + (l - l_2)r}$$

$$\frac{R + S + lr}{S + (l - l_1)r} = \frac{S + R + lr}{R + (l - l_2)r}$$

i.e., $$R + (l - l_2)r = S + (l - l_1)r$$

i.e., $$(S - R) = (l_1 - l_2)r \qquad .....(6.32)$$

From equation (6.32), it is clear that the difference between the standard resistance S and the resistance under test R is obtained as the resistance of the slide wire between the two balance points.

The slide wire of resistance r per unit length of the slide wire can be calibrated by shunting resistance R or S with a known high value resistance. Thus effective value of the standard resistance S reduces to S^1. Now the complete procedure is repeated and two new balance points are obtained at lengths l_1^1 and l_2^1 from the left hand side of the slide wire.

Thus equation (6.32) can be modified as,

i.e., $$(S^1 - R) = \left(l_1^1 - l_2^1\right)r \qquad .....(6.33)$$

Simplifying equation (6.32) and (6.33) by dividing equation (6.32) by equation (6.33) we get.

$$\frac{S - R}{S^1 - R} = \frac{l_1 - l_2}{l_1^1 - l_2^1}$$

$$(S - R)\left(l_1^1 - l_2^1\right) = (S^1 - R)(l_1 - l_2)$$

$$\therefore \quad S\left(l_1^1 - l_2^1\right) - R\left(l_1^1 - l_2^1\right) = S^1\left(l_1 - l_2\right) - R\left(l_1 - l_2\right)$$

$$\therefore \quad S\left(l_1^1 - l_2^1\right) - S^1(l_1 - l_2) = R\left(l_1^1 - l_2^1\right) - R(l_1 - l_2)$$

$$R = \frac{S\left(l_1^1 - l_2^1\right) - S(l_1 - l_2)}{l_1^1 - l_2^1 - l_1 + l_2} \qquad \text{.....(6.34)}$$

From equation (6.34), it is clear that this method compares resistance S and R directly in terms of lengths only and the resistance P, Q and the lead resistance are being eliminated. All high accuracy bridges are with errors because of thermo-electric e.m.f in the arms with low resistances. To eliminate this, the Carey-Forster uses a switch which enables the connections to R and S resistance to be interchanged without handling the coils.

6.10 Methods for Measurement of High Resistances

There are different methods that can be employed for the measurement of high resistance. Some of the important methods are as follows:

1. Direct deflection method
2. Loss of charge method
3. Megohm bridge
4. Megger

Let us discuss above methods.

6.11 Direct Deflection Method

In this method, a high resistance more than 1000 Ω and very sensitive moving coil galvanometer is connected in series with the resistance to be measured along with supply as shown in the Fig. 6.13.

By this method, insulation resistance can be obtained with respect to the deflection of galvanometer. The galvanometer measures current between core and metal sheath. The leakage currents over the surface of the insulating material are carried by the guard wire wound on the insulation. These currents do not flow through galvanometer. This ratio of the voltage applied between core and metal sheath and current flowing between them gives the insulation resistance of the cable. A typical direct deflection method for measuring the insulation resistance is as shown in the Fig. 6.13 (a).

Before testing, the cable is immersed in water at least for 24 hours. The temperature of water is maintained constant. The water enters through the pores if any, and soaks through the defect and at the same time, the cable attains the temperature of water. In the earlier phase of the measurement, the galvanometer must be properly shunted. It includes a series resistance of high value. The true value of insulation resistance can be determined by subtracting the value of series resistance from the observed resistance. In tests on cables, the galvanometer should be short circuited before applying voltage. After the application of voltage short circuit connection across galvanometer is removed. Thus galvanometer is protected from sudden initial spike of current which charge cable which acts as capacitor.

The galvanometer used in this method should be very sensitive and should be of high resistance. To prevent leakage currents, the galvanometer circuit along water, the switch and other circuitry must be well insulated.

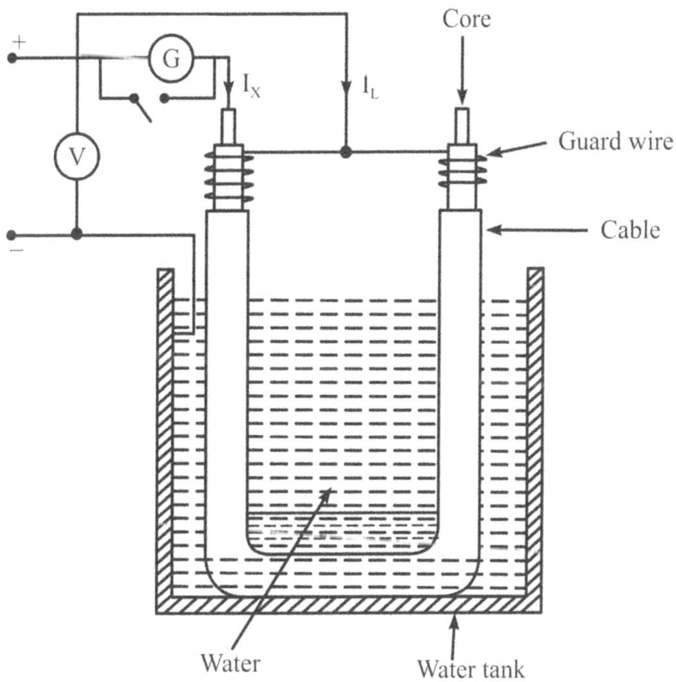

Fig. 6.13 Measurement of insulation resistance without metal sheath

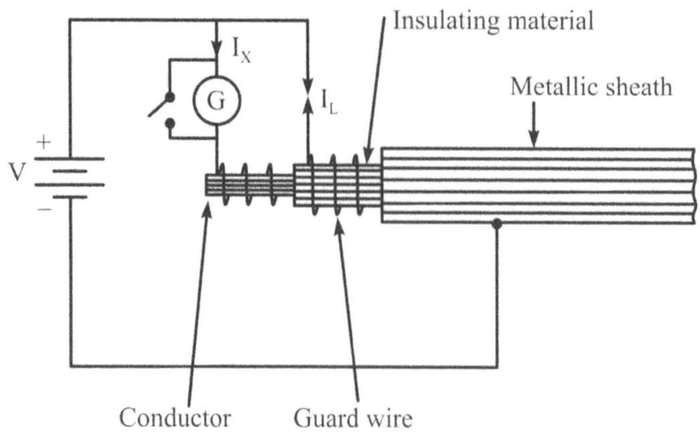

Measurement of insulation resistance with metal sheath

Fig. 6.13 (a)

6.12 Loss of Charge Method

This is very typical method of measuring the insulation resistance of very high value. The typical arrangement of measurement is as shown in the Fig. 6.14.

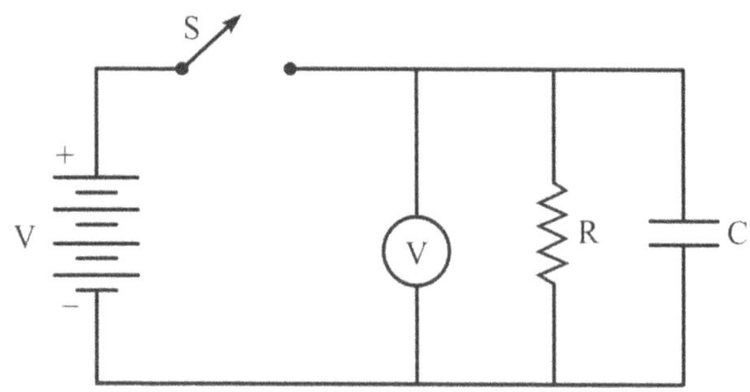

Fig. 6.14 Circuit arrangement of loss of charge method

The resistance to be measured is shunted by a known value capacitor. The voltage across parallel combination is measured using electrostatic voltmeter. The circuit is driven by a d.c. voltage source of value V. This voltage is applied to the circuit through a switch.

Initially switch is kept open. When the switch is closed, at certain instant, the capacitor C start charging. The voltage across C is given by

$$V_C = V\left(1 - e^{-\frac{t}{RC}}\right) \qquad .....(6.35)$$

Then at certain instant say $t = t_1$, switch is opened. Then capacitor 'C' starts discharging through R. Then at instant voltage is given by

$$V_C = Ve^{-\frac{t}{RC}}$$

$$\frac{V_C}{V} = e^{-\frac{t}{RC}}$$

Simplifying

$$\text{Insulation resistance } R = \frac{t}{C l_n \frac{V}{V_C}} = \frac{0.4343\, t}{C \log_{10} \frac{V}{V_C}}$$

The variation of voltage across 'C' and with time shown in Fig. 6.15.

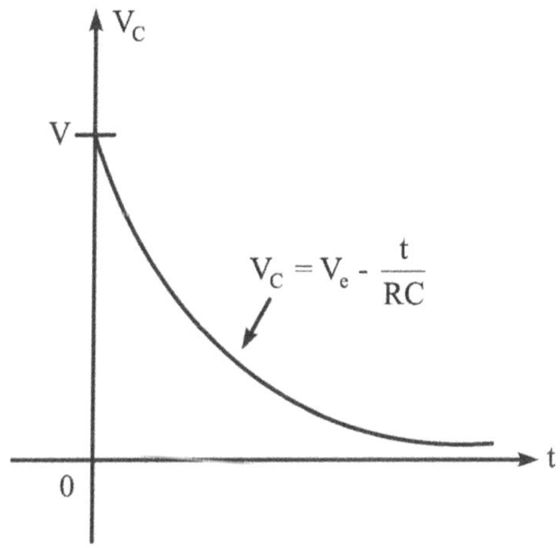

Fig. 6.15 Variation of voltage across 'C' with respect time 't'

If the value of resistance 'R' is very large, then capacitor 'C' requires more time for discharging. In such cases, the process becomes time consuming.

This method can be used effectively for measurement of high resistance, but it needs a capacitor with a high leakage resistance. The typical circuit arrangement is as shown in the Fig. 6.17.

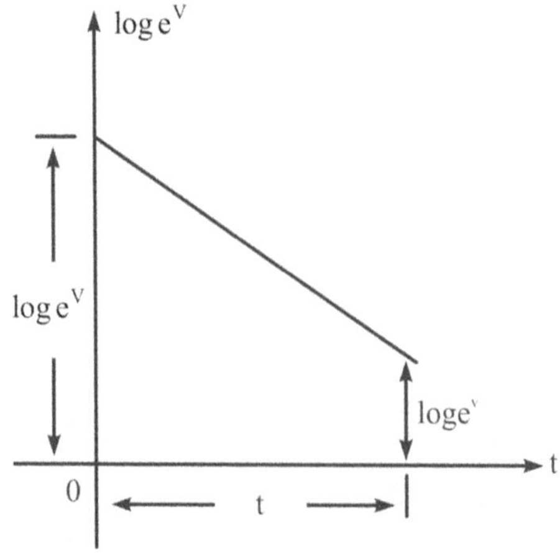

Fig. 6.16 Plot of log e^V verse time

The circuit consists high insulation resistance R to be measured along with capacitor of known value C shunted with electrostatic voltmeter and leakage resistance R_1.

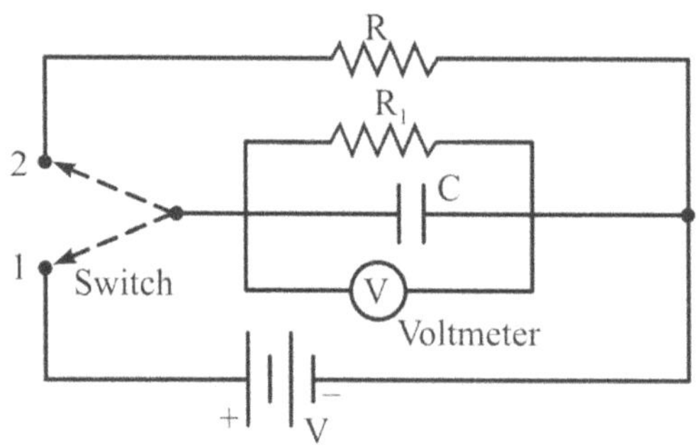

Fig. 6.17 Measurement of high resistance using loss of charge method include leakage resistance of 'C'

Initially, capacitor 'C' is charged to suitable voltage say V_1 by moving switch to position '1'. Then switch is moved to position '2'. The capacitor starts discharging through parallel combination of R and R_1. At certain instant 't' voltage across capacitor 'C' i.e., V_2 is measured using electrostatic voltmeter. Thus in time t, voltage across capacitor drops down from V_1 to V_2. Let the equivalent resistance through which 'C' discharges be denoted by R' where $R \| R_2$.

The expression for current any instant 't' is given by

$$i = -\frac{dq}{dt} = -C\frac{dV}{dt} \qquad .....(6.36)$$

but $$i = \frac{\text{Potential drop across } R'}{R'} = \frac{V}{R_1} \qquad .....(6.37)$$

comparing equation (6.36) and (6.37), we can write,

$$\frac{V}{R^1} = -C\frac{dV}{dt} \quad \text{or} \quad \frac{dV}{V} = -\frac{dt}{RC}$$

Integrating both sides,

$$[\ln V]_{V1}^{V2} = \left[\frac{-t}{R_1C}\right]_0^t$$

i.e., $$\ln\frac{V_2}{V_1} = -\frac{t}{R_1C}$$

$$\therefore \quad V_2 = V_1\, e^{-\frac{t}{R'C}} \qquad .....(6.38)$$

Thus if time t is known, the resistance R' can be obtained by measuring voltages V_1 and V_2. The same test is repeated with unknown resistance R removed. Then 'C' discharge through only R_1. Then expression is given by,

$$V_2 = V_1\, e^{-\frac{t}{R_1C}} \qquad .....(6.39)$$

Thus the value of the leakage resistance of the capacitor can also be found out using this method. Note that in this method, leakage resistance of the voltmeter can be neglected if its value is low.

If it is high, then it must be considered along with R_1.

6.13 Errors in Loss of Charge Method

1. Error due to leakage and absorption
2. Error due to variation in insulation resistance due to variation in temperature
3. Error due to time of application of voltage.

6.13.1 Error due to Leakage and Absorption

Every insulating body will have some leakage current, since we con't get an ideal insulator. The value of this leakage current is quite significant with respect to the current under measurement. Due to this, the results will not be accurate. Hence in order to reduce the error due to leakage, the insulation resistance of the capacitor and voltmeter must be very high. Also every insulation material will have the absorption effect. The net current flowing is the sum of currents due to resistance and absorption. Again due to this effect, the readings will have an error. In order to reduce it, the observation should be taken for a long period of time.

6.13.2 Error due to Variation in Insulation Resistance due to Variation in Temperature

The resistance of any insulating material is a function of temperature. Whenever the temperature increases, the resistance decreases and hence the current flowing through the material increases.

As we know that the temperature is a variable quantity, therefore the resistance as well as current vary along with it. Hence, in order to have accurate results, resistance along with temperature at which the test was performed must be specified.

6.13.3 Error due to Time of Application of Voltage

The effect of absorption is associated with almost all insulating materials. The value of current due to absorption effect is not constant, but varies with the time for which voltage is applied. The value of the absorption current is different at different instance of time. Hence due to this accuracy of the measurement will be lost.

6.14 Megohn Bridge

To avoid the leakage current external to the bridge, the junction of ratio arms R_A and R_B is brought out as a separate guard terminal on the front panel of the instrument. This can be used to connect "three terminal resistance". The three terminal resistance is shown in the Fig. 6.18.

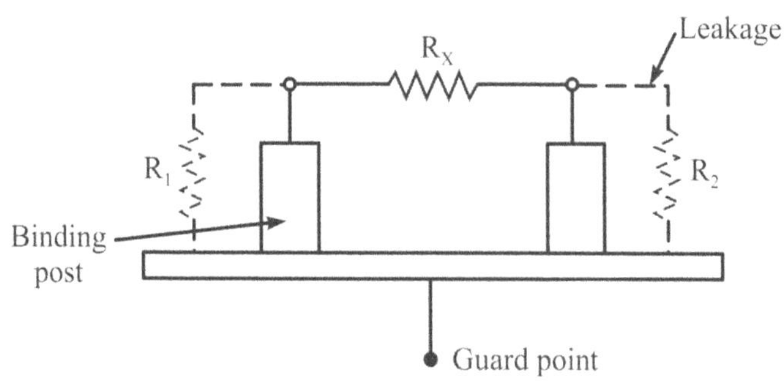

Fig. 6.18 Three terminal resistance

The high resistance is connected between two binding posts which are fixed to metal plate. The two main terminals of the resistor are connected to the R_x terminals in the bridge. The third terminal is the common point of resistance R_1 and R_2, which represents the leakage paths from the main terminal along the insulating post of the metal plate. The guard is connected to the guard terminal in the front panel of the bridge as shown in the Fig. 6.18a.

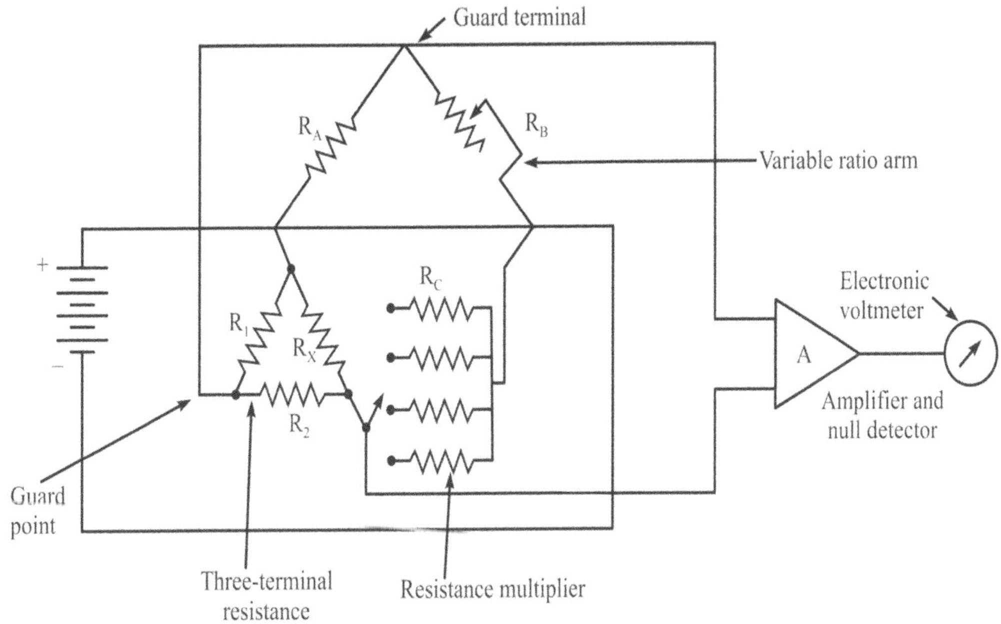

Fig. 6.18(a) Guarded bridge with three terminal resistor

This connection puts R_1 in parallel with ratio arm resistance R_A but since R_1 is very much larger than R_A, its shunting effect is negligiblc. Similarly R_2 in parallel with galvanometer has no effect as 'R_2' is much higher than galvanometer. It slightly reduces the sensitively of the galvanometer. Thus the effect of external leakage path can be removed by using the guard circuit on the three terminal resistance.

If guard circuit is absent, leakage resistance R_1 and R_2 would be placed directly across Rx and giving large error in the measurement.

6.15 Megger

Resistance of the order of 0.1 MΩ and upwards are classified as high resistances. These high resistances are measured by portable instrument known as "Megger". It is also used for "Testing the insulation resistance of cable".

Principle of Operation

It is based on the principle of electromagnetic induction. The Fig. 6.19 shows the construction of Megger. When a current carrying conductor is placed in a uniform magnetic field, it experiences a mechanical force whose magnitude depends upon the strength of current and magnetic field. While its direction depends on the direction of current and magnetic field.

6.15.1 Construction

It consists of a permanent magnet which provides the field for both the generator G and ohmmeter. The moving element of the ohmmeter consists of three coil viz., current or deflection coil, pressure or control coil and compensation coil. These coils are mounted on a central shaft which are free to rotate over a stationary C-shaped iron core.

The coils are connected to the circuit through flexible leads called ligaments which do not produce a restoring torque on the moving element. Consequently, the moving element takes up any position over the scale when the generator handle is stationary. The current

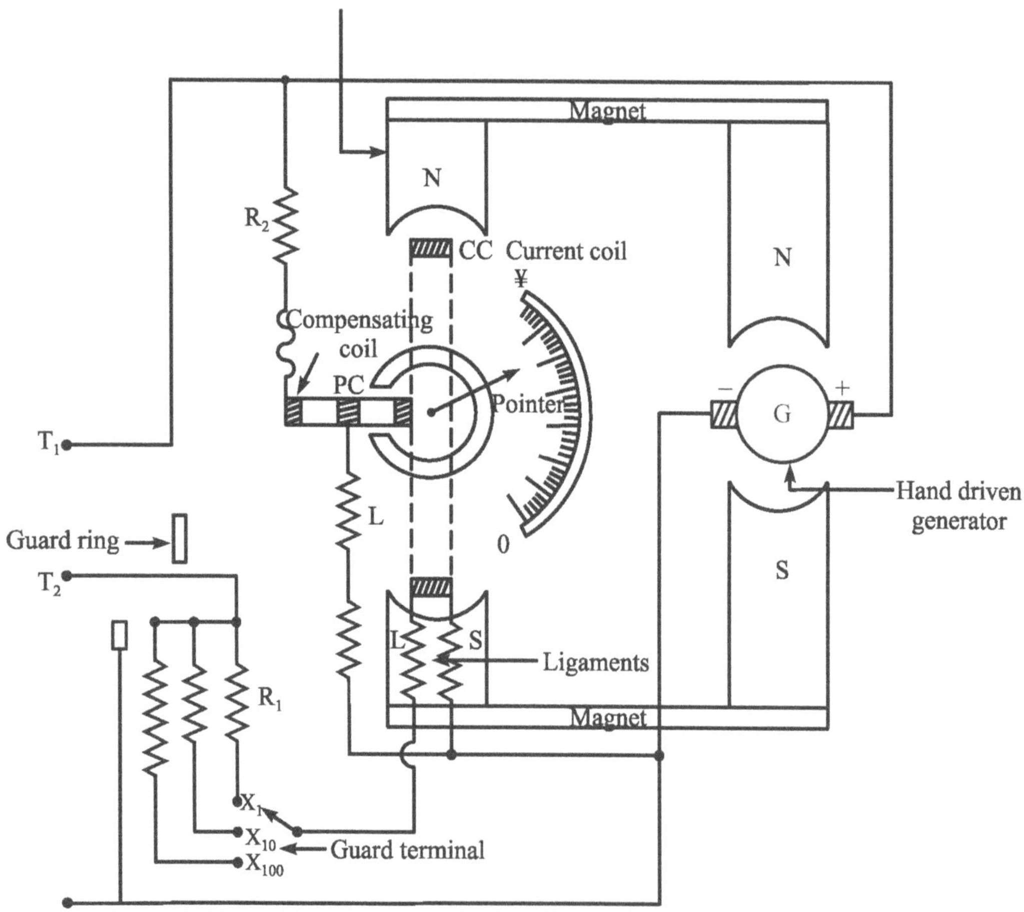

Fig. 6.19 Megger

coil is connected in series with resistance R_1 between one generator terminal and the test terminal T_2. The series resistance R_1 protects the current coil in the event of the test terminals getting short circuited and also controls the range of the instrument. The pressure coil, in series with a compensating coil and protection resistance R_2 is connected across the generator terminals. The compensating coil included in the circuit to ensure better scale proportions. The scale is calibrated reversely; means the normal position of pointer indicates infinity while full scale deflection indicates zero resistance.

6.15.2 Working

When the current flows from the generator, through the pressure coil, the coil tends to set itself at right angle to the field of the permanent magnet.

When the test terminals are open, corresponding to infinite resistance, no current flows through deflection coil. Thus the pressure coil governs the motion of the moving element making it moves to its extreme anticlockwise position. The pointer comes to rest at the infinity end of the scale.

When the test terminals are short circuited i.e., corresponding to zero resistance the current from the generator flowing through the current coil is large enough to produce sufficient torque to overcome the counter-clockwise torque of the pressure coil. Due to this, pointer moves over a scale showing zero resistance. When the high resistance to be tested is connected between terminals T_1 and T_2, the opposing torque of the coils balance each other so that pointer attains a stationary position at some intermediate point on scale. The scale is calibrated in mega ohms so that the resistance is directly indicated by pointer.

The guard ring is provided to eliminate the error, due to leakage current. The supply to the meter is usually given by a hand-driven permanent magnet d.c. generator. Sometimes motor-driven generator may also be used.

6.15.3 Applications

The Megger can be used to determine whether there is sufficiently high resistance between the conducting part of a circuit and the ground. This resistance is called insulation resistance.

The Megger can also be used to test continuity between any two parts. When connected to the two points, if pointer shows full deflection then there is an electrical continuity between them.

6. 16 Bridge in Controlled Circuits

The unbalanced bridge causes a potential difference across the galvanometer. This potential causes current to flow through the galvanometer. If the bridge is used in control circuit, the galvanometer is not used and the potential developed due to un balanced conditions across the two terminals is used to drive some other control circuit. This potential is called "Output voltage (or) error voltage".

While using bridge in such circuit, one arm of the bridge consists of a resistance which is sensitive to the parameter under control. The various parameters are strain, pressure, temperature, light etc, while parameter sensitive elements are strain gauges, temperature sensitive resistor (thermistors), photo resistors etc. Under normal value of the parameter, the sensitive resistance value R_v is such that R_2/R_v is same as R_1/R_3 and the bridge is perfectly balanced. The error signal is zero.

But when the physical parameter changes, then R_v changes. This causes unbalance in the bridge which produces the error voltage. This voltage is used to drive some circuity, which restores the condition of the parameter, thus bringing R_v back to its normal value, maintaining the bridge balance. The entire control action depends on the change in R_v and the production of the error voltage. Hence, Wheatstone bridge used in such circuit is called "error detector". The error voltage is generally very small and needs to amplify before using it for any control purposes. The Wheatstone bridge used as an error detector is show in the Fig. 6.20.

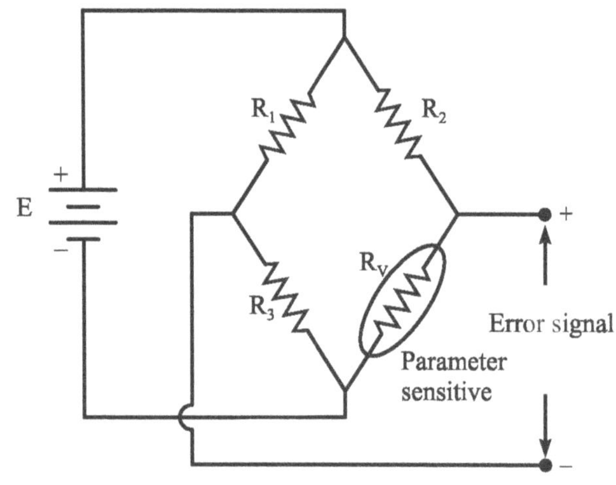

Fig. 6.20 Wheatstone bridge as error detector

6.17 Difficulties in the Measurement of High Resistance

The measurement of high resistance of the order of hundreds and thousands of mega ohms is often required in electrical equipments. The examples are:

1. Leakage resistance of capacitors
2. Insulation resistance of components like machines, cables etc.

3. Resistance of high circuit elements like vacuum tube circuits.

For measuring high resistance the value of current in the circuit will be very small. It is very difficult to sense such small currents in the circuit. Thus, due to this reason difficulties arises. They are follows:

1. Whenever high resistance are to be measured, there is always a time delay between the application of voltage and the observation of the reading, this time delay takes place because the insulation resistance always require certain amount of time for charging and absorbing currents, when the voltage is applied to it.
2. In order to increase the current magnitude in the circuit, very high voltage is supplied to the instrument galvanometer (or) micro-ammeter. But such high voltage is harmful to the instrument. Thus, proper precautions must be taken in order to avoid danger to micro-ammeter or to the galvanometer.
3. Due to the presence of leakage current in the circuit, errors are produced, whose magnitude is equal to the magnitude of measuring current.
4. Electrostatic effect can also cause errors in the measurement i.e., electrostatic effect gives rise to stray changes in the measuring circuit, which may cause errors in the measurement.
5. One point of the circuit may be connected to earth for accuracy in measurements, and
6. Very high voltage is required in order to raise the current magnitude. The galvanometer should be very sensitive and proper steps must be taken to prevent the damage of galvanometer due to high voltages.

In order to minimize or to eliminate the errors, guard circuits are used. These guard circuits are mainly used to eliminate the error, which are caused by leakage currents. The circuit for measurement of high resistance is as shown in Fig. 6.21.

6.18 Use of Guard Circuits

The Fig. 6.21 shows the basic principle of guard circuit.

The high resistance mounted on a piece of insulating material is measured by ammeter-voltmeter method. In Fig. 6.21(a), it can be seen that the microammeter carries the current which is a sun of leakage current I_2 and the resistance current I_R. Hence the reading obtained by such measurement will not be a true reading due to the error caused by leakage current.

For this, a guard terminal is added to the resistance terminal block, as shown in the Fig. 6.21(b).

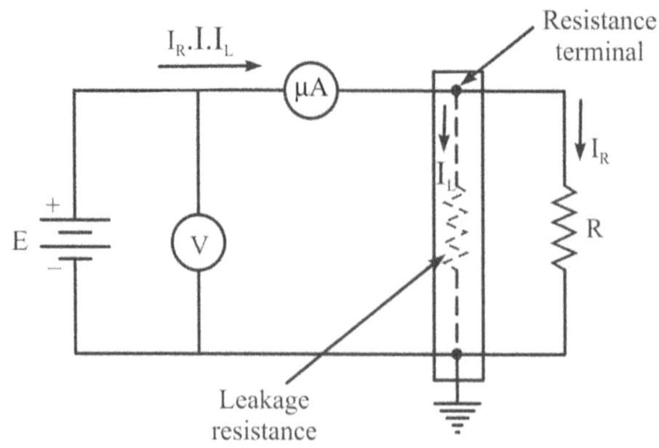

Fig. 6.21 (a) without guard

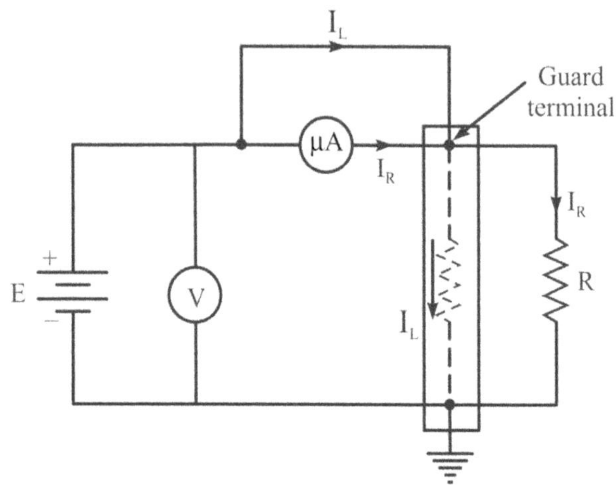

Fig. 6.21 (b) with guard

This terminal surrounds the resistance entirely and is connected to the battery side of the microammeter. The leakage current. I_L has now a separate path and it by passes the microammeter. The current through microammeter is I_R only and hence the high resistance can be determined correctly. The guard and resistance terminals are at the same potential hence no current can flow in between them.

The same principle is used to guard the Wheatstone bridge. This is shown in the Fig. 6.22 without guard, the leakage current I_L along the insulated surface adds to the current I_x causing the error.

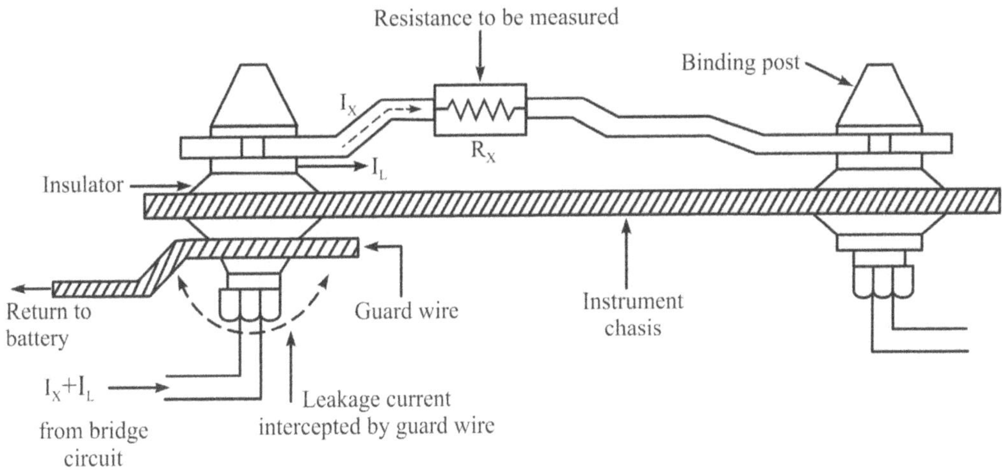

Fig. 6.22 Simple guard circuit

A guard wire shown thick, completely surrounds the surface of the insulated post and abstructs the leakage current and returns it back towards battery. In schematic diagram, the guard around the R_x, binding post is shown by small circle around the terminal. It does not touch to any part of the bridge and is directly connected to positive of battery. The Wheatstone bridge with guard terminal is shown in the Fig. 6.23.

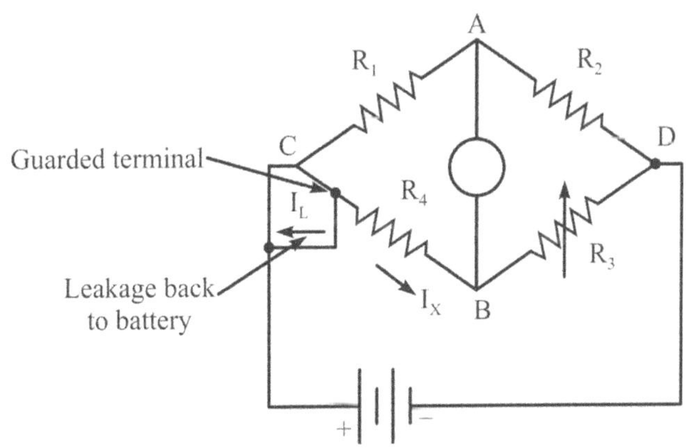

Fig. 6.23 Guarded wheatstone bridge

Problems with Solutions

1. A voltmeter of resistance 500 Ω and a milli-ammeter of 1 Ω resistance are used to measure a resistance by ammeter-voltmeter method. If the volt meter reads 20 V and milli-ammeter 100 mA, calculate the value of measured resistance.

(i) If the voltmeter is put across the resistance and the milli-ammeter connected in series with the unknown resistance and

(ii) If the voltmeter is put across the resistance with ammeter connected on the supply side.

***Solution*:** Given Data

Voltmeter resistance, $R_v = 500\ \Omega$

Ammeter resistance, $R_a = 1\ \Omega$

Voltmeter reading, $V = 20$ V

Ammeter reading $1 = 100$ mA

Let R be the value of unknown resistance, R' and R'' be the measured values of resistance in case (i) and (ii) respectively

(i) Milli-ammeter connected towards the unknown resistance

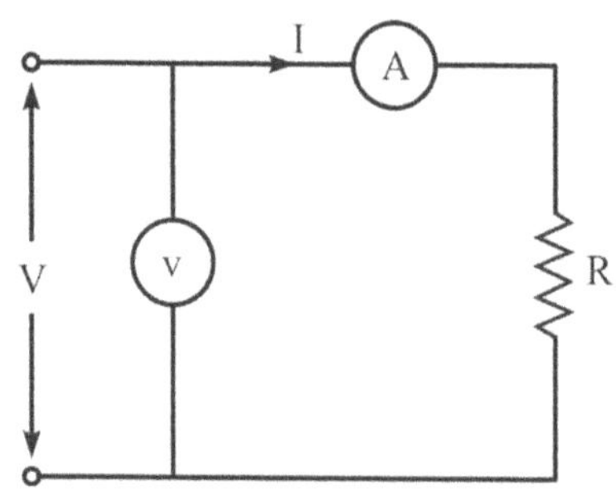

$$R' = R + R_a$$

$$R = R' - R_a$$

$$R' = \frac{\text{Voltmeter reading}}{\text{Ammeter reading}}$$

$$R' = \frac{20}{100 \times 10^{-3}} = 200\ \Omega$$

$$\therefore \quad R = 200 - 1 = 199\ \Omega$$

(ii) Milli-ammeter connected towards the supply side

$$R'' = \frac{RR_V}{R + R_V}$$

$$R'' = \frac{20}{100 \times 10^{-3}} = 200\ \Omega$$

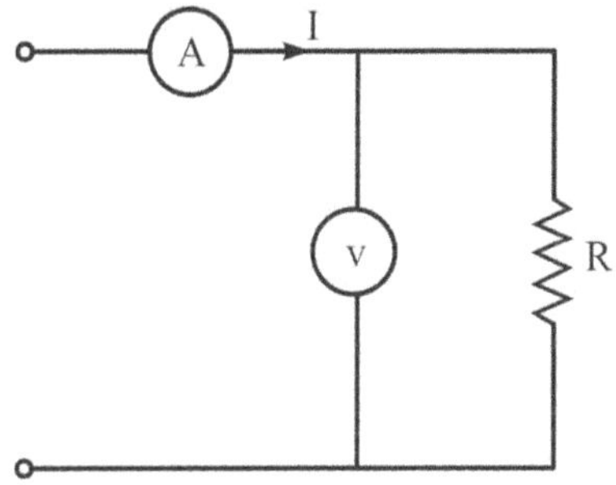

$$\therefore \qquad 200 = \frac{R \times 500}{R + 500}$$

$$200\ R + 100000 = 500\ R$$

$$300\ R = 100000$$

$$R = \frac{100000}{300}$$

$$= 333.33\ \Omega$$

2. In a laboratory, a volt meter of 200 Ω resistance and an ammeter of 0.02 Ω resistance are available. Calculate the value of resistance that can be measured by the ammeter-voltmeter method for which the two different circuit measurements give equal errors.

Solution: Given Data

$R_v = 200\ \Omega$

$R_a = 0.02\ \Omega$

The two different circuits for the measurement of resistance are shown in Fig. (i) and (ii).

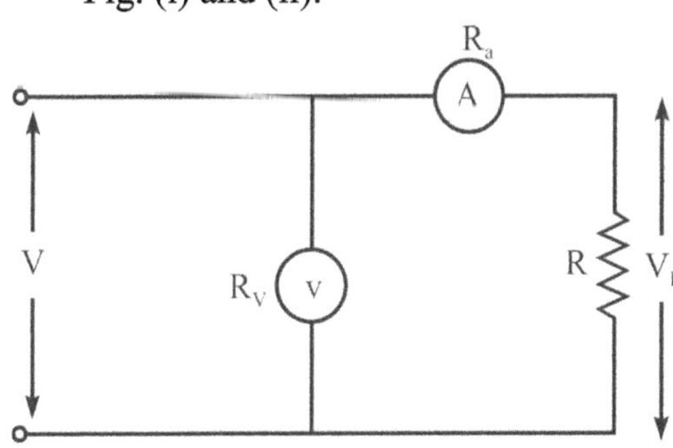

Fig. (i)

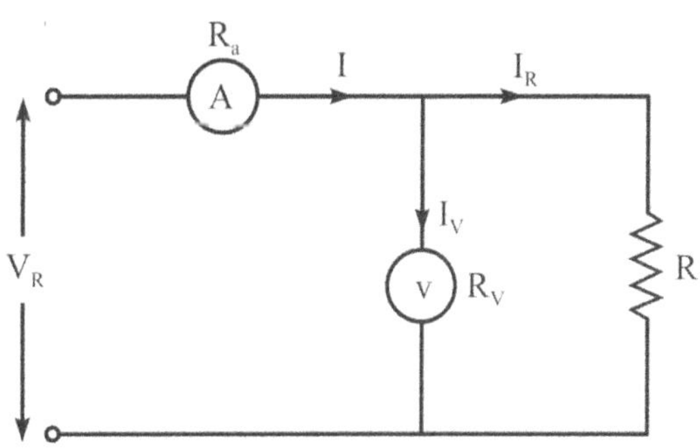

Fig. (ii)

Let,

e' = error with connection (i)

e" = error with connection (ii)

Given

$$e' = e''$$

$$e' = \frac{R_a}{R} \quad \text{and} \quad e'' = \frac{-R}{R_v}$$

$$\therefore \quad \frac{R_a}{R} = \frac{-R}{R_v}$$

$$-R^2 = R_a R_v$$

$$-R^2 = 0.02 \times 200$$

$$-R^2 = 4$$

$$-R = \pm\ 2$$

$$R = -2\ \Omega \quad \text{and} \quad +2\ \Omega$$

∴ R can't be negative

∴ $R = 2\ \Omega.$

3. The four arms of a Wheatstone bridge are as follows: AB = 100 Ω, BC = 10 Ω, CD = 4 Ω, DA = 50 Ω. The galvanometer has a resistance of 20 Ω and is connected across BD. A source of 10 V DC is connected across AC. Find the current through the galvanometer. What should be the resistance in the arm DA for no current through the galvanometer?

***Solution*:** Given Data

Resistance in arm AB, $P = 100\ \Omega$

Resistance in arm BC, $Q = 10\ \Omega$

Resistance in arm CD, $S = 4\ \Omega$

Resistance in arm DA, $R = 50\ \Omega$

Apply potential across AC, $V = 10$ V

Galvanomter, $1G = ?$

The thievinin source generator voltage,

$$V_0 = V\left[\frac{R}{R+S} - \frac{P}{P+Q}\right]$$

$$= 10 \times \left[\frac{50}{50+4} - \frac{100}{100+10}\right]$$

$$= 10 \times \left[\frac{50}{54} - \frac{100}{110}\right]$$

$$= 0.16835 \text{ V}$$

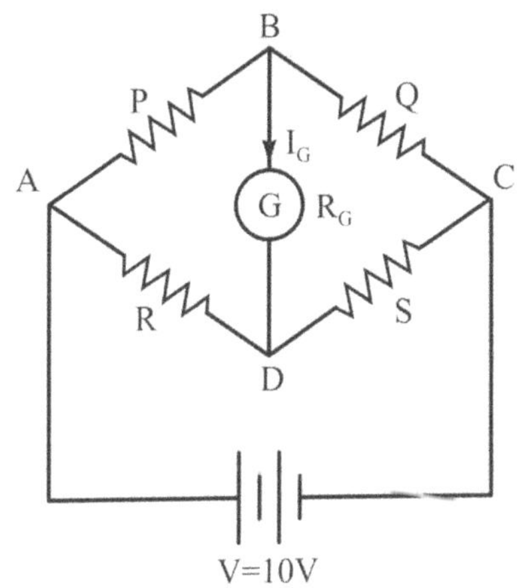

Fig. (i)

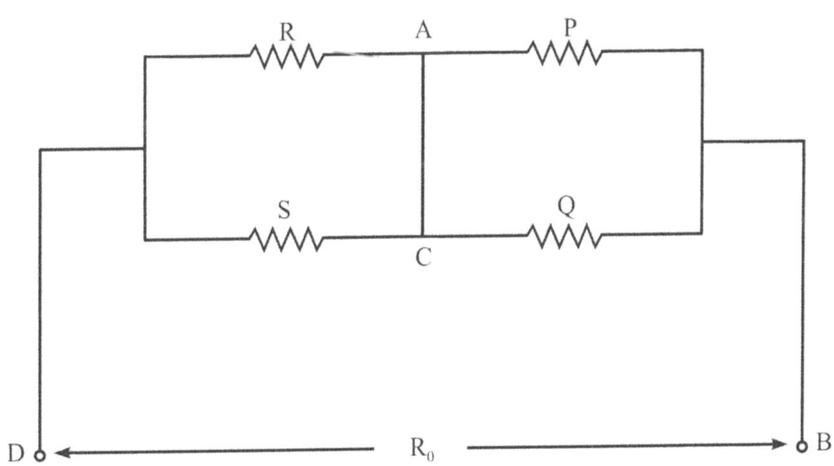

Fig. (ii)

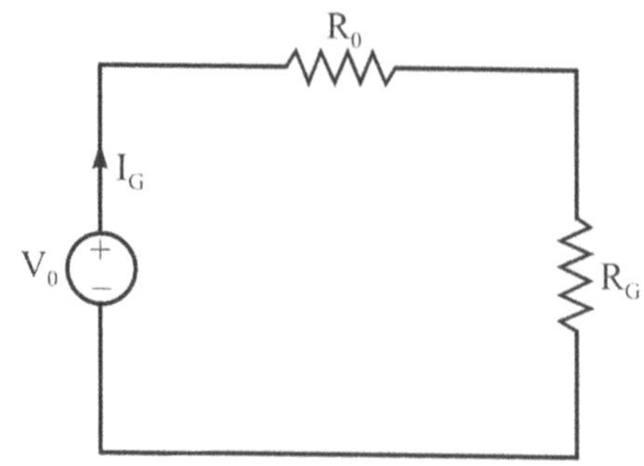

Fig. (iii)

The thievinin's equivalent resistance, R_0 = resistance of circuit looking into terminal D and B with terminals A and C short circuited (as shown in Fig. (ii))

$$R_0 = (R \parallel S) + (P \parallel Q)$$

$$= \frac{RS}{R+S} + \frac{PQ}{P+Q}$$

$$= \frac{50 \times 4}{50+4} + \frac{100 \times 10}{100+10}$$

$$= \frac{200}{54} + \frac{1000}{110}$$

$$\therefore \quad R_0 = 12.794\ \Omega$$

From Fig. (iii), we have,

$$I_G = \frac{V_0}{R_0 + R_G} = \frac{0.16835}{12.794 + 20}$$

$$= 5.1335 \text{ mA.}$$

$\therefore$ current through the galvanometr,

$$I_G = 5.1335 \text{ mA}$$

The bridge is said to be balanced when there is no current through the galvanometer.

We know under balanced condition

$$QR = PS$$

$$(10)\ (R) + (100)\ (4)$$

$$R = 40\ \Omega$$

∴ The value of resistance in the arm DA for the current through the galvanometer.

$$R = 40\ \Omega.$$

4. The four arms of a Wheatstone bridge are as follows: AB = 100 Ω, BC = 1000 Ω, CD = 4000 Ω, DA = 400 Ω. The galvanometer has a resistance of 100 Ω, a sensitively of 100 mm/μA and is connected across AC. A source of 4 V DC is connected across BD. Calculate the current through the galvanometer and its deflection if the resistance of arm DA is changed from 400 Ω to 401 Ω.

***Solution*:** Given Data

Resistance of

Arm AB = 100 Ω

Arm BC = 1000 Ω

Arm CD = 4000 Ω

Arm DA = 400 Ω

Applied voltage, E = 4 V.

Sensitivity of galvanometer, $S_g = 100$ mm/μA

Internal resistance of galvanometer, $R_g = 100\ \Omega$.

(i) Current through the galvanometer

The Wheatstone's bridge according to the given data is drawn as shown in the Fig.

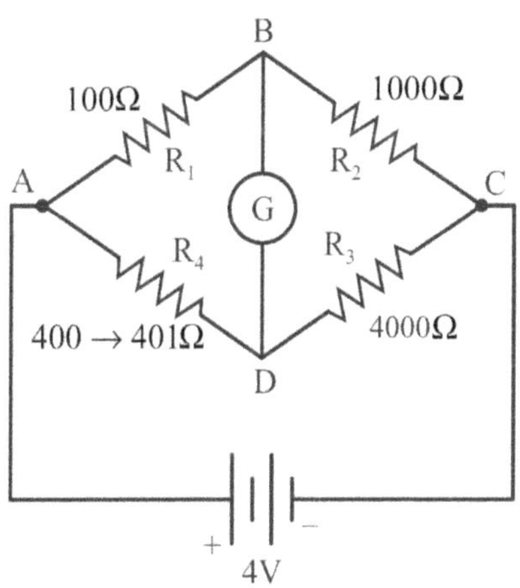

In the above Fig, there is a change in arm DA i.e., R4 is changed from 400 Ω to 401 Ω. Due to this change in R_4, the current will flow through the galvanometer.

Let us find thevinin's voltage

$$V_{Th} = E\left[\frac{R_4}{R_4+R_3} - \frac{R_1}{R_1+R_2}\right]$$

$$= 4\left[\frac{401}{401+4000} - \frac{100}{100+1000}\right]$$

$$= 4\,[0.09011 - 0.09090]$$

$$V_{Th} = 0.84 \text{ mV}$$

$$R_{eq} = \frac{R_1\,R_2}{R_1+R_2} + \frac{R_4\,R_3}{R_4+R_3}$$

$$= \frac{100\times1000}{100+1000} + \frac{401\times4000}{401+4000}$$

$$= 90.909 + 364.462$$

$$= 455.371\ \Omega$$

∴ current through the galvanometer is given as,

$$I_g = \frac{V_{Th}}{R_{eq}+R_g}$$

$$= \frac{0.84\times10^{-3}}{455.371+100}$$

$$= 1.51\ \mu A$$

(ii) Deflection of the galvanometer

Deflection of the galvanometer can be found from the expression.

$$S_g = \frac{\theta}{I_g}$$

$$\therefore \quad \theta = S_g \times I_g$$

$$= 100 \times 1.51$$

$$= 151 \text{ mm}$$

5. A highly sensitive galvanometer can detect a current as low as 0.1 nanoampere. This galvanometer is used in a wheatstone bridge as detector. The resistance of galvanometer is negligible. Each arm of the bridge has a resistance of 1 k Ω. The input voltage applied to the bridge is 20 V. Calculate the smallest change in resistance, which can be detected.

***Solution*:** Given data

For wheatstane bridge each arm has,

Resistance, $R = 1\ k\Omega$

Supply voltage, $V = 20\ V$

Galvanometer can detect a current as low as 0.1 nanoamperes i.e.,

$I_g = 0.1 \times 10^{-9}\ A$

To determine the change is resistance $\Delta r = ?$

For bridge with equal arms

$R_1 = R_2 = R_3 = R_4 = R$, the thevinin's approximate voltage is given as,

$$V_{th} = \frac{VR\Delta r}{4R^2}$$

$$V_{th} = \frac{V\Delta r}{4R}$$

The equivalent resistance is given as

$$R_{eq} = \frac{R^2}{2R} + \frac{R^2}{2R} = R = 1000\ \Omega$$

Also,

$$I_g = \frac{V_{Th}}{R_{eq} + R_g}$$

$$= \frac{\frac{V\Delta r}{4R}}{R_{eq} + R_g}$$

where, R_g is the resistance of galvanometer, which is negligible i.e., $R_g = 0$

$$\therefore\ I_g = \frac{V\Delta r}{4R\ .\ R_{eq}}$$

$$\Rightarrow \quad 0.1 \times 10^{-9} = \frac{20 \times \Delta r}{4 \times 1000 \times 1000}$$

$$\therefore \quad \Delta r = 2 \times 10^{-5}$$
$$= 20\ \mu\Omega$$

6. A Kelvin double bridge is balanced with the following constants: outer ratio arms 100 Ω and 1000 Ω, inner ratio arms 99.92 Ω and 1000.6 Ω, resistance of link 0.1 Ω, standard resistance 0.00377 Ω. Calculate the value of unknown resistance.

***Solution*:** Given Data

Outer ratio arm resistances, A = 100 Ω; B = 1000 Ω

Inner ratio arm resistances, a = 99.92 Ω; b = 1000.6 Ω

Resistance of link, r = 0.1 Ω

Standard resistance, S = 0.00377 Ω

Unknown resistance, x = ?

The value of unknown resistance (x) when the Kelvin's double bridge is given by,

$$x = \frac{A}{B} \cdot S + \frac{b\,r}{a+b+r}\left[\frac{A}{B} - \frac{a}{b}\right]$$

$$\Rightarrow \qquad x = \frac{100}{1000} \times 0.00377 + \left[\frac{1000.6 \times 0.1}{99.92 + 1000.6 + 0.1}\right]\left[\frac{100}{1000} - \frac{99.92}{1000.6}\right]$$

$$= 0.000377 + 0.090912 \times 0.000139$$
$$= 3.897 \times 10^{-4}\ \Omega$$
$$= 0.3897\ \text{m}\Omega$$

∴ The value of unknown resistance

$$x = 0.3897\ \text{m}\Omega.$$

7. A length of cable is tested for insulation resistance by the loss of charge method. An electrostatic voltmeter of infinite resistance is connected between the cable conductor and earth, forming there with a joint capacitance of 600 pF. It is observed that after charging, the voltage falls from 250 V to 92 V in 1 minute. Calculate the insulation resistance of the cable.

***Solution*:** Given Data

C = 600 pF

Battery voltage, V = 250 V

Reduced voltage, V = 92 V

Time, $t = 1$ minute $= 60$ seconds

$$R = \frac{0.4343\ t}{c\ \log\ (V/v)}$$

$$= \frac{0.4343\ \times 60}{600 \times 10^{-12} \log\ (250/92)}$$

$$= \frac{26.058}{600 \times 10^{-12} \times 0.43415}$$

$$= \frac{26.058}{260.49 \times 10^{-12}}$$

$$R = 100 \times 10^{9}\ \Omega$$

8. The following results were obtained by loss of charge method of testing cable, discharged immediately after charging the deflection = 200 divisions.

   Discharged 30 seconds after charging the deflection = 125 divisions.

   Discharged 30 seconds after charging, when in parallel with a resistance of 10 MΩ, the deflection = 100 divisions. Calculate the insulation resistance of the cable.

***Solution*:** Given Data

Voltage across the capacitor, $V_c = 200$ V.

Electrostatic voltmeter reading, $V_1 = 125$ V

Time, $t_1 = 30$ sec

Electrostatic voltmeter reading, $V_2 = 100$ V

Time, $t_2 = 30$ sec

Resistance, $R_1 = 10$ MΩ

The loss of charge method is shown in Fig. (a).

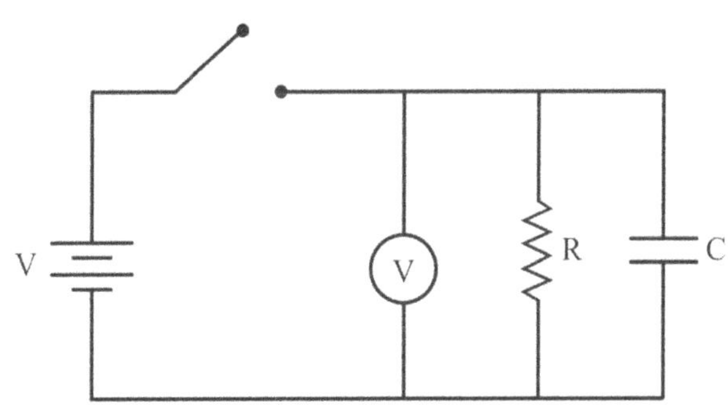

Let us assume,

R is the insulation resistance of the cable.

R^1 is the resultant resistance i.e., R^1 is the parallel combination of insulation and 10 MΩ resistance.

we know that insulation resistance for loss of charge method is given as,

$$R = \frac{0.4343\ t}{C\ \log_{10} V_c / v}$$

$$R = \frac{0.4343\ \times 30}{C\ \log_{10}\ 200/125}$$

$$= \frac{13.029}{0.204\ C} \qquad .....(1)$$

$$R^1 = \frac{0.4343 \times 30}{C\ \log_{10}\ 200/10}$$

$$= \frac{13.029}{0.301\ C} \qquad .....(2)$$

Dividing equation (2) by equating (1), we get,

$$\frac{R^1}{R} = \frac{0.204}{0.301}$$

$$R^1 = 0.677\ R$$

The test is repeated with unknown resistance 'R' disconnected and the capacitor is discharging through R^1.

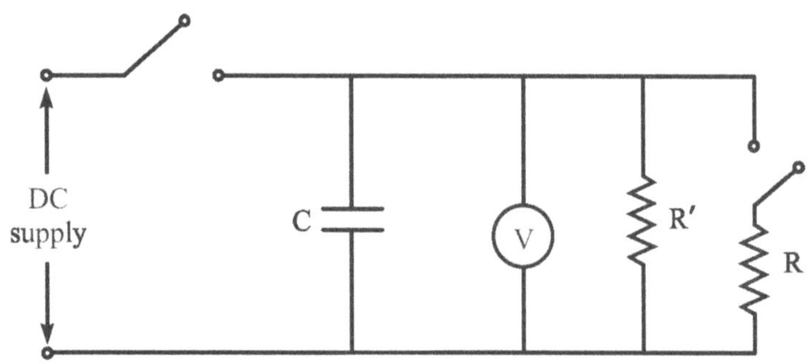

The value of R^1 from this test is given as,

$$R^1 = \frac{R\,R^1}{R + R^1}$$

$$0.677\,R = \frac{R \times 10}{R + 10}$$

$$0.677R^2 + 6.77R = 10R$$

$$0.677R^2 = 3.23\,R$$

$$\therefore \quad R = 4.77\ M\Omega$$

Review Questions

1. Classify the resistance from the point of view of measurements.
2. Describe in brief the different methods used for measurement of medium resistance.
3. Describe the volt-meter method used for measurement of cow resistance.
4. Draw the circuit of a Kelvin's double bridge used for measurement of low resistances. Derive the condition for balance.
5. Explain the potentiometer method used for measurement of low resistance.
6. Describe the substitution method of measurement of medium resistance.
7. Draw the circuit of a Wheatstone bridge and derive the conditions of balance.
8. Derive the expression for bridge sensitivity for a wheatstone bridge with equal arm. Find also the expression for current through the galvanometer for a small unbalance.
9. Explain the measurement errors for the precision measurement of wheatstone bridge.
10. Explain the advantages, applications and limitations of wheatstone bridge.
11. Draw and explain the direct deflection method used for measurement of low resistance.
12. Explain the errors in loss of charge method.
13. Describe the working of a Carey-Foster slide-wire bridge.
14. Explain the loss of charge method for measurements of insulation resistance of cable.
15. Describe the working of a Megohm bridge.

16. What are the difficulties in measurement of high resistance? Explain the use of guard circuits.
17. Draw and explain the construction, working principle of megger.

Quiz Questions

1. For eliminating the effect of earth capacitance from the bridge network we use:
   (a) wagner's earthing device (b) high voltage at low frequency
   (c) low voltage at high frequency (d) campbell-maxwell device
2. Loss of charge method is used to measure:
   (a) low R (b) high R
   (c) low L (d) high L
3. Megger is an instrument used for the measurement of:
   (a) high resistance and insulation resistance
   (b) medium resistance
   (c) low resistance
   (d) leakage current
4. Kelvin double bridge is best suited for the measurement of:
   (a) inductance (b) capacitance
   (c) low resistance (d) high resistance
5. The material used to make standard resistance is:
   (a) manganin (b) aluminium
   (c) nichrome (d) platinum
6. Specific resistance is measured in
   (a) mho (b) ohm
   (c) ohm – cm (d) ohm/cm
7. The material to be used in the manufacture of a standard resistor should be of
   (a) Low resistivity
   (b) High resistivity and low temperature co-efficient
   (c) High temperature co-efficient
   (d) Low resistivity and high temperature co-efficient

8. The best material of use for standard resistor is
   (a) manganin (b) aluminium
   (c) nichronme (d) platinum

9. The insulation resistance of a cable can be measured by
   (a) megger (b) ammeter-voltmeter method
   (c) ohm meter (d) kelvin's bridge

10. The insulation resistance of a T/F winding can be easily measures with
   (a) wheatstone bridge (b) megger
   (c) voltmeter (d) kelvin's bridge

11. Dielectric loss can be measured by
   (a) energy meter (b) wheat stone bridge
   (c) electrostatic wattmeter (d) mega ohm bridge

Questions and Answers

1. Classify resistance.

Ans. Low resistance

Medium resistance

High resistance

2. What is the range of medium resistance?

Ans. Resistance of about 1 ohm to 100 kilo ohms are called medium resistance.

3. Name the methods used for low resistance measurement.

Ans. Ammeter – voltmeter method

Potentiometer method

Kelvin double bridge method

Ohm meter method.

4. Name the methods used for medium resistance measurement

Ans. Ammeter – voltmeter method

Substitution method

Wheatstone bridge method

Carey-Foster bridge method.

5. Define megger.

Ans. The megger is an instrument used for the measurement of high resistance and insulation resistance.

6. Classify the cables according to their sheathing.

Ans. Armoured cables

Unarmoured cables.

7. How resistance is measured by using ohm meter method.

Ans. Series ohm meter method

Shunt ohm meter method.

8. How resistance is measured in loss of charge method.

Ans. In this method a capacitor is charged and discharged for a specific time period and from this resistance is measured.

9. State the advantages of price's guard wire method.

Ans. In this method leakage current does not flows through the meter and therefore it gives accurate reading.

10. Manganin is :__________

Ans. An alloy used to make standard resistors

11. A megger is usually

Ans. Moving coil type instrument

Quiz Questions Key

1. (a)	2. (b)	3. (a)	4. (c)	5. (a)
6. (a)	7. (b)	8. (a)	9. (a)	10. (c)
11. (c)				

CHAPTER 7

A.C Bridges

7.1 Introduction

Alternating current bridge methods are of outstanding importance for measurement of "electrical quantities". Measurement of "inductance", "capacitance", storage factor, loss factor may be made conveniently and accurately by employing A.C bridge networks.

A "bridge circuit" in its simplest form consists of a network for four resistance arms forming a closed circuit. A source of current is applied to two opposite junctions, the current detector is connected to other two junctions.

The bridge circuit use the comparison measurement methods and operate on null-indication principle. The bridge circuit compares the value of an unknown component with that of an accurately known standard component. Thus the accuracy depends on the bridge components and not on the null indicator. Hence high degree of accuracy can be obtained.

In a bridge circuit, when no current flows through the null detector which is generally galvanometer, the bridge is said to be balanced. The relationship between the component values of the four arms of the bridge at the balancing is called "balancing condition" or "balancing equation". This equation gives us the value of the unknown component.

7.2 Advantages of Bridge Circuit and Applications

The various advantages of the bridge circuit are:

1. The measurement accuracy is high as the measurement is done by comparing the unknown value with the standard value.
2. The balance equation is independent of the magnitude of the input voltage or its source impedance. These quantities do not appear in the balance equation expression.

3. The accuracy is independent of the characteristics of a null detector and is dependent on the component values.
4. The balance equation is independent of the sensitivity of the null detector, the impedance of the detector or any impedance shunting the detector.
5. The balance condition remains unchanged if the source and detector are interchanged.
6. The bridge circuit can be used in the control circuits. When used in such control applications one arm of the bridge contains a resistive element that sensitive to the physical parameter like pressure, temperature etc., which is to be controlled.
7. This bridge circuits are commonly used for phase shifting, providing feedback paths for oscillators and amplifiers, filtering out undesirable signals and measuring the frequency of audio signals.

7.3 Types of Bridges

The two types of bridges are:

1. D.C bridge and 2. A.C bridges

The "D.C bridges" are used to measure the resistances while the "A.C bridges" are used to measure the impedences consisting capacitance and inductances. The D.C bridge use the D.C voltage as the excitation voltage while the a.c bridges use the alternating voltage as the excitation voltage.

The two types of D.C bridges are:

1. Kelvin bridge
2. Wheatstone bridge

The various types of A.C bridges are:

1. Inductance comparison bridge
2. Capacitance comparison bridge
3. Maxwell's bridge
4. Anderson bridge
5. Schering bridge
6. Hay's bridge
7. Wien bridge

Let us discuss the various type of the bridges in details

7.4 AC Bridge

An A.C bridge in its basic form consist of four arms, a source of excitation and a balance detector. Each arm consists of an impedance. The source is an A.C supply which supplies A.C voltage at the required frequency. For high frequencies, the electronic oscillators are used as the source. The balance detectors commonly used for A.C bridge are head phone, tunable amplifier circuits or vibration galvanometers. The headphones are used as detectors at the frequencies of 25 Hz to 3 to 4 kHz, while working within single frequency a tuned detector is the most sensitive detector. The vibration galvanometers are useful for low audio frequency range from 4 Hz to 1000 Hz but are commonly used below 200 Hz. Tunable amplifier detectors are used for frequency range of 10 Hz to 100 Hz.

The simple A.C bridge is the outcome of the Wheatstone bridges. The impedance at audio and radio frequency range can be easily determined by such simple A.C wheatstone bridge. It is shown in the Fig. 7.1. This is similar to D.C Wheatstone bridge.

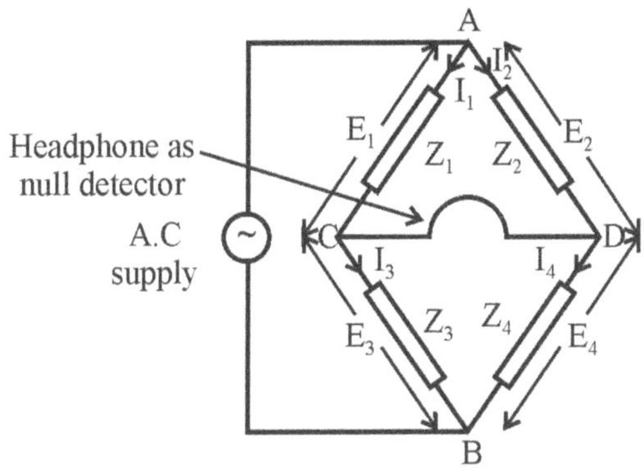

Fig. 7.1 Basic A.C bridge network

The bridge arms are impendence. The bridge is excited by A.C supply and pair of head phones is used as a null detector. The null response is obtained by varying one of the bridge arms.

7.4.1 Bridge Balance Equation

For bridge balance, the potential of point C must be same as the potential of point D. These potentials must be equal in terms of amplitude as well as phase.

The drop from A to C must be equal to drop across A to D, in both magnitude and phase for the bridge balance

$$\therefore \qquad \bar{E}_{AC} = \bar{E}_{AD} \qquad .....(7.1)$$

The vector notation indicates, both amplitude and phase to be considered

$$\overline{I_1 Z_1} = \overline{I_2 Z_2} \qquad .....(7.2)$$

When the bridge is balanced, no current flows through the headphones

$$\therefore \qquad \bar{I}_3 = \bar{I}_1 \text{ and } \bar{I}_4 = \bar{I}_2$$

$$\text{Now} \qquad \bar{I}_1 = \frac{\bar{E}}{\bar{Z}_1 + \bar{Z}_3} \qquad .....(7.3)$$

and $$\bar{I}_2 = \frac{\bar{E}}{\bar{Z}_2 + \bar{Z}_4} \quad .....(7.4)$$

Substituting (7.3) and (7.4) into (7.2) we get,

$$\frac{\bar{E}.\bar{Z}_1}{\bar{Z}_1 + \bar{Z}_3} = \frac{\bar{E}.\bar{Z}_2}{\bar{Z}_2 + \bar{Z}_4}$$

$$\therefore \quad \overline{Z_1Z_2} + \overline{Z_1Z_4} = \overline{Z_1Z_2} + \overline{Z_2Z_3}$$

$$\therefore \quad \overline{Z_1Z_4} = \overline{Z_2Z_3} \quad .....(7.5)$$

The equation (7.5) is the balancing equation in the impedance form. In the "admittance" form the condition can be expressed as,

$$\boxed{\overline{Y_1Y_4} = \overline{Y_2Y_3}}$$

The admittance is the reciprocal of the impedance.

Now in the polar form, the impedances are expressed as,

$$\bar{Z}_1 = Z_1\angle\theta_1,\ \bar{Z}_2 = Z_2\angle\theta_2,\ \bar{Z}_3 = Z_3\angle\theta_3,\ \bar{Z}_4\angle\theta_4 \quad .....(7.6)$$

where Z_1, Z_2, Z_3, Z_4 are the magnitudes and θ_1, θ_2, θ_3 and θ_4 are the phase angles.

Note that the product of the impedance must be carried out in "polar form" where magnitudes get multiplied and phase angles get added.

Substituting in equation (7.3) we get,

$$Z_1\angle\theta_1 \times Z_4\angle\theta_4 = Z_2\angle\theta_2 \times Z_3\angle\theta_3 \quad .....(7.7)$$

The eqn. (7.7) gives the two conditions to be satisfied for the bridge balance. Equating magnitudes of both sides, we get the magnitude condition as,

$$Z_1Z_4 = Z_2Z_3 \quad .....(7.8)$$

Equating phase angles, we get,

$$\theta_1 + \theta_4 = \theta_2 + \theta_3 \quad .....(7.9)$$

Note 1: Thus the products of the magnitude of the opposite arms be equal while sum of the phase angles of the opposite arms must be equal.

Note 2: The phase angles are positive for the inductive impedances and negative for the capacitive impedances.

Thus the bridge must be balanced for both the conditions magnitude as well as phase. The phase angles depend on the components of the individual impedances.

- For capacitive branch; $Z_c = R - j\,X_c = |Z_c| \angle -\theta$
- For inductive branch; $Z_L = R + j\,X_L = |Z_L| \angle +\theta$

where $$X_L = 2\pi\,fL\ \Omega \text{ and } X_c = \frac{1}{2\pi fc}\ \Omega$$

Therefore, from the above equation, the conditions for balancing of in A.C bridges are:

1. The magnitudes of impedance or admittance should satisfy the relation
$$Z_1\,Z_4 = Z_2\,Z_3 \quad \text{or} \quad Y_1\,Y_4 = Y_2\,Y_3$$
2. The phase angles of impedance or admittances should satisfy the relation
$$\theta_1 + \theta_4 = \theta_2 + \theta_3$$

7.4.2 Sources and Detectors

For measurements at low frequencies, the power line may act as the source of supply to the bridge circuit. For higher frequencies electronic oscillators are universally used as bridge source supplies.

These electronic oscillators are used as a source of supply universally, because,

1. The output frequency is very stable
2. The output waveform is very close to sine wave
3. The output power is sufficient to drive the bridge circuits
4. The output frequency is easily determinable with accuracy and also it is easily adjustable

A typical oscillator has a frequency range of 40 Hz to 120 kHz with power output of 7 W.

For the a.c bridges commonly used detectors are as follows.

1. *Vibration galvanometers*: For low audio frequency ranges and power ranges, these detectors are extremely effective. Even though these type of detectors are manufactured to work at various frequency ranges starting from 5 Hz to 1000 Hz, these detectors can be effectively used below 200 Hz with greater sensitivity than the headphones.
2. *Head phones*: These are the most commonly used as detectors in A.C bridges. The frequency range over which headphones can be used as detector effectively is 250 Hz up to 3 to 4 kHz.
3. *Tuneable amplifier detectors*: The transistor amplifier can be tuned electrically to any desired frequency and then it can be made to respond to a narrow bandwidth at a bridge frequency. The output of such amplifier is connected to the indicating instruments. The frequency range for these detectors is 10 Hz to 100 kHz.

7.5 Inductance Comparison Bridge

By this bridge, unknown inductance L_x and its internal resistance 'R_x' can be determined. The schematic diagram of bridge is shown in the Fig. 7.2 The bridge consists of pure resistance R_1 and R_2 in the ratio arms. The third arm consists of a standard inductor L_3 and the variable resistance R_3 while the fourth arm consists of the unknown inductor L_x with its internal resistance R_x.

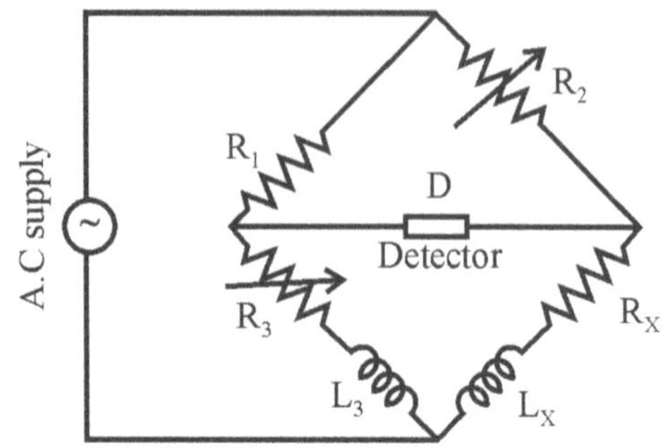

Fig. 7.2 Inductance comparison bridge

Here $Z_1 = R_1$ and $Z_2 = R_2$

$$Z_3 = R_3 + jX_{L3} = R_3 + j(\omega L_3)\ \Omega$$

$$Z_4 = R_x + jR_{Lx} = R_x + j(\omega L_x)\ \Omega$$

From balance condition,

$$Z_1\bar{Z}_4 = Z_2\bar{Z}_3$$

$$R_1[R_x + j\omega L_x] = R_2[R_3 + j\omega L_3]$$

$$R_1R_x + j\omega R_1L_x = R_2R_3 + j\omega R_2L_3$$

Equating real parts, $R_1R_x = R_2R_3$

$$R_x = \frac{R_2R_3}{R_1} \quad .....(7.10)$$

Equating imaginary parts,

$$\omega R_1L_x = \omega R_2L_3$$

$$\therefore \quad L_x = \frac{R_2L_3}{R_1} \quad .....(7.11)$$

By using equation (7.10) and (7.11), unknown elements can be determined. In this bridge, R_2 is selected as inductive balance control and R_3 as resistance balance control. The balance is obtained by alternately varying L_3 or R_3.

7.6 Maxwell's Bridge

Maxwell's bridge can be used to measure inductance by comparison either with a variable standard self inductance or with a standard variable capacitance. These two measurements can be done by using the Maxwell's bridge in two different forms.

7.7 Maxwell's Inductance Bridge

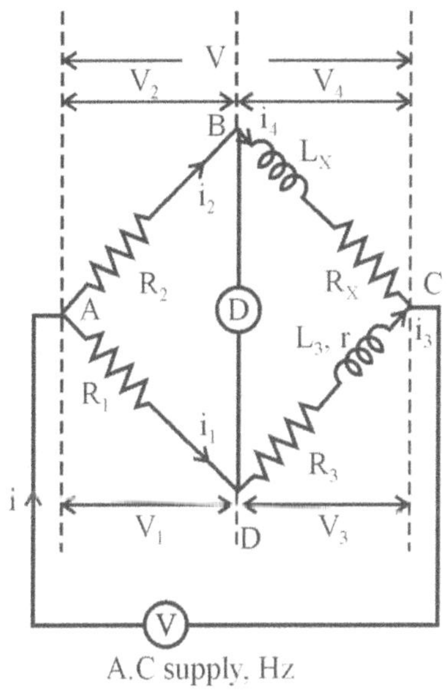

Fig. 7.3(a) Circuit diagram

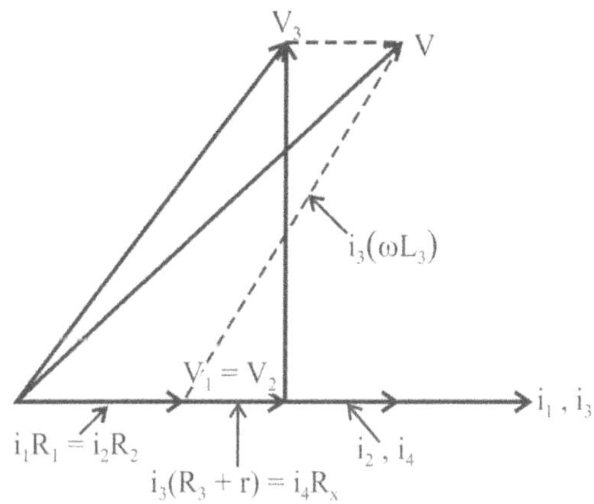

Fig. 7.3(b) Phasor diagram

Using this bridge, we can measure inductance by comparing it with a standard variable self inductance arranged in bridge circuit as shown in Fig. 7.3 (a).

Two branches consist of non-inductive resistance R_1 and R_2. One of the arms consists variable inductance with series resistance r. The remaining arm consists unknown inductance L_x.

At balance, we get condition as

$$\frac{R_1}{\left[(R_3+r)+j\omega L_3\right]} = \frac{R_2}{R_x + j\omega L_x} \quad .....(7.12)$$

$$R_1\left[R_x + j\omega L_x\right] = R_2\left[(R_3+r)+j\omega L_3\right]$$

$$\therefore \quad R_1R_x + j\omega R_1L_x = R_2R_3 + R_2r + j\omega R_2L_3$$

Equating imaginary terms, we can write

$$R_1L_x = R_2L_3 \quad \Rightarrow \quad L_x = \frac{R_2}{R_1}L_3 \qquad .....(7.13)$$

Equating real terms, we can write

$$R_1R_x = R_2(R_3 + r) \quad \Rightarrow \quad R_x = \frac{R_2}{R_1}(R_3 + r) \qquad .....(7.14)$$

Under the balanced condition, the vector diagram for Maxwell's inductance bridge is as shown in the Fig. 7.3(b).

7.8 Maxwell's Inductance Capacitance Bridge

Using this bridge, we can measure inductance by comparing with variable standard capacitor. The bridge circuit diagram is as shown in the Fig. 7.4(a). One of the ratio arms consists of resistance and capacitance in parallel. Hence it is simple to write the bridge equations in the admittance form.

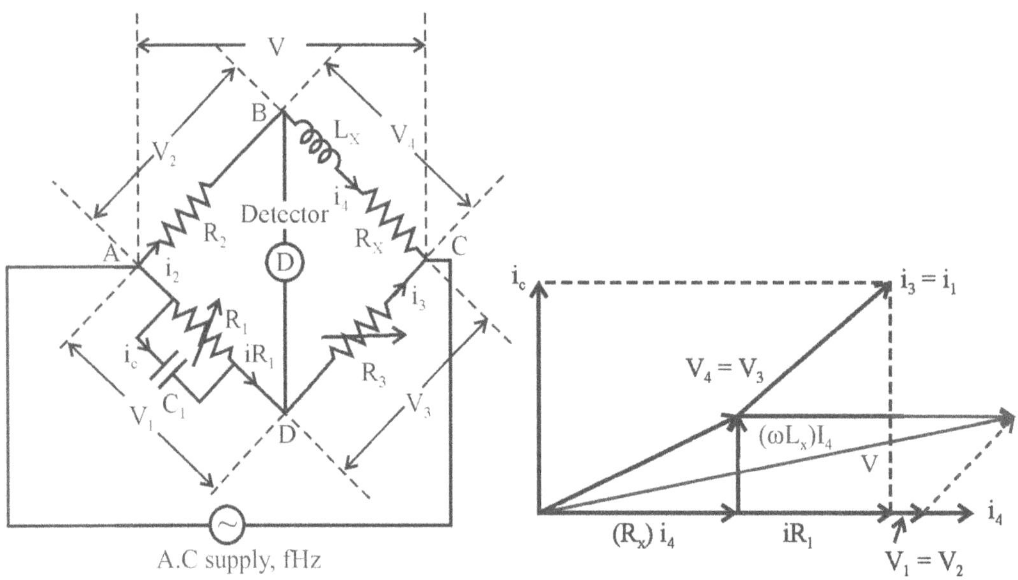

(a) Circuit diagram (b) Phasor diagram

Fig. 7.4 Maxwell's inductance capacitance bridge

The general bridge balance equation is,

$$\overline{Z_1Z_x} = \overline{Z_2Z_3}$$

$$\therefore \overline{Z_x} = \frac{\overline{Z_2 Z_3}}{\overline{Z_1}} = \overline{Z_2 Z_3}\,\overline{Y_1} \quad .....(7.15)$$

where $\overline{Y_1} = \frac{1}{\overline{Z_1}}$ i.e. R_1 in parallel with C_1

$$\overline{Z_2} = R_2$$

$$\overline{Z_3} = R_3$$

$$\overline{Z_x} = R_x + j\omega L_x, \qquad \text{as } L_x \text{ in series with } R_x$$

Now $$\overline{Y_1} = \frac{1}{R_1} + j\omega c_1 \quad .....(7.16)$$

as $$\overline{Z_1} = R_1 \parallel j\left(\frac{1}{\omega c_1}\right) \qquad as\,\frac{1}{j} = -j$$

substituting all the values in (7.15) we get,

$$R_x + j\omega L_x = R_2 R_3\left[\frac{1}{R_1} + j\omega c_1\right]$$

$$\therefore R_x + j\omega L_x = \frac{R_2 R_3}{R_1} + jR_2 R_3 j\omega c_1 \quad ..... (7.17)$$

Equating real parts,

$$\boxed{R_x = \frac{R_2 R_3}{R_1}} \quad ....(7.18)$$

Equating imaginary parts,

$$\omega L_x = R_2\,R_3\,\omega c_1$$

$$\boxed{L_x = R_2 R_3 C_1} \quad .....(7.19)$$

The resistances are expressed in ohms, the inductance in henries and capacitance in farads.

The quality factor of the coil is given by,

$$Q = \frac{\omega L_x}{R_x} = \frac{\omega R_2 R_3 C_1}{\left(\frac{R_2 R_3}{R_1}\right)}$$

$$\boxed{Q = \omega\,R_1 C_1} \quad .....(7.20)$$

The advantages of using standard known capacitor for measurement are:

1. External fields have less effect on a capacitor. The standard inductor requires well shielding in order to eliminate the effect of stray magnetic fields.
2. The standard inductor will not present its rated value of inductance unless current flow through it is precisely adjusted.
3. The capacitors are less expensive than stable and accurate standard inductors.
4. The capacitors are smaller in size.
5. The capacitors are almost loss less.

This bridge is also called as "Maxwell wien bridge".

7.9 Advantages of Maxwell Bridge

The advantages of the Maxwell Bridge are:

1. The stray magnetic fields does not disturb the balance, as the variable element used here is capacitor.
2. The balance equation is independent of losses associated with inductance.
3. The balance equation is independent of frequency of measurement.
4. When the bridge is balanced, the only component in series with coil under test is resistance R_2. If R_2 is selected such that it can carry high current, then heavy current carrying capacity coils can be tested using this bridge.
5. The scale of R_1 can be calibrated to read the Q value directly.
6. The scale of the resistance can be calibrated to read the inductance directly.
7. Very wide range of inductance can be measured not only at power frequencies but also at audio frequencies.

7.10 Disadvantages of Maxwell Bridge

The disadvantages of the Maxwell bridge are:

1. The use of standard variable capacitor makes the bridge highly expensive. However this can be overcome by using a fixed capacitor and varying R_1 and an additional resistor in series with the inductance to be measured.
2. It can not be used for the measurement of high Q values. Its use is limited to the measurement of low Q values from 1 to 10. This can be proved from phase angle balance condition which says that sum of the angles of one pair of opposite arms must be equal.

$$\theta_1 + \theta_4 = \theta_2 + \theta_3$$

but θ_2 and θ_3 are zero, as the corresponding impedance are pure resistance. For high θ values, the angle θ_4 is almost 90°, hence θ_1 must be –90°. But θ_1 gets decided by parallel combination of R_1 and C_1. To get θ_1 as almost –90°, the value of R_1 should be very high. Practically such high resistance is not possible. Hence high θ values can not be measured.

3. There is an interaction between the resistance and reactance balances. Getting the balance adjustment is little difficult.
4. It is unsuited for the coils with low Q values, less than one, because of balance convergence problem.
5. The bridge balance equations are independent of frequency but practically, the properties of coil under test vary with frequency which can cause error.

Commercial Maxwell Bridge measures the inductance from 1-1000 H, with ± 2% error.

7.11 The Different Sources of Errors in A.C Bridges

The six different sources of errors in A.C bridges are:

7.11.1 Eddy Current Errors

The induced eddy currents in the standard resistors and inductors changes the standard value of bridge, causing errors. To avoid such errors, the materials with large conductivity masses should not be placed near the bridge.

7.11.2 Stray Conductance Effects

There is no proper insulation between various components of a bridge circuit. Errors may arise because of leakage currents from one arm to another. This mainly happens for high impedance bridge. To overcome this, variable components and other pieces of apparatus are mounted on insulating stand.

7.11.3 Errors due to Electromagnetic Coupling

If there is an electromagnetic coupling between the source and detector, an e.m.f is induced in the detector directly from the source. The zero indication of detector is obtained only when the bridge supplies the voltage which is equal in magnitude and opposite in phase as that of induced e.m.f.

7.11.4 Errors due to Electrostatic Coupling

Inter-capacitance effect arises if the adjacent branches of a bridge network are at different potential. In other words, the electric field of the branches interacts thus introducing errors. Effective shielding by means of covering elements with earthed metallic shields can minimize this error.

7.11.5 Frequency Errors

The frequency of supply system plays a vital role in causing error in supply system. i.e. the presence of harmonics in supply causes variations in the sinusoidal wave form due to which the appreciable errors are generated.

7.11.6 Residual Errors

Residues means small inherent inductance or capacitance that are used in bridge circuits and are assumed to be non-inductive and non-capacitive but in practice their inductance and capacitance are never be zero. So it is necessary to evaluate these components otherwise these errors will creep in.

7.12 Anderson Bridge

It is another important A.C bridge used for the measurement of self-inductance in terms of a standard capacitor. Actually this bridge is nothing but modified Maxwell's bridge in which also the value of self-inductance is obtained by comparing it with a standard capacitor. This bridge is basically used for the precise measurement of inductance over a wide range of value. The Anderson Bridge is as shown in the Fig. 7.5(a). One arm of the bridge consists of unknown inductor L_x with known resistance in series with L_x. This resistance R_1 includes resistance of the inductor. C is the standard capacitor with R_1, R_2, R_3 and R_4 are non-inductive known resistances.

The bridge balance equations are.

$$i_1 = i_3,\ i_2 = i_4 + i_c, \quad v_2 = i_2R_2,\ v_3 = i_3R_3,$$

$$v_1 = v_2 + i_{cr} \text{ and } v_4 = v_3 + i_{cr},\ v_1 = i_1 R_1 + i_1\omega L_1,$$

$$v_4 = i_4R_4$$

$$V = \overline{V_2} + \overline{V_4} = \overline{V_1} + \overline{V_3}$$

to find balance equation transforming a star formed by R_2, R_4 and r into its equivalent delta as shown in the Fig. 7.5(a) (b) and 7.5 (c) (d).

The element in equivalent delta are given by,

$$R_5 = \frac{R_2 r + R_4 r + R_2 R_4}{R_4}$$

$$R_6 = \frac{R_2 r + R_4 r + R_2 R_4}{R_4}$$

$$R_7 = \frac{R_2 r + R_4 r + R_2 R_4}{r}$$

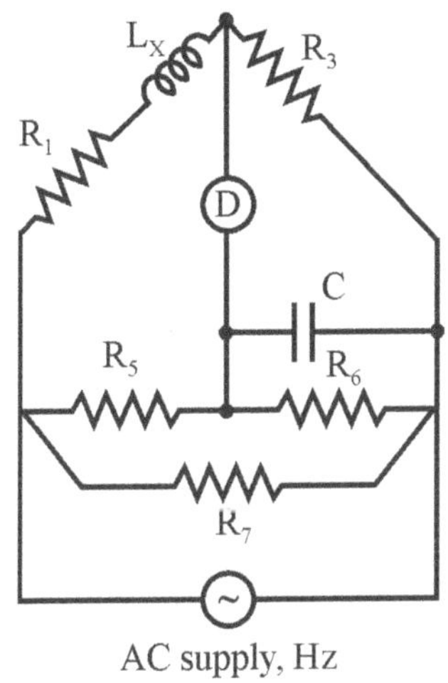

(c)

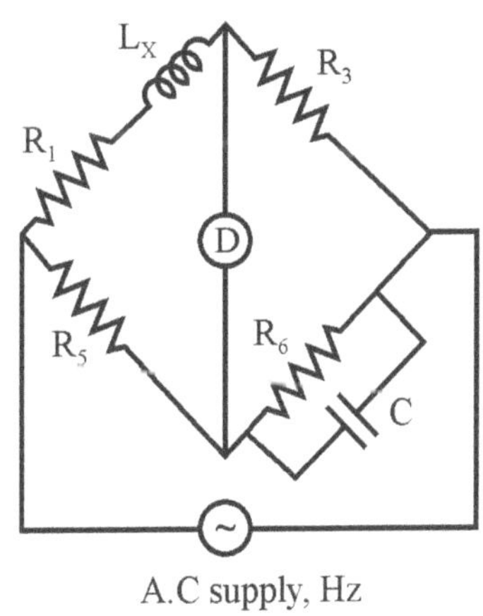

(d)

Fig. 7.5 Transformed of Anderson bridge

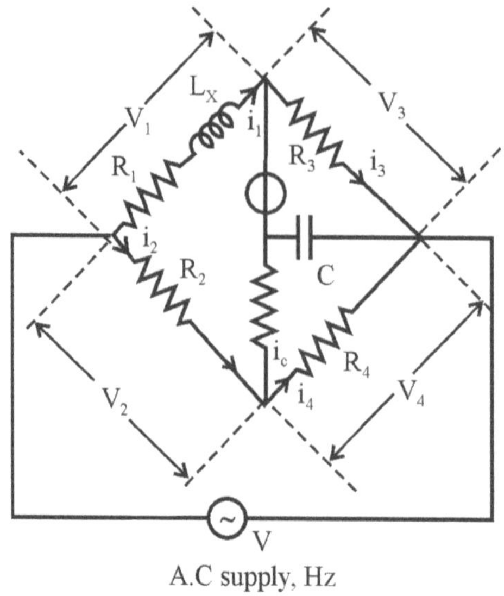

Fig. 7.5(a) Circuit diagram

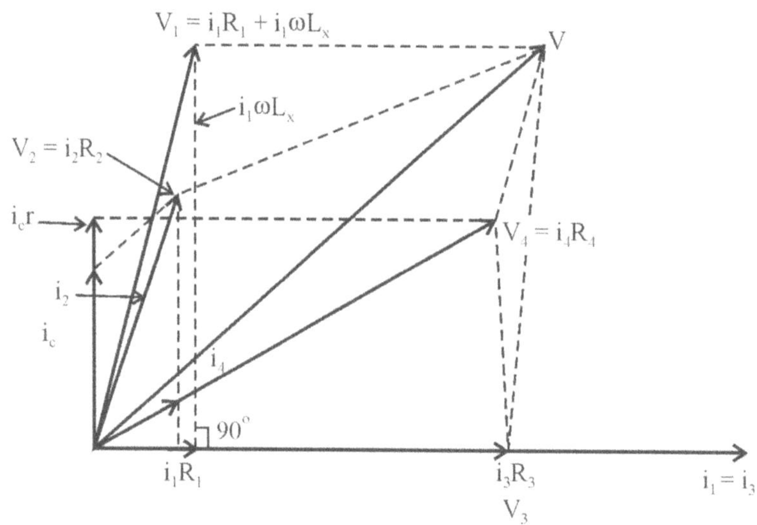

Fig. 7.5(b) Phasor diagram

Now R_7 shunts the source hence it does not effect the balance condition, thus by neglecting R_7 and rearranging a network as shown in the Fig. 7.5 (d), we get a Maxwell inductance bridge.

Thus balance equation are given by,

$$L_x = CR_3R_5 \text{ and } R_1 = R_3\frac{R_5}{R_6}$$

Substituting values of R_5 and R_6, we can write,

$$\boxed{L_x = \frac{CR_3}{R_4}\left[R_2 r + R_4 r + R_2 R_4\right] \text{ and}}$$

$$\boxed{R_1 = \frac{R_2 R_3}{R_4}}$$

If the capacitor used is not perfect, the value of inductance remain unchanged, but the value of R_1 changes.

This method can also be used to measure the capacitance of the capacitor C if a calibrated self inductance is available.

7.12.1 Advantages of Anderson Bridge

The 'advantages' of Anderson Bridge are:

1. The problem of sliding balance condition normally faced with low 'Q' coil is overcomed in the Anderson's bridge. This is because both the variable resistances to be adjusted are independent of each other. Hence, the balance can be obtained easily.
2. Instead of a variable capacitor, a fixed capacitor can be used. This makes the bridge cheaper than Maxwell's bridge.
3. Determination of unknown capacitance in terms of known inductance is also possible.
4. The expression for self-inductance of coil do not change even with the use of imperfect capacitor (i.e. the capacitor with dielectric loss) instead, only the value of coil resistance is effected.
5. For more precise measurements, second balance is obtained by short circuiting the coil and the inductance of coil leads is calculated. Finally the actual self-inductance of the coil is obtained by subtracting the values of inductance obtained in both the measurement cases.
6. It can also be used for accurate measurement of capacitance in terms of inductance.

7.12.2 Disadvantages of Anderson Bridge

The disadvantages of Anderson's bridge are:

1. Use more number of components.
2. It is more complicated than other bridges.
3. Bridge cannot be easily shielded due to additional junction points, to avoid the effect of stray capacitance.
4. Balance equations are also complicated to derive.

7.13 Hay's Bridge

The limitation of Maxwell's bridge is that it cannot be used for high Q values. The "hay's" bridge suitable for the coils having high Q values. The difference in Maxwell's bridge and hay's bridge is that "the Hay's bridge consists of resistance R_1 in series with the standard capacitor C_1 in one of the ratio arms. Hence large phase angles R_1 needed is very low, which is practicable. Hence bridge can be used for the coils with high Q values. The hay's bridge is shown in the Fig. 7.6(a).

Under balanced condition, the phasor diagram is as shown in the Fig. 7.6(b).

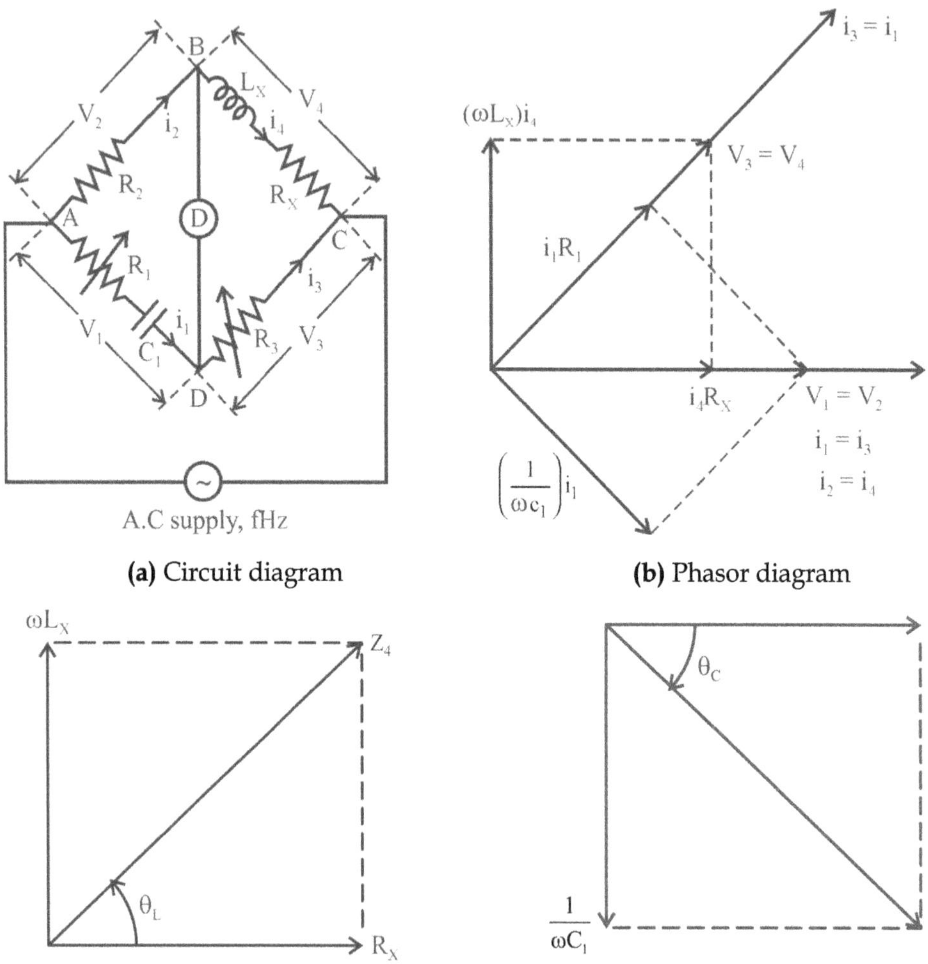

(a) Circuit diagram

(b) Phasor diagram

(c)

(d)

Fig. 7.6 Hay's bridge

The various constants of the bridge are:

$$Z_1 = R_1 - jX_{C1} = R_1 - j\left(\frac{1}{\omega c_1}\right)$$

$$Z_2 = R_2 \text{ and } Z_3 = R_3$$

$$Z_y = Z_x = R_x + j(\omega L_x)$$

At the balance condition,

$$\overline{Z_1 Z_x} = \overline{Z_2 Z_3}$$

$$\therefore \left[R_1 - j\left(\frac{1}{\omega C_1}\right)\right]\left[R_x - j(\omega L_x)\right] = R_2 R_3$$

$$\therefore R_1 - R_x - j\left(\frac{R_x}{\omega C_1}\right) + j\omega r_1 L_x + \frac{L_x}{C_1} = R_2 R_3$$

$$\therefore \left[R_x R_1 + \frac{L_x}{C_1}\right] + j\left[\omega R_1 L_x - \frac{R_x}{\omega C_1}\right] = R_2 R_3 \qquad .....(7.21)$$

Equating the real parts of both sides,

$$R_x R_1 + \frac{L_x}{C_1} = R_2 R_3 \qquad .....(7.22)$$

Equating the imaginary parts of both sides of (7.21).

$$\omega R_1 L_x - \frac{R_x}{\omega C_1} = 0 \qquad .....(7.23)$$

To obtain R_x and L_x, solve equations (7.22) and (7.23) simultaneously

From equation (7.23), $\omega R_1 L_x = \dfrac{R_x}{\omega C_1}$

$$\therefore L_x = \frac{R_x}{\omega^2 R_1 C_1} \qquad .....(7.24)$$

Substituting in (7.22), $R_x R_1 + \dfrac{R_x}{\omega^2 R_1 C_1^2} = R_2 R_3$

$$\therefore R_x\left[R_1 + \frac{1}{\omega^2 R_1 C_1^2}\right] = R_2 R_3$$

$$R_x\left[\frac{1+\omega^2R_1^2C_1^2}{\omega^2R_1C_1^2}\right]=R_2R_3;\ \boxed{R_x=\frac{\omega^2R_1C_1^2R_2R_3}{1+\omega^2R_1^2C_1^2}} \quad .....(7.25)$$

Substituting (7.25) in (7.24) we get,

$$L_x=\frac{\omega^2R_1C_1^2R_2R_3}{\left(1+\omega^2R_1^2C_1^2\right)\omega^2R_1C_1}$$

$$\boxed{\therefore L_x=\frac{R_2R_3C_1}{1+\omega^2R_1^2C_1^2}} \quad .....(7.26)$$

From equation (7.25) and (7.26) the unknown resistance and the inductance can be calculated. The expressions are frequency (ω) sensitive and hence source frequency must be accurately known. The inductance balance equation depends on the losses of the inductor.

From the phase angle balance equation, the opposite sets of phase must be equal as Z_2 and Z_3 are purely resistive, the inductive phase angle must be equal to the capacitive phase angle.

The inductive and capacitive phase angle can be determined from the impedance triangles shown in the Fig. 7.6 (c) (d)

$$\tan\theta_L=\frac{X_{L_x}}{R_x}=\frac{\omega L_x}{R_x}=Q \quad .....(7.27)$$

while

$$\tan\theta_c=\frac{Xc_1}{R_1}=\frac{1}{\omega C_1R_1} \quad .....(7.28)$$

The two angles must be equal at bridge balance

$$\therefore \quad Q=\frac{1}{\omega C_1R_1} \quad .....(7.29)$$

Substituting in the eq. (7.26) we get

$$L_x=\frac{R_2R_3C_1}{1+\left(\frac{1}{Q}\right)^2} \quad .....(7.30)$$

For a large value of Q, $\frac{1}{Q^2}$ becomes small, can be neglected:

$$\therefore \quad L_x=R_2R_3C_1 \quad \text{....... for high 'Q'}$$

This is same as Maxwell's bridge equation.

The commercial Hay's bridge measure the inductances in the range 1μH – 100 H with ± 2% error.

7.13.1 Advantage of Hay's Bridge

1. Hay's bridge is suitable for coil having quality factor greater than 10 (**Q > 10**). Also it gives a simple expression for unknown inductance for high Q coils.
2. It requires very low value resistor R_1 to measure high Q inductance.
3. It gives very simple expression for Q factor in terms of elements in the bridge.

7.13.2 Disadvantage of Hay's Bridge

1. The major draw back of an Hay's bridge is that it cannot be used for measuring coils having a Q-factor less than 10.

We have
$$L_x = \frac{R_2R_3C_1}{1+\left(\frac{1}{Q}\right)^2}$$

For lower values of Q (< 10) the term $\left(\frac{1}{Q}\right)^2$ in the above expression cannot be neglected. Hence this bridge cannot be used for coils which have the Q-factor less than 10. For such coils the Maxwell's bridge can be used.

7.14 Capacitance Comparison Bridge

In the capacitance comparison bridge, the ratio arms are resistive in nature. The impedance Z_3 consists of two known standard capacitor C_3 in series with the resistance R_3. The resistance R_3 is variable, used to balance the bridge. The impedance Z_4 consists of the unknown capacitor C_x and its small leakage resistance R_x. The unknown capacitor C_x is compared with the standard capacitor by using the balance equation, the capacitor and its leakage resistance value is obtained. The bridge is shown in the Fig. 7.7.

Here,
$$Z_1 = R_1 + j0\ \Omega$$
$$Z_2 = R_2 + j0\ \Omega$$
$$Z_3 = R_3 - j\,X_{C_3}$$
$$= R_3 - j\left(\frac{1}{\omega_{C_3}}\right)\ \Omega$$

$$Z_3 = R_x - jX_{C_X}$$

$$= R_X - j\left(\frac{1}{\omega_{C_X}}\right)\ \Omega$$

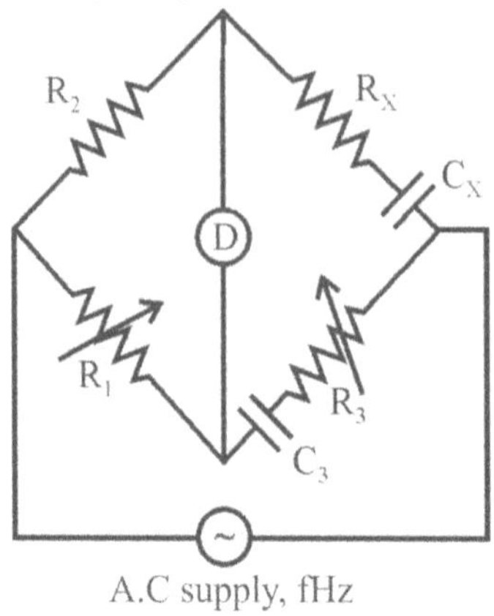

Fig. 7.7 Capacitance comparison bridge

From balance condition,

$$\bar{Z}_1 Z_4 = Z_2 \bar{Z}_3$$

$$\therefore \quad R_1\left[R_X - j\left(\frac{1}{\omega C_X}\right)\right] = R_2\left[R_3 - j\left(\frac{1}{\omega C_3}\right)\right]$$

$$\therefore \quad R_1 R_X - j\left(\frac{R_1}{\omega C_X}\right) = R_2 R_3 - j\left(\frac{R_2}{\omega_{C_3}}\right)$$

Equating real parts,

$$R_1 R_X = R_2 R_3; \qquad R_X = \frac{R_2 R_3}{R_1} \qquad .....(7.31)$$

Equating imaginary parts,

$$\frac{R_1}{\omega_{C_X}} = \frac{R_2}{\omega_{C_3}}; \qquad C_X = \frac{C_3 R_1}{R_2} \qquad .....(7.32)$$

By using eqn. (7.31) and (7.32), the unknown capacitor and its leakage resistance can be determined. By varying R_1 and R_3 simultaneously, true balance can be obtained.

7.15 Schering Bridge

It is one of the most widely used A.C bridges for the measurement of **unknown capacitors, dielectric loss** and **power factor**.

The Fig. 7.8 shows the connections of schering bridge. It can be used for low voltage. The C_X is perfect capacitor to be measured. R_X is series resistance. C_2 is standard air capacitor having very stable value. R_3 and R_4 are non-inductive resistance while C_4 is variable capacitor. From the general balance eqn. $\overline{Z_1Z_4} = \overline{Z_2Z_3}$

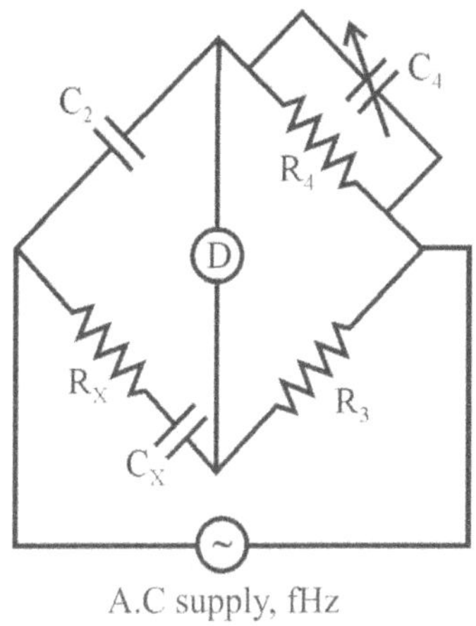

Fig. 7.8 Schering bridge

From the general balance eqn. $\overline{Z_1Z_4} = \overline{Z_2Z_3}$

Now $$Z_1 = R_X - j\frac{1}{\omega C_X}$$

$$Z_2 = -\frac{j}{\omega C_2}$$

$$Z_3 = R_3$$

$$Z_4 = R_4 \,||\, \frac{-j}{\omega_{C4}} = \frac{R_4\left(-\frac{j}{\omega_{C4}}\right)}{\left(R_4 - j\frac{1}{\omega_{C4}}\right)}$$

$\therefore$ $$Z_4 = \frac{-jR_4}{\omega R_4C_4 - j} = \frac{-jR_4(\omega R_4C_4 + j)}{(\omega R_4C_4 - j)(\omega R_4C_4 + j)}$$

$$= \frac{R_4 - j\omega R_4^2C_4}{\omega^2R_4^2C_4^2 + 1}$$

$$\therefore \qquad Z_1 = \frac{Z_2 Z_3}{Z_4} = \frac{\left(-\frac{j}{\omega_{C_2}}\right)(R_3)}{\left[\frac{R_4 - j\omega R_4^2 C_4}{1+\omega^2 R_4^2 C_4^2}\right]} = \frac{\left(1+\omega^2 R_4^2 C_4^2\right) R_3 \left(-\frac{j}{\omega_{C_2}}\right)}{\left(R_4 - j\omega R_4^2 C_4\right)}$$

Rationalising,

$$Z_1 = R_3\left(1+\omega^2 R_4^2 C_4^2\right)\left\{\frac{\frac{-j}{\omega_{C_2}}\left(R_4 + j\omega R_4^2 C_4\right)}{R_4^2 + \omega^2 R_4^2 C_4^2}\right\}$$

$$\therefore \qquad R_X - j\frac{1}{\omega C_X} = \frac{R_3\left(1+\omega^2 R_4^2 C_4^2\right)}{R_4^2\left(1+\omega^2 R_4^2 C_4^2\right)}\left\{\frac{R_4^2 C_4}{C_2} - \frac{jR_4}{\omega C_2}\right\}$$

Equating real and imaginary parts

$$R_X = \frac{R_3}{R_4^2} \times \frac{R_4^2 C_4}{C_2} = \frac{R_3 C_4}{C_2} \qquad .....(7.33)$$

$$-j\frac{1}{\omega C_X} = -j\frac{R_3}{R_4^2} \times \frac{R_4}{\omega C_2} = -j\left[\frac{1}{\frac{R_4}{R_3}\omega C_2}\right]$$

$$\therefore \qquad \omega C_X = \frac{R_4}{R_3}\omega C_2$$

$$\therefore \qquad C_X = \frac{R_4}{R_3} C_2 \qquad .....(7.34)$$

The equations (7.33) and (7.34) gives the required valves of C_X and R_X.

7.16 Power Factor and Loss Angle

7.16.1 Power Factor (P.F)

The P.F of the series RC combination is defined as the cosine of the phase of the circuit.

Thus,

$$P.F = \cos\phi_x = \frac{R_x}{Z_x}$$

For phase angles very close to 90°, the reactance is almost equal to the impedance,

$$\therefore \qquad P.F = \frac{R_X}{X_X} = \frac{R_X}{\left(\dfrac{1}{\omega C_X}\right)}$$

$$P.F = \omega R_X C_X$$

7.16.2 Loss Angle (δ)

For a series combination of R_X and C_X, the angle between the voltage across the series combination and voltage across the capacitor C_X is called "loss angle δ". This is shown in Fig. 7.9(a).

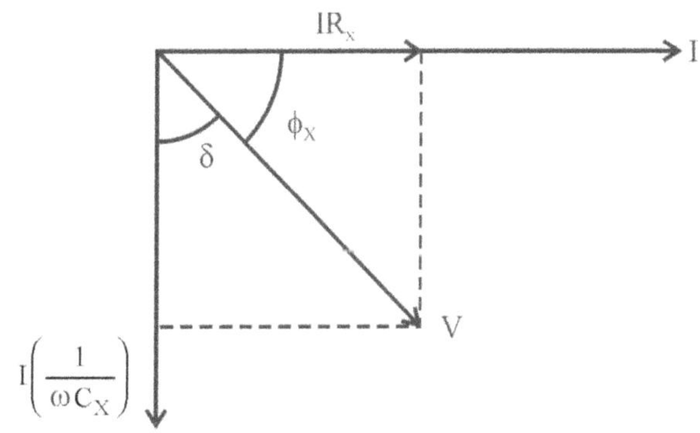

Fig. 7.9(a)

Now
$$\tan\delta = \frac{IR_X}{I\left(\dfrac{I}{\omega C_X}\right)} = \omega R_X C_X$$

$$\therefore \qquad \tan\delta = \omega\left(\frac{R_3 C_4}{C_2}\right)\left(\frac{R_4}{R_3}C_2\right)$$

$$= \omega R_4 C_4$$

$$\therefore \qquad \tan\delta = \omega R_4 C_4;$$

Thus loss angle can be measured, knowing the values of ω, R_4 and C_4.

7.16.3 Dissipation Factor (D)

For $R_X - C_X$ series circuit it is cotangent of the phase angle ϕ_X.

$$\therefore \qquad D = \cot\phi_x = \frac{1}{\tan\phi_x} = \frac{1}{\left[\dfrac{I\left(\dfrac{1}{\omega C_X}\right)}{IR_X}\right]} = \omega R_X C_x = \omega R_4 C_4$$

$$\therefore \qquad D = \cot\phi_X = \omega R_4\, C_4$$

Note: The quality factor $Q = \frac{X}{R} = \frac{1}{\omega} CR$ hence "dissipation factor is reciprocal of quality factor Q" and gives the information about quality of the capacitor.

Thus if the resistance R_4 is fixed, then the dial of capacitor C_4 can be directly calibrated to give dissipation factor D i.e., quality for the capacitor. As the term 'W' is present in the equation, the calibration of C_4 dial holds good for only one particular frequency. The different frequency can be used but a correction should be made to multiply the C_4 dial reading by the ratio of the two frequencies. Similarly, if the resistance ratio is maintained at fixed value, the dial of C_3 can be graduated im terms of direct readings of C_X. Commercial Schering bridge measures the capacitors from 100 pF – 1μF, with $\pm$ 2% accuracy.

Note: The bridge is widely used for testing small capacitors at low voltage with very high precision.

The phasor diagram is shown in the Fig. 7.9(b) at the balance condition.

I_1 is chosen reference. Now V_1 is drop across R_X and C_X. Thus OA = $I_1\ R_X$ and OB = $I_1\left(\frac{1}{\omega C_X}\right)$ and I_1 leads capacitor drop by 90°. Thus OC = V_1 is $\overline{OA} + \overline{OB}$. Now $V_1 = I_2\left(\frac{1}{\omega C_2}\right)$, i.e. drop across C_2. So current I_2 leads V_2 by 90^0. Then $V_2 = I_1\ R_3$ which is OD in phase with I_1, and $\overline{V_1} + \overline{V_2} = \overline{V}$ is supply voltage OE. V_2 is also drop across R_4 and C_4. Let OF is current through R_4 thus OF = $\frac{V^2}{R_4}$ in phase with V_2 while OG is current through C_4 which is $V_2\omega C_4$ and is leading V_2 by 90^0. The OH = I_2 which is vector sum of OF and OG.

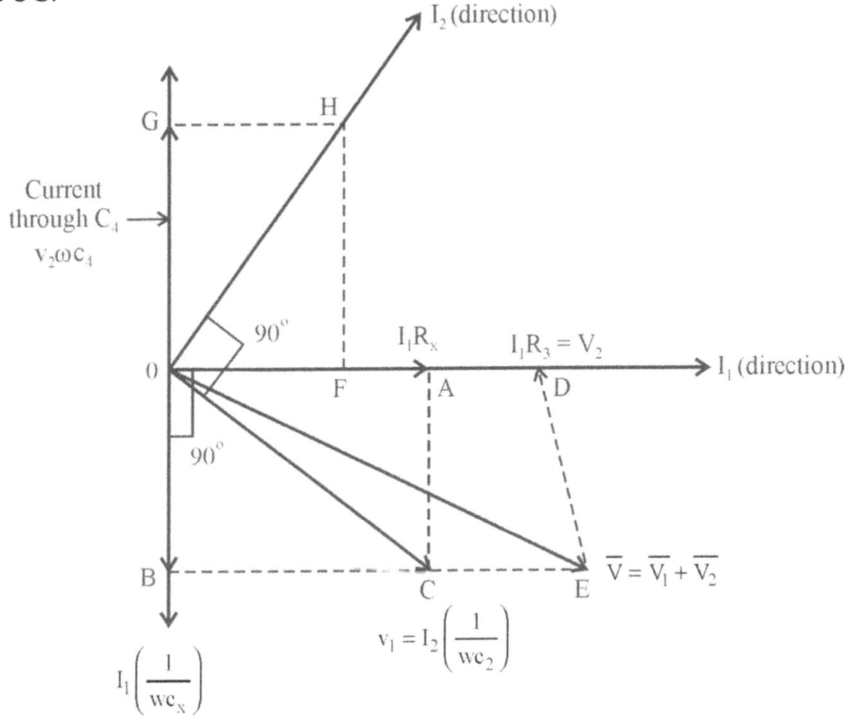

Fig. 7.9(b)

7.17 High Voltage Schering Bridge

When the Schering bridge is used for the measurement of small capacitance at low voltage, then it suffers from errors. To avoid this, high voltage Schering bridge is used. The bridge is shown in the Fig. 7.10. The high voltage is obtained from step up transformer at 50 Hz. The detector used is vibration galvanometer. The arms AB and AD have only capacitors, which are designed for high voltages.

As compared to arms BC and CD, the branches AB and AD have very high impedance. Hence major part of the overall potential drop will be across the arms AB and AD, as compared to the arms BC and CD. The point C is grounded. Thus for very high voltages of the order of 100 KV, the drop across the arms BC and CD is few volts above earth. This is advantageous and safe from the operator point of view, that the detector and controlling part is at low voltage.

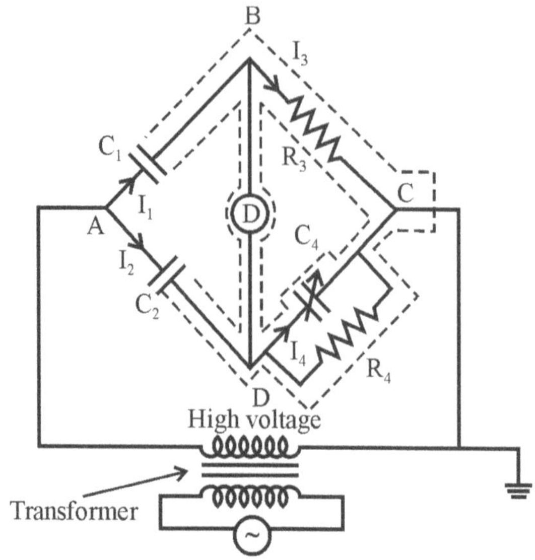

Fig. 7.10 High voltage schering bridge

In case of breakdown of either of the high voltage capacitors, very high voltage may appear across the branches BC and CD. The detector has to be very sensitive because due to high impedance values, the current drawn by the arms is very small.

But power loss is less due to low current level. To avoid the errors due to inter-capacitance between high and low impedance arms, the earthed screens are used. Similarly, to eliminate the effect of earth capacitance on the galvanometer and contact leads, a wagner ground connection may be used.

7.18 Wien Bridge

Basically the bridge is used for the frequency measurement but it is also used for the measurement of the unknown capacitor with great accuracy. Its one ratio arm consists of a series RC circuit i.e. R_1 and C_1. The second ratio arm consists of a resistance R_2. The third arm consists of the parallel combination of resistance and capacitor i.e. R_3 and C_3. The circuit of the wien bridge is shown in the Fig. 7.11.

From the Fig. 7.11, we can write,

$$Z_1 = R_1 - j\left(\frac{1}{\omega C_1}\right)$$

$$Z_2 = R_2$$

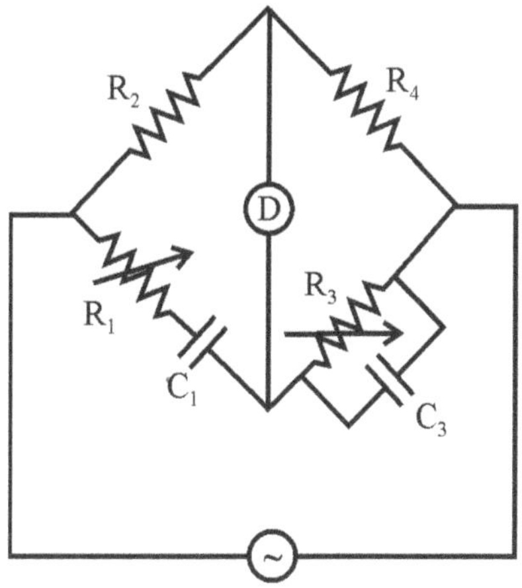

Fig. 7.11

$$Z_3 = R_3 \parallel C_3$$

$$Y_3 = \frac{1}{R_3} + j\omega C_3$$

and $$Z_4 = R_4$$

the balance condition is,

$$\overline{Z_1 Z_4} = \overline{Z_2 Z_3}$$

$$\overline{Z}_2 = \frac{\overline{Z_1 Z_4}}{\overline{Z_3}} = Z_1 \overline{Z_4 Y_3}$$

$\therefore$ $$R_2 = \left[R_1 - j\left(\frac{1}{\omega C_1}\right)\right] R_4 \left[\frac{1}{R_3} + j\omega C_3\right]$$

$\therefore$ $$R_2 = R_4 \left[\frac{R_1}{R_3} + j\omega R_1 C_3 - j\frac{1}{\omega C_1 R_3} + \frac{C_3}{C_1}\right]$$

$\therefore$ $$R_2 = R_4 \left[\frac{R_1}{R_3} + \frac{C_3}{C_1}\right] + jR_4 \left[\omega R_1 C_3 - \frac{1}{\omega C_1 R_3}\right]$$

Equating real parts of both sides,

$$R_2 = \frac{R_4 R_1}{R_3} + \frac{C_3 R_4}{C_1}$$

$$\frac{R_2}{R_4} = \frac{R_1}{R_3} + \frac{C_3}{C_1} \quad .....(7.35)$$

Equating imaginary parts of both sides,

$$\omega R_1 C_3 - \frac{1}{\omega C_1 R_3} = 0$$

$$\therefore \quad \omega^2 = \frac{1}{R_1 R_3 C_1 C_3}$$

$$\omega = \frac{1}{\sqrt{R_1 R_3 C_1 C_3}} \quad .....(7.36)$$

$$f = \frac{1}{2\pi\sqrt{R_1 R_3 C_1 C_3}} \quad .....(7.37)$$

The equation (7.35) gives the resistance ratio while the equation (7.37) gives the frequency of applied voltage.

Generally in Wien bridge, the selection of the components is such that

$$R_1 = R_3 = R$$

and $$C_1 = C_3 = C$$

$$\therefore \quad \frac{R_2}{R_4} = 2 \quad .....(7.38)$$

and $$f = \frac{1}{2\pi RC} \quad .....(7.39)$$

The equation (7.39) is the general equation for the frequency of the bridge circuit.

7.18.1 Applications

The bridge is used to measure the frequency in audio rage, the audio range is [20 – 200 – 2k – 20k] Hz. The resistances are used for the range changing while the capacitances are used for fine frequency control.The bridge can be used for capacitance measurement if the operating frequency is known.

The bridge is also used in a harmonic distortion analyzer, as a notch filter and in audio frequency and radio frequency oscillators as a frequency determining element. The accuracy of 0.5% -1% can be readily obtained using this bridge.

7.19 De Sauty Bridge

The basic application of the De sauty bridge is to compare two capacitances. The circuit diagram of De sauty bridge is as shown in Fig. 7.12(a).

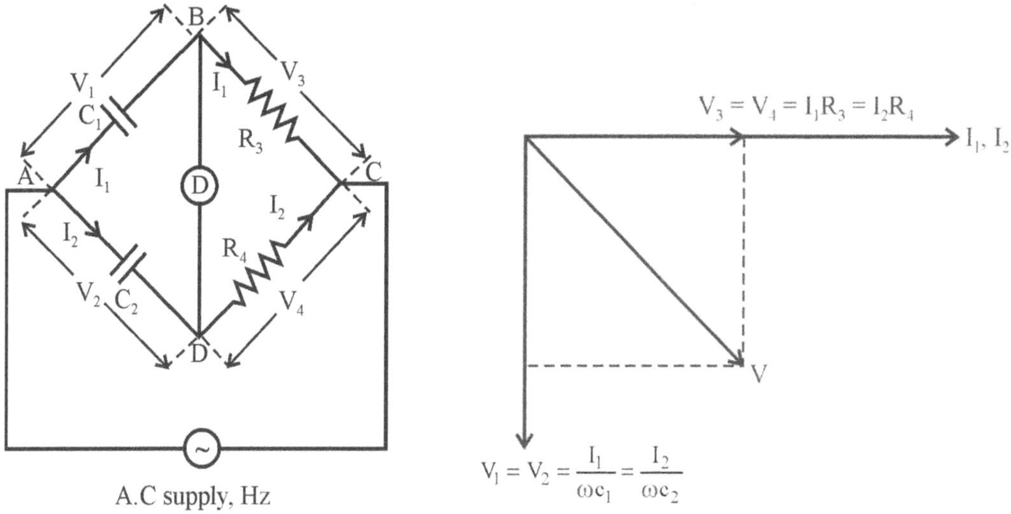

(a) Circuit diagram

(b) Phasor diagram

Fig. 7.12 De sauty bridge and phasor diagram

The De sauty bridge consists a capacitor C_1 which is capacitance under test in branch AB. The branch AD consists a known standard capacitor. The remaining branch BC and CD consists non-inductive resistance R_3 and R_4 respectively.

Let C_1 = capacitor where capacitance is to be measured

C_2 = a standard capacitor

R_3, R_4 = non-inductive resistors.

At balance, we get condition as, $Z_1 Z_4 = Z_2 Z_3$

$$\frac{-jX_{C1}}{R_3} = \frac{-jX_{C2}}{R_4} \quad .....(7.40)$$

$$\therefore \quad \left(\frac{-j}{\omega C_1}\right) R_4 = \left(\frac{-j}{\omega C_2}\right) R_3$$

$$\therefore \quad C_1 = \frac{C_2 . R_4}{R_3} \quad .....(7.41)$$

The balance in the bridge can be achieved by varying either R_3 or R_4. Under the balanced condition the vector diagram for the De sauty bridge is as shown in the Fig. 7.12(b). The main advantage of De sauty bridge is that it is very simple in construction and the measurement of the capacitor is done with simplicity.

Inspite of the simplicity of the measurement, it has one drawback that when the capacitors are with dielectric losses. Then it highly impossible to achieve balance condition. Thus practically De sauty can be used effectively for the capacitors with very low dielectric losses such as air capacitors.

7.19.1 Modified De Sauty Bridge

This bridge is the modification of the basic De sauty bridge. The capacitors with the dielectric losses are called "imperfect capacitors". Prof. Grover modified the basic De sauty bridge and obtained modified De sauty bridge as shown in Fig. 7.13, which is used for the measurement of unknown, imperfect capacitor. The circuit diagram of the De sauty bridge is as shown in the Fig. 7.13(a).

The two capacitors i.e., C_1 and C_2 are connected with non-inductive resistances R_1 and R_2 respectively. Let the loss component of the two capacitors be given by two resistances r_1 and r_2.

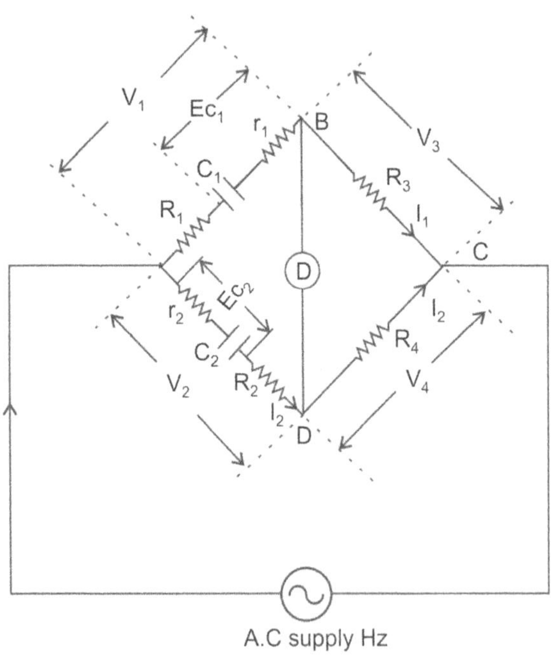

(a) Circuit diagram

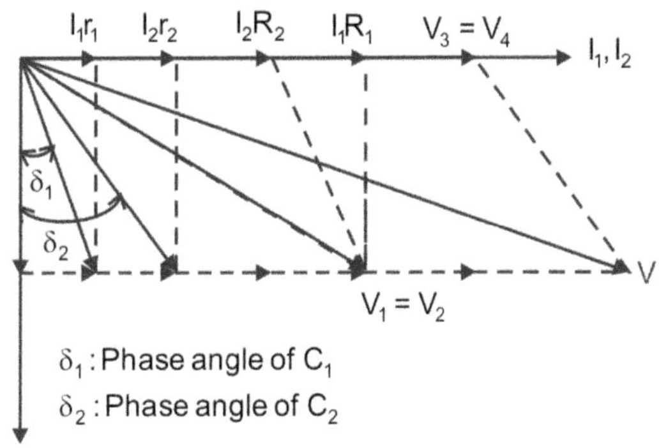

(b) Phasor diagram

Fig. 7.13 Modified De sauty's bridge

At balance condition expression is given by,

$$\frac{\left(R_1 + r_1 + \dfrac{1}{\sin j\omega C_1}\right)}{R_3} = \frac{\left(R_2 + r_2 + \dfrac{1}{j\omega C_2}\right)}{R_3}$$

i.e.,

$$R_4\left(R_1 + r_1 + \frac{1}{j\omega C_1}\right) = R_3\left(R_2 + r_2 + \frac{1}{j\omega C_2}\right)$$

By simplifying equation we get,

$$\frac{C_1}{C_2} = \frac{R_2 + r_2}{R_1 + r_1} = \frac{R_4}{R_3} \qquad .....(7.42)$$

In the modified De sauty bridge, the balance condition can be easily obtained by varying the resistance in the circuit, such as resistance R_1, R_2, R_3 and R_4. The dissipation factor (D) of the capacitor depends on phase angle δ.

Hence we can write,

D_1 = dissipation factor of capacitor $C_1 = \tan \delta_1 = \omega C\, r_1$ .....(7.43)

D_2 = dissipation factor of capacitor C_2

$$= \tan \delta_2 = \omega C_2 r_2 \qquad .....(7.44)$$

Simplifying eq. (7.42) we get,

$$C_1(R_1 + r_1) = C_2(R_2 + r_2)$$

$$C_1R_1 + C_1r_1 = C_2R_2 + C_2r_2; \quad C_2R_2 - C_1k_1 = C_1r_1 - C_2r_2$$

$$\therefore \qquad \omega C_2R_2 - C_1R_1 = \omega C_1r_1 - C_2r_2$$

From equations (7.43) & (7.44) substituting values of factors ωC_2R_2 and ωC_1R_1 as D_2 and D_1 respectively.

$$\omega C_2R_2 - \omega C_1R_1 = D_1 - D_2$$

$$\therefore \qquad D_1 - D_2 = \omega(C_1R_1 - C_2R_2) \qquad .....(7.45)$$

But $\frac{C_1}{C_2} = \frac{R_4}{R_3}$ or $C_1 = C_2\left(\frac{R_4}{R_3}\right)$. Substituting value of C_1 in equation (7.45), we get

$$D_2 - D_1 = \omega\left[C_2\frac{R_4}{R_3}R_1 - C_2R_2\right]$$

$$D_2 - D_1 = \omega C_2\left[\frac{R_1R_4}{R_3} - R_2\right] \qquad .....(7.46)$$

From equation (7.46), it is clear that if the dissipation factor for one of the capacitors is known, we can obtain dissipation factor for the second one.

But the main drawback of the measurement of the dissipation factor is not accurate. It depends on difference of quantities. These quantities are themselves very large but differences are very less. Thus the difference is not known with high degree accuracy. The dissipation factor cannot be accurately measured.

7.20 Owen's Bridge

This bridge can be used to measure the inductance of capacitance. The basic arrangement of the Owen's bridge is as shown in the Fig. 7.14(a)

In branch AB unknown self inductance with resistance R_1 is connected. In branch 'BC' non-inductive resistance R_3 is used. The branch 'CD' consists the fixed value standard capacitor C_4 while the branch DA consists variable non-inductive resistance R_2 in series with variable capacitor C_2

At balance we get the condition as,

$$(R_1 + j\omega L_1)\left(\frac{1}{j\omega C_4}\right) = R_3\left(R_2 + \frac{1}{j\omega C_2}\right)$$

$$\therefore \qquad \left(\frac{R_1 + j\omega L_1}{j\omega C_4}\right) = \frac{R_2R_3(j\omega C_2) + R_3}{j\omega C_2}$$

$$\therefore \qquad R_1 + j\omega L_1 = \left(\frac{C_4}{C_2}\right) R_3 \left[1 + j\omega R_2 C_2\right] \qquad \ldots\ldots(7.47)$$

$$\therefore \qquad R_1 + j\omega L_1 = \left(\frac{C_4}{C_2}\right) R_1 \left[1 + j\omega R_r C_2\right]$$

$$\therefore \qquad R_1 + j\omega L_1 = R_3 \frac{C_4}{C_2} + j\omega\, R_2\, R_3\, R_4$$

Comparing real terms, we get,

$$R_1 = R_3 \frac{C_4}{C_2} \qquad \ldots\ldots(7.48)$$

Similarly comparing imaginary, we get,

$$L_1 = R_2 R_3 C_4 \qquad \ldots\ldots(7.49)$$

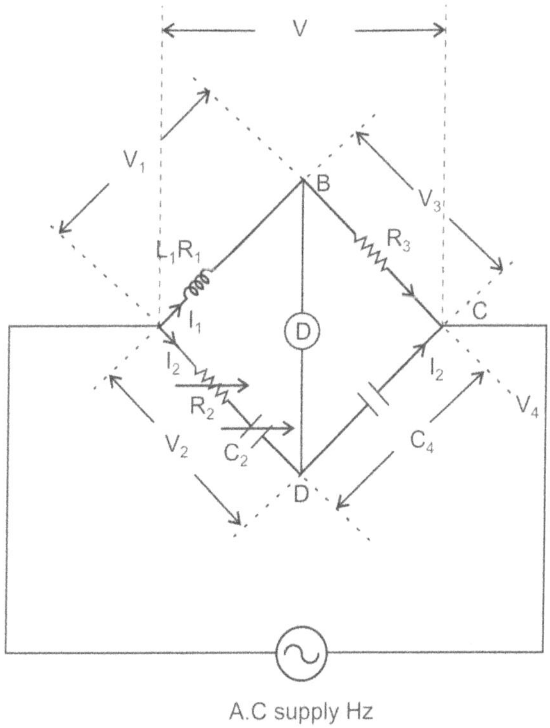

(a) Ckt diagram

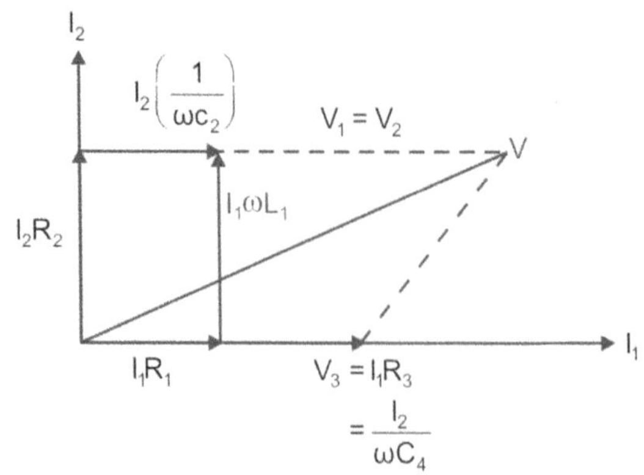

(b) Phasor diagram

Fig. 7.14 Owen's bridge

7.20.1 Advantages of Owen's Bridge

The main advantages of the Owen's bridge are:

1. From the equation of balance conditions, it is clear that both conditions are independent of each other even if R_2 and C_2 are variables. Moreover these two elements are connected in the same arm in series. Hence, it is very easy to achieve balance condition properly.
2. The balance equations are independent of the frequency term.
3. It is possible to use this bridge over a wide range of inductance values.
4 The balance equation are of very simple form.

7.20.2 Disadvantages of Owen's Bridge

Disadvantages are:

1. The use of standard variable capacitor makes the bridge costlier. For measurement of wide range of inductances in terms of capacitance, the capacitor C_2 is required to be variable. This increases the cost of the bridge.
2. With the inductances of large Q-factor values, the value of C_2 becomes very large.
3. Accuracy of the variable capacitor is about 1%.

7.20.3 Application of Owen's Bridge

The main application of the Owen's bridge is to "measure incremental inductance. It is necessary to feed coil from A.C as well as D.C voltage supply. Thus it is necessary to modify original Owen's beidge such that both A.C and D.C sources are included in it. The modified Owen's bridge used for the measurement of the incremental inductance is as shown in the Fig. 7.15. In the bridge circuit, the coil L_1 is fed with A.C source as well as D.C source.

Both the sources are connected in parallel. The capacitor C_b is the blocking capacitor which is used to prevent d.c current from flowing into a.c sources. Similarly a large value inductance L is used in series with d.c supply. It stops the a.c current from flowing into the d.c supply. As this bridge consists both the types of sources, it is necessary that no d.c current should affect any a.c balance conditions. In actual bridge circuit no d.c current can flow through detector as the bridge consists capacitors C_2 and C_4.

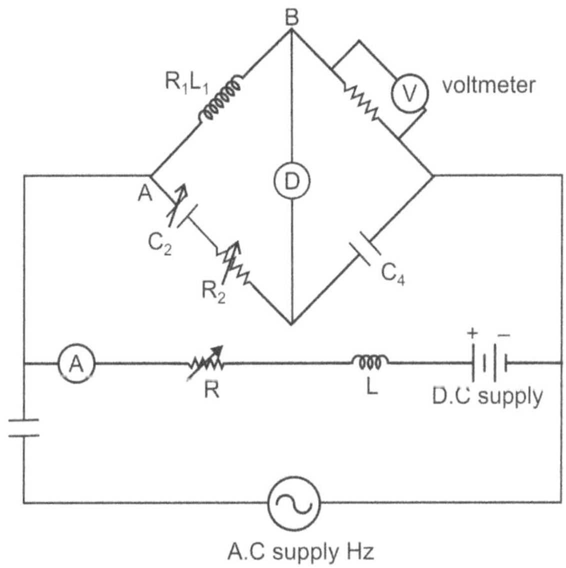

Fig 7.15 Owen's bridge for measurement of an incremental inductance

To measure the D.C current, a moving coil type ammeter is used, while the component of the A.C current can be obtained from the reading on the voltmeter. The voltmeter used is a special type of voltmeter known as "Valve voltmeter" which is insensitive to the d.c voltage. The current obtained from this reading is nothing but the current through R_3 as well as current through coil L_1 at balance.

At balance the incremental inductance as per previous derivation is given by,

$$L_1 = R_2 R_3 C_4$$

and
$$L_1 = \frac{N^2}{\left(\frac{l}{\mu A}\right)} \qquad .....(7.50)$$

Incremental permeability $\mu = \frac{L_1 l}{N^2 A}$

where, l = length of flux path

A = area of flux path

L_1 = incremental inductance

7.21 Heaviside Mutual Inductance Bridge

As the name suggests, Heaviside mutual inductance bridge measures **mutual inductance in terms of self inductances.**

Let M be the unknown mutual inductance between two coils of self inductances of L_1 and L_2. The bridge consists coils L_1, L_2 and inductive resistance R_1, R_2, R_3 and R_4 as shown in the Fig. 7.16 (a).

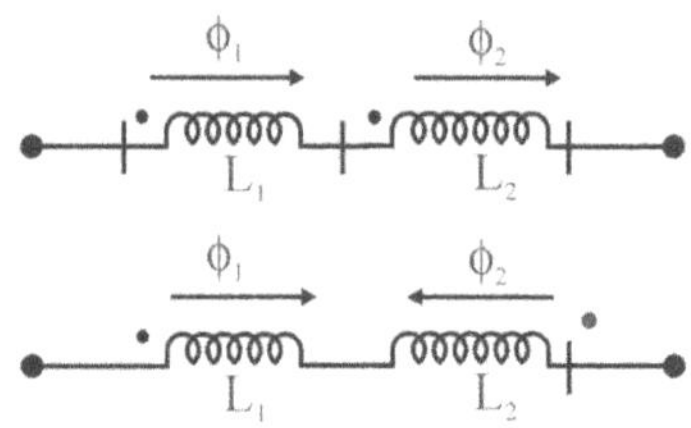

Fig. 7.16 (b) Positive and negative couplings for mutual inductors

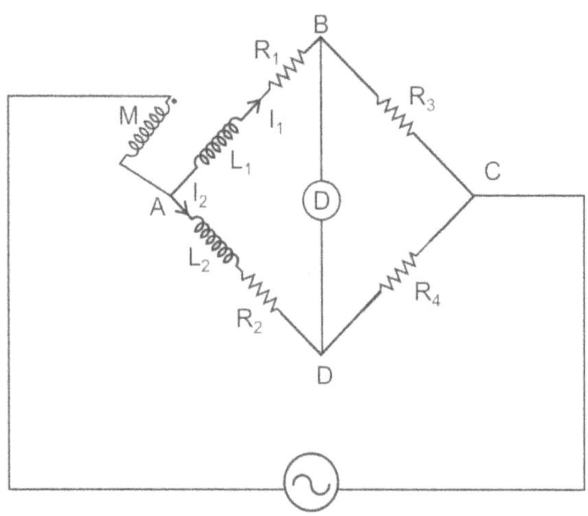

(a) Circuit diagram

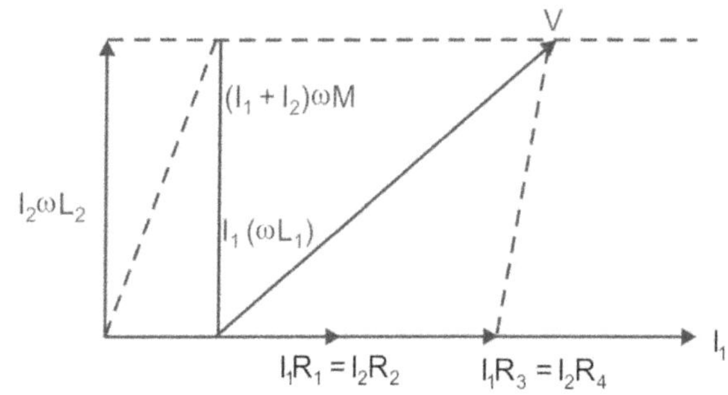

(c) Phasor diagram

Fig. 7.16 Heaviside mutual inductance bridge

Let M = unknown mutual inductance

L_1 = self inductance of secondary of mutual inductance

L_2 = known self-inductance

R_1, R_2, R_3, R_4 = Non-inductive resistors

The balance condition can be written as,

Voltage drop across brach BC = Voltage drop across brach D.C

$$\therefore \quad I_1R_3 = I_2 R_4 \qquad \text{.....(7.51)}$$

Also another balance condition is given by

Voltage drop between A-B-C nodes = voltage drop between A-D-C nodes.

$$\therefore \quad (I_1 + I_2)(j\omega M) + I_1(R_1 + j\omega L_1 + R_3) = I_2(R_2 + j\omega L_2 + R_4) \qquad \text{.....(7.52)}$$

Substituting value of $I_1 = \frac{R_4}{R_3} I_2$ from equation 7.51 in equation 7.52 we get,

$$I_2\left(\frac{R_4}{R_3}+1\right)(j\omega M) + \frac{R_4}{R_3}(R_1 + j\omega L_1 + R_3)L_2 = I_2(R_2 + j\omega L_2 + R_4)$$

$$\therefore \quad j\omega M\left(\frac{R_4}{R_3}+1\right) + \frac{R_4}{R_3}(R_1 + R_3) + \frac{R_4}{R_3} j\omega L_1 = (R_2 + R_4) + j\omega L_2$$

$$\therefore \quad \frac{R_1R_4}{R_3} + R_4 + j\omega\left[L_1\left(\frac{R_4}{R_3}\right) + \left(\frac{R_4}{R_3}+1\right)M\right] = (R_2 + R_4) + j\omega L_2 \qquad \text{.....(7.53)}$$

Equating real turns, we get

$$\frac{R_1R_4}{R_3} + R_4 = R_2R_4$$

$$\therefore \qquad R_1 = \frac{R_3}{R_4}.R_2 \qquad .....(7.54)$$

Equating imaginary terms, we get,

$$\left(\frac{R_4}{R_3}\right)L_1 + \left(\frac{R_4}{R_3}+1\right)M = L_2$$

$$M\left[\frac{R_4+R_3}{R_3}\right] = L_2 - \frac{R_4}{R_3}L_1$$

$$M\left[\frac{R_4+R_3}{R_3}\right] = \frac{L_2R_3 - R_4L_1}{R_3}$$

$$\therefore \qquad = \frac{L_2R_3 - L_1R_4}{R_3+R_4} \qquad .....(7.55)$$

If $R_3 = R_4$, the balance conditions becomes

$$R_1 = R_2 \qquad .....(7.56)$$

and

$$M = \frac{L_2 - L_1}{2} \qquad .....(7.57)$$

7.22 Campbell's Modification of Heaviside Bridge

The heaviside mutual inductance bridge is modified to measure the **self inductance in terms of mutual inductance by Campbell.**

The circuit diagram of campbell's modified Heaviside bridge is as shown in the Fig.7.17. In the modified Heaviside bridge, a balancing coil of inductance L and resistance R is used with inductor L_2 in series. A variable resistor r is used in series with L_1. The balance condition can be obtained by varying M and r. A switch is used across L_2 and R_2. During measurement, readings are taken with S_1 open initially and then S_1 closed. Let the values of M and r be M_1 and r_1 respectively with S_1 open and M_2 and r_2 with switch closed.

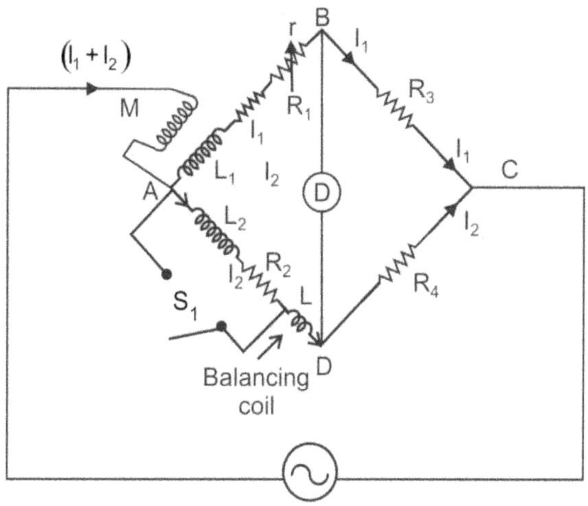

Fig. 7.17 Campbell's modification of heaviside bridge

Let switch S_1 is in open condition, then from equation 7.55, we can write,

$$M_1 = \frac{(L + L_2)R_3 - L_1R_4}{(R_3 + R_4)}$$

$$\therefore \quad L_1 + L_2 = \frac{M_1(R_3 + R_4) + L_1R_4}{R_3} \quad .....(7.58)$$

With switch S_2 in closed condition, we can write

$$L = \frac{M_2(R_3 + R_4) + L_1R_4}{R_3} \quad .....(7.59)$$

From equations in (7.58) and (7.59), the value of unknown inductance L_2 is given by

$$L_2 = (M_1 - M_2)\left(1 + \frac{R_4}{R_3}\right) \quad .....(7.60)$$

Another condition for R_2 can be derived as follows

Consider equation 7.54, with switch S_1 open, we can

$$(R_1 + r_1) = \frac{R_3}{R_4}(R_2 + R) \quad ......(7.61)$$

Similarly, with switch S_1 closed, we can write

$$(R_1 + r_2) = \frac{R_3}{R_4} + (R_2) \quad ......(7.62)$$

Hence, solving equations in 7.61 and 7.62, we get

$$R_2 = (r_1 - r_2)\frac{R_4}{R_3} \qquad .....(7.63)$$

When we use equal ratio arms i.e., $R_3 = R_4$, then

$$L_2 = 2(M_1 - M_2) \qquad .....(7.64)$$

$$R_2 = (r_1 - r_2) \qquad .....(7.65)$$

7.23 Heaviside Campbell Equal Ratio Bridge

The use of balancing coil in the above method reduces the sensitivity of the bridge. Fig. 7.18 shows Heavisible Campbell equal ratio bridge. This is a better arrangement which improves sensitivity and also dispenses with the use of a balancing coil.

In this method, the secondary of the mutual inductor is made up of two equal coils each having a self-inductance. One of the coils is connected in arm AB and the other arm AD. The primary of mutual inductance reacts with both of them. L_2, R_2 is the coil whose self-inductance and resistance is to be determined. The resistance R_3 and R_4 are made equal. Balance is obtained by varying the mutual inductance and resistance r.

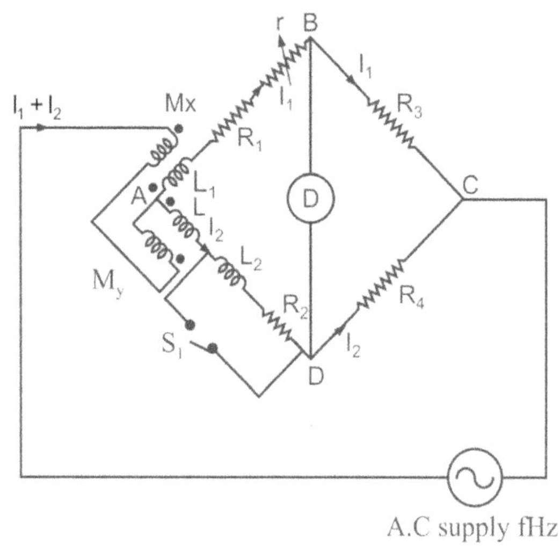

Fig. 7.18 Heaviside campbell equal ratio bridge

At balance $\quad I_1 R_3 = I_2 R_4$

But $\quad R_3 = R_4$

$\therefore \quad I_1 = I_2 = \frac{I}{2} \qquad$ as $I = I_1 + I_2$

Writing the other equation for balance

$$I_1(R_1 + r) + I_1 j\omega L + Ij\omega M_x = I_2R_2 + I_2 j\omega(L_2 + L) - Ij\omega\, C$$

or

$$\frac{R_1 + r}{2} + j\omega\left(\frac{L}{2} + M_x\right) = \frac{R_2}{2} + j\omega\left(\frac{L_2 + L}{2}\right)_{j\omega C}$$

Equating the real and imaginary terms

$$R_2 = R_1 + \psi \qquad .....(7.66)$$

$$L_2 = 2(M_x + M_y) = 2M \qquad .....(7.67)$$

and thus the magnitude of inductance which can be measured with this method is twice the range of the mutual inductor. The values calculated above include the effects of leads etc. In order to eliminate this effect, we take two readings with switch open circuit and another with switch closed. Let M_1, r_1 be the readings of M. r with open circuit and M_2, r_2 with short circuit.

$$R_1 = r_1 - r_2 \qquad .....(7.68)$$

$$L_1 = 2\,(M_1 - M_2) \qquad .....(7.69)$$

7.24 Shielding and Grounding of Bridges

The various bridges discussed up till now consist of the lumped impedances, which do not interact in any way. In practice, the stray capacitance between the various bridge elements and ground, and between the bridge arms themselves exist. Thus stray capacitance shunt the bridge arms and cause considerable error in the measurement.

The stray capacitance are uncertain in magnitude. They often vary with the adjustment of bridge arms and position of operator. "The shielding and grounding" of bridge is one way of reducing the effect of stray capacitances. But this technique does not eliminate the stray capacitance but makes them constant in value and hence they can be compensated.

One very effective and popular method of eliminating the stray capacitances and the capacitances between the bridge arm is using a ground connection called "Wagner ground connection". This connection is shown in the Fig. 7.19.

The circuit is a capacitance bridge where C_1 and C_2 represent the stray capacitances, the Wagner's connection is the use of separate arm consisting of the resistance R_w and the capacitance C_w across the terminals C and D, forming a potential divider. This arm is also called "Guarding arm". The procedure for the adjustment is as follows.

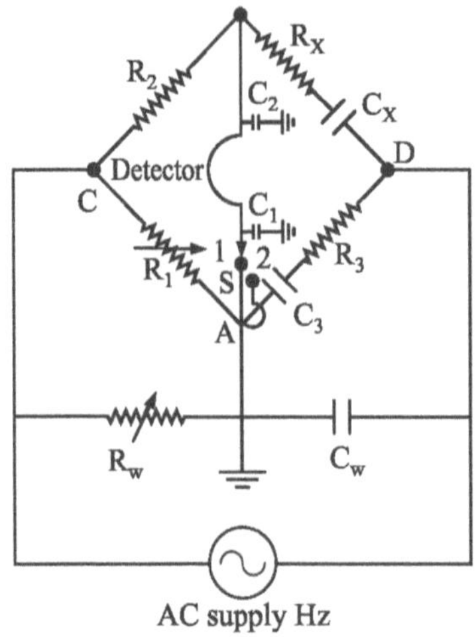

Fig. 7.19 Wagner ground connection

The switch S is connected is series with the detector. The switch is connected to position '1' and R_1 is then adjusted to get null response i.e., minimum sound in head phones. The switch is then thrown to position 2, which connects the detector to the Wagner ground point. The resistance R_w of the Wagner connection is now adjusted to get the minimum sound.

The switch is again thrown back to position 1. There is some inbalance present now. The resistance R_1 and R_3 are then adjusted to get minimum sound.This procedure is repeated till a null is obtained on both the switch position 1 and 2.

This null obtained at both the position indicates that the points 1 and 2 are at the same potential. But the point 2 is at the ground potential due to Wagner ground connection. Hence point 1 and 2 are at the ground potential. Thus the stray capacitance C_1 and C_2 are effectively short circuited. Thus they have no effect on the normal balancing.

There are capacitance existing from points C and D to ground but the addition of the Wagner ground connection eliminates them as current through these capacitances will enter through Wagner ground connection.

Note: As the addition of Wagner ground connection does not affect the balance conditions, the procedure for the measurement remains unchanged.

Basic Bridge Configuration

Series capacitance comparison bridge		
Most common capacitance bridge circuit usually R_3 and R_1 are variable so that bridge reads capacitance and D.	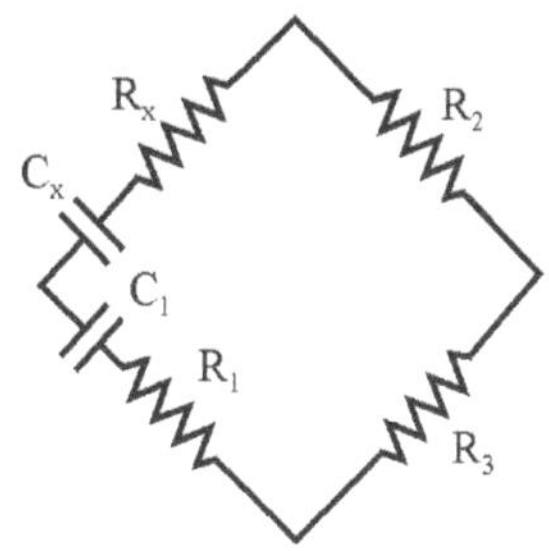	$C_X = C_1.\frac{R_3}{R_2}$ $R_x = R_1.\frac{R_2}{R_3}$ $D = \omega C_1 R_1$ Series components
Parallel capacitance comparison bridge		
Used for parallel capacitance measurements and for high D measurements where R_1 of the series bridge would be on practically high	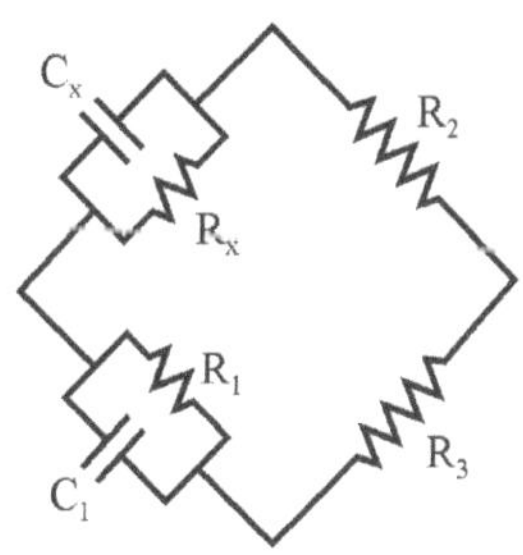	$C_X = C_1.\frac{R_3}{R_2}$ $R_X = R_1.\frac{R_2}{R_3}$ $D = \frac{1}{\omega C_1 R_1}$ Parallel components
Schering bridge		
Used in high-voltage bridge with a high-voltage capacitor as C_1. Used in high frequency bridges because variable capacitors can be used for both adjustments. Used in dielectric measurement because C_2 gives high resolution loss measurement. A series combination of C_2 and R_2 reads parallel capacitance	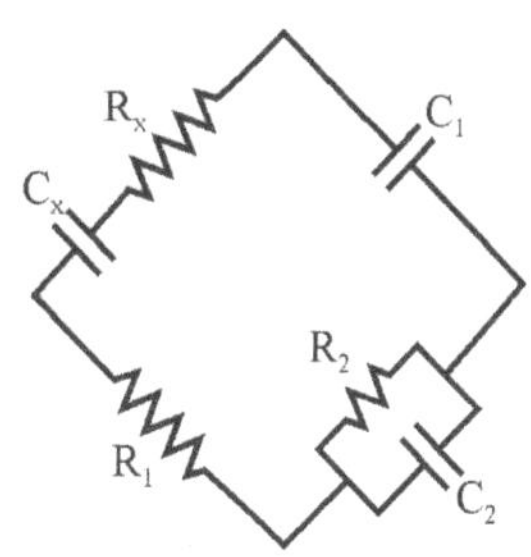	$C_X = C_1.\frac{R_2}{R_1}$ $R_X = R_1.\frac{C_2}{C_2}$ $D_X = \omega C_2 R_2$ Series components

Table *Contd....*

Series resonant bridge		
Rarely used to measure L because of frequency dependence. Parallel vertion has same equations for parallel components, and is used with variable frequency source to measure resonant frequency of coils	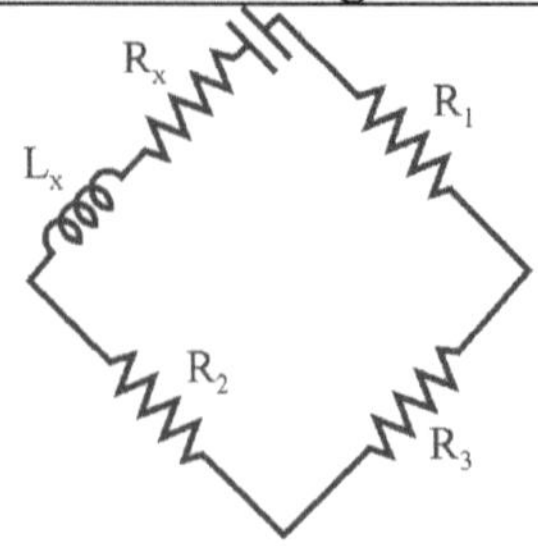	$L_X = \frac{1}{\omega^2 C}$ $R_X = \frac{R_1 R_2}{R_3}$ Series components
Maxwell inductance comparison bridge		
Basic inductance comparison bridge additional resistance added to inductance with lower Q to get resistance balance series circuit shown : could also be parallel	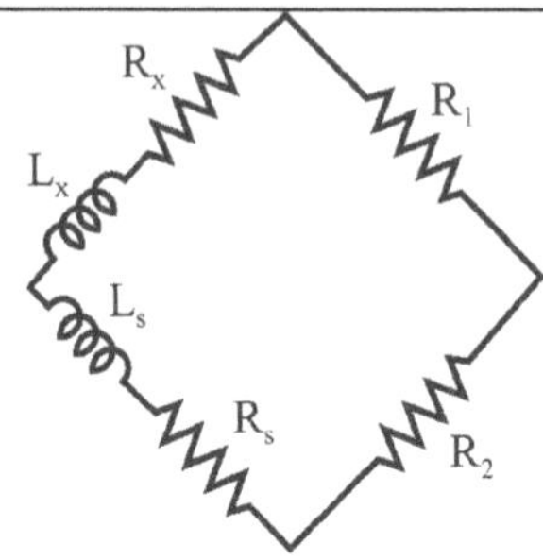	$L_X = L_s.\frac{R_1}{R_2}$ $R_X = R_s.\frac{R_1}{R_2}$ Series components
Maxwell wien bridge		
Most common inductance bridge. If R_2 and R_3 are variable, it reads L and Q. If R_3 and C are variable, it reads L and R.	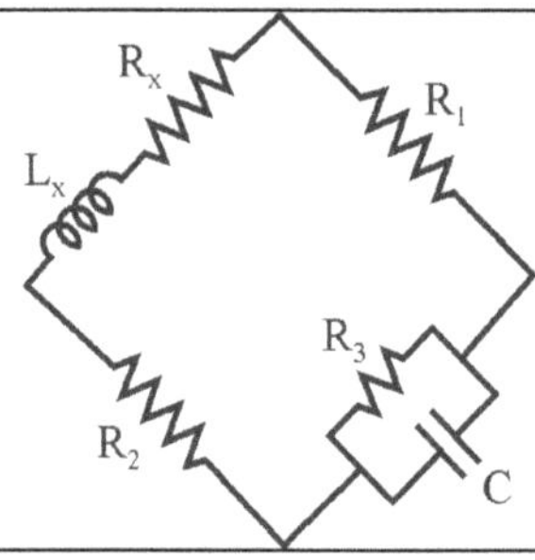	$L_X = R_1 R_2 C$ $R_X = \frac{R_1 R_2}{R_3}$ $Q = \omega C R_3$ Parallel components
Hay bridge		
Useful for parallel inductance measurement of high-Q measurements where R_3, of Maxwell bridge would have to be impractically high. Also used for measurement with d.c current bias because all current applied across bridge flows in L_X.	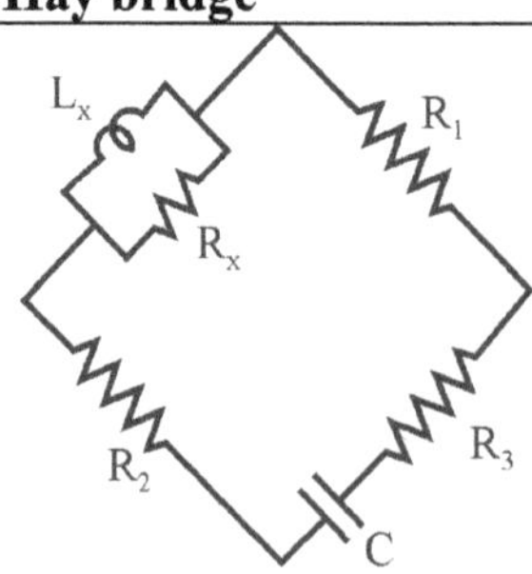	$L_X = R_1 R_2 C$ $R_X = \frac{R_1 R_2}{R_3}$ $Q = \frac{1}{\omega R_3 C}$ Parallel components

Table *Contd....*

Series Owen bridge		
Used in precision inductance bridges where R_2 is a high-resolution decade reading L_x, and C_2 reads G_{x1}, parallel - Lx and G_x is formed if R_2 and C_2 connect in parallel.	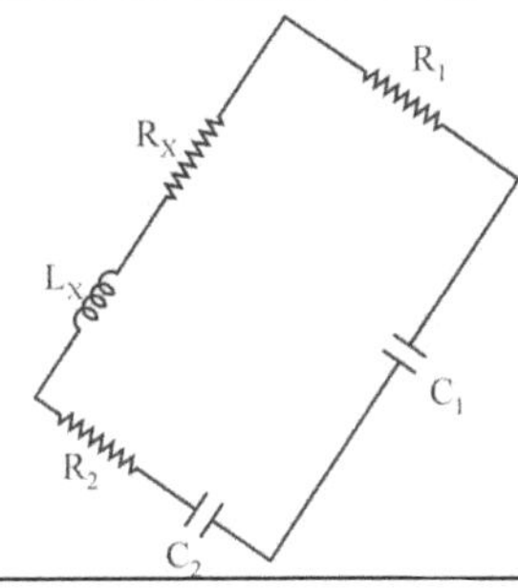	$L_X = R_1 R_2 C_1$ $G_X = \frac{C_2}{R_1 C_1}$ Series components

Problems with Solutions

1. The four arms of a Maxwell's capacitance bridge at balance are, arm 'ab', an unknown inductance L_1 having a resistance R_1, arm 'bc', a non inductive resistance of 1000 Ω, arm 'cd', a capacitor of 0.5 μF in parallel with a resistance of 1000 Ω, arm 'da', resistance of 1000 Ω. Find inductance L_1 of arm 'ab'.

***Solution*:** Given data:

Arm ab: unknown inductance L_1, resistance R_1

Arm bc: $R_3 = 1000\ \Omega$

Arm cd: $C_4 = 0.5\ \mu F$

$R_4 = 1000\ \Omega$

Arm da: $R_2 = 1000\ \Omega$

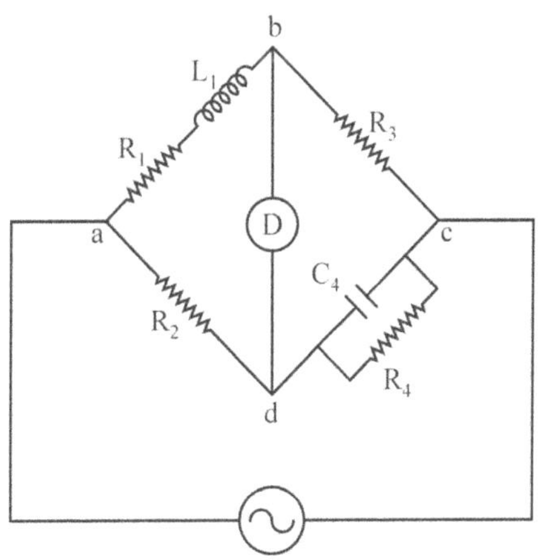

Fig. Maxwells bridge

Now,

$$\boxed{R_1 = \frac{R_2 R_3}{R_4}}$$

$$= \frac{1000 \times 1000}{1000} = 1000\ \Omega$$

$$\therefore \quad \boxed{L_1 = R_2 R_3 C_4}$$

$$= 1000 \times 1000 \times 0.5 \times 10^{-6}$$

$$= 0.5\ H$$

2. A Maxwell's bridge shown in figure has the following constants. Arm ab consist of a coil with inductance L_1 and resistance r_1, in series with a non-inductive resistance R. Arm bc and cd each are a non-inductive resistance of 100 Ω. Arm ad consist of standard variable inductor L of resistance 32.7 Ω. Balance is obtained when L_2 = 47.8 mH and $R = \frac{1}{36}$ Ω. Find the resistance and inductance of the coil in the arm ab.

***Solution*:**

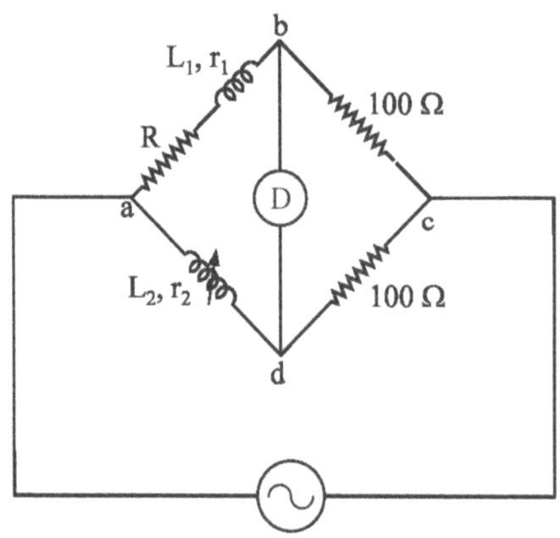

$$\boxed{Z_1 = R + [r_1 + jwL_1]}$$

$$= (R + r_1) + jwL_1$$

$$\boxed{Z_2 = (r_2 + jwL_2)}$$

$$Z_3 = 100$$

$$Z_4 = 100$$

Under balanced condition we have,

$$\boxed{Z_1Z_4 = Z_2Z_3}$$

$$[(R + r_1) + jwL_1] \times 100 = (r_2 + jwL_2) \times 100$$

$$\Rightarrow (R + r_1) + jwL_1 = r_2 + jwL_2$$

Equating real and imaginary parts,

$$\boxed{R + r_1 = r_2,} \qquad \boxed{L_1 = L_2}$$

$$\therefore \qquad r_1 = r_2 - R$$

Given,

$$R_3 = R_4 = 100\ \Omega$$

$$R = \frac{1}{36}\ \Omega = 0.027\ \Omega$$

$$r_2 = 32.7\ \Omega$$

$$L_2 = 47.8\ \text{mH}$$

$$\therefore \qquad r_1 = r_2 - R$$

$$= 32.7 - 0.027$$

$$= 32.67\ \Omega$$

$$L_1 = L_2 = 47.8\ \text{mH}$$

3. The four arms of Hay's bridge are arranged as follows, AB is a coil of unknown impedance, BC is a non-reactive resistance of 100 Ω, CD is a non-reactive resistance of 833Ω in series with a standard capacitor of 0.38 μF, DA is a non-reactive resistance of 16,800 Ω. If the supply frequency is 50 Hz, determine the inductance and the resistance at the balance conditions.

***Solution*:** Given data:

Arm BC: $R_3 = 100\ \Omega$

Arm CD: $C_4 = 0.38\ \mu F$

$R_4 = 833\ \Omega$

Arm DA: $R_2 = 16{,}800\ \Omega$

Arm AB: $R_1 = ?$

$L_1 = ?$

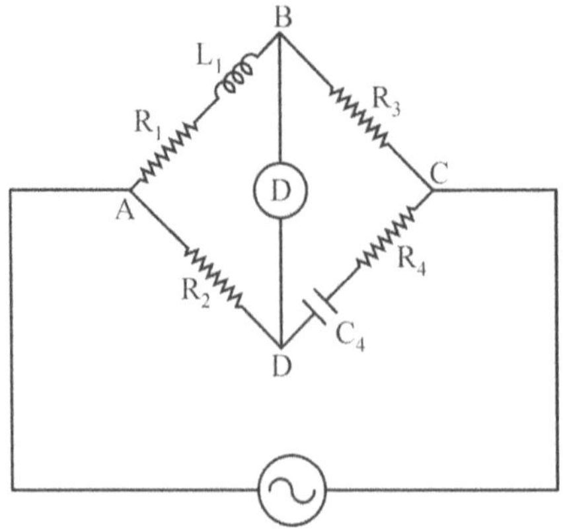

***Solution*:**

Angular frequency,

$$\begin{aligned} w &= 2\pi F \\ &= 2\pi \times 50 \\ &= 100\pi \end{aligned}$$

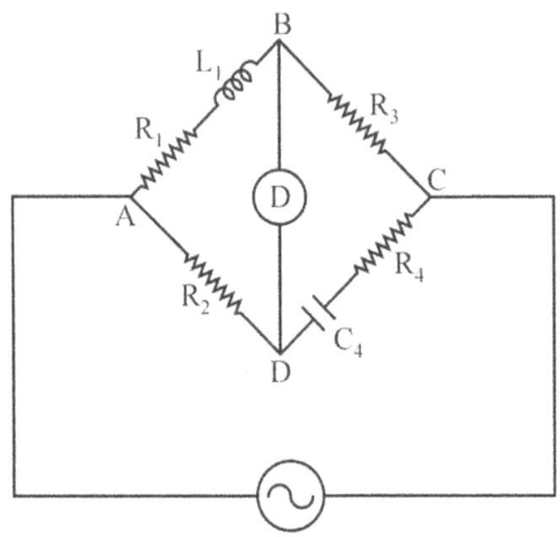

Fig. Hay's bridge

According to Hay's Bridge L_1 and R_1 are given as,

$$\boxed{L_1 = \frac{R_2\, R_3\, C_4}{1 + w^2\, C_4^2\, R_4^2}}$$

$$\boxed{R_1 = \frac{w^2\, C_4^2\, R_2\, R_3\, R_4}{1 + w^2\, C_4^2\, R_4^2}}$$

$$\therefore \quad L_1 = \frac{(16800)\,(100)\,(0.38 \times 10^{-6})}{1+\left[(100\pi)^2\;(0.38\times10^{-6})^2\;(833)^2\right]}$$

$$= 0.63 \text{ H}$$

$$R_1 = \frac{(100\pi)^2\,(0.38\times10^{-6})^2(16800)(100)(833)}{1+[(100\pi)^2(0.38\times10^{-6})^2(833)^2]}$$

$$= 19.75\ \Omega$$

4. In an Anderson bridge for measurement of inductance L_x and resistance R_x in the arm AB, the arm CD and DA have resistance of 600 Ω each and the arm CE has a capacitor of 1 μF. With AC supply at 100 Hz supplied across A and C, balance is obtained with a resistance of 400 Ω in arm DE and 800 Ω in the arm BC. Calculate the value of L_x and R_x.

***Solution*:** Given data:

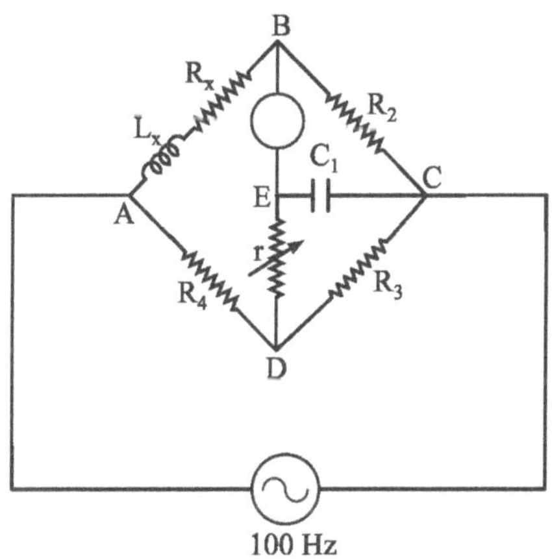

$R_3 = 600\ \Omega$

$R_4 = 600\ \Omega$

$C_1 = 1\ \mu F$

$r = 400\ \Omega$

$R_2 = 800\ \Omega$

$$\boxed{L_x = \frac{C_1 R_2}{R_3}\left[r\,(R_3 + R_4) + R_3 R_4\right]}$$

$$= \frac{1\times10^{-6}\times 800}{600}\;[400\;(600 + 600) + 600 \times 600]$$

$$= 1.33 \times 10^{-6} \; [480000 + 360000]$$
$$= 1.12 \text{ H}$$

$$\boxed{R_x = \frac{R_2}{R_3} \times R_4}$$

$$= \frac{800}{600} \times 600$$
$$= 800\ \Omega$$

5. An owen's bridge is used to measure the properties of a sample of sheet steel at 2 kHz. At balance, arm AB is test specimen; arm bc is $R_3 = 100\ \Omega$; arm CD is $C_4 = 0.1\ \mu F$. Calculate the effective impedance of the specimen under test conditions.

Solution: Given data:

Arm AB: effective resistance R_1, unknown inductance, L_1.

Arm BC: $R_3 = 100\ \Omega$

Arm CD: $C_4 = 0.1\ \mu F$

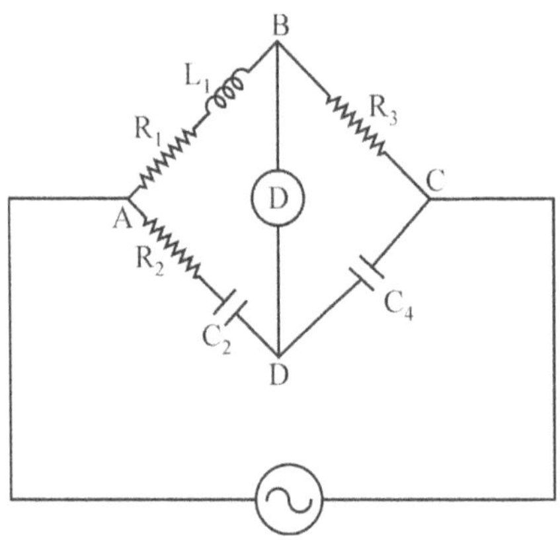

From the above owen's bridge it is observed that the values of R_2 and C_2 are not given. Thus assuming the values under balanced condition,

$$\boxed{(R_1 + J\omega L_1)\frac{1}{j\omega C_4} = R_3\left(R_2 + \frac{1}{j\omega C_2}\right)}$$

$$\frac{R_1}{j\omega C_4} + \frac{j\omega L_1}{j\omega C_4} = R_3R_2 + \frac{R_3}{j\omega C_2}$$

$$-\frac{R}{j\omega C_4} + \frac{L_1}{C_4} = R_3R_2 - \frac{jR_3}{\omega C_2}$$

$$\frac{L_1}{C_4} - \frac{jR_1}{\omega C_4} = R_3R_2 - \frac{jR_3}{\omega C_2}$$

Equating imaginary and real parts, we get

$$\frac{L_1}{C_4} = R_3R_2$$

$$\boxed{L_1 = R_2R_3C_4} \quad .....(1)$$

and $$\frac{-R_1}{\omega C_4} = \frac{-R_3}{\omega C_2}$$

$$\boxed{R_1 = \frac{R_3C_4}{C_2}} \quad .....(2)$$

from equation (1) and (2)

$$\therefore \quad L_1 = (100)\,(834)\,(0.1 \times 10^{-6})$$

$$= 8.34\ \mu H$$

$$R_3 = \frac{(100)\,(0.1\times10^{-6})}{(0.124\times10^{-6})}$$

$$= 80.64\ \Omega$$

Reactance (x), of specimen at 2 kHz is given as,

$$x_1 = 2\pi\times2\times1000\times L_1$$

$$= 2\pi\times2\times1000\times8.34\times10^{-3}$$

$$= 104.8\ \Omega$$

$\therefore$ The effective impedance of specimen under this test condition is given as,

$$\boxed{Z = \sqrt{R_1^2 + x_1^2}}$$

$$= \sqrt{(80.64)^2 + (104.8)^2}$$

$$= 132.23\ \Omega$$

6. A sheet of bakelite 4.5 mm thick is tested at 50 Hz between electrodes 0.12 m in diameter. The Schering bridge employs a standard air capacitor C_2 of 106 pF capacitance, a non-reactive resistance R_4 of $1000/\pi$ ohms in parallel with a variable capacitor C_4, and a non-reactive variable resistance R_3. Balance is obtained with C_4

= 0.5 μF and R_3 = 260 Ω. Calculate the power factor, capacitance and relative permittivity of sheet.

***Solution*:** Given data:

Thickness, d = 4.5 mm

$= 4.5 \times 10^{-3}$ m

Frequency, f = 50 Hz

Diameter, D = 0.12 m

Standard air capacitor, C_2 = 106 pF

$= 106 \times 10^{-12}$ F

Non-reactive resistance,

$$R_4 = \frac{1000}{\pi} \Omega$$

Balanced condition is obtained at,

Variable capacitance,

$C_4 = 0.5\ \mu F$

$= 0.5 \times 10^{-6}$ F

Non-reactive variable resistance,

$R_3 = 260\ \Omega$

Power factor = ?

Capacitance, C_1 = ?

Relative permittivity, E_r = ?

The bridge employed is a Schering bridge.

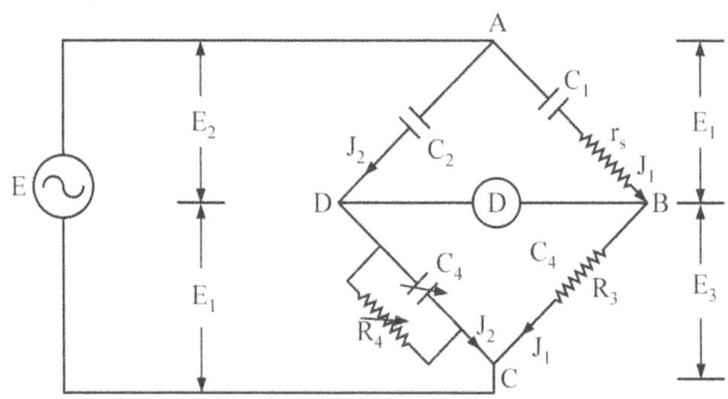

C_1 = required capacitance

r_s = series resistance of the capacitance

From the analysis of the Schering bridge, we have

Series resistance,

$$\boxed{r_s = \frac{R_3 C_4}{C_2}}$$

$$= \frac{260 \times 0.5 \times 10^{-6}}{106 \times 10^{-12}}$$

$$= 1.22 \times 10^{6}$$

$$= 1.22\ \text{M}\Omega$$

Capacitance,

$$\boxed{C_1 = \frac{R_4 C_2}{R_3}}$$

$$= \frac{1000}{\pi} \times 106 \times 10^{-12} \times \frac{1}{260}$$

$$- 1.297 \times 10^{-10}\ \text{F}$$

$$= 129.7\ \text{pF}$$

Now, power factor of the sheet

$$= 2\pi F\ C_1\ r_s$$

$$= 2\pi \times 50 \times 129.7 \times 10^{-12} \times 1.22 \times 10^{6}$$

$$= 0.0497$$

$$= 0.05$$

Area of the sheet,

$$\boxed{A = \frac{\pi}{4}\ (0.12)^2}$$

$$= 0.0113$$

We have, capacitance,

$$\boxed{C_1 = \frac{EA}{d}}$$

$$\Rightarrow \qquad C_1 = \boxed{E_o\ E_r}\frac{A}{d}$$

$$\therefore \quad E_r = \frac{C_1 \, d}{E_o \, A}$$

$$= \frac{129.7 \times 10^{-12} \times 4.5 \times 10^{-3}}{8.854 \times 10^{-12} \times 0.0113}$$

$$= 5.83$$

p.f = 0.05

C_1 = 129 p.f

E_r (Relative permittivity)

E_r = 5.83

7. In a low voltage Shering bridge designed for the measurement of permittivity. The branch consist of two electrodes between which the specimen under test may be inserted; arm BC is a non-reactive resistor R_3 in parallel with a standard capacitor C_3; arm CD is a non-reactive resistor R_4 in parallel with a standard capacitor C_4; arm DA is a standard air capacitor of capacitance C_2. Without the specimen between the electrodes, balance is obtained with the following values, $C_3 = C_4 = 120$ pF, $C_2 = 150$ pF, R3 = $R_4 = 5000\ \Omega$. With the specimen inserted, these values become, $C_3 = 200$ pF; C4 = 1000 pF; $C_2 = 900$ pF, and $R_3 = R_4 = 5000\ \Omega$. In each test w = 5000 rad/sec. Find the relative permittivity of the specimen.

***Solution*:**

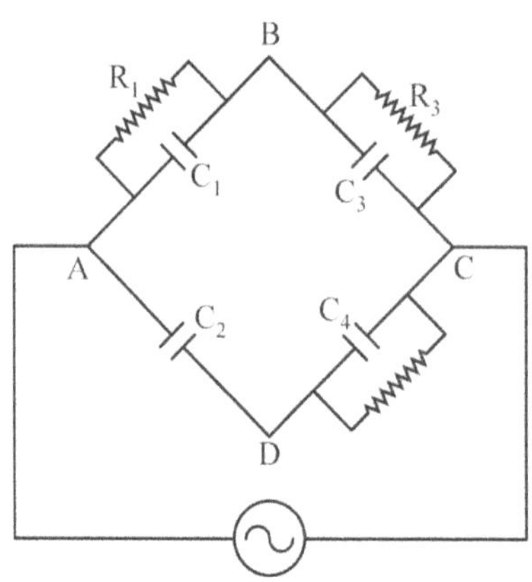

Given data:

From figure,

$$Y_1 = \frac{1}{R_1} + jwC_1$$

$$Y_2 = jwC_2$$

$$Y_3 = \frac{1}{R_3} + jwC_3$$

$$Y_4 = \frac{1}{R_4} + jwC_4$$

Under balanced condition we get,

$$\boxed{Y_1Y_4 = Y_2Y_3}$$

$$\left(\frac{1}{R_1} + j\omega C_1\right)\left(\frac{1}{R_4} + j\omega C_4\right) = JWC_2\left(\frac{1}{R_3} + j\omega C_1\right)$$

$$\left(\frac{1}{R_1R_4} - \omega^2 C_1C_4\right) + j\omega\left(\frac{C_4}{R_1} + \frac{C_1}{R_4}\right) = \frac{j\omega C_2}{R_3} - \omega^2 C_2C_3$$

Equating the real and imaginary parts,

$$\frac{1}{R_1R_4} - \omega^2 C_1C_4 = -\omega^2 C_2C_3 \quad \text{.....(1)}$$

and

$$\frac{C_4}{R_1} + \frac{C_1}{R_4} = \frac{C_2}{R_3} \quad \text{.....(2)}$$

Solving equations (1) and (2), we get

$$C_1 = \frac{C_2\frac{R_4}{R_3} + \omega^2 C_2C_1C_4\, R_4^2}{1 + \omega^2\, C_4^2\, R_4^2} \quad \text{.....(3)}$$

Since, $\omega^2 C_2C_3C_4 R_4^2 <<< \frac{C_2\, R_4}{R_3}$

Also, $\omega^2\, C_4^2\, R_4^2 <<< 1$

Hence, the equation (3) reduces to,

$$\boxed{C_1 = \frac{C_2\, R_4}{R_3}}$$

(i) without the specimen,

Let C_1 be the value of capacitor,

$$\boxed{C_1' = \frac{C_2\, R_4}{R_3}}$$

$C_2 = 150$ pF

$R_4 = 5000\ \Omega$

$R_3 = 5000\ \Omega$

$$\therefore \quad C_1^{'} = \frac{150 \times 10^{-12} \times 5000}{5000}$$

$$= 150 \times 10^{-12} = 150 \text{ pF}$$

(ii) with the specimen

Let $\boxed{C_1^{''} = \frac{C_2 R_4}{R_3}}$

$C_2 = 900$ pF

$R_3 = R_4 = 5000\ \Omega$

$$C_1^{''} = \frac{900 \times 10^{-12} \times 5000}{5000} = 900 \text{ pF}$$

$$\therefore \text{ Relative permittivity, } E_r = \frac{C_1^{''}}{C_1^{'}} = \frac{900}{150} = 6$$

8. The four arms of a bridge are,

Arm ab: an imperfect capacitor C_1 with equivalent series resistance of 'r'

Arm bc: a non-inductive resistance R_3

Arm cd: a non-inductive resistance R_4

Arm da: an imperfect capacitor C_2 with an equivalent resistance of r_2 series with a resistance R_2.

A supply of 450 Hz is given between terminal a and c and the detector is connected between 'b' and 'd'. At balance $R_2 = 4.8\Omega$ $R_3 = 2000\Omega$, $R_4 = 2850\ \Omega$ and $C_2 = 0.5\ \mu F$ and $r_2 = 0.4\ \Omega$. Calculate the value of C_1 and r_1 and dissipating factor.

***Solution*:** Given data:

$f = 450$ Hz

$R_2 = 4.8\ \Omega$

$R_3 = 2000\ \Omega$

$R_4 = 2850\ \Omega$

$C_2 = 0.5\ \mu F$

$r_2 = 0.4\ \Omega$

$C_1 = ?$

$r_1 = ?$

$D = ?$

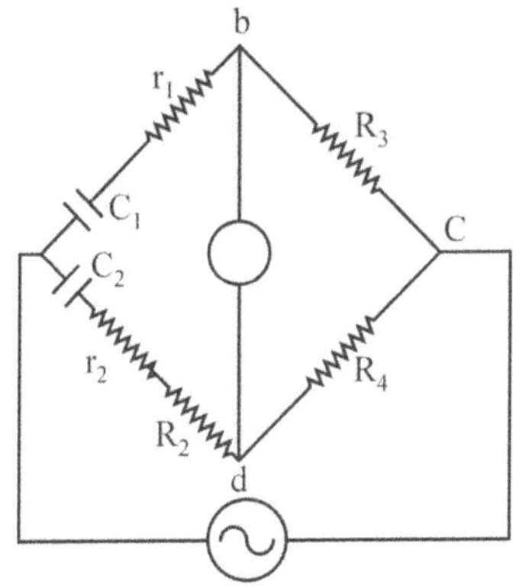

From figure,

$$z_1 = r_1 + \frac{1}{jwC_1}$$

$$z_2 = r_2 + R_2 + \frac{1}{jwC_2}$$

$$z_3 = R_3$$

$$z_4 = R_4$$

Under balanced condition we hence

$$\boxed{Z_1Z_4 = Z_2Z_3}$$

$$\Rightarrow \quad \left(r + \frac{1}{j\omega C_1}\right) R_4 = \left(r_2 + R_2 + \frac{1}{j\omega C_2}\right) R_3$$

$$\Rightarrow \quad r_1 + \frac{1}{j\omega C_1} = \frac{r_2 + R_2}{R_4} \times R_3 + \frac{R_3}{j\omega R_4 C_4}$$

Equating read and imaginary terms on both sides, we get,

$$r_1 = (r_2 + R_2)\,\frac{R_3}{R_4} = (0.4 + 4.8) \times \frac{2000}{2850}$$

$$= 3.649\ \Omega$$

and $\quad \dfrac{1}{C_1} = \dfrac{R_3}{R_4 C_2}$

$$\Rightarrow \quad C_1 = \frac{R_4 C_2}{R_3}$$

$$= \frac{2850 \times 0.5 \times 10^{-6}}{2000}$$

$$= 0.7125 \ \mu F$$

Dissipating factor, $D = \tan \delta$,

$$D = \omega C_1 r_1$$

$$= 2\pi f C_1 r_1$$

$$= 2\pi \times 450 \times 0.7125 \times 10^{-6} \times 3.649$$

$$= 0.00735$$

9. An A.C bridge circuit working at 100 Hz, have its arms as follows. Arm AB in 0.1 microfarad capacitance. Arm BC is a 500 Ω resistance arm, CD contains an unknown impedance and arm DA has a 300 Ω resistance in parallel with a 0.1 μF capacitor. Find the R and L or C constants of arm CD considering it as a series circuit.

Solution: Given data:

Supply frequency, $f = 1000$ Hz

Arm AB: $C_1 = 0.1 \ \mu F$

Arm AD: $R_2 = 300 \Omega$

$C_2 = 0.1 \ \mu F$

Arm BC: $R_3 = 500 \ \Omega$

Arm CD: $C_4 = ?$ $\quad L_4 = ?$ $\quad R_4 = ?$

under balanced condition we have, $Z_1 Z_4 = Z_2 Z_3$

According to the data given, the bridge is as shown in the figure.

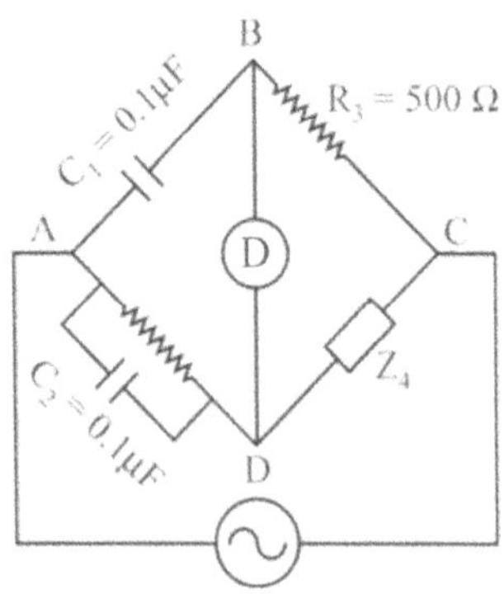

Let Z_1, Z_2, Z_3 and Z_4 be the impedance of arms AB, AD, BC, and CD respectively

Impedance of arm AB is

$$Z_1 = \frac{1}{jwC_1} = \frac{-j}{wC_1} \quad .....(1)$$

Impedance of arm AD is

$$Z_2 = \frac{R_2}{1 + jwC_2R_2} \quad .....(2)$$

Impedance of arm BC is

$$Z_3 = R_3 \quad .....(3)$$

Under balanced condition, we have

$$Z_1\, Z_4 = Z_2\, Z_3 \Rightarrow Z_4 = \frac{Z_2\, Z_3}{Z_1} \quad .....(4)$$

Substituting the equation (1) (2) and (3) in equation (4) we get

$$Z_4 = \frac{\left(\frac{R_2}{1 + jwC_2R_2}\right) R_3}{\frac{-j}{wC_1}} = \frac{jwC_1\, R_2\, R_3}{1 + jwC_2\, R_2}$$

$$= \frac{j\, 2\, \pi\, f\, C_1R_2R_3}{1 + j\, 2\, \pi\, f\, C_2R_2} \quad .....(5)$$

Now substituting the given value in equation (5) we get

$$= \frac{188.5\, \underline{|90}}{1.0176\, \underline{|10.674}}$$

$$= 185.24\, \underline{|79.32}$$

the positive angle for impedance indicates that the branch consists of a series R-L circuit

thus the impedance Z_4 can be written as

$$\therefore \quad Z_4 = 34.33 + j\, 182.03$$

$$\therefore \quad R_4 = 34.33\ \Omega$$

$$XL_4 = 2\pi\, f\, L_4 = 182.03$$

$$\therefore \quad L_4 = \frac{182.03}{2\pi \times 1000} = 0.0289\ H$$

$$= 28.9\ mH$$

10. In a Carey-Foster's bridge a resistance of 1.0125 Ω is compared with a standard resistance of 1.0000 Ω, the slide wire has a resistance of 0.250 Ω in 100 divisions. The ratio arms nominally 10 Ω each are actually 10.05 Ω and 9.95 Ω respectively. How far (on scale divisions) are the balance positions from those which would obtain of arms ratio were true to their nominal value? The slide wire is 100 cm long.

***Solution*:** Given Data

Resistance being compared, R = 1.0125 Ω

Standard resistance, S = 1.0000 Ω

Slide wire resistance, $R_T = 0.250\ \Omega$

Number of divisions, n = 0.250 Ω

Nominal values of ratio arms, P = Q = 10 Ω

Actual values of ratio arms, P = 10.05 Ω and Q = 9.92 Ω

Slide wire length, L = 100 cm

Difference in balance positions when P and Q value, are true and nominal,

$$l_1 - l_1^1 = ?\,, \qquad l_2 - l_2^1 = ?$$

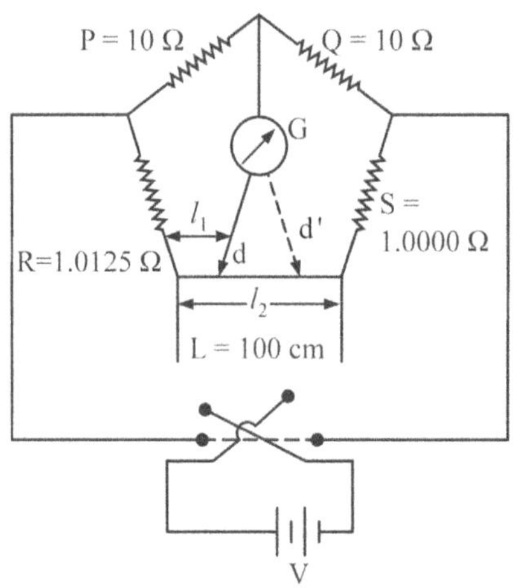

Resistance per with length of slide wire,

$$r = \frac{R_T}{L} = \frac{0.25}{100} = 0.0025\ \Omega/\text{cm}$$

(i) with nominal values of ratio arms:

we know that at balance

$$\frac{P}{Q} = \frac{R + l_1 r}{S + (L - l_1)r} \qquad .....(1)$$

$$\frac{10}{10} = \frac{1.0125 + (l_1 \times 0.0025)}{1 + (100 - l_1) \times 0.0025}$$

$\Rightarrow \quad 1 + 0.25 - 0.0025\, l_1 = 1.0125 + 0.0025\, l_1$

$\Rightarrow \quad 0.005\, l_1 = 0.2375$

$\Rightarrow \quad l_1 = 47.5 \text{ cm} \simeq 48 \text{ cm}$

$= 48$ divisions.

[since 100 divisions are present for 100 cm long slide wire, 1 div = 1cm]
At second balance i.e., after replacing R and S in the circuit, we have,

$$\frac{P}{Q} = \frac{R + l_2 r}{R + (L - l_2)r} \qquad .....(2)$$

$$\frac{10}{10} = \frac{1 + 0.0025\, l_2}{1.0125 + (100 - l_2) \times 0.0025}$$

$\Rightarrow \quad 1.0125 + 0.25 - 0.0025\, l_2 = 1 + 0.0025\, l_2$

$\Rightarrow \quad 0.005\, l_2 = 0.2625$

$\Rightarrow \quad l_1 = 52.5 \text{ cm} \simeq 53 \text{ cm}$

$= 53$ division.

(ii) with true values of ratio arms:

for first balance condition, substituting the known values in eqn. 1, we get

$$\frac{10.05}{9.92} = \frac{1.0125 + 0.0025\, l_1^1}{1 + (100 - l_1^1) \times 0.0025}$$

$$\Rightarrow \quad 1.01005 = \frac{1.0125 + 0.0025\, l_1^1}{1 + 0.25 - 0.0025\, l_1^1}$$

$$\Rightarrow \quad 1.01005 = \frac{1.0125 + 0.0025\, l_1^1}{1 + 0.25 - 0.0025\, l_1^1}$$

$\Rightarrow \quad 1.26256 - 0.00253\, l_1^1 = 1.0125 + 0.0025\, l_1^1$

$$\Rightarrow \quad 0.00503\ l_1^1 = 0.25006$$

$$\Rightarrow \quad l_1^1 = 49.71372 \text{ cm} = 50\text{cm}$$

$$= 50 \text{ divisions}$$

Similarly for second balance conditions, substituting the known values in eqn. 2, we get

$$\frac{10.05}{9.95} = \frac{1 + 0.0025\ l_2^1}{1.0125 + (100 - l_2^1) \times 0.0025}$$

$$\Rightarrow \quad 10.05 \times (1.0125 + 0.25 - 0.0025\, l_2^1) = 9.95 \times (1 + 0.0025\, l_2^1)$$

$$\Rightarrow \quad 12.6881 - 0.025\, l_2^1 = 9.95 + 0.0249\, l_2^1$$

$$0.05\, l_2^1 = 2.7381$$

$$\Rightarrow \quad l_2^1 = 54.762 \text{ cm} = 55 \text{ cm} = 55 \text{ division}$$

$$\therefore \quad \Delta l_1 = \left| l_1^1 - l_1 \right| = |50 - 48| = 2 \text{ division (cm)}$$

$$\Delta l_2 = \left| l_2^1 - l_2 \right| = |55 - 53| = 2 \text{ division (cm)}$$

Hence, the balance positions obtained are 2 cm or division for from those which obtained of arms ratio were true to their nominal value.

11. In a Schering bridge balance, the following values are obtained $C_2 = 100\ \mu F$; $R_3 = 100\ \Omega$; $R_4 = 300\ \Omega$; $C_4 = 0.5\ \mu F$; $f = 50$ Hz. Find out unknown capacitance C_1 and its loss angle δ.

***Solution*:** unknown capacitance $C_1 = C_2 . \dfrac{R_4}{R_3}$

$$= 100 \times 10^{-12} \times \frac{300}{100} = 300 \text{ PF}$$

loss angle $\delta = \tan^{-1} w\, C_4\, R_4$ [since $\tan \delta = WC_4R_4$]

$$= \tan^{-1} 2 \times 3.14 \times 50 \times 0.5 \times 10^{-6} \times 300 = 2.7^{\circ}$$

Review Questions

1. What are the advantages of bridge circuit its applications?
2. Derive the bridge balance equation for the basic A.C bridge
3. Write short notes on sources and detectors used in measurement?
4. What is the inductance comparison bridge? Derive its balance equation?
5. What is the maxwells inductance bridge? Derive its balance equation?
6. Derive balance equation for maxwell's inductance capacitance bridge and what are the advantages of using standard J known capacitor for measurements?
7. What are the advantages and disadvantages maxwells bridge?
8. What are the different sources of errors in A.C bridges explain each in details?
9. Draw the phasor diagram for anderson's bridge under balance conditions?
10. State the advantages and disadvantages of anderson's bridge?
11. What is Hay's bridge? Derive its balance equation?
12. Discuss the advantages and disadvantages of Hay's bridge.
13. What is capacitance comparison bridge? Derive it's balance equation?
14. Explain how capacitance of an imperfect capacitor is measured using A.C bridge and draw the phasor diagram for the balanced bridge?
15. Derive the power factor and loss angle for schering bridge.
16. Define dissipation factor and derive the equation for dissipation factor in case of L.V schering bridge?
17. Explain about high voltage schering bridge?
18. What is wien bridge? Derive its balance equation and what are application of it?
19. Draw the phasor diagram of modified De sauty's bridge and explain in details?
20. Draw a neat sketch and explain operation of owen's bridge?
21. What is compbell's modification of Heaviside bridge. Derive its balance equation?
22. Explain about Heaviside compbell equal ratio bridge.
23. What is the important of shielding and grounding of bridges?

Quiz Questions

1. The insulation resistance of a cable can be measured by
   (a) megger (b) ammeter-voltmeter method
   (c) ohm meter (d) kelvin's bridge
2. Which bridge is used for the measurement of low resistance a
   (a) kelvin (b) wheatstone
   (c) hay's (d) wagner ground bridge
3. Which bridge is used for the measurement of medium resistance
   (a) wheat stone bridge (b) hay`s bridge
   (c) wagner ground bridge (d) anderson's bridge
4. Medium resistances in the range from
   (a) 1Ω to 0.1MΩ (b) 1Ω to 0.5 MΩ
   (c) 1Ω to 0.1kΩ (d) 1.Ω to 0.01MΩ
5. Before connecting an ohmmeter to a circuit, you should first:
   (a) operate the circuit.
   (b) set the range selector to the highest range.
   (c) check the circuit with a voltmeter to make sure the circuit is not powered.
   (d) install new batteries.
6. Oscilloscopes can be used to
   (a) view the shape of voltage waveforms
   (b) measure voltage over a period of time.
   (c) measure resistance.
   (d) both a and b
7. Swaping resistance is added to the moving coil of a meter to
   (a) increase the sensitivity (b) prevent from over voltages
   (c) reduce the temperature error (d) none of the above
8. Maxwell-wien bridge is used for measuring
   (a) inductance (b) capacitance
   (c) voltage (d) none of the above

9. Megger instrument is used for
   (a) testing the insulation resistance of the circuit
   (b) voltage measurement
   (c) current measurement
   (d) none of the above

10. Dielectric loss can be measured by
    (a) energy meter (b) wheat stone bridge
    (c) electrostatic watt meter (d) mega ohm bridge

11. Power frequency measurements the frequency meter used in
    (a) mechanical resonance type (b) weston frequency meter
    (c) ratio meter type (d) electrodynanometer type

Questions and Answers

1. Audio frequency can be determined by ______ _______ ___bridge

Ans. Wien

2. From Maxwell Wien bridge L_X = _________ R_X =_______ Q = ______

Ans. $L_X = R_1 R_2 C,\ R_X = \frac{R_1R_2}{R_3},\ Q = CR_3$

3. Hay's bridge is used for ______

Ans. measurement of high-Q measurements

4. What is the use of Campbell bridge?

Ans. This is used for the measurement of mutual inductance.

5. Define Q-factor of the coil.

Ans. It is the ratio between power stored in the coil to the power dissipated in the coil.

6. State the use of Wein bridge.

Ans. It is used for the measurement of unknown capacitance and frequency.

7. Name the sources of errors in ac bridge.

Ans. Errors due to stray magnetic fields

Leakage errors

Eddy current errors

Residual errors

Frequency and waveform errors.

8. Name the bridge circuits used for the capacitance.

Ans. De Sauty's bridge

Schering Bridge

Wein bridge

9. Name the bridge circuits used for the mutual inductance.

Ans. The Heaviside Campbell bridge

The Campbell bridge.

10. Which type of detector is used in ac bridges?

Ans. Vibration galvanometers are used.

11. Name the ac sources used in ac bridges.

Ans. AC supply with step-down transformer

Motor driven alternator

Audio frequency and radio frequency oscillator.

12. State the use of ac bridges.

Ans. AC bridges are used for the self and mutual inductance and capacitance.

13. State the balance equation used in ac bridges.

Ans. The product of opposite branch impedances are equal.

14. State the advantages of price's guard wire method.

Ans. In this method leakage current does not flows through the meter and therefore it gives accurate reading.

15. The dielectric loss can be measured by

Ans. Electro-static meter

16. Schering bridge is used to

Ans. Determine dielectric loss

17. Which bridge is used to determine frequency?

Ans. Wien bridge

18. The current through the galvanometer and the direction of current in

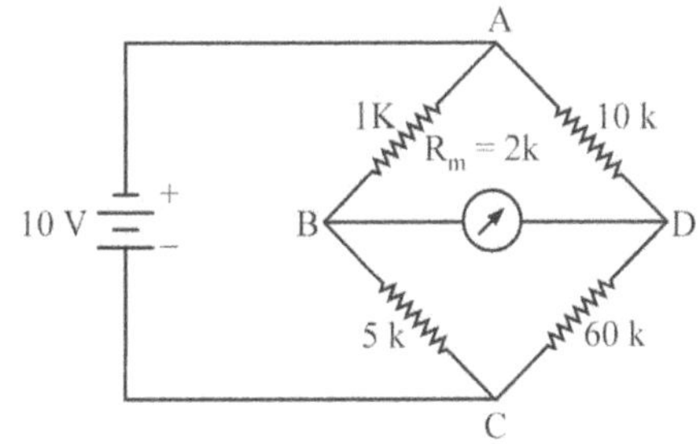

Ans. 21 μA, DB

Ex: $Ig = \frac{E_{TH}}{R_{TH} + R_M} \Rightarrow R_{TH} = \frac{5}{6}K + \frac{600}{70}K = 9.4\ K,$

$E_{TH} = 10\left(\frac{1}{6} - \frac{10}{70}\right) = 10\left(\frac{1}{42}\right),\ I_g = \frac{10}{42} + (9.4\ K + 2\ K) = 11.85\ \mu\ A$

19. In the wheatstone bridge shown in the given figure, if the resistance in each arm is increased by 0.05% then the value of V_{out} will be

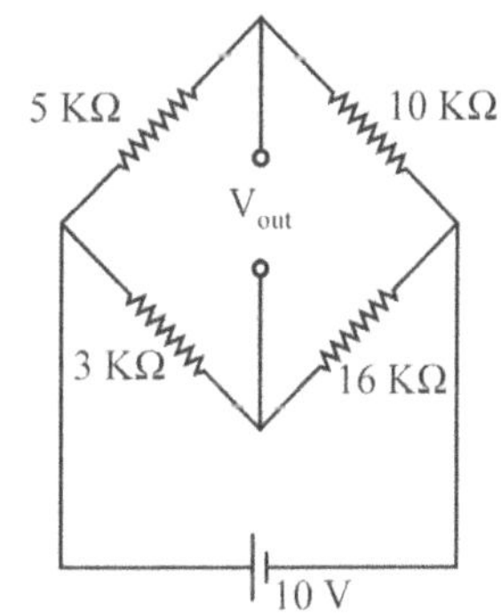

Ans. Zero

20. State the advantages of Wheatstone bridge method.

Ans. Free from errors

The balance is quit independent of source emf

21. Which type of detector is used in ac bridges?

Ans. Vibration galvanometers are used.

22. The dielectric loss can be measured by

Ans. Electro-static meter

23. Schering bridge is used to

Ans. Determine dielectric loss

24. Which bridge is used to determine frequency?

Ans. Wien bridge

25. Draw the diagram owen's bridge?

Ans.

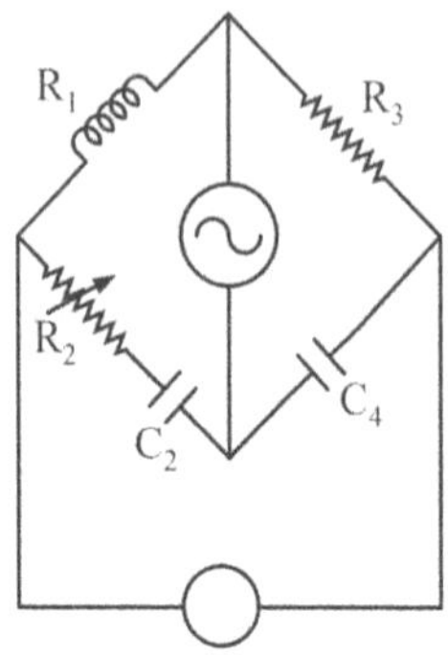

26. The inductance of a high θ inductor can be measured by a

Ans. Hay bridge

Match the Following

1. Match the following measuring equipment with their specific applications

(a)	kelvin double bridge	(p)	low resistance
(b)	wien bridge	(q)	frequency
(c)	schering bridge	(r)	capacitance
(d)	maxwell's bridge	(s)	self inductance

2. Match list I with list II

	List I		List II
(a)	megger	(r)	measurement of insulation resistance
(b)	spectrum analyses	(s)	measurement of harmonies
(c)	schering bridge	(p)	measurement of less angle in a Dieelectrics
(d)	digital counter	(q)	measurement of frequency

Quiz Questions Key

1. (a)	2. (a)	3. (a)	4. (a)	5. (c)
6. (d)	7. (c)	8. (d)	9. (a)	10. (c)
11. (a)				

Match the Following Key

1. a-p, b-q, c-r, d-s
2. a-r, b-s, c-p, d-q

CHAPTER 8

Magnetic Measurements

8.1 Introduction

The measurement of various properties of a magnetic material are called "magnetic measurements".

Electrical and magnetic phenomena are closely associated with each other. The operating characteristics of electrical machines, apparatus and instruments are greaty influenced by the properties of ferro-magnetic materials used for their construction. Therefore, magnetic measurements and a thorough knowledge of characteristics of magnetic material is almost importance in designing and manufacturing electrical equipments.

The magnetic measurements include:

1. Measurement of magnetising force H, producing the flux density B, field strength in air
2. Measurement of flux density B in a specimen of ferro-magnetic material
3. Determination of eddy current and the hysteresis losses
4. Testing of permanent magnets
5. Determination of B-H curve and hysteresis losses

For such magnetic measurements following tests are performed:

8.1.1 D.C Tests of Ballistic Tests

These are used determine B-H curve and hysteresis losses of ferro-magnetic materials. The direct current is used to have variable m.m.f and flux meter is used to measure the flux density. A ballistic galvanometer can be used to measured flux density.

Such tests are also called ballistic tests.

These methods do not directly measure the value of flux density in a magnetic material corresponding to a particular value of magnetizing force but measure instead the changes in flux brought about by changes in magnetising force. These changes in flux density are either measured by a flux meter or a ballistic galvanometer.

8.1.2 A.C Testing

These tests may be carried at power, audio or radio frequencies. They give information about eddy current and hysteresis losses in magnetic materials.

8.1.3 Steady State Tests

These are performed to obtain the steady value of flux density existing in the air gap of a magnetic circuit.

The results of magnetic measurement are not very accurate because of following reasons:

1. The magnetic materials are not homogenous
2. Various groups of test specimens have no uniformity
3. The condition at the time of calculation are different than the conditions existing at the time of testing of magnetic materials.

8.2 Ballistic Galvanometer

The Ballistic galvanometer and the flux meter are the necessary instruments for various types of magnetic measurements. The ballistic galvanometer has a "Search Coil" connected to its terminals. When there is change in the flux linked with search coil, an e.m.f is induced in the search coil. The electricity proportional to this e.m.f is measured by the ballistic galvanometer.

The ballistic galvanometer is usually of D'Arsonal type. The electricity passing through the ballistic galvanometer is in the form of transitory current which is instantaneous in nature. It exists only till the change of flux is associated with the search coil. Thus the ballistic galvanometer gives a 'throw' proportional to the quantity of the current instantaneously passing through it. It does not give steady deflection. The scale of the ballistic galvanometer is calibrated in such a way that from its throw, the quantity of electricity and the change in the flux producing it can be measured.

The throw of the ballistic galvanometer is proportional to the electricity only if the discharge of electricity through it gets completed before my appreciable change in position of the moving system. Hence the moving system of the galvanometer is made to have large moment of interia and due to this, it keeps on vibrating for long time for about 10-15 seconds. The large moment of intertia is achieved by the addition of weights to the moving system. Thus the oscillations of the galvanometer moving system are damped with large time period and small damping ratio. The small damping ensures that the first deflection (throw) is high and large in amplitude. The throw or deflection is measured at the extreme point of the first throw using a lamp and scale.

The dead beat galvanometer is best. It has to have electromagnetic damping such that it can be determined from the constants of the instrument. Air damping should not be present as it is indeterminate. To bring the moving system back rapidly, a key is provided with which galvanometer terminals are shirt circuited.

The special construction of the ballistic galvanometer also includes:

1. The terminals, coil and all the connections are made up of copper to avoid thermoelectric e.m.fs at the junctions.
2. The suspension strip is selected carefully and mounted precisely.
3. In precision instrument, the suspension is non-conducting and current is led in and out of the coil with the help of delicate spirals of very thin strip of copper.
4. The moving coil is free from magnetic material.

8.2.1 The Construction of a Moving Coil Ballistic Galvanometer

The construction of moving coil ballistic galvanometer is similar to that of D'Arsonval galvanometer with small changes in moving system and suspension. The basic difference between them is that ballistic galvanometer is used for "Magnetic Measurement" where as D'Arsonal's galvanometer is used for "Electrical Measurement". The moving coil ballistic galvanometer is shown in Fig. 8.1.

In this, electromagnetic damping in provided rather than air (or) oil damping because the latter dampings does not satisfy the conditions required for deflection system.

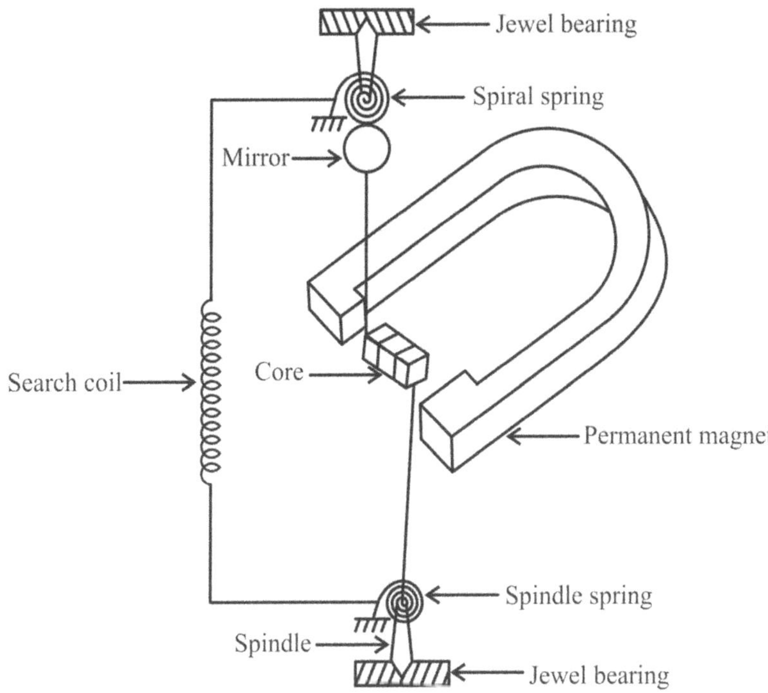

Fig. 8.1 Moving coil ballistic galvanometer

8.2.2 Theory of Ballistic Galvanometer

The quantity of electricity in the galvanometer is instantaneous and for very short period of time. While discharging it gives energy to the moving system which is dissipated gradually in frication and damping.

Thus let θ = deflection in radians

b = damping constant

a = moment of inertia

c = controlling constant

During the actual motion, electricity is not present once gets discharged hence the equation of motion becomes,

$$a\frac{d^2\theta}{dt^2}+b\frac{d\theta}{dt}+c\theta=0 \qquad .....(8.1)$$

The solution of this differential equation is.

$$\theta = A\ e^{m_1 t}+B\ e^{m_2 t} \qquad .....(8.2)$$

where A, B = constants

$$m_1, m_2 = \text{roots of quadratic} = \frac{-b \pm \sqrt{b^2 - 4ac}}{2a}$$

As damping is very small, 'b' is very small hence ($b^2 + 4ac$) term is negative and the roots are imaginary in nature.

The solution of second order differential equation having complex conjugate roots is damped sinusoidal oscillations given by

$$\theta = e^{-\frac{bt}{2a}} b \sin\left[\frac{\sqrt{4ac - b^2}}{2a}.t + \alpha\right]$$

but as 'b' is very small hence $4ac - b^2 \approx 4ac$

$$\therefore \quad \theta = e^{-\frac{bt}{2a}} b \sin\left[\sqrt{\frac{c}{a}}.t + \alpha\right] \quad .....(8.3)$$

From the initial conditions when t = 0, θ = 0 hence α = 0,

$$\theta = e^{-\frac{bt}{2a}}.b.\sin\left[\sqrt{\frac{c}{a}}.t\right] \quad .....(8.4)$$

During the discharge of the electricity through the instrument for the short period of time 'α', the deflecting torque proportional to instantaneous current i exist given by Gi where 'G' is displacement constant.

Hence the torque produced by Gi is represented as,

$$Gi = a\frac{d^2\theta}{dt^2}$$

$$\therefore \quad \int_0^\tau Gi\, dt = \int_0^\tau a\frac{d^2\theta}{dt^2} dt = a\frac{d\theta}{dt} \quad .....(8.5)$$

But $\int_0^\tau i\, dt$ represents the change Q coulombs

$$\therefore \quad QG = a\frac{d\theta}{dt} \quad .....(8.6)$$

where $\frac{d\theta}{dt}$ is angular velocity of the system at the end of the period 'τ' i.e., at the beginning of the first deflection

$$\therefore \qquad \frac{d\theta}{dt} = \frac{G}{a}.Q \qquad .....(8.7)$$

Now differentiating eq. 8.4, we get

$$\frac{d\theta}{dt} = -\frac{b}{2a} e^{-\frac{bt}{2a}}.b\sin\left(\sqrt{\frac{c}{a}}.t\right) + e^{-\frac{bt}{2a}}\, b\cos\left(\sqrt{\frac{c}{a}}.t\right)\sqrt{\frac{c}{a}} \qquad .....(8.8)$$

As the period of discharge is small, at the end of discharge, t = 0, using above

$$\therefore \qquad \frac{d\theta}{dt} = b\sqrt{\frac{c}{a}} \qquad .....(8.9)$$

From eq. in 8.7 and 8.9

$$b.\sqrt{\frac{c}{a}} = \frac{G}{a}.Q$$

$$\therefore \qquad b = \frac{G}{a}\sqrt{\frac{a}{c}}.Q \qquad .....(8.10)$$

Substituting value of 'b' in equation 8.4, we get

$$\theta = c^{-\frac{bt}{2a}}\left[\frac{G}{a}\sqrt{\frac{a}{c}}.Q\right]\sin\left(\sqrt{\frac{c}{a}}.t\right) \qquad .....(8.11)$$

Note: This shows that deflection of ballistic galvanometer is proportional to the charge Q and it is oscillatory on nature.

The waveform of 'θ' is shown in the Fig. 8.2 with angular frequency 'ω'.

$$\therefore \qquad \omega = \sqrt{\frac{c}{a}} = 2\pi f$$

$$\therefore \qquad f = \frac{1}{2\pi}\sqrt{\frac{c}{a}}$$

$$\therefore \qquad T' = \frac{1}{f} = 2\pi\sqrt{\frac{a}{c}} = \text{time period of oscillations} \qquad .....(8.12)$$

In Fig. 8.2 shows successive decreasing maxima at the instants $\frac{T'}{4}, \frac{3T'}{4}, \frac{5T'}{4}$ etc ., using in θ,

$$\therefore \quad \theta_1 = \frac{G}{a}Q\sqrt{\frac{a}{c}}\, e^{-\frac{\pi}{4}\cdot\frac{b}{\sqrt{ac}}} \quad ....t = \frac{T'}{4} = \frac{2\pi}{4}\sqrt{\frac{a}{c}}$$

$$\therefore \quad \theta_2 = \frac{G}{a}.Q\sqrt{\frac{a}{c}}\, e^{-\frac{3\pi}{4}\cdot\frac{b}{\sqrt{ac}}}$$

$$\therefore \quad \theta_n = \frac{G}{a}.Q\sqrt{\frac{a}{c}}\, e^{-\frac{\pi}{4}(2n-1)\frac{b}{\sqrt{ac}}} \qquad .....(8.13)$$

$$\therefore \quad \frac{\theta_1}{\theta_2} = \frac{e^{-\frac{\pi}{4}\cdot\frac{b}{\sqrt{ac}}}}{e^{-\frac{3\pi}{4}\cdot\frac{b}{\sqrt{ac}}}} = e^{+\frac{\pi}{2}\frac{b}{\sqrt{2ac}}} \qquad .....(8.14)$$

$$\therefore \quad l_n \frac{\theta_1}{\theta_2} = \frac{\pi}{2}\cdot\frac{b}{\sqrt{ac}} = \lambda \qquad .....(8.15)$$

The 'λ' is called "Logarithmic decrement' of the galvanometer.

8.2.3 Logarithmic Decrement

The 'Logarithmic Decrement' 'λ' is a constant of the galvanometer which is proportional to the damping and depends upon the resistance of the galvanometer circuit.

From above, we have the ratio of any two successive peaks remains same and given by 'λ'.

$$\therefore \quad \frac{\theta_2}{\theta_3} = \frac{\theta_3}{\theta_4} = ... = \lambda$$

$$\therefore \quad \frac{\theta_1}{\theta_n} = e^{\lambda(n-1)}$$

$$\therefore \quad l_n\left(\frac{\theta_1}{\theta_n}\right) = \lambda(n-1)$$

$$\therefore \quad \lambda = \frac{1}{n-1}\, l_n\left(\frac{\theta_1}{\theta_n}\right) \qquad .....(8.16)$$

If the damping is "Undamped" i.e., ideally with amplitude constant at all the time then damping should be zero. i.e., b = 0. Let the corresponding deflection be 'θ' from eq. 8.13 with b = 0

$$\theta' = \frac{G}{a}.Q.\sqrt{\frac{a}{c}} = \frac{GQ}{\sqrt{ac}}$$

$$\therefore \qquad Q = \frac{\sqrt{ac}}{G}\theta' \qquad \text{.....(8.17)}$$

Note: The quantity of the electricity to be measured is directly proportional to the undamped (zero damping) deflection θ'.

Now $$\theta' = \frac{GQ}{\sqrt{ac}} = \theta_1 e^{+\frac{\pi}{4}\frac{b}{\sqrt{ac}}} \qquad \text{as } \theta_1 = \frac{GQ}{\sqrt{ac}} e^{-\frac{\pi}{4}\frac{b}{\sqrt{ac}}}$$

But $$\frac{\pi}{2}\frac{b}{\sqrt{ac}} = \lambda \text{ i.e., } \frac{\pi}{4}\frac{b}{\sqrt{ac}} = \frac{\lambda}{2}$$

$$\theta' = \theta_1 e^{\lambda/2} = \theta_1\left[1 + \frac{\lambda/2}{1!} + \frac{\lambda/2}{2!} + \ldots\right]$$

Neglecting higher order terms as λ is small due to b,

$$\theta' = \theta_1\left[1 + \frac{\lambda}{2}\right] \qquad \text{.....(8.18)}$$

Substituting in 8.17

$$Q = \frac{\sqrt{ac}}{G} \times \theta_1\left[1 + \lambda/2\right]$$

$$\therefore \qquad Q = K\theta_1 \qquad \text{.....(8.19)}$$

where K = constant of the galvanometer $= \frac{\sqrt{ac}}{G}\left[1 + \frac{\lambda}{2}\right]$

As $$T' = 2\pi\sqrt{\frac{a}{c}}$$

$$\therefore \qquad K = \frac{C}{G}\frac{T'}{2\pi}\left[1 + \frac{\lambda}{2}\right] \qquad \text{.....(8.20)}$$

To eliminate 'C' and 'G', assume that a fixed direct current I_g is passed through the galvanometer producing steady deflection θ then,

$$GI_g = C\theta$$

$$\therefore \qquad \frac{C}{G} = \frac{I_g}{\theta}$$

$$\therefore \qquad K = \frac{I_g}{\theta} \cdot \frac{T'}{2\pi} \left[1 + \frac{\lambda}{2} \right] \qquad \text{.....(8.21)}$$

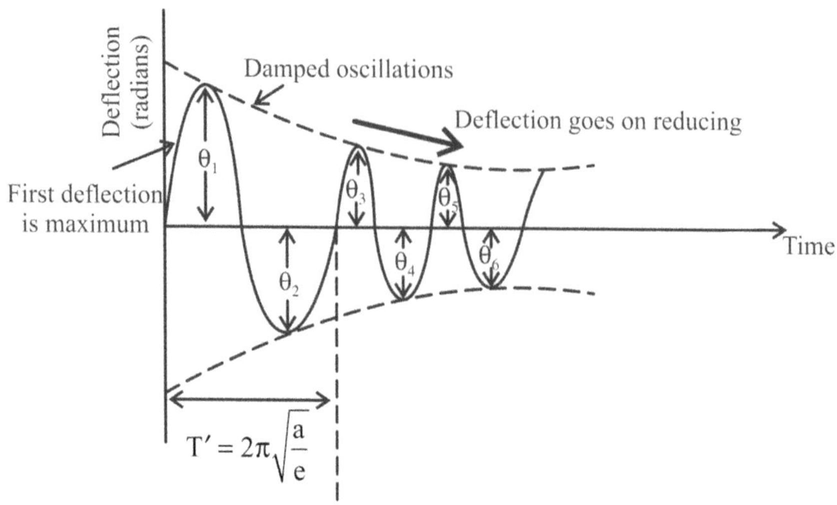

Fig. 8.2 Waveform of 'θ'

The galvanometer scale is calibrated according to the value of K, which is sensitivity of the galvanometer. The charge is measured in microcolumbs or in weber-turns. Q_1 is measured in mm while 'K' is measured in microcoulombs per mm deflection.

8.3 Flux Meter

A special ballistic galvanometer with very small controlling torque and high electro-magnetic damping is used as a flux meter. It consists of a small cross-section carrying a coil C. The cross-section is suspended in the narrow air gap of permanent magnet. The cross-section is suspended with the help of spring support and a single thread of slik. The current is injected to the coil through very thin annealed silver springs as shown in the Fig. 8.3(a).

The pointer is fitted to the moving system of the flux meter and the scale is calibrated in terms of flux turns. Such a flux meter is designed by Grassot and hence it is called Grassot flux meter. The spirals of silver spring keep the controlling torque to minimum.

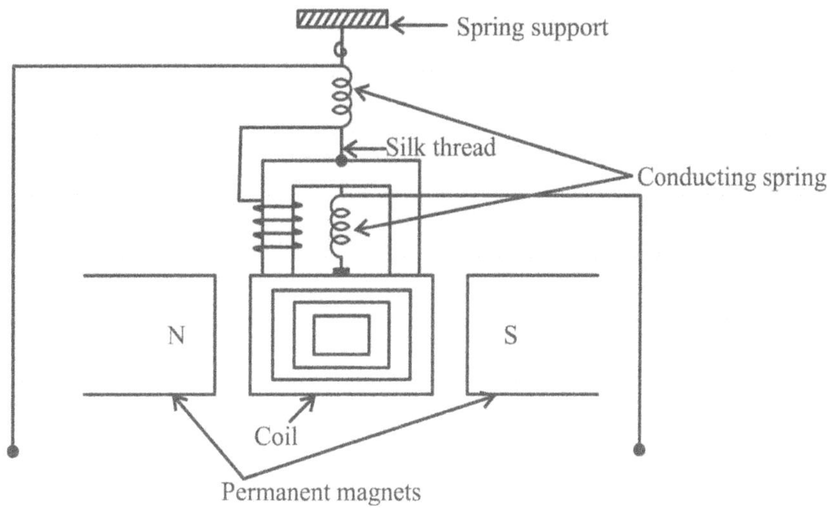

Fig. 8.3(a) Flux meter

The equivalent current of a flux meter is shown in fig. 8.3 (b).

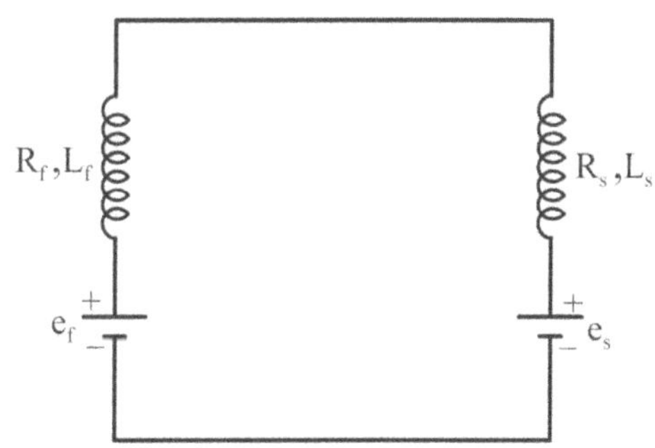

Fig. 8.3(b)

As controlling torque is minimum, pointer takes time to come back to the zero position. But reading may be taken by observing the difference in deflections at the beginning and end of the change in flux, without waiting for pointer to restore its zero position.

The resistance R_s of the search coil connected to the flux meter must be small. The inductance of search coil may be large.

Let e_f = induced e.m.f on flux meter in volts

e_s = induced e.m.f in the search coil in volts

R_f = Resistance of flux meter in ohms

R_s = Resistance of search coil in ohms

L_f = inductance of flux meter in henry

L_s = inductance of search coil in henry

i = instantaneous value of current in the circuit in amps

θ = delfection of the pointer

Let ϕ_1 and ϕ_2 are the interlinking fluxes at the beginning and the end of the change in flux to be measured respectively. The corresponding deflections are θ_1 and θ_2 respectively. Then

$$d\theta = \text{change in deflection} = \theta_2 - \theta_1$$

$$d\phi = \text{change in flux} = \phi_2 - \phi_1$$

and it can be proved that

$$G\, d\theta = N\, d\phi$$

where G = constant of flux meter called displacement constant

N = number of turns on search coil

$$\therefore \quad \boxed{d\theta = \frac{N}{G} d\phi}$$

In modern flux meter, the coil is supported with pivots and mounted in jewelled bearings. The current is passed to the coil through fine ligameters.

The flux meter has an advantage over ballistic galvanometer as the deflection is same irrespective of the time taken for the corresponding change in flux interlinking with the search coil. Another advantage of flux meter is it is portable.

The main differences between flux meter and ballistic galvanometer are:

8.4 Comparision between Flux Meter and Ballistic Galvanometer

S. No.	Flux meter	Ballistic galvanometer
1.	The scale is directly calibrated in weber turns	The scale is directly calibrated in coulombs
2.	In case of flux meter, the electromagentic damping is heavy	The electromagnetic damping is light
3.	The deflection of flux meter depends on the charge in the value of flux	The deflection of galvanometer is proportional to the charge passing through it.
4.	The controlling torque is very small	The controlling torque is quite large

Table *Cont...*

S. No.	Flux meter	Ballistic galvanometer
5.	The sensitivity of flux meter is less	Sensitivity of galvanometer is more
6.	It can be used in null method for determining the flux	It can't used in null method for determining the flux
7.	The time taken for the change in flux producing the deflection need not be small	The time taken for discharging of charges must be very small

8.5 Measurement of Flux Density (B)

Let us study the measurement of flux density in a ring specimen. A coil with sufficient number of turns is wound on a ring specimen. This coil is called search coil or a 'B' coil. This coil is then connected to a flux meter or a ballistic galvanometer, as shown in the Fig. 8.4.

The magnetizing winding carries a current I which produces the flux to be measured. The current I in the magnetising coil is reversed using a reversing switch. Thus flux linkages associated with the search coil also change inducing e.m.f in it. This e.m.f drives a current through a ballistic galvanometer, causing the corresponding delfection.

8.5.1 Theory of the Flux Density Measurement

Let N = No. of turns on search coil

ϕ = flux embraced by the search coil

R = Resistance of the ballistic galvanometer circuit.

t = the time (in seconds) taken to reverse the magnetising current (and hence to reverse the flux ϕ)

The average e.m.f induced in the search coil is given by,

$$e = N\frac{d\phi}{dt} \text{ volts}$$

Initial flux = ϕ. After reversal = $-\phi$

$$d\phi = \phi - (-\phi) = 2\phi$$

$$dt = t$$

$$\therefore \quad e = \frac{N2\phi}{t} \text{ volts}$$

Thus the average current through ballistic galvanometer is

$$i = \frac{e}{R} = \frac{2N\phi}{Rt} \text{ A}$$

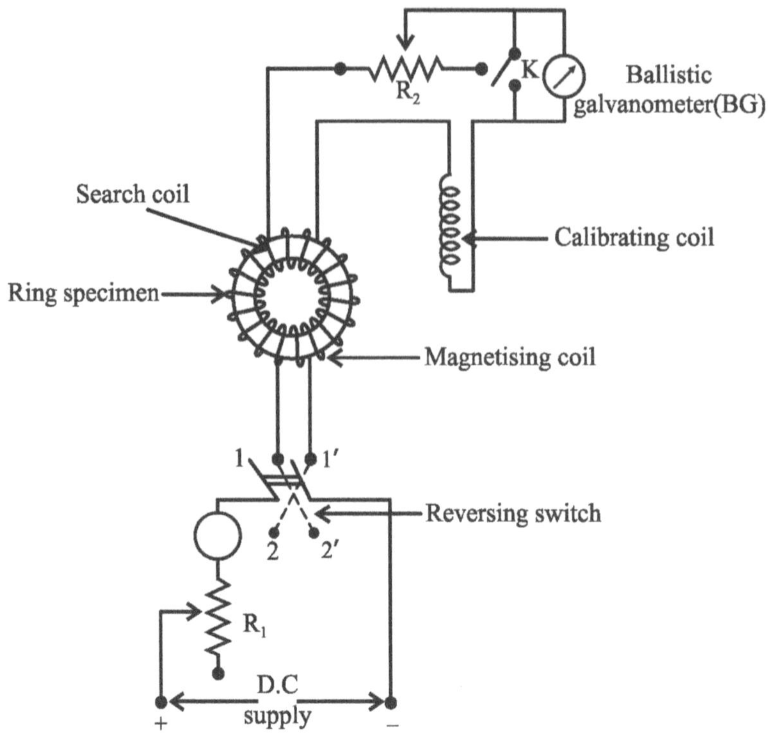

Fig. 8.4 Measurement of flux density

Hence, the charge Q discharged through the galvanometer during t sec is,

$$i = \frac{Q}{t}$$

$$\therefore \qquad Q = it = \frac{2N\phi}{Rt}.t = \frac{2N\phi}{R} \text{ coulombs}$$

The deflection of the galvanometer is proportional to the charge discharged through it,

$$\therefore \qquad Q = K\theta' \qquad \text{where K = Galvanometer constant}$$

$$\therefore \qquad \frac{2N\phi}{R} = K\theta'$$

$$\therefore \qquad \phi = \frac{K\theta' R}{2N} = \frac{RK\theta'}{2N} \text{ wb}$$

Thus if A is the area of cross-section of the specimen, the flux density can be obtained as

$$B = \frac{\phi}{A} = \frac{RK\theta'}{2NA} \text{ wb}/\text{m}^2$$

Thus flux and flux density can be measured from the deflection of the ballistic galvanometer.

8.5.2 Correction of Air Flux

In the above discussion, it is assumed that the flux is uniform throughout the specimen and area of cross-section of search coil and specimen is equal, but practically cross-section of search coil is higher than that of specimen. Hence total flux linking with the search coil is sum of the flux in the specimen and the flux existing on the air gap between search coil and the specimen.

$\therefore$ total flux observed = flux in specimen + Flux in air gap between coil and specimen

Let B' = observed value of flux density

$$B'A = B_t A + \mu_o H(A_c - A)$$

B_t = True value of flux density

A_c = Area of cross-section of search coil

$$B'A = B_t + A + \mu_o H(A_C - A)$$

$$\therefore \quad B_t = B' - \mu_o H\left[\frac{A_c}{A} - 1\right]$$

This is called correction of air flux.

8.6 Measurement of Magnetising Force H

The magnetising force H is also measured by using a search coil and a ballistic galvanometer. The arrangement is shown in the Fig. 8.5.

In such method 'H' cannot be obtained directly but is to be calculated by measuring flux density by the method of current reversal, as described earlier.

The position of search coil shown in the Fig. 8.5 is for the measurement of flux density B_a in air, by reversing current 'I' with the help of reversing switch. The search coil used for such measurement is called 'H' coil. Once B_o is measured then H can be obtained as,

$$H = \frac{B_o}{\mu_o} \text{ A/n}$$

where $\mu_o = 4\pi \times 10^{-7}$ H/m

Thus search coil is placed in the air gap itself if H in the air gap is to be determined.

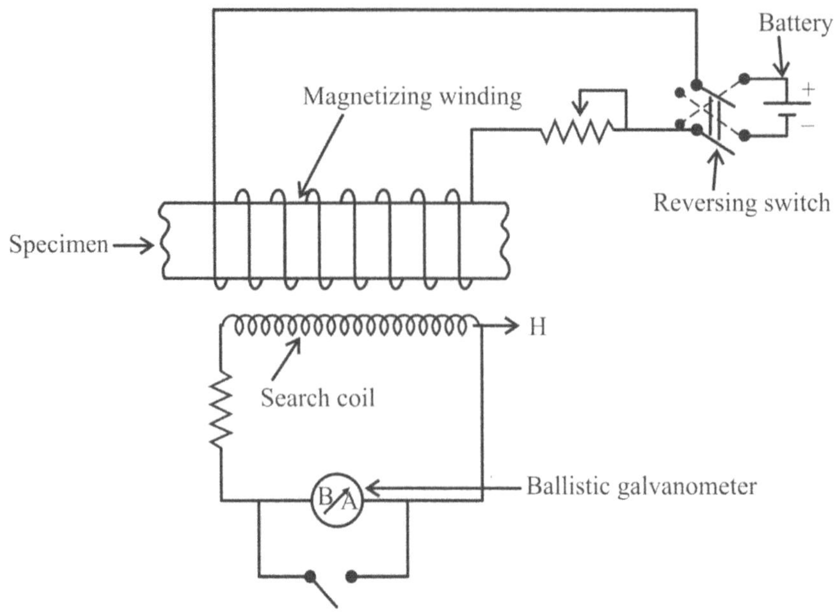

Fig. 8.5 Measurement of 'H'

While if magnetising force N within the ferro-magnetic material specimen is to be obtained then H is measured on the surface of the specimen as the tangential components of field are of equal in magnitude for both the side of the interface.

While the value of H inside a specimen can be calculated as,

$$H = \frac{NI}{l}\ AT/m$$

where l = mean circumference of ring in 'm'

I = current flowing through specimen

N = No. of turns of specimen

8.7 Magnetic Potentiometer

The device used to measure the magnetic potential between any two points in a magnetic field is called magnetic potentiometer. It consists of a uniform helix of a thin wire wound on a thin strip of some flexible insulating and non-magnetic material. This is used along with the ballistic galvanometer for the measurement of magnetic potential.

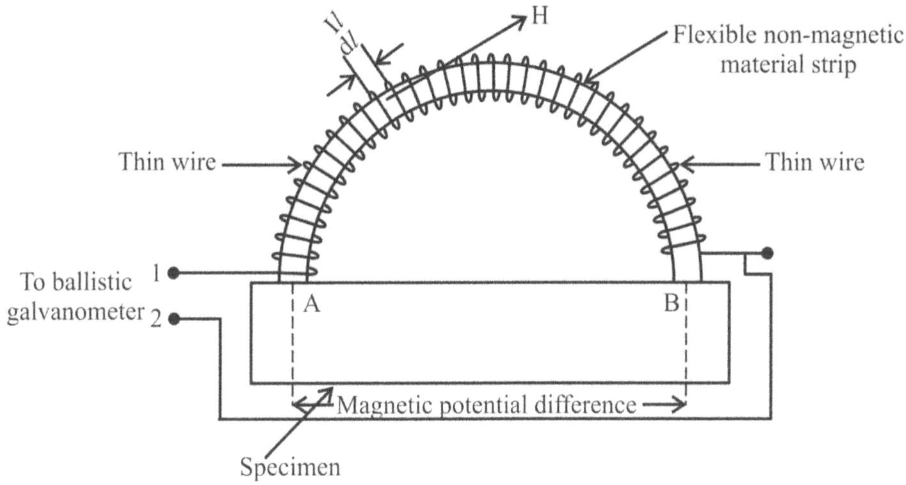

Fig. 8.6 Magnetic potentiometer

The arrangement of magnetic potentiometer is shown in the Fig. 8.6. The coil of thin wire is wound on a flexible non-magnetic material.

The two ends 1-2 of the coil are to be connected to a ballistic galvanometer. The two ends of the non-magnetic material strip are placed at the two points between which the magnetic potential is to be measured. The corresponding deflection of the ballistic galvanometer is to be observed, when the flux through the specimen is reversed.

According to Ampere's circuit law, the line integral of the magnetizing force H produced by a coil of N turns carrying current I is

$$\int H dl = NI \text{ around any closed path linking coil}$$

This principle is used in magnetic potential measurement

R = Resistance of ballistic galvanometer circuit.

A = Area of flexible strip in m^2

n = no. of turns per unit length of the strip

H = Field strength at any point in tangential direction at that point in A/m

Then total flux linkages of a small infinitesimal part of the strip of length dl are

as $$\psi = \text{flux} \times \text{turns} = (\mu_o HA)\, n\, dl$$

$$\phi = BA = \mu_o HA \text{ and turns} = n\, dl$$

Hence, total flux linkages of the strip can be obtained as,

$$\psi_T = \int \mu_o\, HA\, ndl = \mu_o An \int H\, dl$$

The integral is to be taken between the ends of the strip. When the flux is reversed, the total change in the flux linkages is given by

$$= 2\mu_o An \int H \, dl$$

But $\int H \, dl = \Im$ = magnetic potential difference between A – B

$$\therefore \quad \Delta\phi\tau = 2\ \mu_o\ An\ \Im \text{ in time t}$$

Now, $$Q = \text{charge} = i \times t = \frac{e}{R} t$$

$$\therefore \quad Q = \frac{2\mu_o An}{t} \Im . \frac{1}{R} = \frac{2\mu_o An}{R} \Im\ C$$

but the galvanometer deflection θ' is proportional to charge

$$\therefore \quad Q = K\theta'$$

$$\therefore \quad \frac{2\mu_o An\Im}{R} = K\theta'$$

where k in galvanometer constant

$$\therefore \quad \Im = \frac{RK\theta'}{2\mu_o An}$$

K can be determined by the calibrating coil.

Such a potentiometer is used for the measurements of the magnetic potential drop across a given part of a magnetic circuit such as a joint, the measurement of magnetic leakage or measurement of m.m.f around a closed path. The important advantage of this device is that the results are same whether the strip on which the helix is wound is straight or of other shape.

As the method does not give uniform values of 'H' over the entire length AB and hence method is rarely used in practice for direct measurement of H. The potentiometer is also called "Chattock" magnetic potentiometer.

8.8 Measurement of Leakage Factor with Flux Meter

In an electrical machine the leakage factor is defined as:

$$\text{Leakage factor } (\lambda) = \frac{\text{total flux}}{\text{useful flux}} \text{ wb}$$

Useful flux = flux in armature (flux crossing the air gap) wb

Total flux = flux in the pole bodies

= useful flux + leakage flux existing in the pole body at its root wb.

Thus in order to measure the leakage factor, we must measure the flux on the pole bodies and the flux in the armature. The flux measurements are done by a flux meter. A ballistic galvanometer cannot be used here as the field winding of electric machines are highly inductive and therefore the flux changes very slowly when the voltage is impressed across the field winding.

It is clear from Fig. 8.7 that the yoke carries half the total flux and therefore it is possible to measure the value of total flux per pole by using two search coils on the yoke and connecting them in series across the flux meter.

The armature is kept stationary and another search coil is put on it. This coil is so positioned that it embraces the useful flux pole. This search coil is then connected to the flux meter and this way the useful flux is measured. Leakage factor can thus be calculated from the two readings of the flux meter. Search coils of usually one turn are used so that the flux meter reading gives the flux directly. In large machines, the value of flux is quite large and therefore it becomes necessary to use a shunt along with the flux meter in order that its range may be increased.

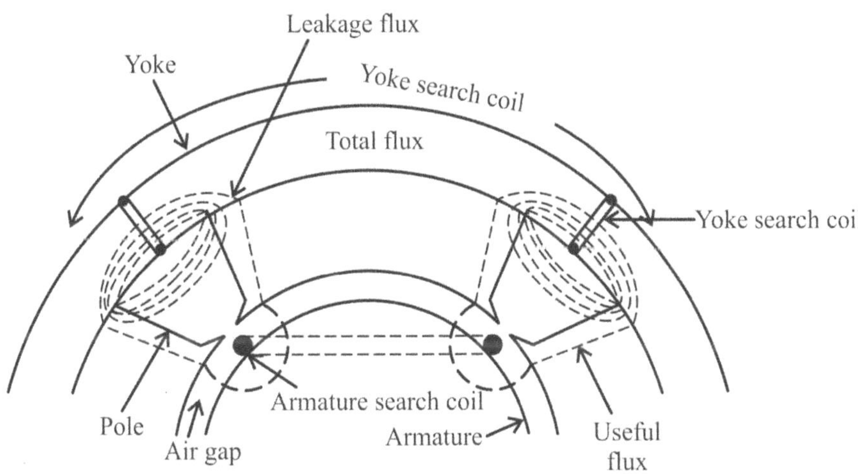

Fig. 8.7 Measurement of leakage factor

8.9 Determination of B-H Curve

There are two methods by which B-H curve can be obtained for the magnetic material specimen.

1. Step by step method
2. Method of reversals

8.9.1 Method of Reversals

It is one of the methods of determining B-H curve. A ballistic galvanometer is used to measure the flux corresponding to different value of 'H'. The value of flux so obtained is then divided with the area of cross-section of the specimen to obtain the value of B. Using these values of B and H graph is plotted. Fig. 8.8 shows the connection diagram of the set up to plot the B-H curve.

In order to ensure satisfactory operation, it is necessary to remove the residual magnetism present in the specimen. This can be done by demagnetising the specimen using the short circuiting key 'S' of the ballistic galvanometer. A current is passed through the magnetising winding by closing the key 'S'.

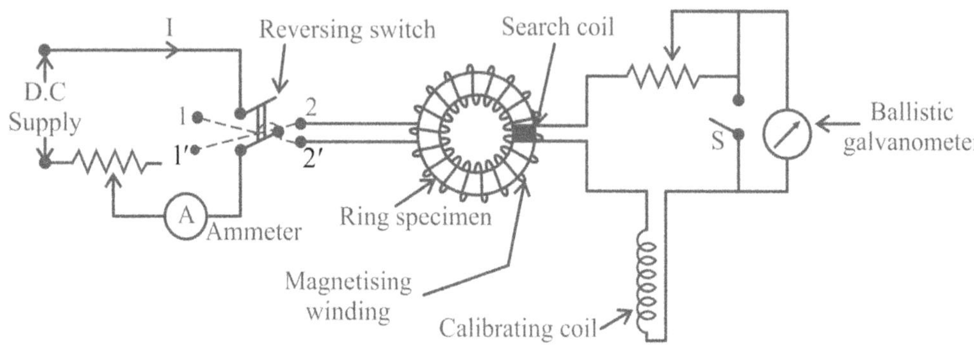

Fig. 8.8 Method of reversals

The value of current is such that it gives the value of H greater than highest value to be used in the test. The reversing switch is then repeatedly reversed so as to reduce the current in the magnetizing winding to zero. Due to this, the specimen will undergo number of magnetising cycles. The current in the magnetising winding is now increased so that it gives a lowest value of H. This value of H corresponds to a part on graph. After the completion of denagnetization, the test is started by assuming the smallest value of magnetising current. The switch 's' is thrown forward and backword at least 20 to 30 times. After this the switch s is opened and the readings of the ballistic galvanometer is noted. This value is the value of flux corresponding to the value of H. The value of flux density, B corresponding to this value of H is obtained by dividing the flux measured with the area of the specimen. The above procedure is repeated for different value of H up to its maximum value and the B-H curve is plotted Fig. (8.9).

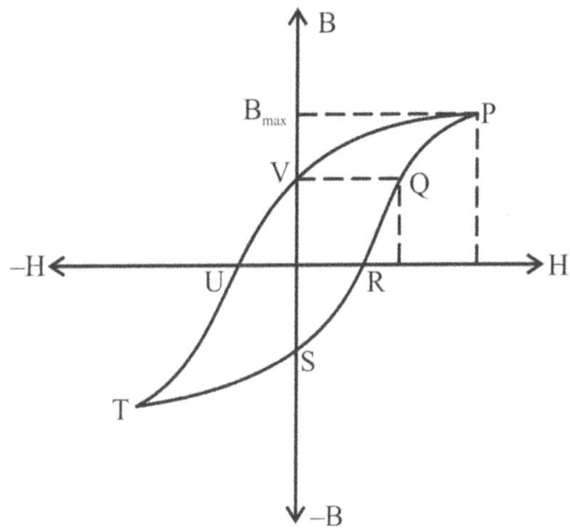

Fig. 8.9

Hysteresis loss calculation

The area of the hysteresis loop (obtained by the above method) is determined. This area gives the hysteresis loss per cycle of magnetization i.e., the hysteresis loss per unit volume per cycle and the units are J/m^3/Hz.

∴ Hystersis loss in the specimen = Area of PQRSTUVP in J/m^3 Hz

$$H = \frac{NI_1}{l}$$

where N = no. of turns on the magetising winding

I_1 = corresponding magnetising current

l = mean circumference length of specimen in m

8.9.2 Sources of Errors and Method of Reducing

1. The distribution of m.m.f will be non-uniform. If the magnetising winding is not wound properly and will give rise to errors. Hence in order to have uniform distribution of mmf magnetising winding should be uniformly distributed.
2. Bar and rod type specimen suffer from magnetic leakage and other magnetic effects due to this, the test results will not be accurate. Hence in order to reduce these errors ring type specimen should be used which are free from these effects.
3. In case of ring specimen, due to the difference in the inner and outer circumference, the distribution of H and B will be non-uniform and will cause error in the measurement. In order to reduce this error, the ratio of mean diameter to radial width should be more.

4. Residual magnetism present in the specimen will make the measurement incorrect. Hence, in order to have accurate results, residual magnetism must be removed in each step by demagnetization.

8.9.3 Step by Step Method

In this method, reversal of magnetising current is not used. The magnetising current in the winding is supplied through a potential divider as shown in the Fig. 8.10(a).

The potential divider has number of tappings. The tappings are arranged in such a way that the magnetising force H increases in suitable number of steps, up to the required maximum value.

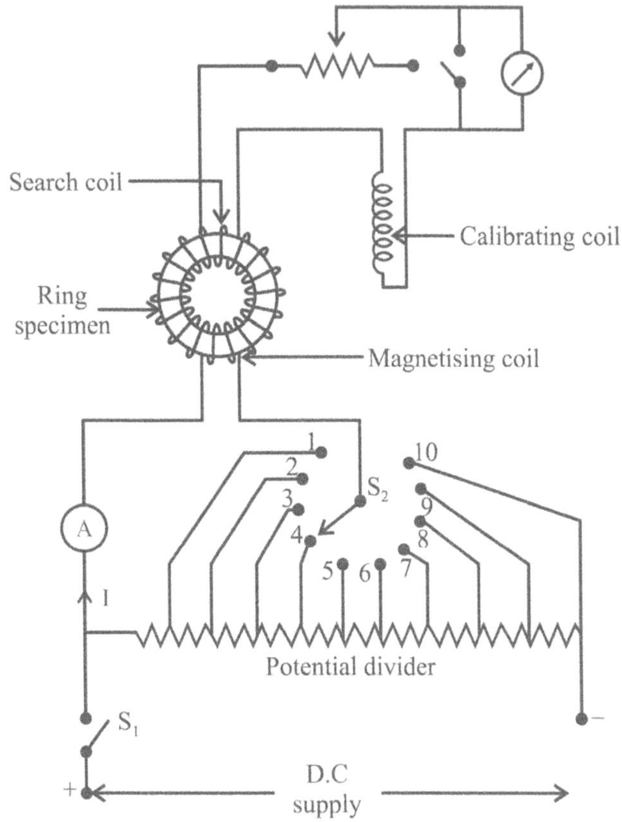

Fig. 8.10(a) Step by step method

The specimen is completely demagnetised before starting the test. The switch S_1 is closed with switch S_2 at tapping 1. Due to this, there will be some change in the flux hence there will be increase in the flux density from 0 to B_1. This value can be obtained by observing the deflection of the ballistic galvanometer. Recording the value of corresponding magnetising current, the corresponding value of the magnetising force H_1 can be obtained. The switch S_2 is instantaneously changed to tapping 2, which increases the magnetising force to H_2. Due to this, there is increase in flux and hence flux density by the amount ΔB. This can be determined from the galvanometer throw hence B_2 at, H_2 can be obtained as $B_1 + \Delta B$. The procedure is repeated for various tappings till maximum value of H is achieved. The graph of B against 'H' is then plotted which is nothing but the B-H curve for the specimen under test. This is shown in Fig. 8.10(b).

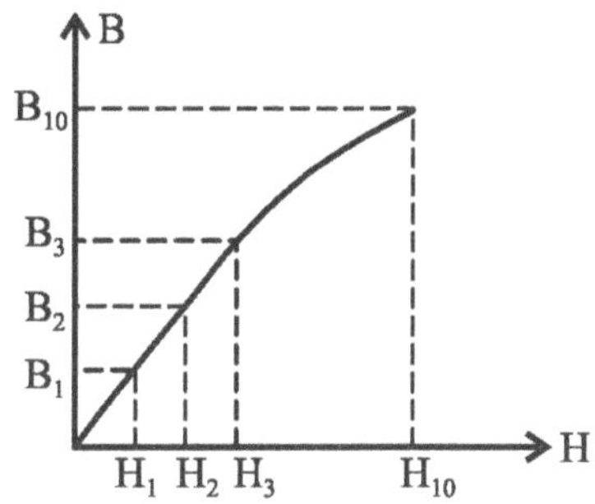

Fig. 8.10(b) B-H curve from step by step method

8.9.4 Advantages of Method of Reversals Over Step-by-Step Method

1. In method by reversals, the value of flux density at each point is calculated by keeping the previous and next value of flux densities at its maximum level (i.e., B_{max}). As a result, the cyclic state of magnetization is preserved; this is necessary for accurate results. Whereas, in case of step by step method, no preservation of the cyclic state of magnetization.
2. In method of reversals, the number of points on the loop can be obtained as many as we require by simply adjusting resistance R. By this the shape of the loop obtained is accurate. Whereas, in step by step method, the number of points required to form the loop are limited by the number of tappings of the potential divider used. Hence, in case of step by step method, the accuracy of the shape of loop is limited, which may result into a serious error when the hysteresis loss for that particular specimen is calculated for higher volumes of that material.

8.10 Magnetic Testing under A.C Conditions

Whenever a piece of magnetic material is subjected to alternating current, it goes under a cycle of magnetisation and demagnetisation. There exists a hysteresis, due to which power loss occurs in the form of hysteresis loss and eddy current loss. This loss is called "Iron Loss". The knowledge of iron loss in ferro-mangetic material plays an important role for the designers. The iron loss depends on:

1. Maximum value of flux density B_M.

2. Frequency of alternating field to which it is subjected.

In practice, for various materials, the curves are obtained at typical frequency giving the variation between iron loss per kg against the maximum flux density. Such curves are called "Iron loss curves". The curves help the designers to select the proper material for the proper application. The Fig. 8.11 shows typical iron loss curves at a frequency of 50 Hz.

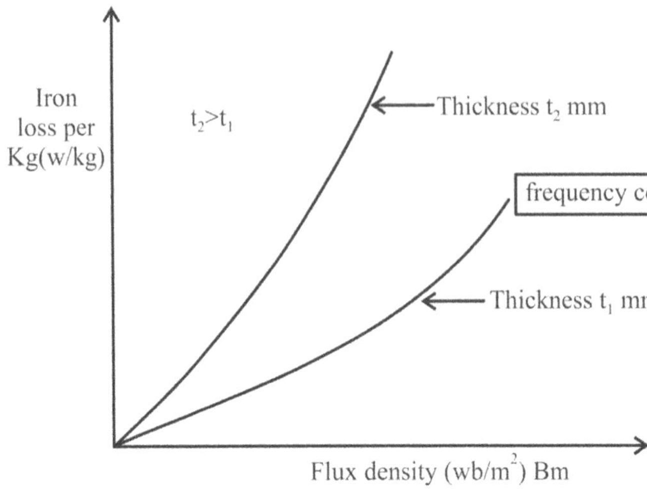

Fig. 8.11 Typical iron loss curves

The total iron loss has two components:

1. Hysteresis loss and 2. Eddy current loss

8.10.1 Hysteresis Loss

For a given volume and thickness of laminations, these losses depend in the operating frequency and maximum flux density in the core. Basically hysteresis loss per unit volume is the area of the hysteresis loop of that material.

Practically Steinmetz has given the formula for the hysteresis loss per unit volume as, equation is

$$P_h = n\, f\, B_m^k \quad \text{watts/m}^3$$

where B_m = maximum flux density

K = Steinmetz coefficient varies between 1.6 to 2

n = hysteresis coefficient

f = frequency

Practically K is taken as 1.67.

For a given specimen of thickness t and certain value, it is given by

$$\boxed{P_h = K_h f\ B_m^{1.67}, w} \qquad .....(8.22)$$

where K_h = Hysteresis loss constant

8.10.2 Eddy Current Loss

The eddy current loss per unit volume for given lamination is given by,

$$P_e = \frac{4K_f^2\ f^2\ B_m^2\ t^2}{3\rho}\ \text{watts} / m^3$$

where K_f = form factor of alternating supply used

t = thickness of lamination

ρ = resistivity of material in Ω-m.

For a given volume thickness and resistivity of the material, the loss is expressed as,

$$P_e = K_e\ K_f^2\ f^2\ B_m^2\ w \qquad .....(8.23)$$

where K_e = Eddy current loss constant

$P_i = P_h + P_e$ = total iron loss

8.11 Separation of Iron Losses

It can be seen from the equations 8.22 and 8.23 that both the losses have different laws of variation with frequency f and form factor K_f, if maximum flux density B_m is maintained constant. Thus by varying f or K_f, the losses can be separated.

8.11.1 Variation of Frequency

Assume that K_f and B_m are maintained constant. The frequency is varied then total Iron loss is,

$$P_i = A_1 f + A_2 f^2$$

where $A_1 = K_h\ B_m^{1.67}$

and $A_2 = K_e B_m^2 K_f^2$

$$\therefore \quad \frac{P_i}{f} = A_1 + A_2 f$$

The graph of $\left(\frac{P_i}{f}\right)$ against f is a straight line as shown in the Fig.8.12. The intercept of line with Y-axis gives the value of A_1 while the slope of the straight lines gives the value of A_2.

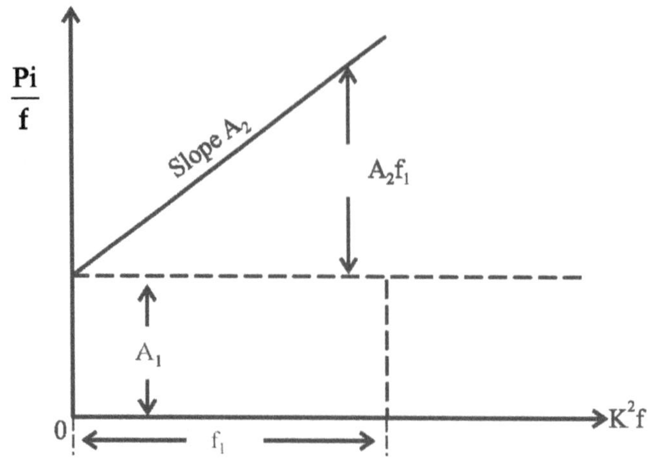

Fig. 8.12 Iron loss against frequency

8.11.2 Variation of Form Factor

Assume that the frequency and B_m are maintained constant. The specimen can be subjected to different waveforms sine, square etc. Whose form factors are different.

When the K_f is varied then total iron loss

$$P_i = B_1 + B_2 K_f^2$$

where $$B_1 = K_h f\, B_m^{1.67},\quad B_2 = K_e f^2 B_m^2$$

The graph of P_i against K_f^2 can be obtained as shown in the Fig. 8.13. Thus the intercept of line on Y-axis gives the value of B_1, while the slope of the straight line gives the value of B_2.

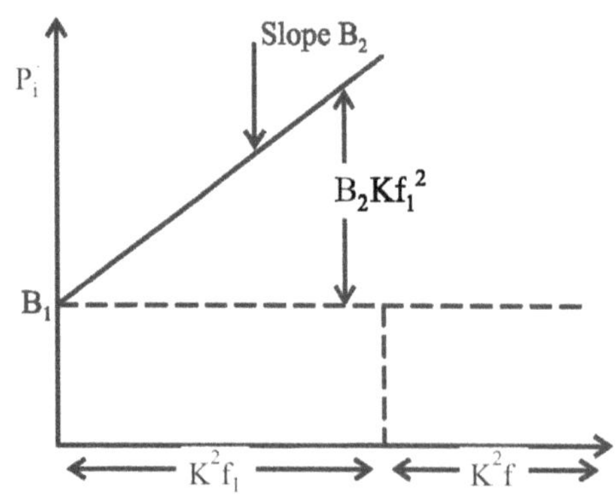

Fig. 8.13 Iron loss against K_f^2

8.12 Methods of Measuring Iron Losses

There are mainly four types of methods used for measurement of iron loss in ferro-magnetic materials.

1. Wattmeter meter
2. Bridge method
3. Potentiometer method
4. Oscillographic method

Let us discuss these methods in detail.

8.12.1 Wattmeter Method

This is most commonly used and simple method of measuring Iron loss in a strip steel material. The material is arranged in a form closed magnetic circuit in a shape of a "Square".

The square carries the magnetising and the search coils. There are two types of such magnetic squares:

1. Epstein square
2. Lioyed fisher square

8.12.1.1 Epstein square

It consists of four stacks of strips which are bound and taped. All the strips are in the plane of the square. The strips are insulated from each other. The corner joints are interleaved. The strips are inserted into four magnetising coils and strips are projected beyond the coils. The arrangement is shown in Fig. 8.14.

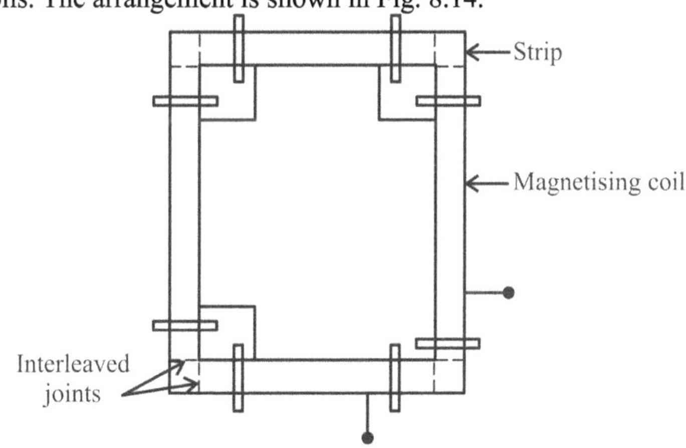

Fig. 8.14 Epstein square

8.12.1.2 Lioyed fisher square

In this square, the strips are built up into four stacks. Each stack consists of two types of strips, one cut in the direction of rolling and other cut perpendicular to the direction of rolling. The stacks are inserted inside four magnetising coils having large cross-sectional area. All these coil are connected in series to form a primary winding.

Below each magnetising coil, there are two similar single layer coils are placed; these are secondary coils. The four coils are connected in series and such two groups exits. Thus two separate secondary windings exist.

The strips are projected beyond the coil and arranged in such a way that each is perpendicular to the plane of the square. The right angled corner pieces are used at the corner to form the corner joints. The corner pieces and strips are overlapped due to which the cross-section of the iron is doubled at the corners. "Hence a correction is required to be applied for the measured value of iron loss".

The Fig. 8.15 shows an arrangement at Lioyed-Fisher square.

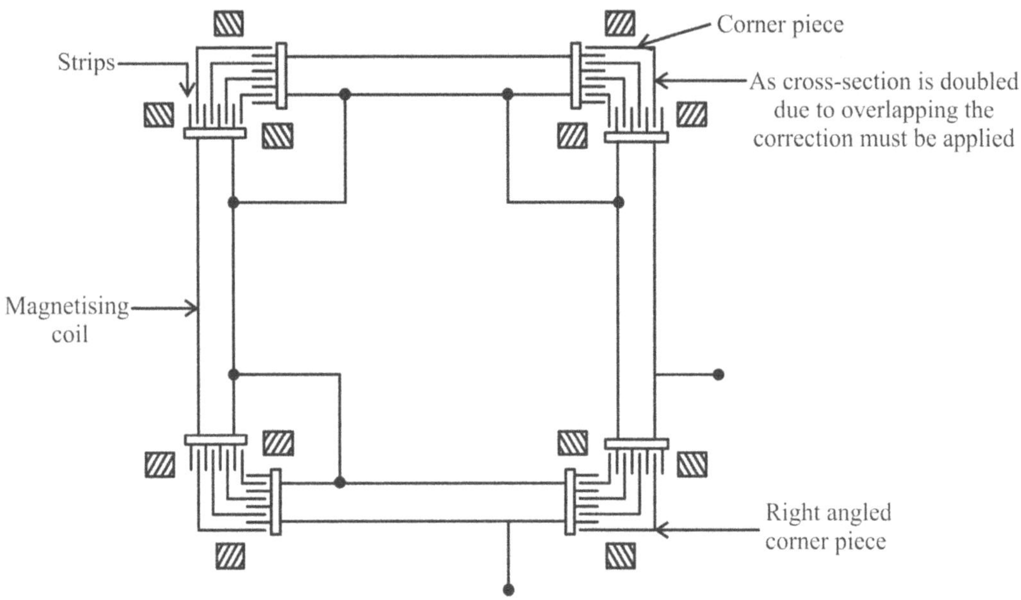

Fig. 8.15 Lioyed fisher square

The correction is required for this value of flux density as S_2 encloses the flux in air gap between specimen and the coil, in addition to the flux in the specimen.

$$\therefore \quad B_m = B_m' - \mu_o H_m \left[\frac{A_c}{A_s} - 1 \right]$$

where B_m = Maximum flux density required

A_c = Cross-section of coil in m^2

A_S = Cross-section of specimen in m^2

H_m = Magnetising force required to develop maximum flux density in A/m

The H_m can be obtained from B-H curve for the specimen.

The wattmeter reading (w) consists of,

w = total iron loss + copper loss in secondary circuit

V_{PC} = voltage across pressure coil in 'V'

I_{PC} = current through pressure coil in 'A'

E = voltage induced in S_2 = voltmeter reading

r_p = resistance of pressure coil in Ω

r_s = resistance of secondary winding in Ω.

The voltage induced in S_1 is same as S_2 as the number of turns are same.

$\therefore$ Voltage induced in S_1 = E

Neglecting leakage reactance of S_1 and leakage reactance of pressure coil as it is highly resistive.

$$E = I_{PC}\, r_p + I_{PC}\, r_s = I_{PC}\,(r_p + r_s) \qquad \text{.....(8.24)}$$

$$V = I_{PC}\, r_p$$

$\therefore$ total iron loss (P_i) in specimen + total copper loss

$$\text{in secondary circuit} = \frac{WE}{V} \qquad \text{.....(8.25)}$$

$$\text{the copper loss in secondary circuit} = I_{pc}^2\left(r_p + r_s\right) = \frac{E^2}{\left(r_p + r_s\right)}$$

$$P_i = \frac{WE}{V} = \frac{E^2}{(r_s + r_p)} = \frac{WI_{PC}(r_p - r_s)}{I_{PC} r_p} - \frac{E^2}{r_p + r_s} \qquad \text{.....(8.26)}$$

By conducting test at different frequencies iron loss can be separated into hysteresis and eddy current loss.

$$P_i = W\left[1 + \frac{r_s}{r_p}\right] - \frac{E^2}{(r_p + r_s)} - \frac{E^2}{r_p + r_s} \text{ watts}$$

8.12.1.3 Comparision between Epstein Square and Lioyed-Fisher Square

	Epstein square	Lioyed-Fisher square
1.	The performance is poor for anistropic materials as the direction of flux at the corners is partially perpendicular to the flux path on other parts of strip	Suitable for anistropic material as the corner pieces eliminates the difficulty of direction of flux existing in Epstein square
2.	The value of flux density is not same in corner and in square sides. The correction for this is difficult to apply. Hence result is not accurate	The correction is required for doubling of cross-section at the corners. This is easy to apply. Once applied, the results are highly accurate and reliable
3.	The strips are in the plane of the square	The strips are perpendicular to the plane of the square
4.	Not preferred due to its inaccuracy	Very commonly used due to its accuracy

8.12.1.4 Measurement setup of Wattmeter Method

Construction: The Fig. 8.16 shows that connection diagram for the measurement of total iron loss by a wattmeter method. The weight and cross-section of the test specimen is determined before the test. The four magnetising coils connected in series together to

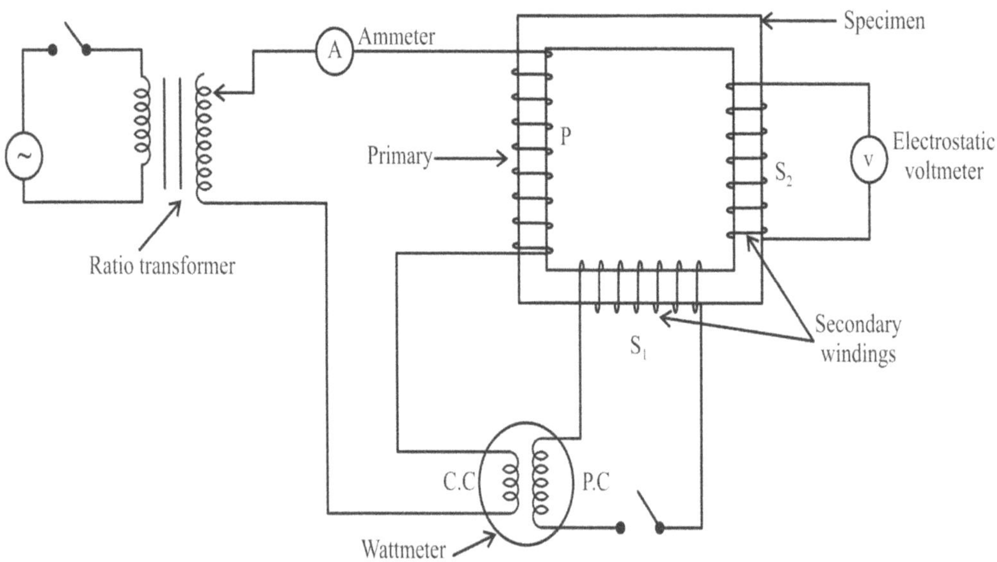

Fig. 8.16 Measurement of iron loss using wattmeter method

form a primary winding. It is connected to the transformer secondary through the current coil (C.C) of the wattmeter, for the alternating supply. The pressure coil (PC) is connected to one of the secondary windings. The other secondary winding is connected to the very high impedance electrostatic voltmeter. The supply frequency is adjusted to the required value. The wattmeter is a low power factor device having power factor of about 0.2.

Theory: The supply voltage is adjusted by the tappings, on the secondary of the ratio transformer, till the required maximum flux density (B_m) is reached in the specimen. The voltage induced in the secondary S_2 is measured on voltmeter while the wattmeter reading is observed.

The voltage measured by voltmeter is given by

$$E = 4K_f(B_m^1 A_S) f N_2 \qquad \ldots \phi_m = B_m' A_S$$

where K_f = form factor = 1.11 for sinusoidal flux

B_m' = apparent value of maximum flux density in wb/m^2

N_2 = number of turns in the winding S_2

f = frequency in Hz

A_S = cross-section area of specimen in m^2

$$\therefore \quad B_m' = \frac{E}{4K_f A_s f N_2} \qquad \ldots(8.27)$$

8.12.2 Bridge Methods

The wattmeter method is not sensitive for the laminated core materials working at low flux densities. Such material are used frequently in communication equipments. The a.c bridge methods are suitable for testing such materials.

The advantages of using bridge methods are:

1. Suitable for materials working at low flux densities
2. Test can be carried out at commercial or audio frequencies.
3. Measurement of Iron loss and a.c permeability is possible
4. Suitable even when only a small quantity of material is available for testing.

8.12.2.1 Maxwell's Bridge Method

The Maxwell's bridge used for testing ring specimen is shown in Fig. 8.17(a).

The arm a – b consists of the specimen

Thus $\qquad R_{ab} = R_s$

$$L_{ab} = L_s$$

At the bridge balance, detector current is zero hence,

$$I_1 R_3 = I_2 R_4 \qquad \text{.....(8.28)}$$

$$I_1(R_s + j\omega L_s) = I_1(R_2 + r + j\omega L_2) \qquad \text{.....(8.29)}$$

Dividing the two equations,

$$\frac{R_s + j\omega L_s}{R_3} = \frac{R_2 + r + j\omega L_2}{R_4}$$

Equating real and imaginary parts,

$$R_s = \frac{R_3}{R_4}(R_2 + r) \text{ and } L_s = \frac{R_3}{R_4} L_2 \qquad \text{......(8.30)}$$

Now R_s = effective resistance of a – b including winding resistance R_w and the resistance representing core losses R_C

$\therefore$ $I_1^2 R_s$ = Iron loss + copper loss in winding = Pi + $I_1^2 R_w$

$$P_i = I_1^2 [R_s - R_w] \qquad \text{......(8.31)}$$

Now $I = I_1 + I_2$

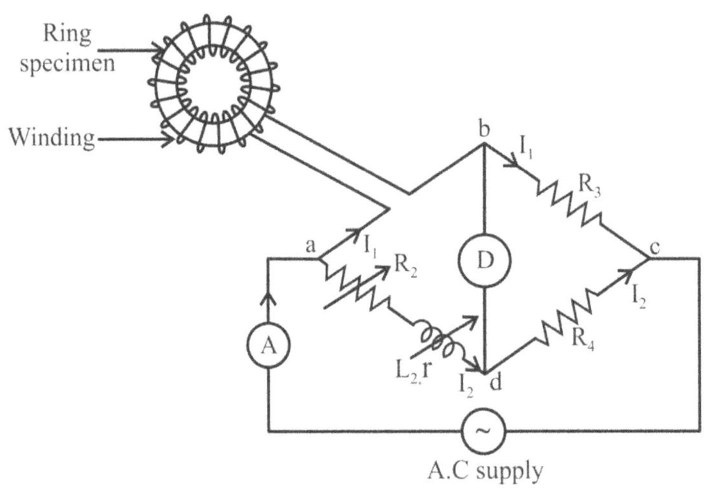

Fig. 8.17(a) Maxwell's bridge method

From equation 8.28, $I_1 R_3 = (I - I_1) R_4$

$$I_1 = I\left[\frac{R_4}{R_3 + R_4}\right] \quad .....(8.32)$$

$$\therefore \quad P_i = I^2\left[\frac{R_4}{R_3 + R_4}\right]^2 [R_s - R_w] \quad .....(8.33)$$

The current I is measured on ammeter and R_s, R_w can be measured. Thus Iron loss can be obtained

The L_s obtained by equation 8.30 can be expressed as

$$L_s = \frac{N^2}{S}$$

where $S = \text{reluctance} = \dfrac{l_s}{\mu_o \mu_r A_s}$

l_s = length of mean flux path in specimen

A_s = cross-sectional area of specimen

μ_r = A.C relative permeability of the specimen

$$\therefore \quad L_s = \frac{N^2}{\dfrac{l_s}{\mu_o \mu_r A_s}}$$

N = number of turns of magnetising coil

$$\therefore \quad \mu_r = \frac{l_s L_s}{N^2 \mu_o A_s} = \frac{R_3 l_s L_2}{N^2 R_4 \mu_o A_s} \quad .....(8.34)$$

8.12.2.2 Comphell's Bridge Method

The circuit of comphell's bridge method used for measurement of Iron loss is shown in the Fig. 8.17(b).

The specimen has two windings. The primary has N_1 turns and secondary has N_2 turns. The cross-sectional area of both the winding is same. M is the variable mutual inductance while R_1, R_2 are the adjustable resistance used to balance the bridge.

Let e_1 and e_2 are the inductanced voltages in primary and secondary windings while e_m is the induced voltage in the secondary of the mutual inductance, ϕ is the flux in the specimen. The phasor diagram at balance condition is shown in the Fig. 8.17(c). Due to Iron loss, the current I leads the flux ϕ by the small angle. When phasor sum of e_m and e_2 is equal and opposite to the drop IR_2, the balance is obtained.

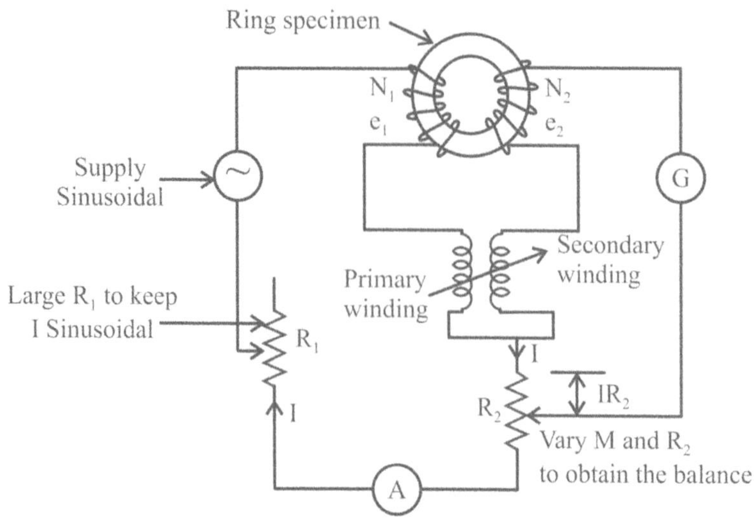

Fig. 8.17(b) Comphell's bridge method

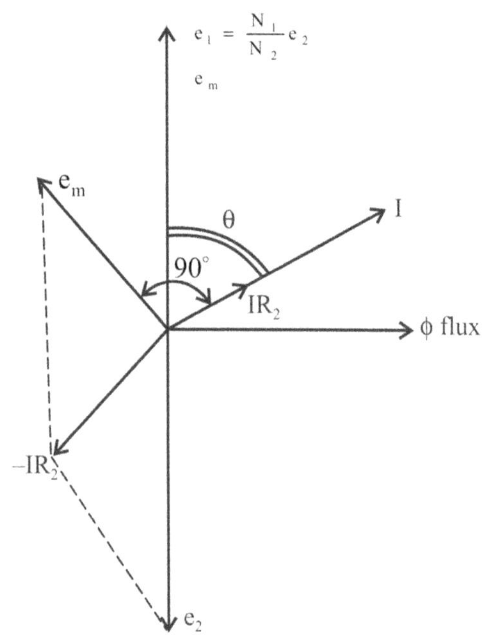

Fig. 8.17(c) Phasor diagram at balance condition

Let M' = mutual inductance between primary and secondary winding on specimen

$\therefore \quad e_2 = \omega M I$

$$e_m = \omega MI$$

At balance, $e_2 = e_m$ i.e., $M' = M$

Now, $L_1 = \dfrac{N\phi}{I}$ and $L = \dfrac{N_1^2}{S} = \dfrac{N_1^2}{\dfrac{l}{\mu_o\mu_r A_s}}$

$\therefore$ $\phi = \dfrac{LI}{N_1} = \dfrac{N_1^2\,\mu_o\mu_r A_s I}{N_1 l} = \dfrac{N_1\mu_o\mu_r A_s I}{l}$ flux

while $M = \dfrac{N_2\phi}{I} = \dfrac{N_1N_2\mu_o\mu_r A_s}{l}$

$\therefore$ $\mu_r = \dfrac{Ml}{N_1N_2\mu_o A_s}$ A.C. permeability

$P_i = e_1 I\cos\theta = e_1 I \times \dfrac{IR_2}{e_2} = I^2R_2\left(\dfrac{e_1}{e_2}\right)$ Iron loss

But $\dfrac{e_1}{e_2} = \dfrac{N_1}{N_2}$

$\therefore$ $P_i = I^2R_2\dfrac{N_1}{N_2}$

Thus by measuring current I by ammeter and knowing R_2, N_1 and N_2, the iron loss in the specimen can be obtained.

8.12.3 A.C Potentiometer Method

The A.C co-ordinate type of potentiometer used for the measurements of Iron loss at low flux densities, as shown in Fig. 8.18.

The ring specimen has two windings, primary with N_1 turns and secondary with N_2 turns. The primary is supplied from A.C supply through a regulating transformer. This also supplies the two slide wire circuits of the two potentiometers.

The primary winding has a regulating resistance. Another standard resistance R is also connected in the primary circuit. By measuring drop across R_1 current through primary can be obtained. The alternating a.c supply must have sinusoidal waveform.

When supply is given to the primary, the voltage E_2 is induced in the secondary.

According to transformer equation, it is given by

$$E_2 = 4.44\,\phi_m f N_2$$

$$= 4.44\,B_m A f N_2$$

$$\therefore \qquad B_m = \frac{E_2}{4.44\,AfN_2} \qquad .....(8.35)$$

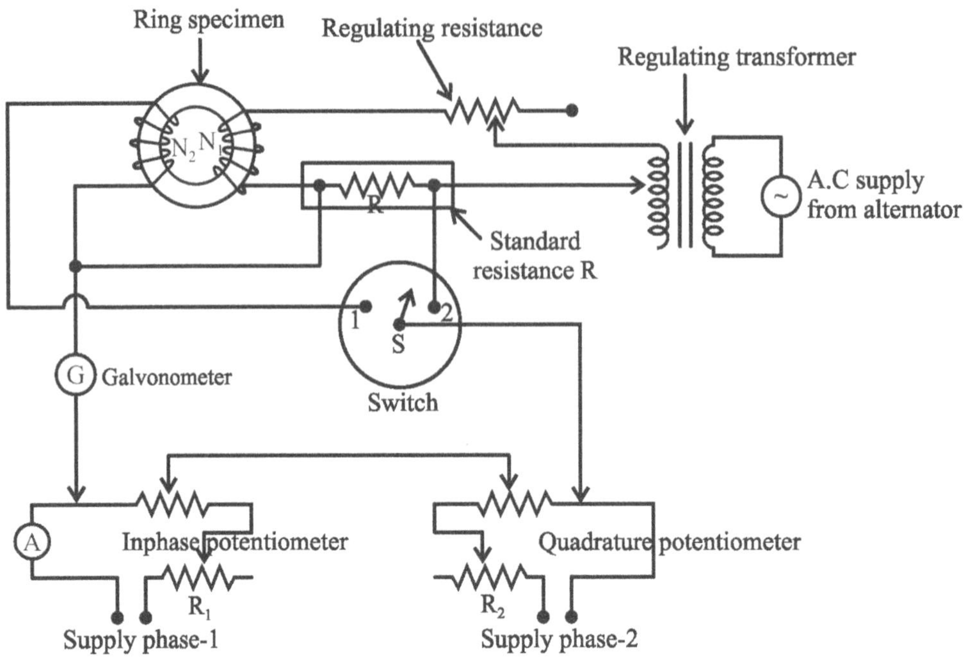

Fig. 8.18 Measurement of iron loss using A.C potentiometer method

For sinusoidal waveform $K_f = 1.11$

The E_2 can be measured as:

1. Put the switch S to position 1.
2. Set the quadrature potentiometer at zero.
3. Adjust the in-phase potentiomenter till galvanometer G shows zero deflection
4. The setting of in-phase potentiometer for balance gives the value of E_2.

Note that the potentiometer must be standardised first then put the switch to position 2 and adjust both the potentiometers till galvanometer shows zero deflection.

The total current has two parts as core loss component I_C and magnetising component I_m.

The in-phase potentiometer reading gives the drop I_CR while the quadrature potentiometer reading gives the drop I_mR.

$$\therefore \qquad I_C = \frac{\text{Reading of in-phase potentiometer}}{R}$$

$$I_m = \frac{\text{Reading of quadrature potentiometer}}{R}$$

$$\therefore \qquad P_i = \text{Iron loss} = I_C\ E_2\left(\frac{N_1}{N_2}\right)$$

The method is very effective as both core loss and magnetising components of currents are obtained separately.

Potentiometer method is very suitable for a.c magnetic testing at low flux densities since magnetising and Iron loss components of exciting current are measured separately.

8.12.4 Oscillographic Method

The magnetic measurement can be done by using cathode ray oscilloscope (C.R.O). The Fig. 8.19 shows the arrangement of CRO for such measurements.

The ring specimen has two windings, magnetising winding having N_1 turns and search coil having N_2 turns. In series with magnetising winding an ammeter and non-inductive resistance R_1 are connected.

The magnetising circuit is excited with a supply from a autotransformer. The magnetising current I_1 flows through the circuit, causing a drop I_1R_1 across R_1.

$\therefore \qquad V_1 = I_1R_1 \propto$ magnetic field in test specimen

The voltage V_1 which is directly proportional to the magnetic field in the test specimen in connected to X-plate amplifier representing horizontal axis on C.R.O.

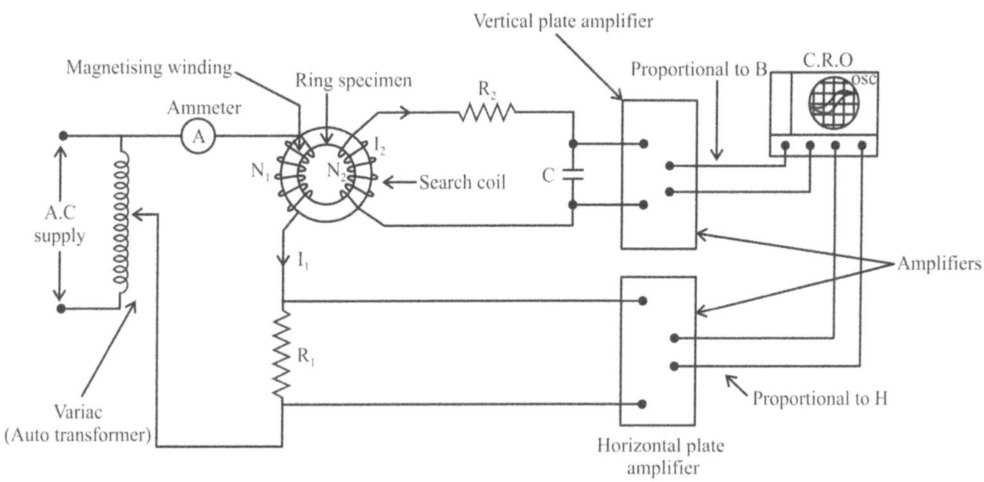

Fig. 8.19 Testing of magnetic material using oscillographic method

The e.m.f gets induced in the search coil, which is shunted by a capacitor C and high resistance R_2.

$$e_2 = N_2 \frac{d\phi}{dt} = N_2 A_S \frac{dB}{dt} \qquad \boxed{\phi = A_S B}$$

Now,
$$I_2 = \frac{e_2}{R_2 - j \times C}$$

But R_2 is selected very high such that $R_2 >> X_C$ and X_C can be neglected.

$$\therefore \quad I_2 = \frac{e_2}{R_2} = N_2 \frac{A_S}{R_2} \frac{dB}{dt}$$

The voltage across capacitor is,

$$V_C = \frac{1}{C}\int I_2 dt = \frac{1}{C}\int N_2 \frac{A_S}{R_2}\frac{dB}{dt} dt = \frac{1}{C}\int N_2 \frac{A_S}{R_2} dB$$

$$\therefore \quad V_C = \frac{N_2 A_S}{R_2 C} B$$

Thus voltage across capacitor is directly proportional to flux density in test specimen. This is given to Y-plate amplifier representing vertical axis on C.R.O. Thus on C.R.O, the graphs can be obtained from which following measurement can be made:

8.12.4.1 Plotting Hysteresis Loop

As $X \propto H$ and $Y \propto B$, directly hysteresis loop can be obtained.

8.12.4.2 Plotting B-H Curve

By varying the peak flux density B_m in the specimen slowly and continuously, by varying the excitation of the alternator supplying magnetising current, the locus of the tips of the hyseresis loops gives the B-H curve of the specimen.

8.12.4.3 Determination of Hysteresis Loss

The hysteresis loss for the material may be estimated from the area of the loop either by comparison with a loop for a standard specimen or by calculation from the 'X' and Y-sensitivities of the cathode-ray oscilloscope and the circuit constants.

Further, it must be realized that the current flowing through resistance contains a components associated with the eddy current loss in the specimen. Thus the loop obtained is really a 'total-loss' loop whose area represents the total iron loss. The method is normally used for such purposes as acceptance tests, where the loop obtained may be compared with a standard loop.

8.13 Ewing Double Bar Permeameter

Ewing double bar permeameter consists of two similar bar specimens made up of the magnetic material whose relative permeability is to be measured. Two similar yokes made up of annealed soft Iron (so that it have a negligible reluctance) and four magnetising coils, in which one pair covers a length 'l' and another pair covers only length '$l/2$'. Both the pairs of coils are wound such that they have same number of turns per unit length. The bars are carefully fitted into the yoke. The complete arrangement is shown in the Fig. 8.20.

Initially the magnetising coils wound for entire length 'l' (i.e., C_1 and C_1') are excited with a current I_1 and the apparent magnetising force is calculated as

$$H_1 = \frac{\text{m.m.f}}{\text{length}} = \frac{nlI_1}{l} = nI_1$$

where n = number of turns of magnetising coils per unit length

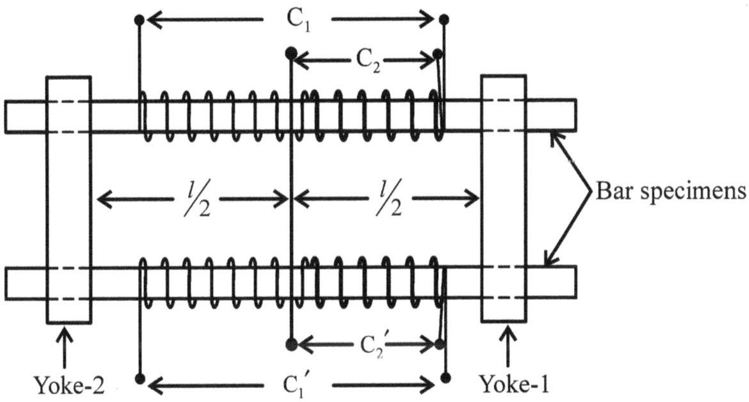

Fig. 8.20 Ewing double bar parmeameter

Now, yoke-2 is moved to the position shown by dotted lines and the magnetising coils wound for length $l/2$ (i.e., C_2 and C_2') are excited by current I_2 and again apparent magnetizing force is calculated as,

$$H_2 = \frac{n \times \frac{l}{2} \times I_2}{l/2} = nI_2 \qquad \left[\because l = \frac{l}{2}\right]$$

The total applied m.m.f will be the sum of the m.m.f required for the specimen and m.m.f required for the yokes and joints.

Assuming that the m.m.f required for yokes and joints in both the cases is equal or same, let this value of m.m.f be $\Im$.

$\therefore$ we have

[(Apparent magnetising force) × (Length)] = [True magnetising force (H) × (length)] + $\Im$

$$(H \times l) = (H \times l) + \Im \qquad .....(8.36)$$

and

$$H_2 \times \frac{l}{2} = N \times \frac{l}{2} + \Im \qquad .....(8.37)$$

From equations in (8.36) and (8.37) we get

$$H_1 l - \frac{H_2 l}{2} = Hl - \frac{Hl}{2}$$

$$H_1 - \frac{H_2}{2} = H - \frac{H}{2}$$

$$H = 2H_1 - H_2 \qquad .....(8.38)$$

Now, using a ballistic galvanometer the flux density (B) corresponding to the obtained true magnetising force (H) is measured.

Finally, the relative permeability of the specimen is calculated as,

$$\mu_r = \frac{1}{\mu_o}\frac{B}{H} \qquad \left[\because \mu = \mu_o \mu_r = \frac{B}{H}\right]$$

8.13.1 Advantage

1. The only main advantage of this method is that, it is a simple method.

8.13.2 Disadvantages

1. This method requires two exactly similar bar specimens.
2. It is a time consuming method.
3. The assumption made is an inpractical one i.e., the reluctance of yokes and joints will not be same for both the position of yoke 2.

8.14 Illiovici Permeameter

The construction of an illovici permeameter consisting of two windings is as shown in Fig. 8.21.

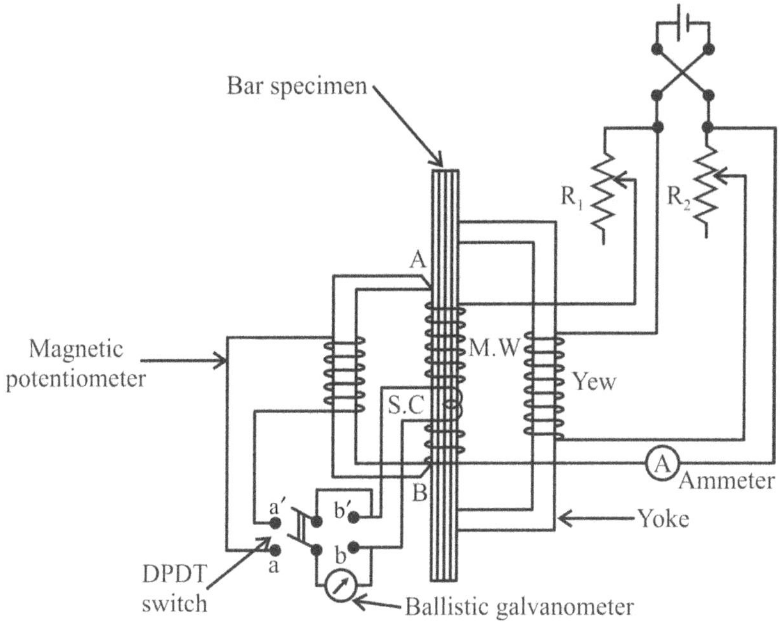

Fig. 8.21 Illiovici permeameter

An illiowici permeameter consists of a bar specimen on which a magnetising is uniformly distributed. This bar specimen is clamped against a yoke to which a compensating winding is wounded in parallel with the magnetic winding across a reversing switch. The search coil wounded on the bar specimen along with the ballistic galvanometer which is connected across the magnetic potentiometer are employed for measuring the flux density on the bar specimen. An ammeter is connected in series with the magnetising winding.

The current in the magnetising winding and the compensating winding are so adjusted that on their reversal, the ballistic galvanometer reads a zero value in turn causing no flux passing through the magnetic potentiometer. When there is no flux passing through the magnetic potentiometer which is connected to a section AB of the bar specimen, there is no magnetic potential difference between the point A and B. The magnetising force is therefore given by

$$H = \frac{NI}{l}$$

Where N = number of turns in the magnetising winding

I = current in the magnetising winding

l = length of section AB

The major disadvantage of an illionic permeameter is that the results obtained are not satisfactory which is due to the lack of symmetry of arrangement.

8.15 Burrow's Permeameter

The constructional features of a burrows permeameter is shown in Fig. 8.22 (a).

The burrow permeameter consists of two bar specimens S_1 and S_2 possessing same magnetic properties wounded by uniformly distributed magnetising windings M_1 and M_2 respectively. The bar specimen which is to be tested is S_1 and specimen S_2 acts as an auxiliary bar. A number of compensating coils or windings C_1, C_2, C_3, C_4 connected in series are used to eliminate the leakage affects at the joints between the yoke and the bar specimen. Three search coils X, Y and Z are connected across the ballistic galvanometer. Search coil 'X' surrounds the middle portion of the test specimen S_1. Search coil 'Y' divided into two equal parts are located at the ends of the test specimen S_1. Search coil 'Z' is wounded in the middle of the auxiliary bar specimen S_2.

Where C_1, C_2, C_3, C_4 = compensating windings

S.C (X, Y, Z) = search coils

M_1, M_2 = magnetising windings

S_1, S_2 = Bar specimens

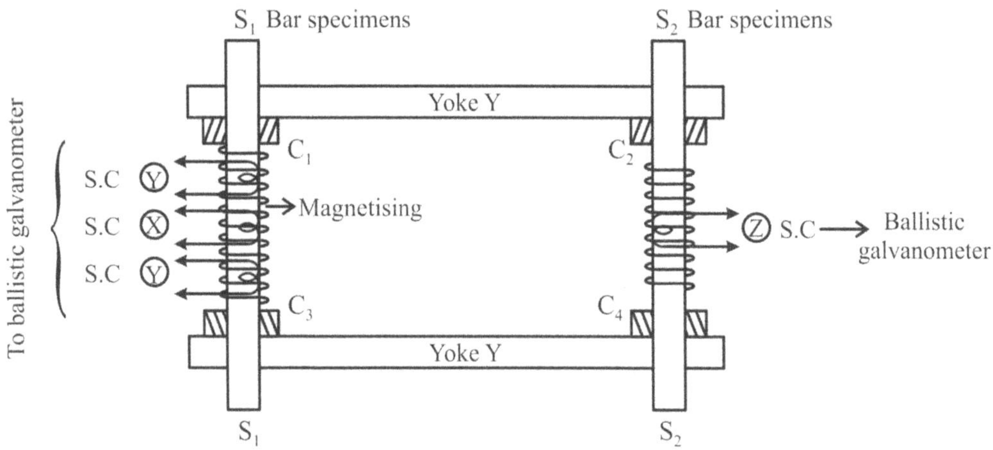

Fig. 8.22(a)

The compensating windings and the magnetising windings are supplied from separate batteries. A selector switch 's' is connected between the ballistic galvanometer and three search coils. The current in the compensating winding and the magnetising winding 'M_2' are so adjusted that the flux linking will at all the search coils should be same for a given value of current in the magnetising winding M_1. The test specimen is demagnetized and the values of current in the magnetising winding M_1 is set according to the required test value. The behaviour of the selector switch at different positions is explained as follows:

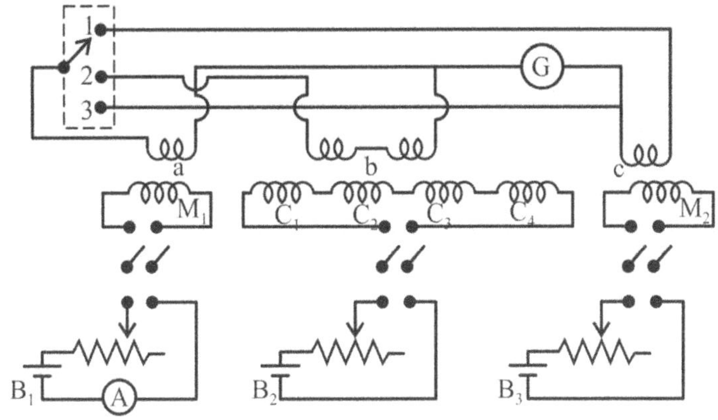

Fig. 8.22(b)

When the switch is at position 1

In this position search coil 'X' and 'Z' are connected in series opposition to a ballistic galvanometer (Fig. 8.22 (b)). The ballistic galvanometer should read a zero value for the reversal of current in the magnetising windings M_1 and M_2. This is achieved by adjusting the current in the magnetising windings M_2. Now equal fluxes are linking with search coils X and Z.

When the switch is at position 2

The search coils 'X' and 'Y' are connected in opposition with ballistic galvanometer. The reading of the ballistic galvanometer should be zero when the currents in the compensating winding and magnetising winding M_1 are reversed. This zero value is achieved by adjusting the current in the compensating winding. Now equal fluxes are linking with the search coil 'X' and 'Y'.

When the switch is at position 3

In this position search coil 'X' is connected across the ballistic galvanometer. The current in the magnetising windings M_1, M_2 and the compensating windings C_1, C_2, C_3, C_4 are reversed and the deflection in the ballistic galvanometer is noted. The flux density 'B' corresponding to the magnetising force 'H' in the magnetising winding M_1 can be found out.

Magnetising force

$$H = \frac{NI}{l}$$

where N = number of turns in the magnetising winding M_1

I = current in the magnetising winding M_1

l = length

8.15.1 The Advantages of Burrow's Permeameter with Respect to Others

Among the various methods used for magnetic testing, Burrow's permeameter method is advantageous over the others in the following ways.

The test specimen which is a straight long bar in this method, processes the conditions which do not vary along the whole length of the bar. This is not so in the case of other methods, where in non-uniform condition arise along the specimen.

Some of the methods for magnetic testing employs the specimens servicing as a standard reference. The Burrows permeameter method can also be used for the calibration of such specimens.

Various kind of magnetic materials can be tested using the Burrow's permeameter methods whereas the other methods restricts the magnetic testing to only few type of materials.

Problems with Solutions

1. The coil of a ballistic galvanometer has 115 turns of mean area 25×10 mm^2. The flux density in the air gap is 0.12 W/m^2 and the moment of inertia is 0.5×10^{-6} kg-m^2. The stiffness constant of springs is 45×10^{-6} Nm/rad. What current must be passed to give a deflection of 100° and what resistance must be added in series with the movement to give critical damping?

***Solution*:**

Given data

$N = 115$ turns

$J = 0.5 \times 10^{-6}$ Kg-m^2

$B = 0.12$ Wb/m^2

$A = 25 \times 10 \times 10^{-6}$ m^2

$K = 45 \times 10^{-6}$ Nm/rad

$$\theta = 100^{\circ}$$
$$= 1.475 \text{ radians}$$

Displacement constant

$$G = NBA$$
$$= 115 \times 0.12 \times 25 \times 10 \times 10^{-6}$$
$$= 3.45 \times 10^{-3} \text{ Nm/A}$$

$$i = \frac{K\theta}{G} = \frac{45 \times 10^{-6} \times 1.475}{3.45 \times 10^{-3}}$$
$$= 19.239 \text{ mA}$$

For critical damping

$$D = 2\sqrt{KJ}$$
$$= 2\sqrt{45 \times 10^{-6} \times 0.5 \times 10^{-6}}$$
$$= 9.48 \times 10^{-6} \text{ Nm/rad S}^{-1}$$

Neglecting the resistance of the galvanometer coil, the external resistance required for critical damping is

$$R_{ext} = \frac{G^2}{D} = \frac{(3.45 \times 10^{-3})^2}{9.48 \times 10^{-6}} = 1.255\ \Omega$$

2. The flux meter is connected to a search coil having 500 turns and a mean area of 500 mm^2. The search coil is placed at the centre of a solenoid 1 metre long, wound with 800 turns. When a current of 5 A is reversed there is a deflection of 25 scale divisions. Calculate the flux linkages per scale division.

Solution:

Given data

$N_2 = 500$ turns

$A = 500 \text{ mm}^2$

$= 500 \times 10^{-6} \text{ m}^2$

$l = 1$ m

$I = 5$ A

$N_1 = 800$ turns

Field strength of solenoid,

$$H = \frac{N_1 I}{l}$$

$$= \frac{800 \times 5}{1}$$

$$= 4000 \text{ A/m}$$

Flux density

$$B = \mu_o H$$

$$= 4\pi \times 10^{-7} \times 4000$$

$$= 5.026 \text{ mwb/m}^2$$

Flux linkages of flux meter search coil

$$= N_2 BA$$

$$= 500 \times 5.026 \times 10^{-3} \times 500 \times 10^{-6}$$

$$= 1.256 \text{ mwb-turns}$$

Change in flux linkages

$$= 2 \times (1.256 \times 10^{-3})$$

$$= 2.512 \times 10^{-3} \text{ wb-turns}$$

Flux linkages per scale division

$$= \frac{2.512 \times 10^{-3}}{25}$$

$$= 0.10048 \times 10^{-3} \text{ wb-turns}$$

3. A flux density = 0.05 W/m^2, turns on moving coil = 40, area of moving coil = 750 mm^2. If the flux linking with a 10 turn search coil of 20 mm^2 area connected to the flux meter is reversed in a uniform field of 0.5 W/m^2. Calculate the deflection of the flux meter.

***Solution*:**

Given data

No. of turns of moving coil, $N_1 = 40$

Flux density $B = 0.05 \text{ w/m}^2$

Area of the moving coil, $A_{MC} = 750 \text{ mm}^2$

$$= 750 \times 10^{-6} \text{ m}^2$$

Area of search coil, $A_{SC} = 20\ mm^2$

$$= 20 \times 10^{-6}\ m^2$$

No. of turns on search coil, $N_2 = 10$

To determine the deflection of the flux meter

The constant of flux meter is given as

$$G = NBA_{MC}$$
$$= 40 \times 0.05 \times 750 \times 10^{-6}$$
$$= 1.5 \times 10^{-3}$$

Now, flux linking with search coil,

$$= BA_{SC}$$
$$= 0.5 \times 20 \times 10^{-6}$$
$$= 1 \times 10^{-5}\ Wb$$

Change in flux linking the coil

$$\phi = 2 \times 10^{-5}$$

If ϕ is the change in flux and θ is the change in flux meter deflection then we have,

$$\phi = \frac{G\theta}{N_1}$$

$$\therefore \quad \theta = \frac{\phi N_1}{G} = \frac{2\times10^{-5}\times40}{1.5\times10^{-3}}$$
$$= 0.53333\ rad$$
$$= 30.5^{o}$$

4. A flux meter is connected to a search coil having 1000 turns and a mean area of 4 cm^2. The search coil is placed at the center of a solenoid 1.2 metres long wound with 1200 turns. When a current of 5 A is reversed, there is a deflection of 25 scale divisions on the flux meter. Determine the flux meter constant.

***Solution*:**

Given data

$N_1 = 1200$ turns

$N_2 = 1000$ turns

$A = 4\ cm^2 = 4 \times 10^{-4}\ m^2$

$L = 1.2$ m

$\theta = 25$ scale divisions

Field strength of solenoid

$$H = \frac{N_1 I}{l} = \frac{1200 \times 5}{1.2}$$

$$= 5000 \text{ A/m}$$

Flux density

$$B = \mu_o H$$

$$= 4\pi \times 10^{-7} \times 5000$$

$$= 6.28 \times 10^{-3} \text{ Wb/m}^2$$

Flux linkages of flux meter search coil,

$$= N_2 BA$$

$$= 1000 \times 6.28 \times 10^{-3} \times 4 \times 10^{-4} = 2.512 \times 10^{-3} \text{ Wb-turns}$$

Change in flux linkages

$$= 2 \times (N_2 BA)$$

$$= 2 \times (2.512 \times 10^{-3})$$

$$= 5.024 \times 10^{-3} \text{ Wb turn}$$

Flux meter constant,

$$K = \frac{\text{Change in flux linkages}}{\text{Deflection}}$$

$$= \frac{5.024 \times 10^{-3}}{25}$$

$$= 0.20096 \times 10^{-3} \text{ Wb turns/scale division}$$

5. A ring having a mean diameter of 0.3 m and a cross-sectional area 400 mm^2 has a primary winding of 80 turns wound uniformly. The secondary winding of 30 turns is connected to a flux meter having a constant of 0.12×10^{-3} Weber turn per division. A deflection of 46 division is observed when a current of 2 A is reversed in the primary winding. Calculate the relative permeability of iron.

***Solution*:**

Given data

Mean diameter, $D = 0.3$ m

Cross-sectional area $A = 400 \text{ mm}^2$

$$= 400 \times 10^{-6} \text{ m}^2$$

No. of turns in the primary winding, $N_1 = 80$

No. of turns in the secondary winding, $N_2 = 30$

Flux meter constant

$$K_\phi = 0.12 \times 10^{-3} \text{ Weber turns/division}$$

Deflection in the flux meter = 46 divisions

Current, $I = 2$ A

Relative permeability of iron, $\mu_r = ?$

Change in flux linkages of search coil

$$\begin{aligned} \Delta\psi &= \text{No. of divisions} \times \text{flux meter constant} \\ &= 46 \times 0.12 \times 10^{-3} \\ &= 5.52 \times 10^{-3} \text{ Weber turns} \end{aligned}$$

This flux linkages is caused by the reversal of flux

$\therefore$ Flux linkages of coil

$$\begin{aligned} \psi &= \frac{\Delta\psi}{2} \\ &= \frac{5.52\times 10^{3}}{2} \\ &= 2.76 \times 10^{-3} \text{ Weber turn} \end{aligned}$$

Flux linking with search coil

$$\begin{aligned} \phi &= \frac{\text{Flux linkages of coil}}{\text{Turn of coil } (N_2)} \\ &= \frac{2.76\times 10^{-3}}{30} \\ &= 9.2 \times 10^{-5} \text{ Weber} \end{aligned}$$

$\therefore$ Flux density in the specimen

$$\begin{aligned} B &= \frac{\phi}{A} \\ &= \frac{9.2\times 10^{-5}}{400\times 10^{-6}} = 0.23 \text{ Wb/m}^2 \end{aligned}$$

Length of flux path in the ring

$$\begin{aligned} l &= \pi D \\ &= \pi \times 0.3 = 0.942 \text{ m} \end{aligned}$$

Magnetising force,

$$H = \frac{N_1 I}{l}$$

$$= \frac{80 \times 2}{0.942}$$

$$= 169.85 \text{ A/m}$$

\ relative permeability of the specimen

$$\mu_r = \frac{B}{\mu_o H} = \frac{0.23}{4\pi \times 10^{-7} \times 169.85}$$

$$= 1077.5$$

6. The iron loss in a sample is 300 W at 50 Hz. With eddy current loss component 5 times as big as the hysteresis loss component. At what frequency will the iron loss be double if the flux density is kept the same.

***Solution*:**

Given data

Iron loss $P_i = 300$ W

Frequency $f = 50$ Hz

Eddy current loss component

$$P_e = 5 \times \text{hysteresis loss component } (P_h)$$

i.e., $P_e = 5\, P_h$

Flux density, B = constant

Now frequency $f' = ?$

At which iron loss $P_i = 2P_i$

$= 600$ W

We know that

$$P_e = K_e f^2 B_m^2$$

$$P_h = K_h f\ B_m^k$$

$$P_i = P_e + P_h$$

Since B is constant

$$P_e = K_e f^2$$

$$P_h = K_h f$$

Where $K_1 = K_e B_m^2$

$$K_2 = K_h B_m^k$$

$\therefore$ $$P_i = K_e f^2 + K_h f$$

$$300 = K_e(50)^2 + K_h(50)$$

$$6 = 50K_e + K_h \quad .....(1)$$

Given that

$$P_e = 5P_h$$

$$K_e f^2 = 5 \times K_h f$$

$$K_e(50)^2 = 5 \times K_h(50)$$

$$K_e \times 50 = 5K_h$$

$$K_e = 0.1K_h \quad .....(2)$$

Substituting eq. 2 in eq. 1, we get

$$K_h = \frac{6}{6} = 1$$

$\therefore$ $$K_e = 0.1 \times 1 = 0.1$$

At frequency f'

Iron loss $$P_i = K_e f'^2 + K_h f'$$

$$600 = 0.1 \times f'^2 + 1 + f'$$

$$0.1f'^2 + f' - 600 = 0$$

$\therefore$ $$f' = \frac{-1 \pm \sqrt{1^2 - [4 \times 0.1 \times (-600)]}}{2 \times 0.1}$$

$$= \frac{-1 \pm \sqrt{1 + 240}}{2 \times 0.1}$$

$$= 72.62 \text{ (or)} - 82.62$$

As negative value of frequency is not feasible

$$f' = 72.62 \text{ Hz}$$

$\therefore$ the frequency at which the iron loss doubles by keeping constant flux density is,

$$f' = 72.62 \text{ Hz}$$

7. A iron ring of 350 mm^2 cross–sectional area with a mean length of 1 m is wound with a magnetizing winding of 100 turns. A secondary coil with 200 turns of wire is connected to a ballistic galvanometer having a constant of 1 μc per scale division, the total resistance of the secondary circuit being 200 Ω. On reversing a current of 10 A in the magnetizing winding, the galvanometer shows a deflection of 100 scale divisions. Calculate the flux density in the specimen and the value of permeability at this flux density.

***Solution*:**

total mmf of coil $= N_1 I_1 = 100 \times 10 = 1000$ A

$$\text{Magnetizing force} = H = \frac{N_1\ I_1}{l} = \frac{1000}{1} = 1000 \text{ A/m}$$

Charge through the galvanometer $Q = K_{an}\ \theta_1 = 1 \times 100 = 100$ μc supposing ϕ is the flux through the ring

Flux linkage of search coil $\Psi = N_2\ \phi = 200\ \phi$

Change in flux linkage of search coil due to reversal of current

$$\Delta\Psi = 2\Psi = 400\phi$$

Average emf induced in the search coil

$$e = \frac{\Delta\Psi}{\Delta t} = \frac{400\ \phi}{\Delta t} \qquad \text{where} \quad \Delta t = \text{time of reversal}$$

$$\therefore \text{ current through search coil } i = \frac{e}{R} = \frac{400\ \phi}{\Delta t \times R} \Delta t$$

$$= \frac{400\ \phi}{R} = \frac{400\ \phi}{2000}$$

$$\text{thus } \frac{400\ \phi}{2000} = 100 \times 16^{-6} \quad \text{or} \quad \phi = 500 \times 10^{-6} \text{ wb}$$

$$\text{flux density in the specimen } B = \frac{\text{flux}}{\text{area}} = \frac{500 \times 10^{-6}}{350 \times 10^{-6}}$$

$$= 1.429 \text{ wb/m}^2$$

$$\text{Permeability of specimen } \mu = \frac{B}{H} = \frac{1.429}{1000} = 1.429 \times 10^{-3}$$

∴ Relative permeability of specimen

$$\mu_r = \frac{\mu}{\mu_0} = \frac{1.429 \times 10^{-3}}{4\pi \times 10^{-7}} = 1137$$

8. A flux meter is connected to a search coil having 200 turns and a mean area of 500 mm^2. Resistance of the search coil is 1 Ω and flux meter has 120 scale divisions. The search coil is placed at the center of a solenoid 1 m long wound with 700 turns. When a current of 4 A is reversed there is a deflected of 25 scale division. Calculate the resistance of shunt required for use search coil if the flux meter is used to measure 0.065 Wb.

Solution:

$$H = \frac{NI}{l} = \frac{700 \times 4}{1} = 2800 \text{ AT/M}$$

$$F = BA = 4\pi \times 10^{-7} \times 2800 \times 0.5 = 175.929 \times 10^{-5} \text{ Wb}$$

$$\text{weber – turns} = \phi \times N = 175.929 \times 10^{-5} \times 200 = 351.858 \times 10^{-3}$$

change in the number of weber – turns on the reversal of solenoid current

$$= 2 \times 351.888 \times 10^{-3} = 703.716 \times 10^{-3}$$

this produces a deflection of 25 scale divisions

calibration constant of the instrument is

$$K = \frac{703.716 \times 10^{-3}}{25} = 0.02815 \text{ Wb-turns/scale division}$$

For flux meter without shunt $\frac{K}{N} = \frac{\phi}{\theta}$

$$\phi = \frac{K\theta}{N} = \frac{0.02815}{200} \times 120 = 0.017$$

for flux meter with shunt $\phi = \frac{R_s + R_c}{R_s} \frac{k\theta}{N}$

$$0.065 = \frac{R_s + 1}{R_s} \times 0.017$$

$$R_s (0.065 - 0.017) = 0.017$$

$$\therefore \quad R_s = \frac{0.017}{0.048} = 0.354\ \Omega$$

9. In loss tests on a sample of Iron laminations the following results were recorded

(a) at 60 Hz, 250 V total iron loss = 200 W

(b) at 40 Hz, 100 V total iron loss = 40 W

Calculate the eddy current and hysteresis loss for each test. The stein metz index is 1.6.

***Solution*:** Given data:

In first loss test, frequency $f_1 = 60$ Hz

Voltage $E_1 = 250$ V

Iron power loss $P_{i_1} = 200$ W

In second loss test, frequency $f_2 = 40$ Hz

Voltage, $E_2 = 100$ V

Iron power losses $P_{i_2} = 40$ W

eddy current loss is given by the expression

$$P_e = K_e B_m^2 f^2 \quad .....(1)$$

hysteresis loss is given by the expression

$$P_h = K_h f B_m^{1.6} \quad .....(2)$$

∴ Total iron loss

$$P_i = P_e + P_n$$

$$= K_e B_m^2 f_1^2 + K_n f B_m^{1.6}$$

From the given tests, B_m is not same for both

$$\therefore \quad (Pi)_1 = K_e B_{m_1}^2 f_1^2 + K_n f_1 B_m^{1.6} \quad .....(3)$$

$$(Pi)_2 = K_e B_{m_2}^2 f_2^2 + K_n f_2 B_m^{1.6} \quad .....(4)$$

we know that, the flux density varies directly with voltage and inversely with frequency i.e.,

$$B_m \propto \frac{E}{f}$$

$$\therefore \quad \frac{B_{m_1}}{B_{m_2}} = \left(\frac{E_1}{f_1}\right)\left(\frac{f_2}{E_2}\right) = \left(\frac{250}{60}\right)\left(\frac{40}{100}\right)$$

$$B_{m_2} = \frac{B_{m_1}}{1.666} \quad .....(5)$$

Substituting the given data in eq. 5 in eq. 3 and 4, we get

$$200 = K_e B_{m_1}^2 60^2 + K_n B_{m_1}^{1.6} 60 \quad .....(6)$$

and $$40 = K_e \left(\frac{B_{m_1}}{1.666}\right)^2 40^2 + K_n \left(\frac{B_m}{1.6666}\right)^{1.6} 40 \qquad .....(7)$$

Assume

$$K_e B^2_{m_1} = K_1 \quad \text{and} \quad K_{b_1} B^{1.6}_{m_1} = K_2$$

Now equation (6) and (7) reduces to

$$200 = 3600\ K_1 + 60\ K_2 \qquad .....(8)$$

$$40 = 576.4\ K_1 + 17.67\ K_2 \qquad .....(9)$$

on solving equation (8) and (9), we get

$$K_1 = 0.039 \text{ and } K_2 = 0.99$$

i.e., $$K_e\ B^2_{m_1} = 0.039 \text{ and}$$

$$K_n\ B^2_{m_1} = 0.039 \text{ and}$$

∴ from first test, eddy current loss.

$$P_{e_1} - K_e\ B^2_{m_1}\ f_1^2$$

$$= (0.039)\ (60)^2$$

$$= 140.4 \text{ W}$$

Hysteresis loss

$$P_h = k_h\ Bm_1^{1.6}\ f_1^2$$

$$= 0.99 \times 60$$

$$= 59.4 \text{ W}$$

for second test, eddy current loss

$$P_{e_2} = K_e\ B^2_{m_2}\ f_2^2$$

$$= K_e \left(\frac{B_{m_1}}{1.6666}\right)^2 40^2$$

$$= \frac{0.039}{(1.6666)^2} \times 40^2 = 22.48 \text{ W}$$

Hysteresis loss,

$$P_{h_2} = K_h\, B^{1.6}_{m_2}\, f_2$$

$$= K_n \left(\frac{B_{m_1}}{1.6666}\right)^{1.6} \times 40$$

$$= \frac{0.99}{(1.6666)^{1.6}} \times 40$$

$$= 17.49 \text{ W}$$

Review Questions

1. What is magnetic measurement? Which tests are necessary for the magnetic measurement?
2. Write short notes on Ballistic galvanometers.
3. Explain the constructional features of moving will Balistic galvanometer?
4. Explain theory and working principle of Ballistic galvanometer?
5. Derive an expression for equation of motion of a ballistic galvanometer?
6. Explain the construction and working principle of flux meter with a neat sketch.
7. Explain the comparision between flux meter and Ballistic galvanometer?
8. Explain the method of measuring flux density?
9. What is correction of air flux why it is necessary?
10. Explain theory of measurement of magnetising force 'H'?
11. Explain the measurement of leakage factor with flux meter?
12. How the B-H curves determined using method of Reversals?
13. Explain the advantages of method of reversals over step-by-step method.
14. Explain B-H is determined using step by step method?
15. Explain with the help of a neat diagram a method of the determination of B-H curve of a magnetic sample; point out the various source of errors and the methods of minimizing them?
16. Write a note on hysteresis loss and eddy current loss.
17. How to obtain separation of iron losses for a given specimen?
18. What are the methods of measuring iron loses? Explain each in details?
19. State the comparision between Epstein square and lioyed-fisher square?

20. Explain the method for measurement of iron loss using wattmeter method?
21. How the value of A-C permeability of magnetic material is determined by using maxwell's bridge?
22. Explain the method for measurement of Iron loss using A.C potentiometer?
23. Explain the method for testing of magnetic material using oscillograhic method?
24. Explain in detail about ewing double bar permeameter?
25. Explain in detail about illiowici permeameter?
26. Explain with the help of a neat diagram constructioned features of Burrow's permeameter?
27. State the advantage of Burrow's permeameter with respect to others?

Quiz Questions

1. Ballistic tests are used in magnetic measurements for
   (a) determination of hysteresis loop of the specimen only
   (b) determination of B-H curve of the specimen only
   (c) determination of flux density, magnetizing force and B-H curve and hysteresis loop of the specimen
   (d) finding out iron losses in the specimen
2. Burrow's permeameter is preferred to other permeameters as it
   (a) is very accurate and can serve as reference standard
   (b) incorporates arrangement for uniform distribution of mmf over the entire length of the specimen
   (c) Incorporates arrangements for elimination of leakage flux at the joints
   (d) All the above
3. The reason for using ring type specimens for ballistic tests is
   (a) they are from end effects
   (b) they required less labour to prepare as compared with bar type specimens
   (c) they can be used for large quantities of material of very small permeability
   (d) all of the above

4. The true value of flux density in a ring specimen's

(a) $B = B' - \mu_o H\left(\frac{A_C}{A_S} - 1\right)$

(b) $B = B' + \mu_o H\left(\frac{A_C}{A_S} - 1\right)$

(c) $B = B' - \mu_o H\left(\frac{A_S}{A_C} - 1\right)$

(d) $B = B' - \mu_o H\left(\frac{A_S}{A_C} - 1\right)$

5. Permanent magnets are tested by
   (a) ballistic methods
   (b) betteridge apparatus
   (c) potentiometeric methods
   (d) using an electric circuit having a mutual inductance

6. Magnetising force is measured by nuclear magnetic resonance method, the expression for magnetising force is

(a) $H = \frac{2\pi f}{V_p}$

(b) $N = \frac{f}{V_p}$

(c) $H = \frac{2\pi f}{n}$

(d) $H = \frac{fh}{4\pi}$

Questions and Answers

1. Siemens is a unit for measuring

Ans. Conductance

2. To measure flux, devices used are based on

Ans. Lorenz method

3. Siemens is a unit for measuring

Ans. Conductance

Quiz Questions Key

1. (c) 2. (d) 3. (d) 4. (a) 5. (b)
6. (a)

CHAPTER 9

Electronic Instruments

9.1 Introduction

"Instrument" may be defined as a device for measuring the value or magnitude of a quantity or variable. "Measurement" involves using an instrument as a physical means of determining a quantity or variable.

For any measuring instrument, the primary requirements are:

1. Power consumption should be as small as possible.
2. No alteration in the circuit conditions with the introduction of instrument in the circuit.

Most of the scientific and industrial instruments require very fast responses. Such requirements be met with by mechanical and electronic devices, since the only movement involved is that of electrons, the response time is extremely small owing to very small inertia of electrons, with the use of electronic devices very weak signals can be detected by using pre-amplifiers and amplifiers.

The electronic instruments are based on electrical or electronic principles for their mcasurcmcnt.

These instruments are either indicating or recording type instruments just like the electrical measuring instruments. In addition to indicating and recording a quantity, some instruments also control a quantity e.g., tachogenerator may be used for indicating or recording the speed as well for controlling the speed in a speed control motor.

The Advantages of an Electronic Measurement are:

1. The measured signals can be transmitted easily over long distances with the help of cables or radio links ect, without any loss of the Data.
2. An electrical or electronic signal can be filtered, amplified, multiplexed, sampled and measured. Electronic circuits can detect and amplify very weak signals and can measure the events of very short duration as well. Higher sensitivity, low power consumption and a higher degree of reliability are the important features of electronic instruments and measurements.
3. The measurement can easily be obtained in or converted into digital form for automatic recording and analysis. And also Electronic measurement makes possible to build analog and digital signals. The digital signals are very much required in computers. The modern development in science and technology are totally based on computers.

9.1.1 True R.M.S Voltmeter

Fig. (9.1) shows one-time of commercial true R. M. S instrument:

- In this instruments the input signal is amplified and fed into the directly-heated filament of high vaccum diode valve, thus furnishing part of the filament heating power. The filament must have a small thermal time constant if the instrument is to respond to rapid changes in signal level.

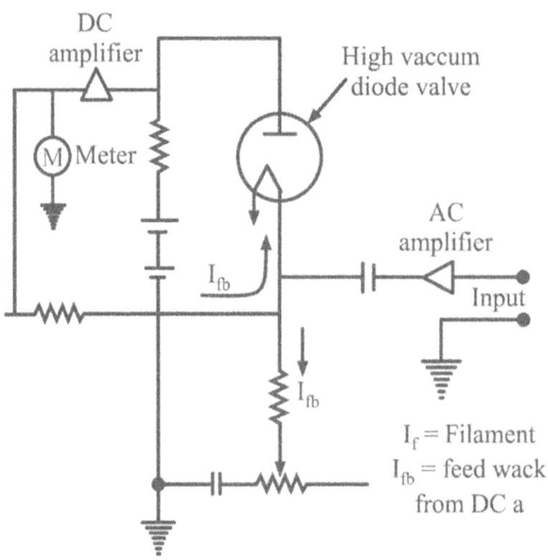

Fig. 9.1 True R.M.S. voltmeter

- After amplification of A.C input signal a cathode follower is used to obtain an impedance match between the filament temperature. Which will cause the anode current to rise and the anode-to-cathode voltage of fall. This drop in voltage is amplified by the D.C amplifier. The output of this amplifier supplies the component of current I_{fb} (feedback current) to the filament.
- The feedback circuit is adjusted so that the decrease in $If_b^2\ R_f$ equals the signal power, $I_s^2\ R_f$, in the filament, where I_S is the R.M.S signal current.
- The meter reads the output voltage of the amplifier, which is proportional to the feedback current I_{fb}. It can be shown that the feedback current I_{fb} is proportional to the square of the R.M.S signal current. The meter reading is, therefore, proportional to the R. M. S signal current.
- The meter will have a liner power scale.

9.1.2 Electronic Multimeter

The solid-state electronic multimeter or VOM (Voltage-ohm-meter) is one of the most versatile general-purpose shop instruments capable of measuring D.C and A.C voltages as well as current and resistance.

An electronic multimeter generally consists of the following elements.

1. Balanced bridge D.C amplifier and indicating meter
2. Input attenuator or Range switch. To limit the magnitude of the input voltage to the desired value
3. Rectifier section to convert an A.C input voltage to a proportional D.C value.
4. Internal battery and additional circuitry to provide the capability of resistance measurement.
5. "Function Switch" to select the various measurement functions of the instrument such as voltage current or resistance.

The instrument is also usually provided with abuitt-in power supply for operation on A.C main and, in most cases, one or more butteries for operation as portable test instrument Fig. (9.2) shows the schematic diagram of a balanced – bridge D.C amplifier using field effect-transistors (FETs).

This circuit also applies to a bridge amplifier with ordinary bipolar transistors (BJTs:

Q_1 and Q_2 are the two FETs: These should be reasonably well matched for current gain to ensure thermal stability of the circuit.

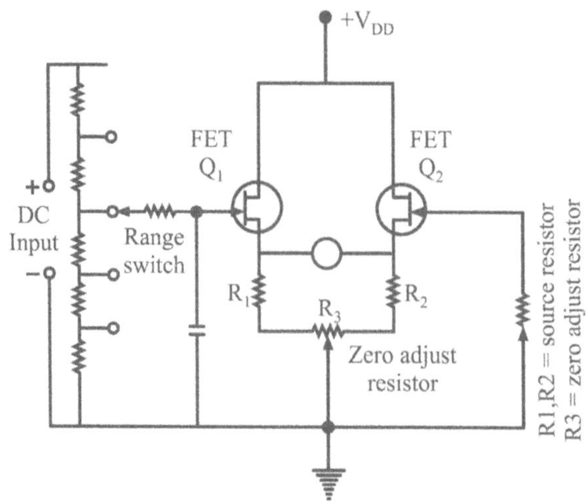

Fig. 9.2 Balanced-bridge D.C amplifier with input attenuator and indicating meter

The two FETs: Q_1 and Q_2 constitute the upper arms of a bridge circuit, whereas source resistor R_1 and R_2, together with "zero" adjust resistor R_3 from the lower bridge arms.

The PMMC meter is connected between the source terminals of the FETs, representing two opposite corners of the bridge.

9.1.3 Voltage Measurement

Where there is no input signal, the gate terminals of the FETs are at ground potential and the transistors operate under identical quiescent conditions. Under such a condition the bridge is balanced and meter reads zero. However in practice, small difference in the operating characteristics of the transistors, and slight tolerance differences in the various resistors, cause a certain amount of unbalance in the drain currents, and the meter shows a small deflection from zero. This current is reduced to zero by adjust control resistor R_3 and now the bridge is balanced.

On the application of the voltage to the gate of input FET Q_1 its drain current increases which causes the voltage at the source terminal to rise. The resulting unbalance between Q_1 and Q_2 source voltage is indicated by the PMMC meter, whose scale in calibrated to agree with the magnitude of the applied input voltage.

The maximum voltage which can be applied to the gate of Q_1 is determined by the operating range of FET and is usually on the order of a few volts. The range of input voltage can easily be extended by an input attenuator a "Range Switch" as shown in Fig.9.2. The unknown D.C input voltage is applied through a large resistor in the probe body to a resistive voltage divider. Thus as shown in Fig. 9.3.

with the "Range Switch" in 3 V position, the voltage at the gate of the input FET is developed across 8 MΩ of the total resistance of 11.3 MΩ and the circuit is so arranged that the meter deflects full scale with 3 V applied to the tip of the probe.

With the "Range Switch" in the 12 V position, the gate voltage is developed across 2 MΩ of the total divider resistance of 11.3 MΩ and an input voltage of 12 V is required to cause the same full-scale meter deflection.

9.1.4 Resistance Measurement

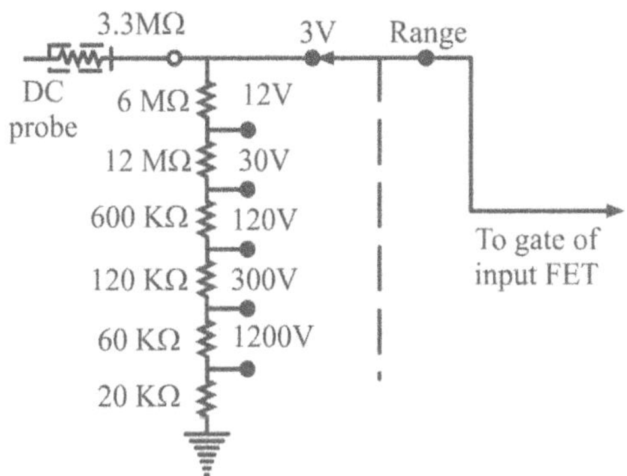

Fig. 9.3 Typical input voltage attenuator for a multimeter

In an electronic multimeter, when the function switch is placed in the "OHMS" position the unknown resistor is connected in series with an internal battery, and the PMMC meter simply measures the voltage drop across the unknown.

Fig. 1.4 shows a typical circuit, where a separate divider net work, used only for resistance measurements, provides for a number of different resistance ranges.

When an unknown resistor R_X is connected to the OHMS terminals of the multimeter, the 1.5 V battery supplies current through one of the range resistors and unknown resistor to ground.

Voltage drop V_X (across R_X) is applied to the input of the bridge amplifier which causes a deflection on the meter. Since the voltage drop across R_X is directly proportional to its resistance, the meter scale can be calibrated in terms of resistance.

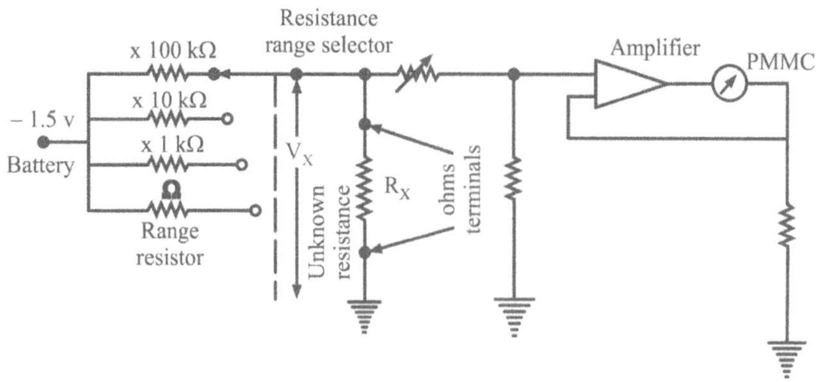

Fig. 9.4 Resistance range selector circuit of an electronic multimeter

It may be noted that the resistance scale of multimeter reads increasing resistance from left to right, opposite the way resistance scale reads on conventional meters. It is expected because electronic multimeter reads a larger resistance as a higher voltage, whereas the ordinary multimeter indicates a higher resistance as a small current.

Advantages

1. Less costly
2. Simple circuit
3. Less electric noise
4. High input impedance
5. High frequency range
6. Rugged construction

Disadvantages

1. Size is not compact
2. Accuracy not high
3. Poor reliability and repeatability
4. Poor resolution
5. Interfacing of the output with external devices is difficult

9.2 Digital Voltmeters (DVM)

9.2.1 Introduction

A digital voltmeter (DVM) displays measurements of D.C or A.C voltage as discrete numerals instead of pointer deflection on a continuous scale as in analog instruments.

It is a versatile and accurate instrument that is employed in many laboratory measurement applications.

Owing to the development and perfection of IC modules, the size, power requirements and cost of the digital voltmeter have been drastically reduced and, therefore, DVMs can actively compete with conventional analog instruments, both in price and portability.

The various varieties of DVM differ in the following ways

1. Number of digits
2. Number of measurements
3. Accuracy
4. Speed by reading
5. Digital output of several types

9.2.2 Characteristic Features of DVMs

1. **Input range:** From $\pm 1.000\,V$ to $\pm 1000\,V$ with automatic range selection and overload indication
2. **Absolute accuracy:** As high as $\pm\ 0.005\%$ of the reading
3. **Stability:** Short-term, 0.002% of the reading for a 24-hr period long term, 0.008% of the reading for a 6 month period.
4. **Resolution:** 1 part in million ($1\mu V$ reading can be read or measured on 1V range)
5. **Input capacitance:** 40 PF

   Input resistance: Typically 10 MΩ
6. **Calibration:** Internally from stabilised reference sources, independent of measuring circuit.
7. **Output signals:** In BCD (Binary-coded-decimal) form, for print output and further digital processing
8. **Optional features:** May include additional circuitry to measure current, ohms and voltage ratio.

9.2.3 Application of DVMs

DVM are often used in "data processing system" or "data logging system"

In such systems a number of analog input signals are scanned sequentially by an electronic system and then each signal is converted to an equivalent digital value by the A/D (analog-to-digital) converter in the DVM.

The digital value is then transmitted to a printer along with the information about the input one from which the signal has been derived.

The whole data is then printed out

In this way, a large number of input signals can be automatically scanned or processed and their values either printed or logged.

9.2.4 Classification of Digital Voltmeters (DVMs)

The digital voltmeters can broadly be classified as follows:

1. Raup - type DVM
2. Dual - slope integrating type DVM
3. Integrating type DVM (voltage-to-frequency conversion)
4. Successive - approximation DVM

9.2.5 Advantages of DVMs

The advantages of DVM over other voltmeters are

1. Higher accuracy and resolution
2. Greater speed
3. No parallax
4. Reduced human error
5. Compatibility with other digital equipment for further processing and recording

9.2.6 Q-meter

The overall efficiency of the coil and capacitors intended for radio frequency (RF) applications is best evaluated using the Q value.

Q-meter is an instrument designed to measure some electrical properties of coils and capacitors. The principle of the Q-meter is based on 'series resonance', the voltage drop across the coil or capacitor is Q times the applied voltage (where Q is the ratio of reactance to resistance, $\frac{X_L}{R}$).

If a fixed voltage is applied to the circuit, a voltmeter across the capacitor can be calibrated to read Q directly.

At resonance : $X_L = X_C$: $V_L = IX_L$:$V_C = IX_C$ $V = IR$

where, X_L = Inductive reactance,

X_C = Capacitive reactance,

R = Coil resistance

I = Circuit current,

V_L = Inductive voltage

V_C = Capacitor voltage and

V = Applied voltage

therefore, $$Q = \frac{X_L}{R} = \frac{X_C}{R}\left(= \frac{1}{\omega CR}\right) = \frac{V_C}{V} \quad .....(9.1)$$

In the above equ (9.1), if V is kept constant, the voltage across the capacitor can be measured by voltmeter calibrated to read directly in terms of Q. Fig. (1.5) shows the basic Q-meter circuit:

A wide-range oscillator with frequency range from 50 KHz to 50 MHz is used as a power supply to the circuit, the output of the oscillator is shorted by a low-value resistance, R_{sh} usually of the order of 0.02 Ω. So it introduces almost no small or almost negligible internal resistance. The voltage across the low-value shunt resistance V_C is measured by an electronic voltmeter.

The inductance (L) of the coil can be determined by connecting it to the test terminals of the instrument. The circuit is tuned to resonance by varying either the capacitance or the oscillatory frequency. If the capacitance is varied, the oscillator frequency is adjusted to a given frequency and resonance is obtained.

If the capacitance is pre-set to a desired value, the oscillator frequency is varied until resonance occurs. The Q reading on the output meter must be multiplied by index setting or the "Multiply Q by" to obtain the actual value.

The inductance of the coil can be calculated from known values of the coil frequency and resonating capacitor (C).

$$X_L = X_C \text{ or } 2\pi f_L = \frac{1}{2\pi f_C} \text{ or } 4\pi^2 f^2 LC = 1 \text{ or } L = \frac{1}{(2\pi f)^2 C} \quad .....(9.2)$$

The Q indicated is not the actual Q_1 because the losses of the resonating capacitor, voltmeter and inserted resistance are all included in the measuring circuit. The actual Q of the measured coil is somewhat greater than the indicated Q. This difference is negligible except where the resistance of the coil is relatively small compared to the inserted resistance R_{sh}.

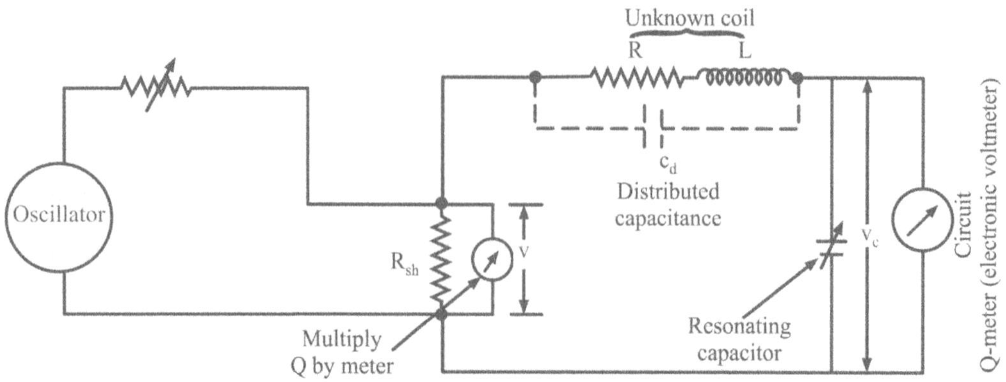

Fig. 9.5 Circuit diagram of a Q-meter

Review Questions

1. With a block diagram explain the working of a True RMS responding voltmeter
2. With a block diagram explain the working of a Ramp type DVM.
3. List the elements of the basic circuit of an electronic multimeter.
4. What is a Q meter? Discuss how the unknown components can be connected to its test terminals J.
5. Explain with the help of block diagram the function of integrating type digital voltmeter.
6. Explain the principle of operation of electronic multimeter.
7. Explain with block diagram any one type of digital voltmeter.
8. What are the advantages of using electronic measuring instruments?
9. Explain the operation of a electronic multimeter to measure current, voltage and resistance. What is the working principle of Q-Meter? How can the distributed capacitance of the coil be
10. Mention the salient features of digital voltmeter.
11. Discuss the different practical method of connection the unknown components to the test terminals of a Q meter.
12. What are the advantages of using electronic measuring instruments.

Quiz Questions

1. The major advantage of electromagnet type meter over a permanent magnet meter is
   (a) The electromagnet meter costs much less
   (b) The permanent magnet meter has a more sluggish coil
   (c) The electromagnet meter need not be aligned with the earth's magnetic field
   (d) The electromagnet meter is more rugged
2. A thermocouple
   (a) Gets warm when dc flows through it
   (b) Is a thin, straight, special wire
   (c) Generates ac when heated
   (d) Generates dc when exposed to visible light
3. An average-reading digital multimeter reads 10 V when fed with a triangular wave, symmetric about the time-axis. For the same input an rms-reading meter will read
   (a) $20/\sqrt{3}$ (b) $10/\sqrt{3}$
   (c) $20\sqrt{3}$ (d) $10\sqrt{3}$
4. A true rms reading voltmeter uses two thermocouples in order
   (a) to increase sensitivity
   (b) that the second thermocouple cancels out the non-linear effects of the first thermocouple
   (c) to prevent drift in the d.c amplifier
   (d) all of the above
5. Electronic voltmeters which use rectifiers employ negative feedback. this is done
   (a) to increase the overall gain
   (b) To improve the stability
   (c) to overcome nonlinearity of diodes
   (d) None of the above

Questions and Answers

1. Why ordinary voltmeter cannot be used in Electronic circuit?

Ans. A moving coil voltmeter draws a large current and hence a large power from the test circuit during measurement. Generally current drawn by the voltmeter does not make much difference in electrical circuit but when measurement is involved in electronic circuits, current drawn by the voltmeter load the test circuit and therefore voltmeter shows an error in the reading.

2. What is the difference between DC voltmeter and AC voltmeter ?

Ans. DC voltmeter consist of a dc amplifier succeded by a conventional DC meter while an ac voltmeter consist of an additional rectifier either before (or) after the amplifier.

3. For what type of measurement is true RMS reading voltmeter required?

Ans. For measuring rms value of a non-sinusoidal waveform a true RMS voltmeter is required.

4. What is Q-meter?

Ans. The Q-meter is an instrument designed for measurement of Q-factor of a coil as well as for the measurement of electrical properties of coils and capacitors.

5. On what principle does a Q-meter works?

Ans. On series resonance principle i.e., under resonant condition of an ac series circuit voltage across the capacitor is equal to the applied voltage times of Q of the circuit.

6. What is the Q-factor of the coil?

Ans. The ratio of inductive reactance to the effective reactance of the coil is called Q-factor of the coil.

7. What are the different parameters that can be measured using Q-meter?

Ans. Inductance, effective resistance, self capacitance, Bandwidth and capacitance.

8. Why the actual Q-factor of the coil is somewhat larger than the calculated Q-factor.

Ans. Because Q-measurement includes the losses of the resonating capacitor, voltmeter and the shunt resistance.

Quiz Questions Key

1. (d) 2. (a) 3. (a) 4. (b) 5. (c)

CHAPTER 10

Dual Trace Oscilloscope

The comparison of two or more voltages is invariably necessary in the analysis and study of many electronic circuits and systems. This is possible by using more than one oscilloscope but in such a case it is difficult to trigger the sweep of each oscilloscope precisely at the same time. A common and less costly method to solve this problem is to use dual trace or multitrace oscilloscopes. In this method, the same electron beam is used to generate "two traces" which can be deflected from two independent vertical sources. Dual trace oscilloscopes are used extensively by industrial firms and research laboratories.

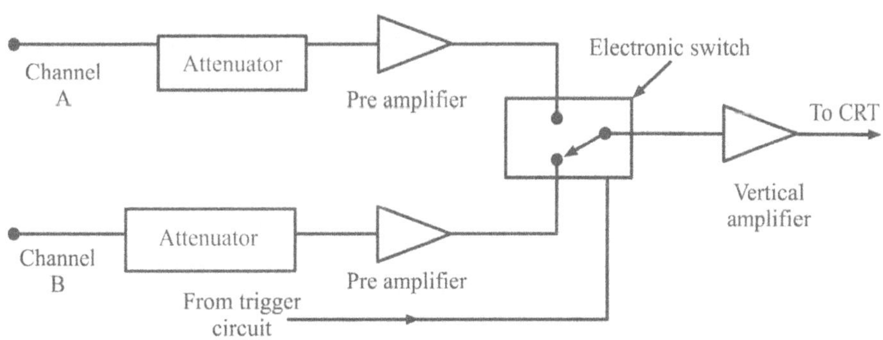

Fig. 10.1 Block diagram of a dual trace oscilloscopes

These oscilloscopes produce a dual-trace display by means of electronic switching of two separate input signals.

Fig. 10.1 Shows the block diagram of a dual trace oscilloscope

There are two vertical input circuits marked channels A and B with identical preamplifiers. The outputs of these preamplifiers are fed to an electronic switch which alternately connects them to the main vertical amplifier of the oscilloscope.

In this type of oscilloscope there in only one electron beam. There are two common operating modes for the electronic switch, called alternate and chop; these are selected from the Instrument's front panel.

In alternate mode the electronic switch alternates between channels A and B, letting each through for one cycle of the horizontal sweep. The display is blanked during the fly back and hold-off periods, as in conventional oscilloscope provided the sweep speed is much greater than the decay time of the CRT phosphor, the screen will show a steady display of the both the waveforms at channels A and B. This mode cannot be used for displaying very low frequency signals.

In the chop mode, the electronic switch free runs at a high frequency of the order of 100 kHz to 300 kHz. The result is the small segments from channels A and B are connected alternately to the vertical amplifier, and displayed on the screen. Provided the chopping rate is much faster than the horizontal sweep rate, the display will show a continuous line for each channel. If the sweep rate approaches the chopping rate then the individual segments will be visible and the alternate mode should now be used.

10.1 Voltage Measurement

The voltage to be measured is applied to y-deflection plates through the vertical amplifier. Generally, a coaxial cable is used for connection.

The X-deflection plates are exicited by the time-base generator.

To aid the measurement, a set of squares is marked on the transparent screen cover. Using these marking and the gain of the vertical amplifier, the peak-to-peak (P-P) value (and hence rms value) of the voltage wave can be measured.

After noting down the selection in $\frac{\text{Volts}}{\text{division}}$ from the front panel (selected for measurement) the peak-to-peak value, amplitude and rms value of sinusoidal signal can be obtained as given below:

1. Peak-to-peak value,

$$V_{P-P} = \left(\frac{\text{Volts}}{\text{division}}\right) \times (\text{number of divisions})$$

the volts/div. is nothing but deflection on sensitivity of CRO

2. Amplitude,

$$V_{max} = \frac{V_{P-P}}{2}$$

3. R.M.S value,

$$V_{rms} = \frac{V_{P-P}}{2\sqrt{2}} \qquad \text{(for sinusoidal signal only)}$$

It must be remembered that an oscilloscope is very complex device and that the accuracy of measurements is not better than $\pm 3\%$.

It is basically a display device so that we can observe the waveform in detail and find out the extent of harmonics in the wave.

***Example*:** For the waveform shown in Fig. below if vertical attenuation is 3 mv/div determine:

1. Peak-to-Peak value
2. Amplitude, and
3. R.M.S value of the signal.

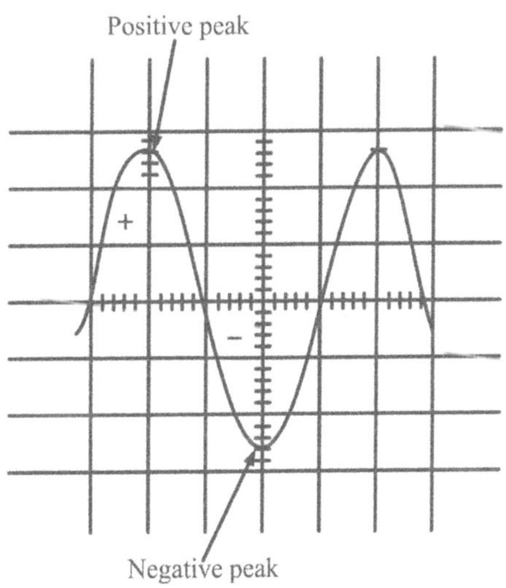

***Solution*:**

Vertical attenuation = 3 mv/div

From the figure, it is evident that the screen is divided such that one part is subdivided into 5 units.

$$1 \text{ sub division} = \frac{1}{5} = 0.2 \text{ units} \quad \text{(See Fig.)}$$

1. Peak-to-Peak value, $V_{P\text{-}P}$:

It can be observed that positive peak of the signal corresponds to two full divisions and three subdivisions

Hence Positive Peak is

$$2 + 3 \times 0.2 = 2.6 \text{ units}$$

While the negative peak also corresponds to 2.6 units.

$$\therefore \quad V_{p\text{-}p} = 2.6 + 2.6 = 5.2 \text{ divisions}$$

Also, $$V_{p\text{-}p} = \frac{\text{volts}}{\text{division}} \times \text{No. of division}$$

$$= 3 \times 5.2 = 15.6 \text{ mv}$$

2. Amplitude, V_{Max}:

$$V_{max} = \frac{V_{p-p}}{2} = \frac{15.6}{2} = 7.8 \text{ mv}$$

3. R.M.S Value, V_{rms}:

$$V_{rms} = \frac{V_{max}}{\sqrt{2}} = \frac{7.8}{\sqrt{2}} = 5.515 \text{ mv}$$

10.1.1 Lissajous Figures/Patterns

Lissajous patterns (named in the honour of French scientist who first obtained them geometrically and optically) are formed when two sine waves are applied simultaneously to the vertical and horizontal deflecting plates of CRO. The two since waves may be obtained from two audio oscillators. Obviously, in this case, a sine wave sweeps a sine-wave input signal.

The shape of the lissajous pattern depends on the frequency and phase relationship of the two sine waves.

Two sine waves of the same frequency and amplitude may produced a line, an ellips or a circle depending on their 'phase difference' as shown in Fig. 10.2.

In general, the shape of Lissajous figures depend on

1. Amplitude
2. Phase difference
3. Ratio of frequency of two waves.

10.1.2 Uses

The Lissajous figures are used for the following purposes:

1. To determine an unknown frequency by comparing it with a known frequency.
2. To check audio oscillator with a known frequency signal
3. To check audio amplifiers and feedback networks for phase shift.

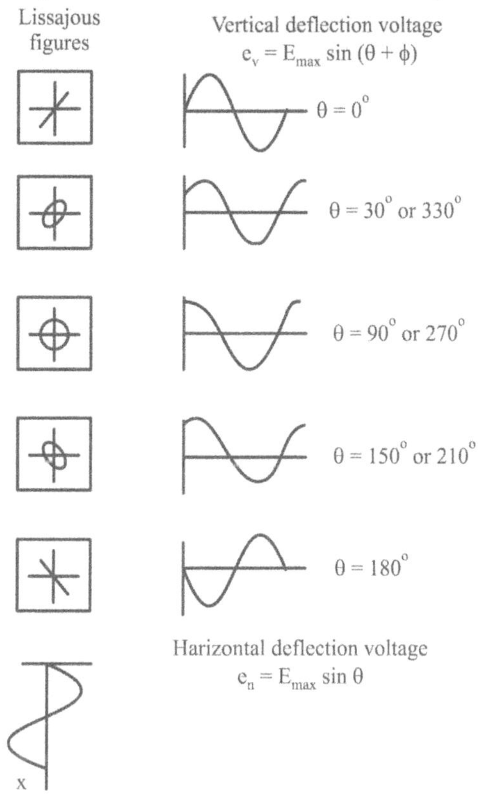

Fig. 10.2 Lissajous patterns for same frequency different phase shift

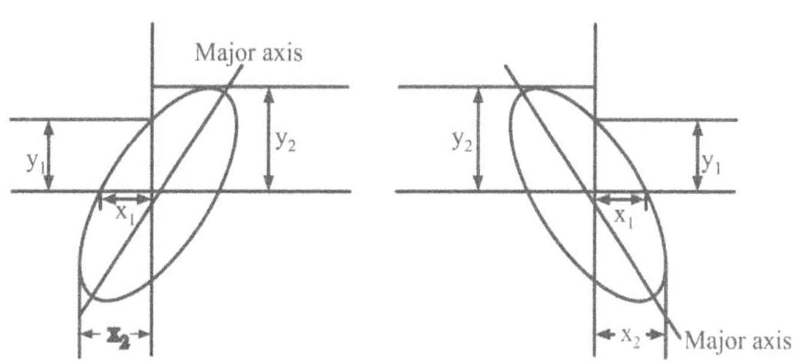

Fig. 10.3 Measurement of phase angles by using CRO

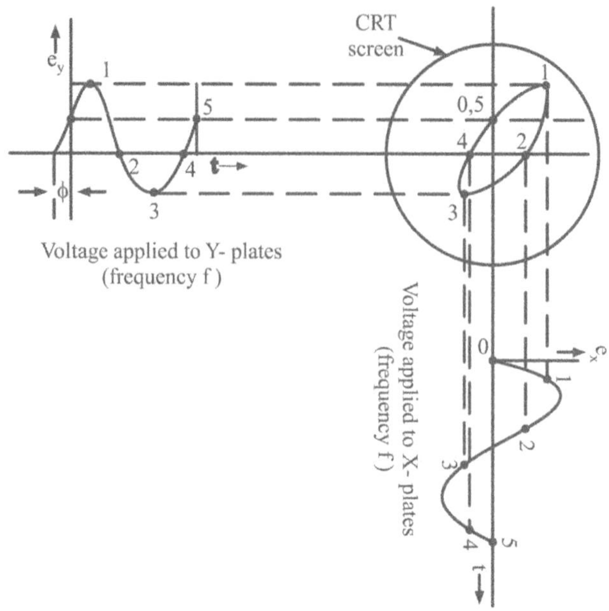

Fig. 10.4 Lissajous patterns with two equal voltage of same frequency and phase shift of ' ϕ '

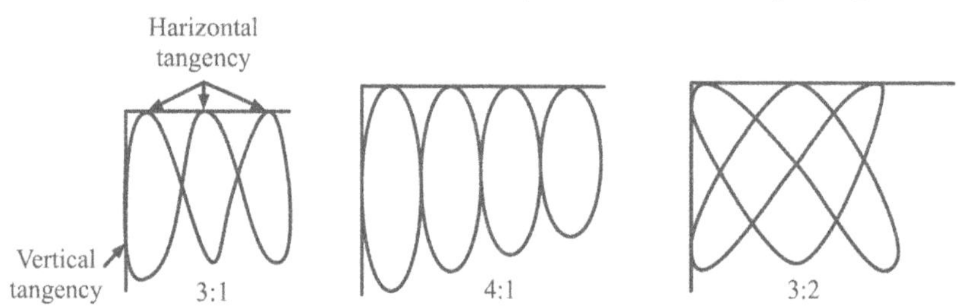

Fig. 10.5 Lissajous patterns with different frequency ratios

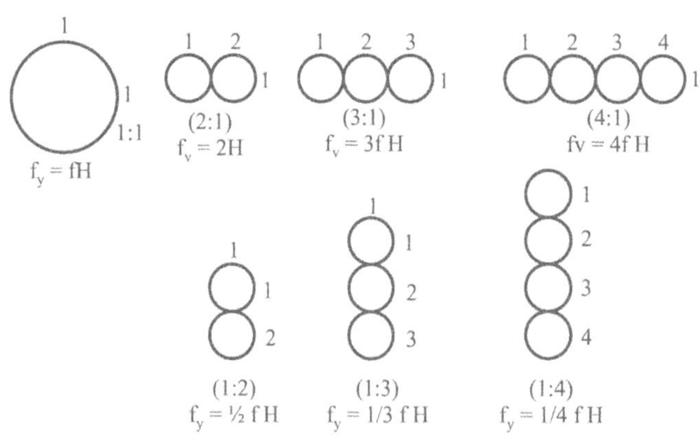

Fig. 10.6(a) For integral frequencies

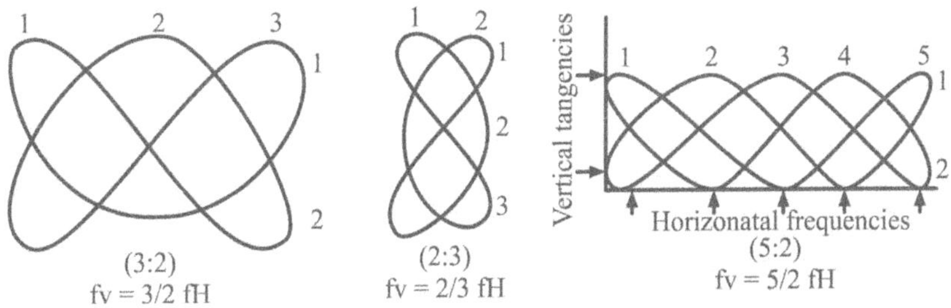

Fig. 10.6(b) For non-integral frequencies **Fig. 10.6(c)** Lissajous patterns

10.1.3 Phase-angle Measurement

An oscilloscope can be used to find the phase angle between the two sinusoidal quantities of the same frequency. One of the signals is fed to the Y - plates. The time-base generator is switched out and the second signal is fed to the X-plates. It is necessary that X and Y are of equal magnitudes. This fact can be checked by measuring the amplitudes separately (as in voltage measurement).

If the two signals are in phase, the display would be a straight line at 45° to the horizontal. If the phase angle is 90°, the display would be a circle. For any other phase difference, the display would be an ellipse.

If the phase difference is between 0° and 90^s or between 270° and 360°, the ellipse would have its major axis in the first and third quadrants. For difference between 90° and 180° or 180° and 270°, the major axis would be in the second and fourth quadrants.

Consider the Lissajous figure obtained on CRO with an unknown phase difference ϕ as shown in Fig 10.2.

The frequency and amplitude of the two wave's are same'

The parameters y_1, y_2, or x_1, x_2 can be measured [see fig 2(a)] the phase angle then can be obtained as,

$$\phi = \sin^{-1}\left(\frac{y_1}{y_2}\right) = \sin^{-1}\left(\frac{x_1}{x_2}\right)$$

If the pattern obtained is as shown in fig. (10.3) then the phase angle

$$\phi = 180^\circ - \sin^{-1}\left(\frac{y_1}{y_2}\right)$$

10.1.4 Frequency Measurement

The unknown frequency can be measured by the following methods:

1. Lissajous method
2. Spot wheel method
3. Gear wheel method

1. Lissajous Method: In this method of measurement a standard frequency is applied to one set of deflection plates (X-plates) of the CRT tube while the unknown frequency (of approximately the same amplitude) is simultaneously applied to the other set of plates (Y-plates). The resulting patterns depend on the integral and phase relationship between the two frequencies. The horizontal signal is designated as f_H and the vertical signal as f_v.

Typical Lissajous pattern are shown in Fig. (10.5) & (10.6) for sinusoidal frequencies which are equal, integral and in ratio.

10.1.5 Measurement Procedure

1. Setup the oscilloscope and switch off the internal sweep (change to ext).
2. Switch off sync. Control.
3. Connect the signal source as shown in fig. (10.7).
4. Set the horizontal and vertical gain control for the desired width and height of the pattern.
5. Keep frequencies f_v constant and vary frequency f_H noting that pattern spins in alternate directions and changes shape. The pattern stands still whenever f_v and f_H are in an integral ratio (either even or odd).
6. When $f_v = f_H$ pattern stands still and is a single circle or ellipse.
7. When $f_v = 2\ f_H$, a two loop horizontal pattern is obtained as shown in fig. (10.5).

For determining the frequency from any Lissajous Pattern, count the number of horizontal 100 Ps in the pattern, divide it by the number of vertical loops and multiply this quantity by f_H (known or standard frequency).

If the two frequencies being compared are not equal, but are fractionally related, a more complex stationary pattern results, whose form is dependent on the frequency ratio and the relative phase between the two signals.

The fractional relationship between the two frequencies is determined by counting the number of cycles in the vertical and horizontal.

$$f_v = (\text{Fraction}) \times f_H$$

or,

$$\frac{f_v}{f_H} = \frac{\text{No.of horizontal tangencies}}{\text{No.of vertical tangencies}}$$

Limitations

The Lissajous method of frequency determination has limitations and is being discarded gradually because Low-cost digital frequency counters are becoming increasingly available in market. Following are the two limitations of this method:

1. The numerator and denominator of the frequency ratio must be whole number.
2. The maximum ratio of frequencies that can be used is 10:1. Beyond that, the Lissajous patterns become too complex to analyse.

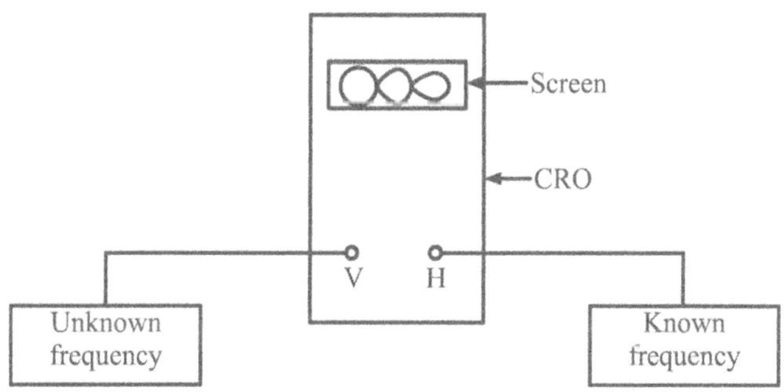

Fig. 10.7 Basic circuit for frequency measurement with Lissajous Patterns

10.1.6 Digital Storage Oscilloscope

The "digital storage oscilloscope" is a superior method of trace storage. In this technique the waveform to be stored is digitized, stored in a digital memory, and retrieved for display on the storage oscilloscope. The stored display can be displayed indefinitely as long as power is applied to the memory, which can be supplied with a small battery. Fig. 10.8 shows the block diagram of a digital storage oscilloscope.

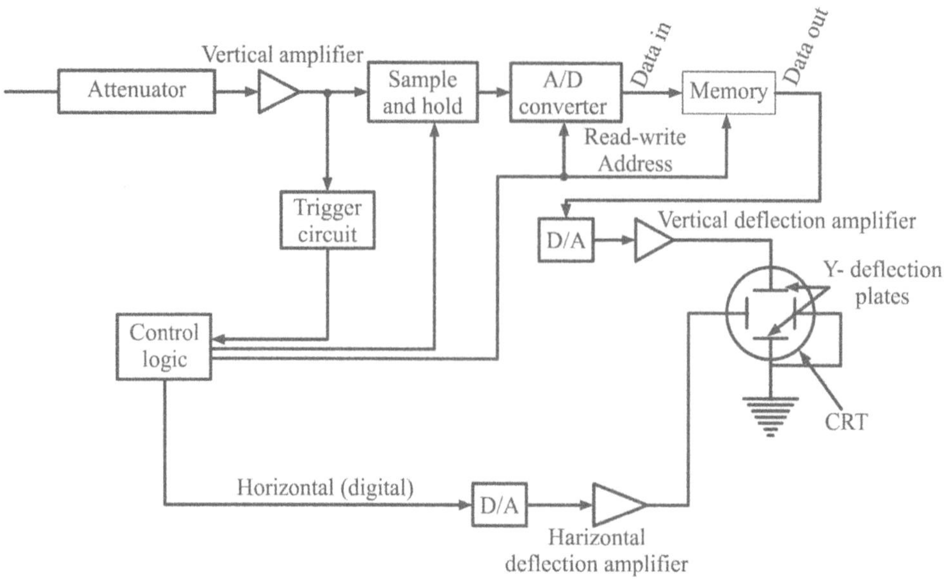

Fig. 10.8 Digital storage oscilloscope

The digital storage oscilloscope uses the same types of input circuitry and oscilloscope probes as conventional oscilloscope, and many digital storage oscilloscope can operate in a conventional mode, by passing the digitizing and storing features.

The output of the input signal amplifiers feeds an A/D (Analog-to-Digital) converter. The main requirement of the A/D converter is its speed, while accuracy and resolution are of secondary independence.

The digitized output is in the uinary form and not BCD form. Generally, successive approximation type of ADC is used in digital storage oscilloscope. If a 12 bit converter is used then accuracy of about 0.1% and resolution of about 0.025% is obtained, while if a 10 bit ADC is used then the resolution of about 0.1% is obtained.

Digitizing the analog signal means to take samples of the input signal at periodic intervals of time. The sampling rate should be greater than twice the highest frequency present in the input signal, so that the sampling theory is satisfied. This ensures that aliasing effect will not occur and there is no loss of information. The sampling rate may be as high as 100,000 samples per second. For such high sampling rates a fast conversion A/D convertor is required. 'Flash' A/D converters are used at the cost of decreased resolution with increase in sampling rate.

The size of the memory is related to the an own of horizontal segments of the trace that can be divided into one sweep of the time base, e.g, if mean by size is 1 k then the trace will be divided into 1000 horizontal segments. If a 10 bit ADC is used then the frequency response of 25 kHz is obtained; the total memory storage capacity is 4096 for a single channel, 2048 for two channels, 1024 for four channels each.

The selection of the sampling rate and memory size depends on the type of waveform being recorded. As the analog signal is sampled at suitable rates, the data points are readable on the memory. Once the signal is stored, the required manipulations can be done.

A continuous storage oscilloscope consist of a feature called "Pretrigger View". This indicates that the events that occurred before the trigger input signal was applied, are displayed. This selection is a percentage selection.

This mode is useful when failure occurs.

'Single shot events' one sometimes transient in nature and hence they are quickly lost. Unless the waveform is stored the observer cannot see such events. Such events are stored in the memory of the digital storage oscilloscope and memory can be rapidly read on the screen; a continuous waveform can be displayed.

The digital storage oscilloscope has the following three modes of operation:

1. Roll Mode: This is used to observe the fast varying signal.
2. Store Mode: It is called as "refresh" mode and is most commonly used.
3. Hold or Save Mode: This mode is used to lock the previously saved signal.

Advantages

1. Easy to operate
2. Infinite storage time
3. Cursor measurement is possible
4. Signal processing is possible
5. It is capable of displaying X- Y plots, p-v diagrams and B-H curve.

10.1.7 Applications

1. It can be used to measure:
    - AC as well as DC, voltage and currents,
    - Frequency, time period, time interval between two signal etc,
    - Inductance, capacitance.

2. It is used to give the visual representation for a target of radar such as aeroplane, ship etc.
3. In medical fields, it is used to display cardiograms that are useful for diagnosis of heart of the patient.
4. It is used to observe the radiation pattern generated by the transmitting antenna.
5. It can be used to check the faulty components in various circuits. It can also be used to analyse TV waveforms

Review Questions

1. With a neat block diagram explain the working of a digital storage oscilloscope
2. Explain the significance of lissajous pattern
3. Explain the panel details of a dual trace oscilloscope
4. Explain the Applications of Digital storage oscilloscope
5. Explain with the help of block diagram of dual trace oscilloscope
6. Explain the working of digital storage oscilloscope

Quiz Questions

1. The most useful transducer for displacement sensing with excellent sensitivity, linearity and resolution is?

   (a) an incremental encoder (b) an absolute encoder

   (c) LVDT (d) a strain gauge

2. One of the following, which is not a transducer in the true sense, is?

   (a) thermocouple (b) piezoelectric pick up

   (c) photo-Voltaic cell (d) LCD

3. In a digital storage oscilloscope, the input signals is

   (a) directly applied to the oscilloscope

   (b) multiplexed, converted to digital form and stored and applied to oscilloscope

   (c) multiplexed, converted to digital form and stored, converted to analog form and applied to oscilloscope

   (d) applied to amplifier, stored as analog signals, Multiplexed, converted to digital form, stored in digital form, converted to analog form and applied to CRO through an amplifier.

4. The pattern used to measure phase and frequency with a cathode ray oscilloscope we called.
   (a) lissajous patterns (b) faraday's patterns
   (c) ohm's Patterns (d) philip's patterns
5. Lissajous patterns appearing on the screen of a CRT, when two sinusoidal voltages of equal frequencies which are phase each others are applied to horizontal and vertical plate is
   (a) a straight line (b) a circle
   (c) an ellipse (d) a parabola
6. When two sinusoidal waveforms of same frequency are applied to CRO to form a Lissajous pattern, a circle is formed, when the two signals have phase difference between them. (b)
   (a) either 0° or 180° (b) 90° 270°
   (c) 0° 120° (d) 150° 330°
7. The Lissajous pattern on a CRO screen is shown in Fig.
   (a) 3 : 2 (b) 5 : 1
   (c) 1 : 5 (d) 2 : 3
8. A dual trace oscilloscope uses
   (a) 1beam (b) 2 beams
   (c) 3 (d) 4
9. A double beam oscilloscope has
   (a) 2 electron guns (b) two screens
   (c) 2 different phosphor coatings (d) one waveform divide into 2 parts
10. In a dual beam oscilloscope
    (a) there are 2 separate vertical inputs and 2 separate horizontal inputs
    (b) there are 2 separate vertical inputs and 1 separate horizontal inputs
    (c) there are 1 separate vertical inputs and 2 separate horizontal inputs
    (d) there are 1 separate vertical inputs and 1 separate horizontal inputs
11. Which of the following is not correct for a storage type oscilloscope
    (a) the storage target is a conductive mesh covered with magnesium flouride
    (b) secondary emission $\overline{e}$ s etch a positively charged pattern
    (c) the flood gun used for display, emit high velocity electrons
    (d) the flood guns are placed between the deflection plates and storage target

Questions and Answers

1. Determine the frequency of this waveform, as displayed by an oscilloscope with a vertical sensitivity of 2 volts per division and a timebase of 0.5 milliseconds per division:

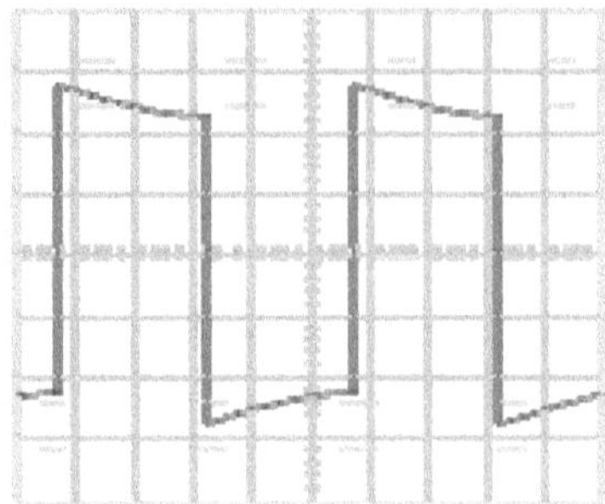

Ans: 400 Hz

2. What is a dual trace oscilloscope?

Ans. Dual trace oscilloscope employs a single electron gun and provides two traces by switching the deflection plates from one input to another which implies that Y-channel is time shared by two signals. The display of the input is continuous and simultaneous to the eyes although it is sampled display.

3. What is the difference between dual trace and dual beam CRO?

Ans. Dual beam CRO uses two electron beams and deflection systems mounted in the same CRI to display two different electron beams simultaneously where as dual trace oscilloscope alternatively displays two signals on one screen by time sharing one electron gun and deflection system.

4. How is dual trace oscilloscope superior to dual beam oscilloscope?

Ans. Dual beam CROs are not as versatile as dual trace oscilloscope because the later can operate in a number of model.

5. What is the difference between Analog storage oscilloscope and digital storage oscilloscope?

Ans. Digital storage oscilloscope uses digital storage techniques whereas analog storage oscilloscope uses special CRT called storage tube.

6. What are the limitations of oscilloscope?

Ans. Loading effect, Hum and noise pick up, oscilloscope errors, Parabola errors, calibration error, frequency response error, loading error.

7. What is a Lissajous pattern?

Ans. It is a pattern that results from applying periodic signals to the deflection plates of a CRO.

Quiz Questions Key

1. (c)	2. (d)	3. (d)	4. (a)	5. (a)
6. (b)	7. (b)	8. (b)	9. (b)	10. (b)
11. (c)				

CHAPTER 11

Transducers

11.1 Introduction

An electronics instrumention system consist of a number of components to perform a measurement and record its results. As explained a generalized measurements system consist of three major components.

1. An input device
2. An output device
3. A signal conditioning or processing device.

The input device receives the measured or the quantity under measurement and delivers a proportional or analogous electrical signal to the signal conditioning device; here the signal is amplified, attenuated fitered, modulated, or otherwise modified in format acceptable to the output device.

The input quantity for most instrumentation system is a "Non-electrical Quantity". In order to use electrical methods and techniques for measurement manipulation or control, the Non-electrical Quantity is generally converted into an electrical form by a device called a "Transducer", we can define a Transducer as a device which, when actuated transforms energy from one form to another.

The broad definition of transducer includes, for example, devices which convert mechanical force into an electrical signal.

The first stage of a measurement system may simply be called a transducer stage instead of detector transducer stage by redefining a transducer.

'A Transducer, in general form, may be defined as a device which converts energy from one form to another, however, this definition has to be restricted, many a time especially in the field of electrical instrumentation. Keeping this restriction in view, a 'Transducer' may be defined as a device which converts a physical quantity or a physical condition into an electrical signal. Another name for a Transducer is 'Pick Up'.

11.2 Classification of Transducers

The transducers can be classified

1. On the basis of Transduction form used.
2. As primary and secondary Transducers,
3. As passive and active transducers,
4. As Analog and digital transducers,
5. As transducers and Inverse Transducers,

Classification based upon principle of Transduction

The transducer can be classified on the basis of principle of Transduction as resistive, inductive, capacity etc. Depending upon how they convert the input quantity into resistance, inductance or capacitance respectively. They can be classified as piezoelectric, thermoelectric, magnet orestrictive, electro kinetic and optical.

Primary and Secondary Transducers

Let us consider the case of Bourdon's tube as shown in Fig. 11.1. The Baurdon tube acting as a primary detector senses the pressure and converts the pressure into a displacement of its free end. The displacement of the free end moves the core of a linear variable differential transformer (L.V.D.T) which produces an output voltage which is proportional to the movement of the core, which is proportional to the displacement of the free end which in turn is proportional to the pressure thus there are two stages of Transduction.

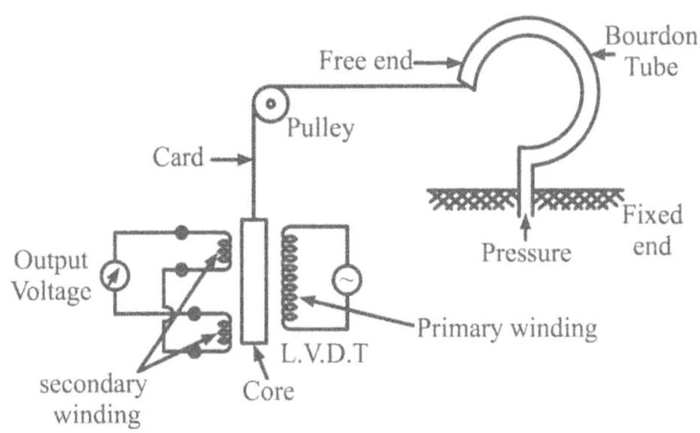

Fig. 11.1 Measurement of pressure using bardon to be and L.V.D.T (example of primary and secondary transducer)

Firstly the pressure is converted into a displacement by Baurdon tube then the displacement is converted into an analogous voltage by L.V.D.T. The Bourdon tube is called a "Primary Transducer" while the L.V.D.T is called a "Secondary Transducer".

Passive and Active Transducers

Passive transducers derive the power required for Transduction from an auxiliary power source. They also derive part of the power required for conversion from the physical quantity under measurement. They are also knows as "Externally powered transducer" Typical examples of passive transducers are resistive, inductive and capacitive transducers

Active transducers are those which do not require an auxiliary power source to produce their output. They are also known as 'self generating type' since they develop their own voltage or current output. The energy required for production of output signal is obtained from the physical quantity being measured velocity, temperature, light intensity and force can be transduced with the help of active Transducers.

Analog and Digital Transducers

The transducers can be classified on the basis of the output which may be continuous function of time or the output may be in discrete steps

Digital Transducers

These transducers convert the input quantity into an electrical output which is in form of pulses.

Analog Transducer

These transducers convert the input quantity into an analog output which is a continuous function of time. Thus a strain gauge an L.V.D.T., a thermocouple or a thermistor may be called as 'Analog Transducers" as they give an output which is a continuous function of time.

Transducers and Inverse Transducers

Transducers

A Transducer can be broadly defined as a device which converts a non-electrical quantity into an electrical quantity.

Inverse Transducers

An 'Inverse Transducer' is defined as a device which converts an electrical quantity into a non-electrical quantity. It is a precision actuator which has an electrical input and a low power non-electrical output. A piezoelectric crystal acts as an inverse transducer because when a voltage is applied across its surfaces. Its change in its dimensions causing a mechanical displacement.

11.3 Selection of Transducers

The most transducers are not sensitive to just one quantity unfortunately. If measurements are to be made under conditions where there is likehood of two or more input quantities influencing the transducer, it is desirable to select a transducer which is sensitive to the desirable quantity and insensitive to the un wanted input quantity. The following is the summary of the factors influencing the selection of a transducer for measurement of a physical quantity

1. **Operating Principle:** The transducers are many a times selected on the basis of operating principle used by them. The operating principles used may be resistive, inductive, capacitive, optoelectronic, piezo electronic etc,
2. **Accuracy:** High degree of accuracy is assured if the transducer does not require frequent calibration and has a small value for repeatability. It may be emphasised that in most industrial applications, repeatability is of considerably more importance than absolute accuracy.
3. **Sensitivity:** The transducer must be sensitive enough to produce detectable output.
4. **Operating Range:** The transducer should maintain the range requirements and have a good resolution over its entire range. The rating of the transducer should be sufficient so that it does not breakdown while working in its specified operating range.
5. **Errors:** The transducer should maintain the expected input-out relationship as described by its transfer functions so as to allied errors.
6. **Loading Effects:** The transducer should have a high input impendence and a low output impendence to avoid loading effects.
7. **Usage and Ruggedness:** The ruggedness both of mechanical and electrical intensities of transducer versus its size and weight must be considered while selecting a suitable transducer.
8. **Insensitivity to Unwanted Signals:** The transducer should be minimally sensitive to unwanted signals and highly sensitive to desired signals.
9. **Static Characteristics:** Apart from low static error, the transducers should have a low non-linearity, low hysteresis high resolution and a high degree of repeatability. The transducer selected should be free from load alignment effects and temperature effects. It should not need frequent calibration, should not have any component limitations and should be preferably small in size.
10. **Electrical Aspects:** The electrical aspects that need consideration while selecting a transducer include the length and type of cable required. Attention also must be paid to signal to noise ratio in case the transducer is to be used in conjunction with amplifiers. Frequency response limitation must also be taken into account.

11. **Transient and Frequency Response:** The transducer should meet the desired time domain specifications like peak overshoot, rise time, settling time and small dynamic error. It should ideally have a flat frequency response curve. In practice, however there will be cut off frequencies and higher cut off frequency should be high in order to have a wide band width.

12. **Stability and Reliabity:** The transducer should exhibit a high degree of stability to be operative during its operation and storage life. Reliability should be assured in case of failure of transducer in order that the functioning of the instrumentation system continues uninterrupted.

13. **Environmental Compatibility:** It should be assured that the transducer selected to work under specified environmental conditions, maintains its input-output relationship and does not break down. For example, the transducer should remain operable under its temperature range. It should be able to work in corrosive environments (if the application so requires), should be able to withstand pressure and shocks and other interactions to which it is subjected to.

11.4 Strain Gauges

If a metal conductor is stretched or compressed, its resistance changes on account of the fact both length and diameter of conductor change, also there is a change in the value of resistivity of the conductor when it is strained and this property is called "piezoresistive effect". Therefore, resistance strain gauges are also know as "Piezoresitive gauges". The strain gauges are used for measurement of strain and associated stress in experimental stress analysis. Secondly, many, diaphragm type pressure gauges, temperature sensors, accelerometers and flow meters, employ strain gauges as secondary transducers.

11.5 Theory of Strain Gauges

The change in the value of resistance by straining the gauge may be partly explained by the normal dimensional behaviour of elastic material if a strip of elastic material is subjected to tension. As shown in fig. (11.2) or in other words positively strained, its longitudinal dimension will increase while there will be a reduction in the lateral dimension. So when a gauge is subjected to a positive strain, its length increases while its area of cross-section decreases as show in fig. (11.2). Since the resistance of a conductor is proportional to its length and inversely proportional to its area of cross-section, the resistance of the guage increases with positive strain. The change in the value resistance of strained conductor is more than what can be accounted for an increase in resistance due to dimensional changes; the extra change in the value of resistance is attributed to a change in the value of resistivity of a conductor when strained. This property, as described earlier, is know as "Peizoresitive effect".

Let us consider a strain guage made of circular wire; the wire has the dimensions:

Length = L, area = A, diameter = D before being strained

The material of the wire has a resistivity l.

$\therefore$ Resistance of unstrained gauge $R = lL/A$

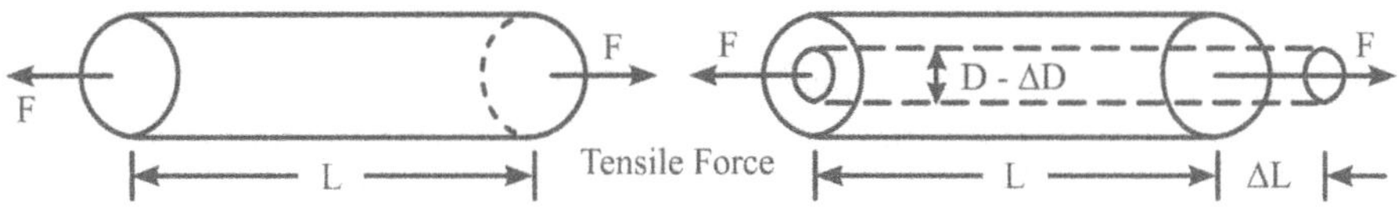

Fig. 11.2 Change in dimensions of a strain gauge element when subjected to a tensile force

Let a tensile stress 'S' be applied to the wire this produces a positive strain causing the length to increase and area to decrease as shown in above Fig (11.2). Thus when the wire is strained there are changes in its dimension

Let

ΔL = change in length; ΔD = change in diameter

ΔA = change in area; ΔR = change in resistance

In order to find how ΔR depends upon the material physical quantities, the expression for R is differentiated with respect to stress 'S'.

Thus we get:

$$\frac{dR}{ds} = \frac{p}{A}\frac{\partial L}{\partial S} - \frac{pL}{A^2}\frac{\partial A}{\partial S} + \frac{L}{A}\frac{\partial P}{ds} \quad .....(11.1)$$

Dividing eqn (11.1) throughout by resistance $R = \partial L/A$, we have

$$\frac{1}{R}\frac{dR}{dS} = \frac{1}{L}\frac{\partial L}{\partial S} - \frac{1}{A}\frac{\partial A}{\partial S} + \frac{1}{l}\frac{\partial P}{\partial S} \quad .....(11.2)$$

It is from eqn (11.2) that per unit change in resistance is due to

(i) Per unit in length $= \frac{\Delta L}{L}$

(ii) Per unit change in area $= \frac{\Delta A}{A}$

(iii) Per unit change in resistivity $= \frac{\Delta e}{e}$

Area $A = \frac{\pi}{4}D^2$ $\quad \therefore \frac{\partial A}{\partial S} = 2.\frac{\pi}{4}D.\frac{\partial D}{ds}$ .....(11.3)

$$\frac{1}{A}\frac{dA}{dS} = \frac{\left(\frac{2\pi}{4}\right)D}{\left(\frac{\pi}{4}\right)D^2}\frac{\partial D}{\partial S} = \frac{2}{D}\frac{\partial D}{\partial S} \quad .....(11.4)$$

∴ eqn (11.2) can be written as:

$$\frac{1}{R}\frac{dR}{dS} = \frac{1}{L}\frac{\partial L}{\partial S} - \frac{2}{D}\frac{\partial D}{\partial S} + \frac{1}{P}\frac{\partial P}{\partial S} \quad .....(11.5)$$

Now, Poissson's ratio $V = \frac{\text{lateral strain}}{\text{longitudinal strain}} = -\frac{\partial D / D}{\partial L / L}$ .....(11.6)

$$\frac{\partial D}{D} = -V \times \frac{\partial L}{L}$$

$$\therefore \quad \frac{1}{R}\frac{dR}{dS} = \frac{1}{L}\frac{\partial L}{\partial S} + V\frac{2}{L}\frac{\partial L}{\partial S} + \frac{1}{e}\frac{\partial P}{\partial S} \quad .....(11.7)$$

For small variations, the above relationship can be written as'

$$\frac{\Delta R}{R} = \frac{\Delta L}{L} + 2v\frac{\Delta L}{L} + \frac{\Delta P}{P} \quad .....(11.8)$$

The gauge factor is defined as the ratio of per unit change in resistance to per unit change in length gauge factor

$$G_f = \frac{\Delta R / R}{\Delta L / L} \quad .....(11.9)$$

$$\frac{\Delta R}{R} = G_f \frac{\Delta L}{L} = G_f \times \in \qquad \text{where } \in = \text{strain} = \frac{\Delta L}{L} \quad ......(11.10)$$

The gauge factor can be written as:

$$= 1 + 2v + \frac{\Delta P / P}{\in} \quad ......(11.11)$$

$$= 1 + 2v + \frac{\Delta e / e}{\in}$$

[Resistance change due to change of length]

[Resistance change due to change in area]

[Resistance change due to piezoresistive effect]

$$G_f = \frac{\Delta R / R}{\Delta L / L} = 1 + 2v + \frac{\Delta e / e}{\Delta L / L}$$

∴ The strain is usually expressed in terms of microstrain

I microstrain = 1 μm/m

If the change in the value of resistivity of material when strained is neglected the gauge factor is:

$$G_f = 1 + 2v \qquad .....(11.12)$$

Eqn (11.12) is valid only when "Piezoresistive effect" in resistivity due to strain is almost negligible.

***Example*:** A strain gauge is bonded to a beam 0.1 m long and has a cross-sectional are 4 cm^2 Young's modulus for steel is 207 GN/m^2. The strain gauge has an unstrained resistance of 2402 and a gauge factor of 2.2. When a load is applied, the resistance of gauge changes by 0.013 Ω. Calculate the change in length of the steel beam and the amount of force applied to the beam.

***Solution*:** we have gauge factor $G_f = \dfrac{\Delta R / R}{\Delta L / L}$

and therefore change in length $\Delta L = \dfrac{\Delta R / R.L}{G_f} = \dfrac{(0.013 / 2402)}{2.2} \times 0.1$

$$= 2.46 \times 10^{-7} \text{ m}$$

Stress $\qquad S = E \in = \dfrac{207 \times 10^9 \times 2.46 \times 10^{-7}}{0.1}$

$$= 5.092 \times 10^5 \text{ N/m}^2.$$

Force $\qquad F = SA = 5.092 \times 10^6 \times 4 \times 10^{-4}$

$$= 2.037 \times 10^3 \text{ N}.$$

11.6 Linear-Variable-Differential Transformer (LVDT)

LVDT is a passive inductive transducer and is commonly employed to measure force (or weigth, pressure and acceleration etc which depend on force) in terms of the amount and direction of displacement of an object.

Construction: It consists of one primary winding (P) and two secondary windings (S_1 and S_2) which are placed on either side of the primary mounted on the same magnetic core. The magnetic core is free to move axially inside the coil assembly and the motion being measured is mechanically coupled to it.

The two secondaries S_1 and S_2 have equal number of turns but are connected in series opposition so that e.m.f_s, (E_1 and E_2) induced in them are 180° out of phase with each other and

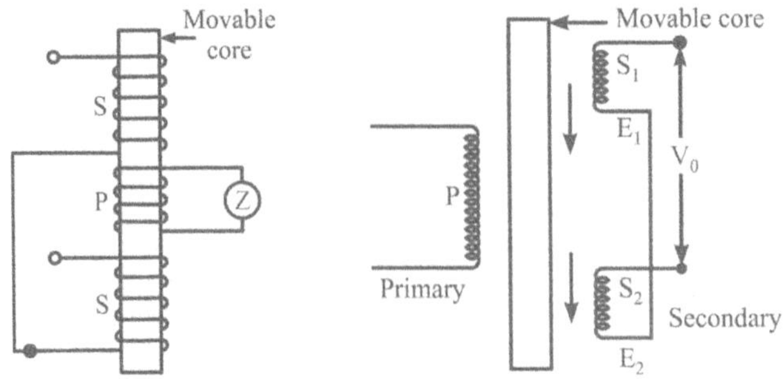

Fig. 11.3 Linear-variable - differential transducer (LVDT)

hence, cancel each other out [see Fig. (11.3)]

The primary is energised from a suitable A.C source

Working

When the core is in the centre (called reference position) the induced voltages E_1 and E_2 are equal and opposite. Hence they cancel out and the output voltage V_0 is zero. When the external applied force moves the core towards coil S_2, E_2 is increased but E_2 is decreased in magnitude though they are still antiphase with each other the net voltage available is $(E_2 - E_1)$ and is in phase with E_2.

Similarly when the movable core moves towards coil S_1, $E_1 > E_2$ and $V_0 = E_1 - E_2$ and is in phase with E_1.

Thus, from above discussion, we find that the magnitude of V_0 is a function of the distance moved by the core and its polarity or phase indicates as to in which direction it has moved.

If core is attached to moving object, the magnitude of V_0 gives the position of that object. Fig. (11.4) shows the pressure measurement by LVDT.

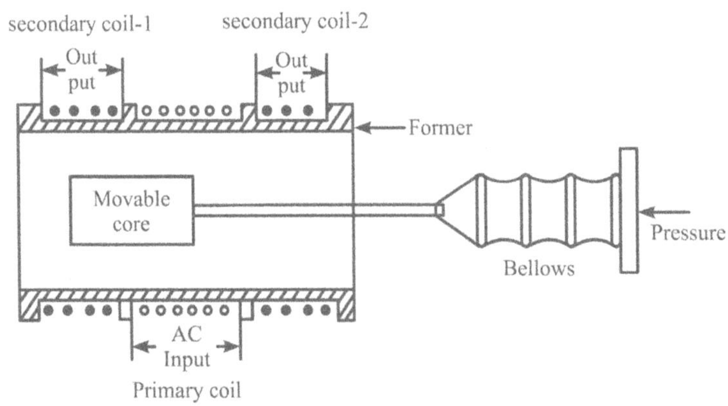

Fig. 11.4 Pressure measurement by LVDT

Advantages

1. It gives a high output and therefore many a times there is no need for intermediate amplification devices
2. The transducer possesses a high sensivity as high as 40 v/mm.
3. It shows a low hysteresis and hence repeaterbility is excellent under all conditions
4. Most of the LVDTs consume a power of less than 1 W.
5. Less friction and less noise (due to absence of sliding contancts)
6. These transducers can usually tolerate a higher degree of shock and vibration without any adverse effects.
7. It can operate over a temperature range from –265°C to 600°C
8. It is available in radiation-resistant design for operation in nuclear reactors.

Disadvantages

1. These transducers are sensitive to stray magnetic fields but shielding is possible. This is done by providing magnetic shields with longitudinal slots.
2. Relatively large displacements are required for appreciable differential output.
3. The receiving instrument must be selected to operate on A.C. signals or demodulator network must be used if a D.C output is required.
4. Several times, the transducer performance is affected by vibrations.
5. The dynamic response is limited mechanically by the mass of core and electrically by the frequency of applied voltage. The frequency of the carrier should be at least ten times the highest frequency component to be measured.

11.7 Applications

1. Measurement of material thickness in not strip or slab steel mills.
2. In accelerometers
3. Jet engine controls in close proximity to exhaust gases.

11.8 Temperature Transducers

Optical fibers can be used for measurement of temperature. This is because the temperature induces changes in refractive index of the fiber. The core and claddings of the fibres have different indices of refraction because of the difference in their-composition. The refractive index is a function of both composition and temperature, and it changes at different rates in response to a temperature change.

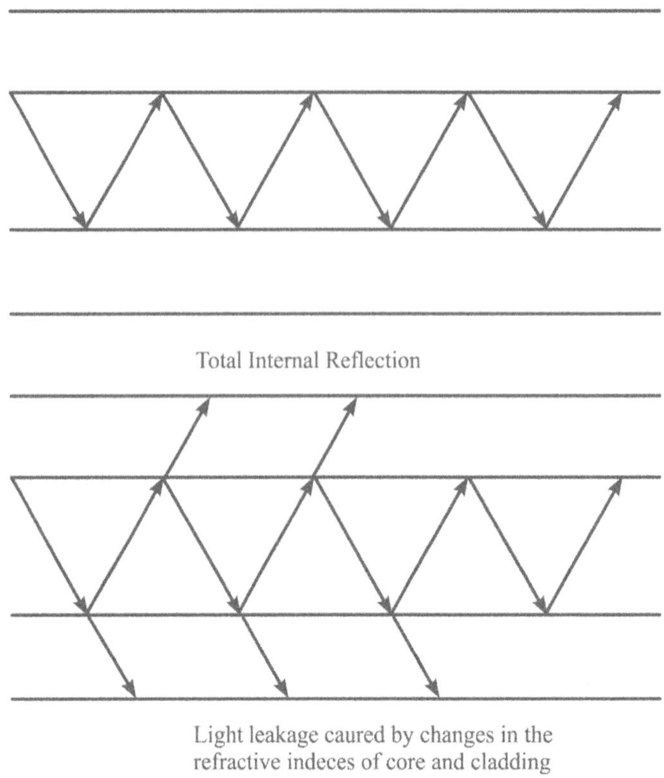

Fig.11.5 Fiber optic temperature transducer

A higher temperature change causes the critical angle of the fiber to change slightly. Which in turn changes the amount of light lost through leakage in the cladding as shown in Fig. (11.5). Total internal reflection occurs at angles of incidence greater than the critical angle. At angles of incidence less than the critical angle, some light is lost due to refraction.

11.9 Photoconductive Cells

Electrical conduction in semiconductor materials occurs when free charge carriers, e.g., electrons, are available in the material when an electric field is applied. In certain semi conductors, light energy falling on them is of the correct order of magnitude to release charge carriers which increase flow of current produced by an applied voltage. The increase of current with increase in light intensity with the applied voltage remaining constant means that the resistance of semi conductors decreases with increase in light intensity. Therefore, these semiconductors are called "Photo conductive cells" or photo resistors or sometimes "Light Dependent Resistors, (LDR), since incident light effectively varies their resistance.

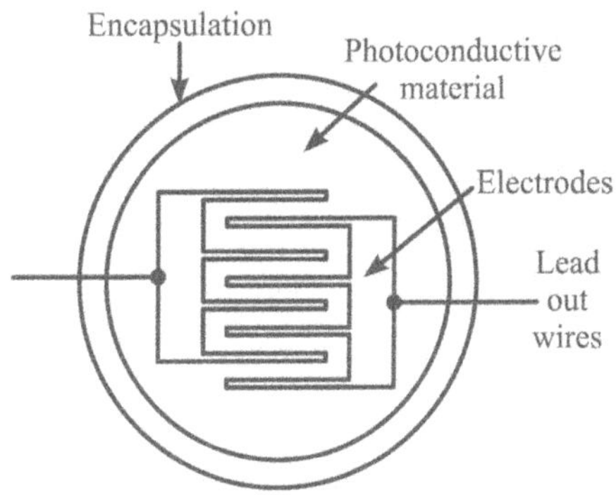

Fig. 11.6 Photoconductive cell

The two most commonly used photo conductive semi conductor materials are cadmium sulphide (Eds) with a bond gap of 2.42 ev and cadmium selenide (cd se) with a band gap of 1.74 ev. The photo conductive cell use a special type of construction which minimises resistance cell use of special type of construction which minimises resistance while providing maximum surface. Photo conductive cell are made by chemically sintering the required powder (cd S) or (cd se) into tablets of the required shape, and enclosing them in a protective envelope of glass or plastic. Electrodes are deposited on the tablet surface and are made of materials which give an ohmic contact, but with low resistance compared with that of the photoconductor gold is typically used.

The electrodes are usually inter-digital i.e., in the form of inter locked fingers or combs shown in above Fig. 11.6.

11.10 Photovoltaic Cell

This is an important class of photo-detectors. They generate a voltage which is proportional to EM radiation intensity. They are called photo votaic cell because of their voltage generating characteristics. They, in fact, convert the EM energy into electrical energy. They are passive transducers i.e., they do not need an external source to power them.

A photovoltaic cell is shown above Fig. (11.7). The cell is a giant diode, constructing a PN junction between appropriately doped semiconductors. Photons striking the cell pass through the thin P-doped upper layer and are absorbed by electrons in the lower N-layer, causing formation of conduction electrons and holes. The depletion zone potential of the PN junction then separate there conduction holes and electrons causing a difference of potential to develop across the junction.

The open circuit voltage is given by

$$E_0 = E_c \log_e E_e V \qquad ....(11.13)$$

where E_c = calibration voltage; V,

E_e = radiant incidence; w/m^2

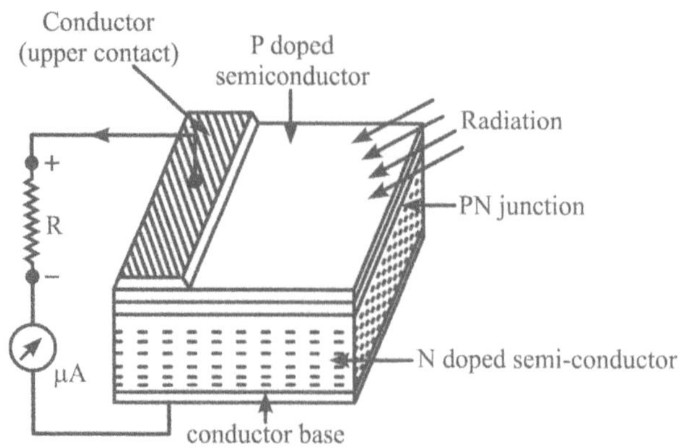

Fig. 11.7 Photovoltaic cell

The photo voltaic cells can operate satisfactorily in the temperature range of –100 to 125°C. The temperature changes have little effect on short circuit current but affect the open circuit voltage considerably. These changes may be of the order of a few mv/°C in output voltage. Table 11.1 below gives the list of some typical materials used for photo voltaic cells and their characteristics.

Table 11.1 Typical photo voltaic cell characteristics

Cell material	Time constant	Spectral Band
Silicon (si)	20 μs	0.44 μm – 1 μm
Selenium (se)	2 ms	0.3 mm – 0.62 μm
Germanium (Ge)	50 μs	0.79 μm – 1.8 μm
Indium Arsenide (InAs)	1 μs	1.5 μm – 3.6 μm (colled)
Indium Antimonide (Insb)	10 μs	2.3 μm – 7 μm (colled)

The tremendous advantage of the photovoltaic cell, as its name, supplies are its ability to generate a voltage without any form of bias and its extremely fast response. This means that it can be used as an energy converter directly. One use of this is in photo graphic exposure meters which require no battery. The cell voltage operates the meter directly, or may be directly linked to the iris to control the aperture.

Review Questions

1. Explain LVDT?
2. What are the various classification of gauges?
3. Explain working of RTD.
4. What are strain gauges?

5. Explain the Electrical transducers and selecting a transducer?
6. Explain the Resistive transducer and Resistive position transducer?
7. Explain the Thermistor Resistance thermometer?
8. Explain the Inductive transducer?
9. Explain the Differential output transducers and LVDT?
10. With a neat sketch explain the unbonded strain gauge
11. Briefly explain the working of LVDT used in displacement measurements. Why is a phase sensitive detector employed along with the LVDT
12. What is a transducer? Briefly explain the procedure for selecting a transducer?
13. Briefly explain photoconductive and photovoltaic cells?
14. What is the principle of electric resistance strain gauge? Derive an expression for the gauge factor in terms of the Poisson's ratio
15. Explain the principle of displacement measurements using two differential transformers in a closed loop servo system
16. Explain the principle of operation of LVDT in translating a linear motion into an electrical signal
17. What are the advantages and disadvantages of LVDT?
18. Explain the classification of electrical transducers
19. What is the principle of electric resistance strain gauge? Explain the unbounded resistance wire strain
20. Explain the operation of a LVDT and anyone application of it
21. Explain the advantage of electric transducer. Also describe the classification of transducer
22. Differentiate between a sensor and a transducer? What are the factors affecting the choice of transducers?
23. Explain the principle and working of LVDT
24. Explain the classification of transducers with the help of examples
25. Derive the expression for gauge factor for a strain gauge
26. Explain the classification of transducers with the help of examples
27. What is a transducer? Briefly explain the photoconductive and photovoltaic cells

Quiz Questions

1. The most useful transducer for displacement sensing with excellent sensitivity, linearity and resolution is?

   (a) an incremental encoder (b) an absolute encoder
   (c) LVDT (d) a strain gauge

2. One of the following, which is not a transducer in the true sense, is?

   (a) thermocouple (b) piezoelectric pick up
   (c) photo-Voltaic cell (d) LCD

3. In a digital storage oscilloscope, the input signals is

   (a) directly applied to the oscilloscope
   (b) multiplexed, converted to digital form and stored and applied to oscilloscope
   (c) multiplexed, converted to digital form and stored, converted to analog form and applied to oscilloscope
   (d) applied to amplifier, stored as analog signals, Multiplexed, converted to digital form, stored in digital form, converted to analog form and applied to CRO through an amplifier.

4. The transducers that converts the input signal into the output signal, which is a discrete function of time is known as ___________ transducer.

   (a) digital (b) analog
   (c) active (d) pulse

5. A transducers that converts measured into the form of pulse is called

   (a) active transducer (b) analog transducer
   (c) digital transducer (d) pulse transducer

6. Which of the following is a digital transducers (b)

   (a) strain gauge (b) encoder
   (c) thermistor (d) LVDT

7. Strain gauge, LVDT and thermocouple are examples of

   (a) active transducers (b) passive transducers
   (c) analog transducers (d) primary transducers

8. An inverse transducer is an device which converts

   (a) an electrical energy into Non electrical energy
   (b) electrical quantity into mechanical quantity
   (c) electrical energy into thermal energy
   (d) an electrical quantity into a non electrical quantity

9. A strain gauge is a passive transducer and is employed for converting
   (a) mechanical displacement into a change of resistance
   (b) presser into a change of resistance
   (c) force into a displacement
   (d) presser into a displacement
10. Resolution of a transducer depends on:
   (a) material of wire
   (b) length of wire
   (c) diameter of wire
   (d) excitation voltage
11. The sensitivity factor of strain gauge is normally of the order of:
   (a) 1 to 1.5
   (b) 5 to 10
   (c) 0.5 to 1.0
   (d) 1.5 to 2.0
12. Bonded wire strain gauges are
   (a) exclusively used for construction of transducer
   (b) exclusively used for stress analysis
   (c) used for both stress analysis and construction of transducer
   (d) pressure measurement
13. The drawback of strain gauges are:

   S1 : Low fatigue life

   S2 : they are expensive, brittle and sensitive to temperature

   S3 : Poor linearity

   (a) S1 and S2
   (b) S2 and S3
   (c) S1 and S3
   (d) S1 only
14. LVDT windings are wound on:
   (a) copper
   (b) aluminium
   (c) ferrite
   (d) steel sheets
15. The size of air cored transducer in comparision with iron core part is
   (a) smaller
   (b) larger
   (c) same
   (d) unpredictable
16. The principle of operation of LVDT is based on the variation of
   (a) mutual induction
   (b) self induction
   (c) reluctance
   (d) permanence

17. LVDT is an a ________________ transducer
    (a) magnetostriction (b) inductive
    (c) resistive (d) eddy current

18. S_1 : Transducer is a device which converts physical into electrical quantity

    S_2 : Transducer is also called as sensor
    (a) S_1 (T) and S_2 (False) (b) S_2 (T) and S_1 (F)
    (c) Both S_1 and S_2 (T) (d) Both S_1 and S_2 (F)

19. In a LVDT the two secondary voltages
    (a) are independent of the core position
    (b) vary unequally depending on the core positon
    (c) Vary equally depending on the core position
    (d) are always in phase equadrature

20. Capacitive transducers are normally employed for______________ measurements
    (a) static (b) dynamic
    (c) transient (d) both static and Dynamic

21. The principle of operation of variable resistance transducer is
    (a) deformation leads to change to resistance
    (b) displacement of contact slider on a resistance
    (c) coupling of two coils changes with displacement
    (d) movement of magnetic field produces variation in resistance of material

22. The Application of LVDT is
    (a) joint motion (b) finger movement
    (c) limb movement (d) heat wall motion

23. Photo conductive cells consist of a thin film of
    (a) quartz (b) lithium sulphate
    (c) barium titunate (d) selenium

24. Self generated type transducer are ________________transducers
    (a) active (b) passive
    (c) secondary (d) inverse

25. The transducer that converts the input signal into the output signal which is continuous function of time is known as____________
    (a) active (b) passive
    (c) analog (d) digital

26. A transducer that converts measured into the form of pulse is called the ________________ transducer
    (a) active
    (b) analog
    (c) digital
    (d) pulse
27. The lower limit of useful working of range of a transducer is determined by
    (a) minimum used input level
    (b) transducer error and Noise
    (c) cross-sensitivity
    (d) dynamic response
28. Which of the following transducer is classified as an active transducer
    (a) metallic strain gauges
    (b) capacitive microphone
    (c) LVDT
    (d) piezo-electric transducer
29. Pair of active transducer
    (a) thermistor, solar cells
    (b) thermocouple, thermistor
    (c) thermocouple, solar cell
    (d) Solar cell, LVDT
30. Which one of the following gives gauge factor of a strain gauge
    (a) $\frac{\Delta L}{L} \Big/ \frac{\Delta R}{R}$
    (b) $\frac{\Delta R}{R} \Big/ \frac{\Delta L}{L}$
    (c) $\frac{\Delta R}{R} \Big/ \frac{\Delta D}{D}$
    (d) $\frac{\Delta R}{R} \Big/ \frac{\Delta P}{P}$
31. In wirewound strain guages, the change in resistance under strained condition is mainly on account of
    (a) change in diameter of wire
    (b) change in length of wire
    (c) change in both length and diameter of wire
    (d) change in resistivity
32. The strain gauges should have low
    (a) guage factor
    (b) resistance temperature coefficient
    (c) resistance
    (d) all of the abovc
33. The carrier material employed with strain gauge at room temperature is
    (a) impregnated paper
    (b) bake lite
    (c) epoxy
    (d) aluminium foil
34. LVDT windings we wounded on
    (a) steel sheets
    (b) aluminium
    (c) ferrite
    (d) copper

35. LVDT can be used for
    (a) vibration measurement (b) angular velocity measurement
    (c) force measurement in a beam (d) load measurement on a column
36. The principle operation of LVDT is based on variation of
    (a) self inductance (b) mutual inductance
    (c) reluctance (d) permeance
37. Sensitivity of LVDT is mainly due to
    (a) magnetic shielding of the core
    (b) permeability of the core
    (c) exact cancellation of secondary voltages
    (d) insulation used on the windings
38. In an LVDT, the output quantity
    (a) is algebraically summed to zero
    (b) is difference of the two currents flowing in two secondaries
    (c) depends upon its rating
    (d) none of the above
39. Hall effect transducers can be used for measuring
    (a) magnetic field (b) current
    (c) electric field (d) pressure
40. Which of the following transducers can be used for magnetic flux density measurement (a)
    (a) LVDT (b) synchro
    (c) Hall effect transducer (d) thermocouple
41. A photo-electric transducers converts
    (a) electric current to voltage
    (b) kinetic energy of electrons into potential energy
    (c) light intensity to voltage
    (d) none of the above
42. Dummy strain guage is used in conjunction with the main strain gauge to
    (a) calibrate the system (b) compensate temperature effect
    (c) improve sensivitity (d) reduce strain on the main gauge

43. In measurement using two strain gauges, the second gauge is provided for
   (a) temperature compensation
   (b) increasing sensitivity
   (c) stability
   (d) linearity
   (e) both (a) and (b)

44. Sensonic transducer is used for measuring
   (a) linear velocity
   (b) angular velocity
   (c) acceleration
   (d) All of the above

45. The electrical transducer used for continuous recording and control of humidity are
   (a) resistive hydrometer
   (b) electrolyte hydrometers
   (c) aluminium oxide hydrometers
   (d) all of the above

Questions and Answers

1. Determine the frequency of this waveform, as displayed by an oscilloscope with a vertical sensitivity of 2 volts per division and a timebase of 0.5 milliseconds per division:

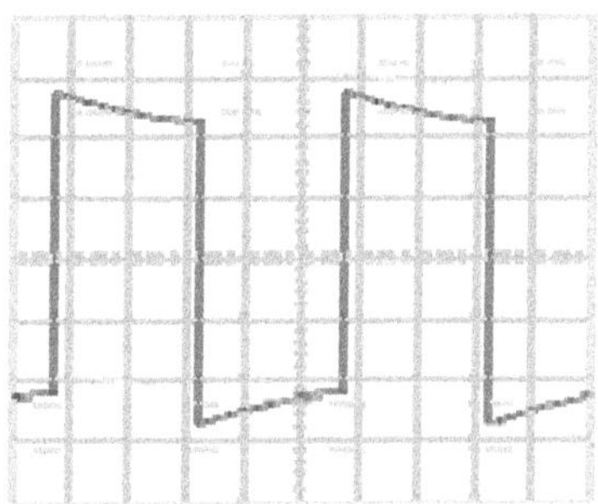

Ans: 400 Hz

2. What is PV?

Ans. PV stands for 'Photovoltaic', the production of electricity from light. The photovoltaic effect is an aspect of the photo electric effect. It is defined as the conversion of electromagnetic radiation to electrical voltage by a specific material.

3. What is a Solar Cell?

Ans. A solar cell is a single piece of photo conductive material for incorporation into a larger module.

4. What types of photo voltaic cells are there?

Ans. (i) Crystalline
(ii) Polycrystalline
(iii) Amorphous
(iv) thin film
(v) multijunction

5. What is a transducer?

Ans. Transducer is a device that converts information energy to be transmitted into electrical signal.

6. What is transducer?

Ans. It is defined as a device that converts energy (or) information from one form to another. These are widely used in measurement work because not all quantities that need to be measured can be displayed as easily as others.

7. What is an electrical transducer?

Ans. It is a sensing device by which a physical, mechanical or optical quantity to be measured is transformed directly with a suitable mechanism, into an electrical signal, current, voltage (or) frequency.

8. Give the three basic requirements of transducer?

Ans. (i) Ruggedness

(ii) Linearity

(iii) No Hysteresis

9. Distinguish between active and passive transducers?

Ans. Self generating type transducers i.e., the transducers, which develop their output in the form of electrical voltage (or) current without any auxiliary source, are called the active transducers. On the other hand the transducers in which electrical parameters (resistance, inductance, capacitance) changes with the change in input signal are called passive transducers.

10. What is difference between Analog and Digital transducer?

Ans. Analog transducers convert input signal to output signal which is a continuous function of time such as thermistor, strain guage, LVDT etc. On the other hand digital transducer converts input signal into the output signal of the form of pulse. *E.g.* discrete output.

11. What is Strain gauge?

Ans. It is a device basically used for measuring mechanical surface strain and is one of the most extensively used transducer. Its working is based on the fact that when stress is applied on the metal conductor its resistance changes owing to change in length and cross sectional area of the conductor.

12. Why are strain gauges called the piezo-resistive strain gauges?

Ans. Resistance of the conductor under stress is also changed due to change in resistivity of the transducer. This property is called the Piezo-resistive effect and the strain gauges are called the Piezo-resistive.

Quiz Questions Key

1. (c)	2. (d)	3. (d)	4. (a)	5. (d)
6. (b)	7. (a)	8. (d)	9. (a)	10. (c)
11. (d)	12. (c)	13. (b)	14. (c)	15. (b)
16. (a)	17. (d)	18. (c)	19. (b)	20. (b)
21. (b)	22 (c)	23. (d)	24. (a)	25. (c)
26. (c)	27. (c)	28. (d)	29. (c)	30. (b)
31. (c)	32. (b)	33. (a)	34. (c)	35. (c)
36. (b)	37. (c)	38. (b)	39. (a)	40. (a)
41. (c)	42. (b)	43. (e)	44. (c)	45. (a)

CHAPTER 12

Display Devices and Signal Generators

12.1 Introduction

An instrument system is an aggregation or assembly of device united by some form of regular interaction of an interdependence. It is a group of devices units or device so combined by nature or by an art to form an integral whole and to function, operate or move in unison and often in obedience to some form of control.

Data acquisition system via analog signals are used in communication, electronic and medical application. Conversion to digitized system is widely used today to implement. In addition there is rapid growth of micro processors and micro computers to perform difficult measurement and control functions.

The demands for more and faster information from physical systems and devices used for space and nuclear research and development, and for industrial automation, have produced many new and effective instrumentation system for data acquisition and reduction.

Data-acquisition systems are used to measure and record analog signals in basically two different ways:

1. Signals which originate from direct measurement of electrical quantities. These signals may be d.c or a.c voltage, frequency, or resistance etc.
2. Signals which originate from use of transducers.

Type of Instrumentation system

1. **Analog System:** These systems deal with information in analog form. An analog signal may be defined as continuous function, such as a plot of voltage versus time or displacement versus force.
2. **Digital System:** A digital quantity may consist of a number of discrete or discontinuous pulses whose time relationship contains information about the magnitude and the nature of the quantity under measurement.

Display Devices

The most commonly used displays in the Digital electronics field, include Light Emitting Diode (LED), Cathode Ray Tube (CRT), Liquid Crystal Display (LCD), gas discharge plasma displays, liquid vapour displays, incandescent displays, electro luminescent displays etc.

Classifications

(i) Based on conversion of electrical signal into the visible light

(a) Active Displays: CRT, Gas discharge plasma display, LED

(b) Passive Displays: LCD, Electrophoresis image displays

(ii) Based on application

(a) Digital Displays: Nixie tubes, LED, Alphanumeric display.

(b) Analog Displays: CRT, Bar graph Display

(iii) Based on display format

(a) Direct View type(Flat panel): Dot matrix, Segmental display.

(b) Stacked Non-planar type: Nixie tubes

(iv) Based on the basis of resolution

(a) Simple single element indicator.

(b) Multi element display.

(v) Based on the size and physical dimensions

(a) Console Displays: LED, CRT

(b) Large Screen Displays: Enlarged projectors

(c) Symabolic Displays: LED, Alphanumeric, Nixie tube.

Components of an analog-data Acquisition System

1. Transducers
2. Signal conditioning equipment
3. Multiplexer
4. Integrating equipment
5. Calibrating equipment
6. Visual display device

7. Analog recorders
8. Analog computers
9. High speed cameras and T.V equipment

Digital data acquisition system are in general, more complex than analog system, both in terms of instrumentation involved and the volume and complexity of the data they can handle.

Digital data Acquisition System

A generalized diagram of a digital data acquisition systems is shown in Fig.12.1 below.

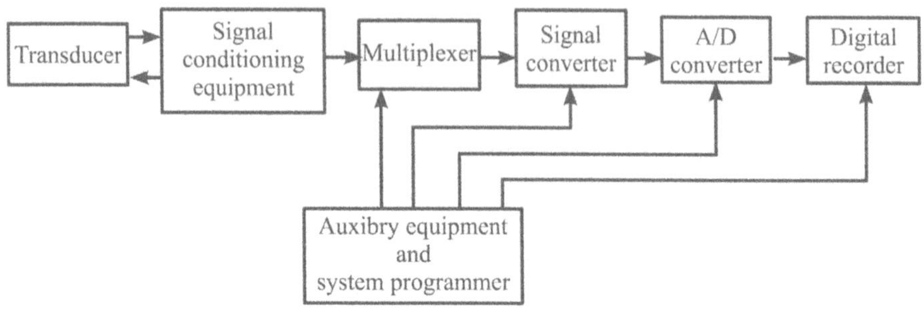

Fig. 12.1 Digital data acquisition system

A digital data acquisition system may include some or all of the components shown in Fig.12.1. The essential functional operations of a digital data acquisition system are:

1. Handling of analog signals,
2. making the measurement,
3. converting the data to digital form and handling it, and
4. internet programming and control.

Modern Digital Data Acquisition System

The electronic devices that perform the interfacing function between the analog and digital world are the analog-to-digital converter abbreviated as A/D converters or ADC, and digital to analog converter abbreviated as D/A converters or DAC.

A/D and D/A converters, data acquisition system may employ one or more of the following circuit functions:

1. Transducers
2. Amplifiers
3. Filters
4. Non linear analog function

5. Analog multiplexers
6. Sample-holds

The inter connection of these components is shown in Fig.12.2, the data-acquisition portion of a computerized feedback control system.

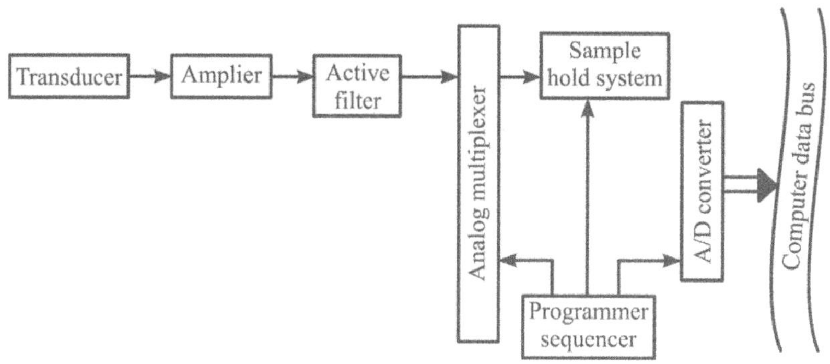

Fig. 12.2 Digital data-acquisition system

12.2 X-Y Recorders

A X-Y recorders is an "Instrument which gives a graphic record of the relationship between two variables".

This system has a pen which can be positioned along the two axis with writing paper remaining stationary. There are two amplifier units. One amplifier actuates the pen in the Y-direction as the input signal is applied, while the second amplifier actuates the pen in X-direction. The movements of the pen in X and Y-directions are automatically controlled by means of a motor, pulleys and a linear potentiometer. Obviously, trace of the marking pen will be due to the combined effects of two signals applied simultaneously. In these recorders "an e.m.f is plotted as a function of another e.m.f".

There are many variations of X-Y recorders with the help of these recorders and appropriate transducers a physical quantity may be plotted against another physical quantity.

Fig. 12.3 shows a block diagram of a typical analog X-Y recorder

A signal enters in each of the two channels.

The signals are "attenuated" to the inherent full scale range of the recorder (often 0.5 mv). The signal then passes to 'a balance circuit' where it is compared with an internal reference voltage.

The "error signal" (i.e the difference between the input signal voltage and the reference voltage) is fed to a "chopper" which converts dc signal to an ac signal.

The signal is then "amplified" in order to actuate a "servomotor" which is used to balance the system and hold it in balance as the value of the quantity being recorded changes.

The action described above it takes place in both the axes simultaneously thus we get a record of one variable with respect of another.

Advantages

1. The instantaneous relationship between two physical quantities can be recorded.
2. The relationship between either electrical or non electrical quantities can be recorded.
3. In modern types of recorders, zero offset adjustments are available.

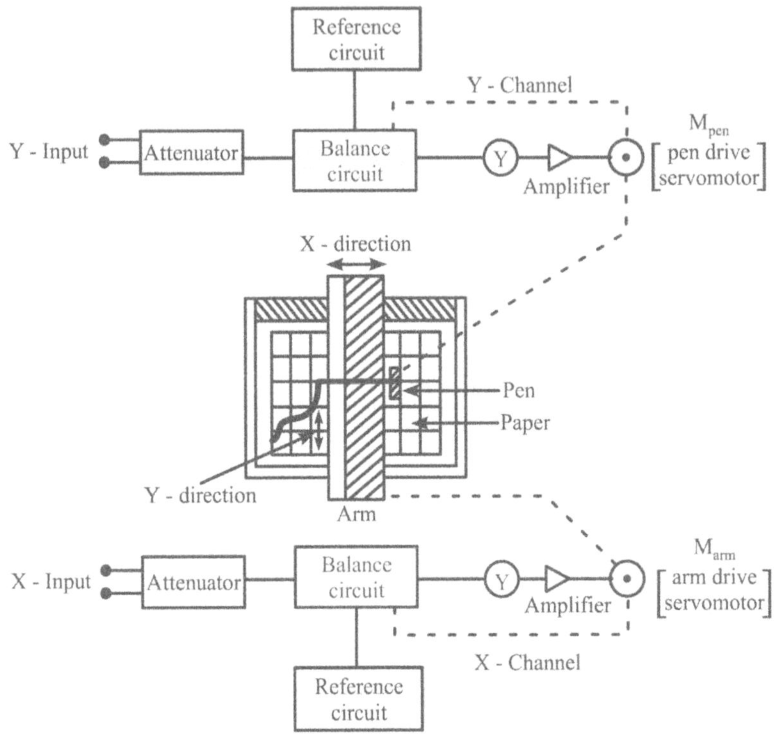

Fig. 12.3 X-Y recorder

12.3 Applications

1. Plotting of stress-strain curves, hysteris curves and vibrations amplitude against swept frequency,
2. Pressure-volume diagrams for I.C engines,
3. Pressure-flow studies for lungs,

4. Lift drag wind tunnel tests,
5. Electrical characteristics of materials such as resistance versus temperature and plotting the output from electronic calculators and computers,
6. Regulation curves of power supplies,
7. Plotting of characteristics of vacuum tubes zener diades, rectifiers and transistors etc,
8. Speed-torque characteristics of motor, etc.

12.4 Digital X-Y Recorders Advantages

1. It is possible to have simultaneous storage of number of input signals.
2. The recorder can record or draw grids, axes.
3. The data can be plotted using multi-pen plotting system.
4. The hardware and software provide better capabilities.

Nixie Tube

The basic construction of a digital indicator tube is shown in Fig. 12.4(a). It is a cold cathode glow discharge tube, which is popularly known as "Nixie" which is the trade mark of M/S Burrough's corporation U.S.A. The display works on the principle that when a gas breaks down, a glow discharge is produced. A gauge electrode with a positive voltage supply functions as an anode, and there are 10 separate wire cathodes. Each in the shape of a numeral from 0 to 9. The electrodes are enclosed in a glass filled envelope with connecting pins at the bottom. Neon gas is usually employed and it gives an oranged-red glow when activated. However, other colours are available when different gases are used.

There is one anode is formed which lights the selected digit.

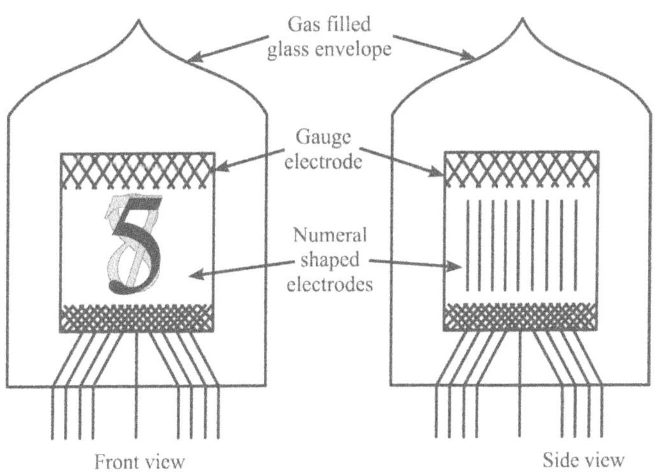

Fig. 12.4 Nixie tube

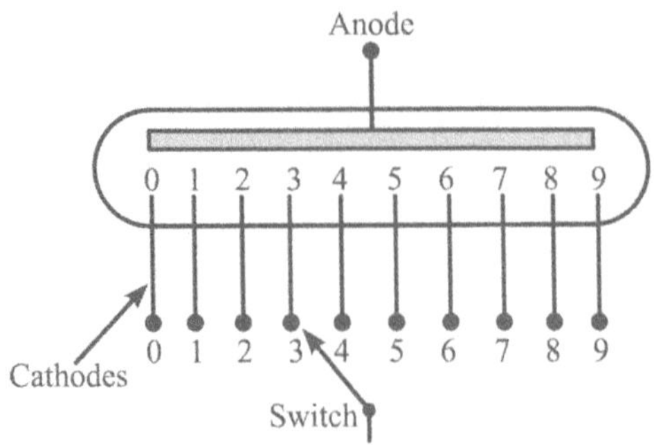

Fig. 12.5 Schematic symbol of nixie

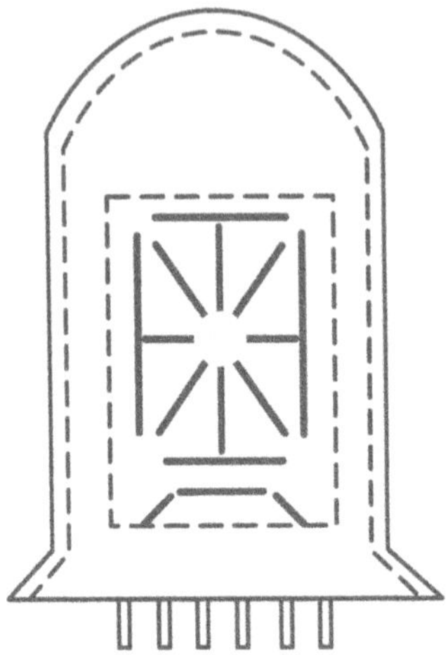

Fig. 12.6 Fifteen nixie

A transistor gate is usually employed at each cathode so that the desired numeral can be switched on. The circuitry driving the Niseie tubes is simpler than that seven segment displays. However, high voltage (150-220 v) are required to produce glow discharge. The current required is of the order of 1-5 mA. The Nixie tubes are bulkier in size than the seven segmental displays.

Modern Nixies

Modern Nixies vary in construction. The original Nixie had 10 cathodes representing numerals 0-9 but additional cathodes displaying decimal symbol and + and – sign are being used.

Special Nixies (Burroughs B7971), shown in Fig (12.6) have 15-cathodes constituting 15 segments that can be used to produce numeric as well as alphanumeric characters.

12.5 Liquid Crystal Displays (LCDs)

LCDs are passive displays characterised by very low power consumption and good contrast ratio.

A liquid crystal is a material, usually an organic compound, which flows like a liquid at room temperature.

Its molecular structure has some properties normally associated with solids (e.g. cholesterol nonanoate and P-azoxyanisole)

When light is incident on an activated layer of a liquid crystal, it is either absorbed or else is scattered by the disoriented molecules.

A liquid crystal "cell" (Fig.12.7) consists of a thin layer (about 10 μm) of a liquid crystal sandwiched between two glass sheets with transparent electrodes deposited on their inside faces.

When both glass sheets are transparent, the cell is known as 'transmittive type cell'.

When one glass is transparent and the other has a reflective coating, the cell is called "reflective type".

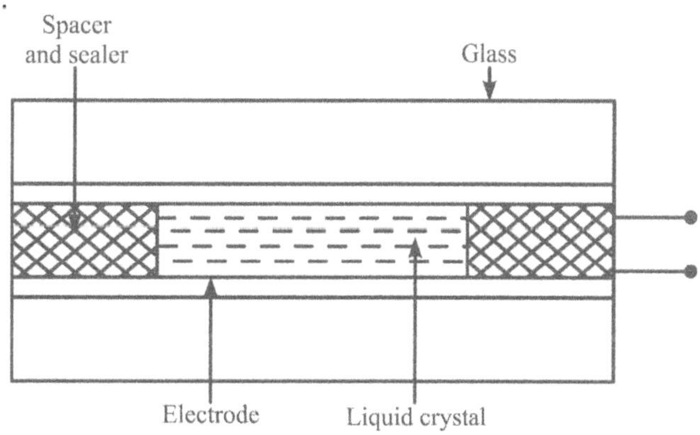

Fig. 12.7 A liquid crystal cell

12.6 Features of LCDs

1. They are light scattering.
2. They can operate in a reflective or transmissive configuration.
3. They do not actively generate light and depend for their operation on ambient or back lighting.

4. The power required by an LCD to scatter or absorb light is extremely small of the order of a few μω/cm. They operate at low voltage ranging from 1 to 15 V.
5. The response time of LCDx typically range between 10 and 100 ms.

Disadvantages

1. LCDs are very slow device, they have a turn-on time of a few milli-seconds, and a turn-off time of tens of milliseconds.
2. They accupy a large area.
3. Poor visibility in low ambient temperature.
4. Limited temperature range of operation (0 – 60°C).
5. Low reliability.

Advantages

1. Extremely low power requirement.
2. Low cost.
3. Long life time about 50,000 hours.

Uses

1. Cellular phone display
2. Desk top LCD monitors
3. Note book computer display
4. Watches and Portable Instruments.
5. Pocket T.V. receiver

12.7 Light Emitting Diode (LED)

A P-N Junction can absorb light energy and produce electric current. The opposite process is also possible, that is a junction diode can emit light. The emission of light occurs under forward bias condition due to recombination of electrons and holes. The emitted light may be visible or invisible.

A P-N junction diode, which emits light when forward biased is known as light emitting diode (LED). The amount of light output is directly proportional to the forward current. Thus, higher the forward current, higher is the light output. Fig.(12.8) shows the schematic symbol of a light emitting diode. The arrows pointing away from the diode symbol represent the "light", which is being transmitted away from the junction.

Fig. 12.8(b) shows the basic structure of a LED

In a forward biased P-N junction, free electrons from N-type material diffuse into P-region. Once in P-region these free electrons encounter holes and eventually recombine thus the free conduction electron fills a vacancy in covalent structure and thus becomes a valence electron. In doing so the electron loses a certain amount of energy as it jumps form conduction band to the valence band. In Si or Ge diode, the energy that recombining electrons lose is dissipated in the form of heat.

But if other semi conductor material such as gallium arsenide and gallium phosphide are used to form P-N diode, the energy lost by recombining electrons is given off in the form of light energy".

The buffer requirement for CRT are couplese and have to fulfil many functions such as:

1. The buffer must translate data from digital to analog form, which determines the beam position and intensity.
2. The buffer must refresh the image on the tube face in excess of thirty times a second to avoid filicker.
3. The buffer must maintain sufficient accuracy to hit any one place out of about five million possible places on the tube face.
4. To reduce the amount of data needed from the computers to drive the roster and vector CRTs, the buffer must be able to fill in details of the intensity and movement needed for the continual control of the beam and to coubine this with data needed to refresh the image.

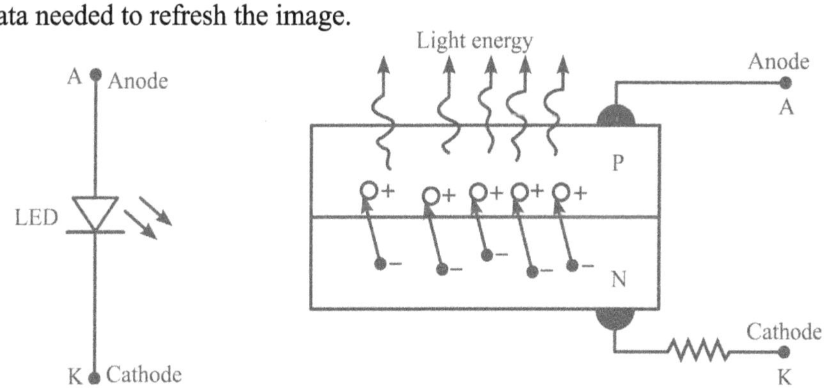

(a) Symbol (b) Basic structure

Fig. 12.8 Light emitting diode (LED)

Diodes made from gallium arsenide (GaAs) emit infrared radiation invisible to eyes and such diodes are referred to as IRED-Infrared emitting diodes.

Fig.12.9 shows two curves used to determine LED operating characteristics.

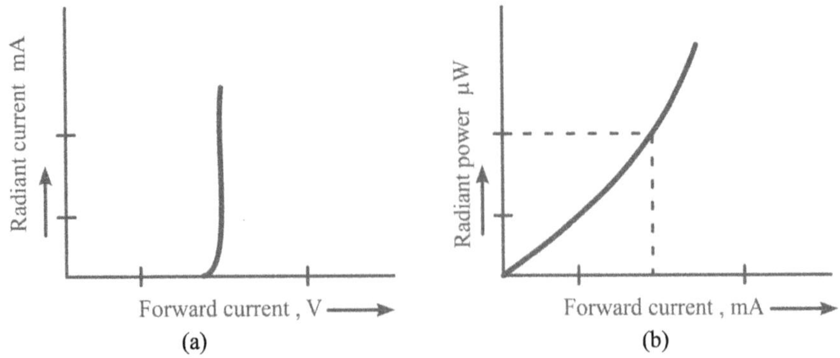

Fig. 12.9 Operating characteristics-LED

Fig.12.9(a) is forward bias V-I curve for typical IRED; the type used in burglar alarms. Forward bias of around IV is required to produce significant forward current.

Fig.12.9(b) is a plot of radiant output power Vs forward current. The radiant output power is rather small (μw) and indicates a very low efficiency of electrical to radiant energy conversion.

They have fast switching characteristics.

Advantages

1. LEDs are miniature in size and they can be stacked together to form numeric and alphanumeric display in high density matrix.
2. The light output from an LED is a function of current flowing through it. Therefore, intensity of light emitted from LEDs can be smoothly controlled.
3. LEDs have a high efficiency as emitted of electromagnetic radiation. They require moderate power for their operation. They are useful where miniaturization in the low d.c. power are important.
4. They are very useful where dynamic operation of large number of arrays is involved.
5. Economical and high degree of reliability.
6. Being rugged, they can withstand shocks and vibrations.
7. Long life.
8. They occupy small area.
9. Available in different colours like red, yellow, green and amber.

Disadvantage

LEDs are not suited for large areas display, primarily because of their high cost.

12.8 Applications

Since LEDs operate at voltage levels 1.5 v to 3.3 v, they are highly compatible with solid state circuitry their uses include the following.

1. Panel indicator
2. Digital watches
3. Multimeters
4. Calculators
5. Intercoms
6. Switch boards
7. Burglar-alarm systems
8. Optical fibre communication system etc.

12.9 Signal Generators

A signal generator is an instrument that provides a controlled output waveform or signal for use in testing or aligning, or in measurements on other circuits or equipments.

The signal generators provide a variety of waveforms for testing electronic circuits, "Usually at low powers."

The term "Oscillator" is used to describe an instrument that "provides only a sinusoidal output signal," and the term "Function generator" to describe an instrument that provides several output waveforms including "since wave, square wave, triangular wave and pulse train, as well as an amplitude modulated waveform".

12.10 Characteristics of Signal Generators

1. The frequency of the signal should be well known and stable.
2. The amplitude of the output signal should be controlled from very low values to relatively large values.
3. The signal should be free from distortion.

12.11 Applications

1. To measure the characteristic of antennas and transmission lines.
2. To provide appropriate signal for calibration, testing and trouble shooting of the amplifier circuits used in communication, electronics, such as radio and television amplifiers.

3. To be employed as power source for the measurement of gain, bandwidth, signal-to-Noise (S/N) ratio and standing wave ratio (SWR) and other properties of circuits.

12.12 Classification of Signal Generators

1. Audio generator
2. Function generators
3. Pulse generators
4. RF generators
5. Frequency synthesizers
6. Other signal generators.

12.13 Conventional Standard Signal Generator

The Fig.12.10 shows the block diagram of a conventional standard signal generator:

The carrier frequency (sine wave) is produced by a stable RF oscillator using LC tank circuit. The frequency of oscillation is shown on frequency control and the vernier dial setting.

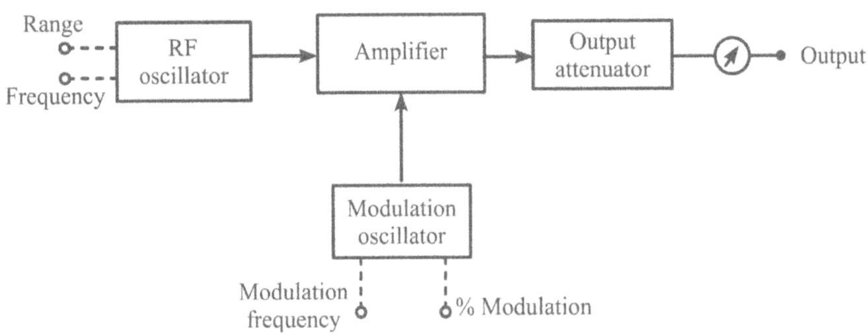

Fig. 12.10 Block diagram of a conventional standard signal generator

Amplitude modulation is obtained by an Internal or an External fixed frequency oscillators.

Modulation is done in output amplifier circuit which delivers its output (i.e., the modulated carrier) to an attenuator.

The output voltage is observed is an "output meter" and the attenuation output setting.

Advantages

1. The output voltage of the conventional standard signal generator can be controlled as per the requirement.
2. The output obtained is stable.

Disadvantages

1. The stability of the frequency is limited, due to LC talk circuit.
2. This type of generator takes some time to stabilise at new frequency when the range is changed.

12.14 Modern Signal Generator

The order to improve the frequency stability, a single master oscillator is optimally designed for the highest frequency range and frequency dividers are switched in to produce lower ranges. The master oscillators is made insensitive to temperature variations and to the influence of the succeeding stages by careful circuit design. Temperature compensation device are used for any temperature changes.

Fig. 12.11 shows the block diagram of a modern signal generator [Laboratory type].

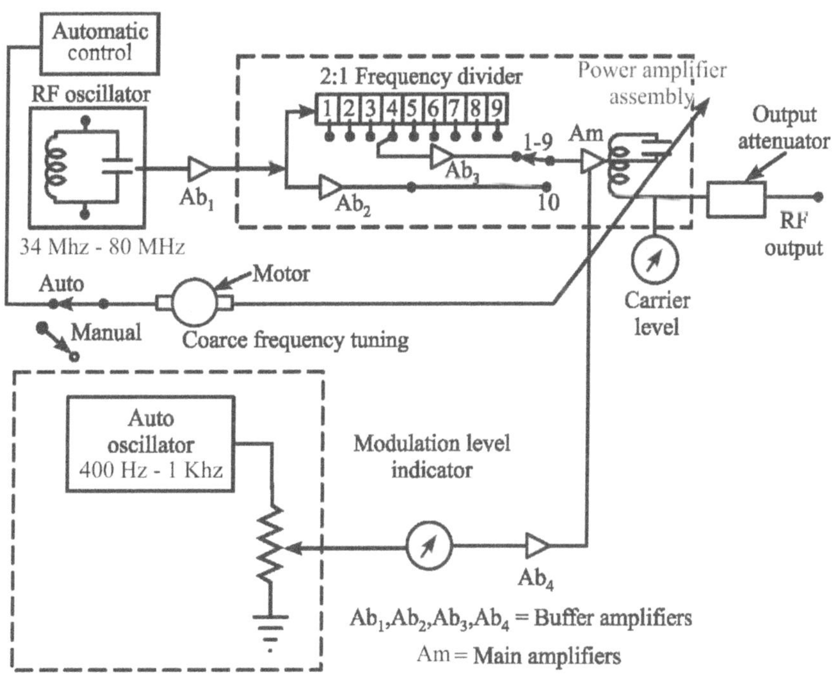

Fig. 12.11 Block diagram of a modern signal generator

RF Oscillator (consisting of LC tank circuit) is used as 'a master – 80 m Hg. This signal is connected to untuned buffer amplifier A_{b_1}. The output of this amplifier is given to frequency divider circuit. The circuit consists of a frequency dividers. The frequency division can be done by using flip-flops to get ratio of 2:1.

The LC tank circuit and frequency divider provides a carrier signal.

The additional buffer amplifiers Ab_2 and Ab_3 provides a very high degree of isolation between the master oscillator and the power amplifier, and almost eliminates all the frequency effects (distortion) between the input and output circuits caused by loading.

The modulating signal is provided by an audio oscillator; the frequency of this oscillator is in the range of 400 Hz – 1 KHz. The modulation is done at the power amplifier stage. The level of modulation can be adjusted up to 95% by using a control device.

The lowest frequency range produced by the cascaded frequency divider is the highest frequency divided by 2^9 or 512. Thus, the frequency stability of the highest range is imported to the lowest frequency ranges.

The master oscillator is turned automatically or manually oscillator, a motor diven variable capacitor is used. This system is extensively used in programmable automatic frequency control devices. The oscillator can be fine tuned by means of a large rotary switch with each division corresponding to 0.01 % of main dial setting.

1 MHz crystal oscillator provides the internal calibration.

The small power consumption of the instruments makes output with very low ripple.

The supply voltage of the master oscillator is regulated by temperature compensated reference circuit;

The output of the main amplifier (A_m) is given to an output attenuator. The "attenuator" controls the amplitude level and provides the required stable RF output.

Advantages

1. Very low power consumption.
2. The instrument provides excellent Q stability (Q-factor/quality factor) with very low ripple.
3. Due to the provision of temperature compensation, the temperature variation does not change output.
4. As the same oscillator is used in all the bands, range switching effects are eliminated.
5. High degree of isolation eliminates distortion between the input and output circuits.

6. Since the same master oscillator is used to get various frequency ranges, the stability is improved even at the highest frequency range.

Disadvantage

Increased cost (as compared to conventional signal generator) due to complex circuitry.

12.15 Function Generator

A "function generator" is an instrument that has the capability of producing different types of waveform as its output signal. The most common output waveforms are:

- Sine waves;
- Triangular waves;
- Square waves;
- Saw tooth waves.

The frequencies of above waves may be adjusted from a fraction of a hertz to several hundred KHz. (A very common frequency range for function generators is 0.01 Hz to 3 MHz).

The triangular-wave and saw tooth wave outputs of function generators are commonly used for those applications which need a signal that increase or reduces at a specific linear rate. They are also used in driving sweep oscillators in oscilloscopes and the X-axis of X-Y recorders.

One of the useful feature of the function generator is that it can be phase locked to an external signal source. One function generator can be phase locked with second function generator; the two output signals can be displaced in phase by an adjustable amount. Furthermore, the fundamental frequency of onc generator can be phase locked to a harmonic of another generator by adjusting the amplitude and phase of the harmonic, almost any waveform can be generated.

When a function generator is phase locked to a standard frequency of the source, then all the output waveforms of the generator will have same accuracy and stability as that of a standard source.

Fig. 12.12 shows the block diagram of a typical function generator.

In this generator the frequency is controlled by varying the magnitude of current that drives the integrator.

The frequency controlled voltage regulates two current supply source. The upper current supply source supplies constant current to the integrator whose voltage rises linearly with time according to output signal equation

$$V_{out} = \frac{1}{c}\int_0^t i\,dt$$

An increase or decrease in the current increase or decreases the slope of the output voltage and thus controls the frequency.

The voltage comparator multivibrator changes state at a predetermined maximum level of the integrator output voltage. This change cuts-off the current supply from the upper supply source and switches to the lower supply source. The lower supply source supplies a reverse current to the integrators so that its output drops linearly with time. When the output attains a predetermined level, the voltage comparator again changes state and switch 'ON' to upper supply source.

The output of the "Integrator" is a "triangular-wave" whose frequency depends on the upper constant current source. The 'voltage comparator' delivers square wave output voltage of the same frequency as that of input triangular wave form. The triangular wave is synthesized into sine wave using diode resistance network. In this shaper circuit, the slope of triangular wave is changed as its amplitude changes; this results in a sine-wave with less than 1percent distortion.

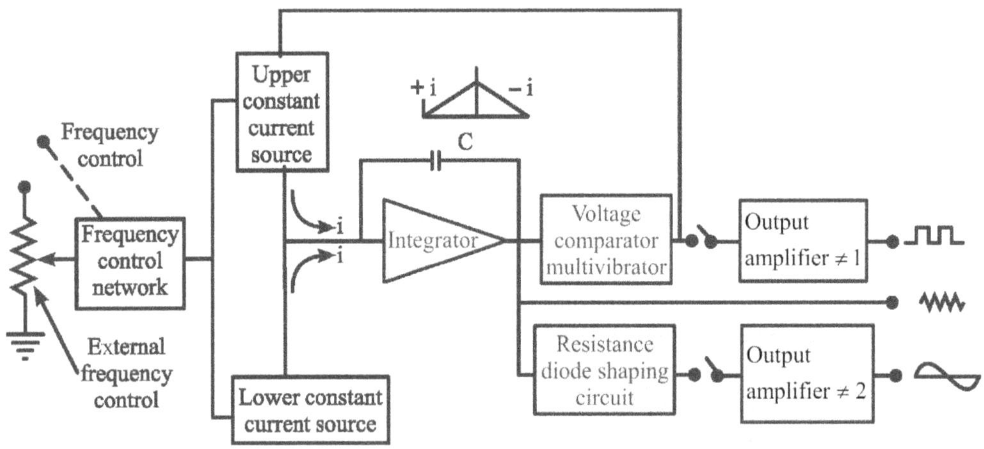

Fig. 12.12 Block diagram of a typical function generator

12.16 Application

Function generator is used for the following purposes:

1. To test the bandwidth of the audio frequency amplifier this method is commonly known as square wave testing.
2. To be used as an important instrument in troubles shooting of different analog and digital circuits and instruments.
3. To act as a source for the alignment of receivers.

Review Questions

1. Write a short notes on: data acquisition systems.
2. What is meant by signal conditioning? is it necessary ?
3. Explain d.c. signal conditioning system with the help of block diagram.
4. Draw block diagram of signal conditioning system. Explain briefly.
5. Explain with block diagram, the essential functional operations of a digital data acquisition system. Compare the digital and analog forms of data acquisition systems
6. Explain with block diagram the essential functional operation of a digital data acquisition system
7. Write a note on digital to analog multiplexing
8. What is the function of instrumentation amplifier? What are its characteristic features?
9. Briefly explain the instruments used in computer controlled instrumentation
10. Write short notes on the following
    (a) objective of DATA acquisition system
    (b) LCD display
11. With a neat sketch, explain the working of a X-Y Recorder
12. Write short note on LCD display
13. Write short note on the various display devices viz. LED, LCD, Nixie tube
14. Explain the classification of displays.
15. Write a note on LED and LCD display.
16. Explain the working of signal generator with the help of neat diagram

Quiz Questions

1. A typical frequency counter
   (a) works by indirectly measuring voltage
   (b) is accurate to six digits or more
   (c) works by indirectly measuring current
   (d) has an analog readout
2. A VU meter is never used to get a general indication of
   (a) sound intensity (b) visible light intensity
   (c) power in an audio amplifier (d) decibels

3. The major advantage of FETVM over a conventional voltmeter is that FETVM
   (a) can withstand higher voltages safely
   (b) draws less current from the circuit under test
   (c) can measure lower voltages
   (d) is sensitive to ac voltage as well as to dc voltage
4. When a signal of 10 mV at 75 MHz is to be measured then which of the following instrument can be used.
   (a) VTVM (b) cathode ray oscilloscope
   (c) moving iron voltmeter (d) digital multimeter
5. The rate at which charge carriers flow is measured in
   (a) coulombs (b) amperes
   (c) watt-hour (d) watts
6. The main problem with bar-graph meters is that
   (a) they are not very sensitive
   (b) they are unstable
   (c) they cannot give very precise readings
   (d) they can display only peak values
7. The simple sine wave generator consists of
   (a) only one oscillator (b) only an attenuator
   (c) oscillators, attenuator (d) frequency divider attenuator
8. A Digital voltmeter has a read out range from 0 to 9, 999 counts. Determine resolution
   (a) 1 μv (b) 1 μv
   (c) 10 μv (d) 100 mv
9. Which of the following is not a static characteristics
   (a) drift (b) dead zone
   (c) sensitivity (d) fidility
10. ____________ is defined as the difference between the largest and smallest reading of instrument
    (a) span (b) range
    (c) dead space (d) resolution

11. ___________ refers to the degree of closeness (or) conformity to the true value of quantity under measurement
    (a) precision (b) accuracy
    (c) drift (d) resolution
12. ___________ defined as ability of the instrument to reproduce a certain set of readings within a given accuracy
    (a) accuracy (b) precision
    (c) reproducibility (d) static error
13. The smallest measurable input change is called___________
    (a) resolution (b) threshold
    (c) dead zone (d) drift
14. The smallest measurable input is called
    (a) resolution (b) threshold
    (c) dead zone (d) drift
15. The addition of ___________ converts the basic D'Arsonal movement into a D.C voltmeter.
    (a) parallel resistor (b) series capacitor
    (c) series resistor (d) parallel capacitor
16. Essentially the reed frequency is
    (a) oscillatory measuring instrument (b) vibration measuring instrument
    (c) deflection measuring instrument (d) recording measuring instrument
17. In reed type measuring instrument all the reeds
    (a) have some natural frequency
    (b) have different natural frequency
    (c) have different natural frequency but difference in natural frequencies of adjacent reeds
    (d) have same natural frequency but difference in natural frequencies of adjacent reeds
18. Vibrating reeds are employed in
    (a) frequency metre (b) p.f meter
    (c) synchroscope (d) megger
19. Signal generator is generally used for generating the following waveform
    (a) sawtooth waveform (b) square waveform
    (c) a periodical waveform (d) sinusoidal waveform

20. In signal generators
    (a) energy is created
    (b) energy is generated
    (c) energy is converted from a simple dc source to ac energy at some specific frequency
    (d) all of the above
21. The maximum output power obtainable from a general purpose audio signal generator is usually
    (a) quite low of the order of μw (b) fairly low of the order of mw
    (c) low of the order of W (d) high of the order of kw
22. The recoder used for plotting B-H curves for magnetic materials, current voltage curves for transistors and speed-time curves for electric moters are ________________ recoders
    (a) LVDT type (b) circular chart
    (c) X-Y (d) potentiometer type
23. X-Y Recorders
    (a) records one quantity on X-axis with reference to time on Y-axis
    (b) records one quantity on Y-axis with reference to time on X-axis
    (c) record one quantity with reference to another quantity
    (d) none of the above
24. An X-Y recorder has X-input and Y input as 1sin 314 t and 1 sin $(314t + \frac{\pi}{2})$ respectively. The trace generator is a
    (a) straight ling (b) parabola
    (c) circle (d) ellipse

Questions and Answers

1. Sensor is a defined as________________________________

Ans. detecting element. Used to locate the presence of matter or energy (energy in the form of heat, light, sound, electrical, pressure, velocity)

2. Video pattern generator employs of multi vibrators

Ans. Two stable chains

3. What is the different between Analog and Digital Multimeter?

Ans. Digital Multimeter have an LCD Display, while an analog meter has a needle that physically moves around.

4. How to read a digital multimeter?

Ans. To read digital multimeter, follow the given instructions. Never test voltage with the meter set to any of the current (amp) settings.

5. How D.M.M works?

Ans. Our digital multimeter may have various functions to take reading on our circuit, we can make a diagnosis when the readings differ from manufacture specific.

6. How to construct a digital multimeter.

Ans. Building a digital multimeter from a kit requires good soldering technique and attention to detail. Kit multimeters are accurate enough for most uses and by calibrating the meter.

7. (a) How to calibrate a D.M.M.

   (b) What are the advantages of D.M.M.

Ans. It provides the functions of dc voltmeter, dc ammeter, Ohm meter and sometime ac current meter, all in one portable package.

8. What are the disadvantages of D.M.M.

Ans. An Analog meter will respond to linearly as current is present. Digital meters cannot do that not until the conversion is complete. In essence an analog meter is human friendly.

9. What is the principle of D.M.M ?

Ans. A YOM digital requires to have a ADC analog to digital converter whereby the Analog input will be displayed as number representation.

10. What is purpose of D.M.M?

Ans. A D.M.M is an instrument which is used to measure voltage, current, resistance also it can measure frequency of periodic waveforms. It mainly used to test equipment.

11. What is a signal generator?

Ans. It is a device that simulates RF signals of accurately known characteristics.

12. What is the purpose of using signal generator?

Ans. It is used for testing amplifiers and radio receivers and transmitters, It is often used in measurement of gain, SW Ratio, BW, standing wave ratio and other properties.

13. How does signal generator differs from ordinary oscillator?

Ans. The signal generator, like an ordinary oscillator is a source of sinusoidal signals but is also capable of modulating its sinusoidal output signal with other signals.

14. Why use of regulated power supply is essential for signal generators?

Ans. Frequency of output from a signal generator changes with the change in supply voltage so use of regulated power supply is essential.

15. Why temperature compensating devices used in signal generators.

Ans. In signal generators, oscillator output frequency varies with variation in temperature so temperature compensating devices are used.

16. What is the use of function generator?

Ans. A function generator is used to generating various forms of the waveforms like as sine wave, square wave, triangular wave.

17. Difference between the signal generator and function generator

Ans. A function generator generates only basic signals, and provides control over its frequency.

18. What is the function of a function switch in a signal generator?

Ans. It is used to select the type of input we want to give to any of our circuits. Inputs such as sine, square, triangular, pulse...etc.

19. How can the range of vibrating reed frequency needs be doubled?

Ans. By polarizing its reeds.

20. On what principle does Weston frequency meter operate?

Ans. It is a moving iron instrument which operates on the principle of variation of impedance in an inductive circuit with the variation in supply frequency.

21. What is the advantage of saturable core frequency meter?

Ans. It has the advantage of wide range and it is well suited for use in tachometer systems.

Quiz Questions Key

1. (b)	2. (d)	3. (b)	4. (b)	5. (b)
6. (c)	7. (c)	8. (a)	9. (d)	10. (a)
11. (b)	12. (b)	13. (a)	14. (b)	15. (c)
16. (d)	17. (c)	18. (a)	19. (d)	20. (c)
21. (b)	22. (c)	23. (c)	24. (c)	

CHAPTER 13

Cathode Ray Oscilloscope

13.1 Introduction

An oscilloscope is a voltage sensing electronic instrument that is used to visualize certain voltage waveforms. An oscilloscope can display the variations of a wave form on the oscilloscope screen.

13.2 Analog and Digital Oscilloscopes

Electronic equipments can be divided into two types analog and digital. Analog equipments works with continuously variable voltage, while digital equipments works with binary no's (Is and zeros), oscilloscopes also come in analog and digital.

Analog oscilloscope works by directly applying a voltage being measured to an electron beam moving across the oscilloscope screen. The voltage deflects the beam up and down proportionally tracing the waveform on the screen. This gives an immediate picture of the waveform.

Digital oscilloscope samples the waveform and uses an (ADC) Analog to digital converter to convert the voltage being measured into digital information.

The oscilloscope is useful for making measurements in both the A.C and D.C. It allows us to view the input waveforms and verify their validity. The device works essentially by plotting (or) tracing the amplitude of the input waveform against time on a cathode Ray Tube (CRT) display. The waveform is plotted many times per second so it is important that the signal is in some way periodic so that the signal seen on the display remains static.

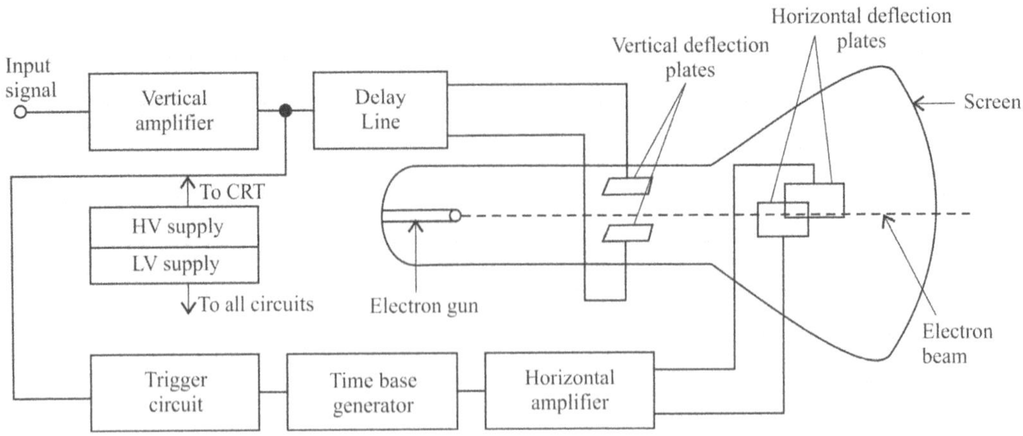

Fig. 13.1 Block diagram of a general purpose CRO

A Basic CRO consist of:

- Cathode Ray tube
- Time Base (sweep) generator
- Horizontal amplifier
- Trigger circuit
- Vertical amplifier
- Delay line

1. **Time Base Generator:** A Time Base Generator is used to generate the saw tooth voltage required to deflect the beam in the horizontal section. The circuit used to generate the sawtooth is called the Continuous Sweep Generator.
2. **Horizontal Amplifier:** The Horizontal Amplifier is used to amplify the saw tooth voltage before it is applied to the Horizontal section.
3. **Trigger Circuit:** A trigger circuit is used to convert the incoming signal into trigger pulses, so that the i/p signal and the sweep frequency can be synchronized.
4. **Vertical Amplifier:** It is a wideband amplifier used to amplify (or) attenuate the input signal to make it seen properly on the screen. This process is controlled by "volt/div" control that is used to select the suitable vertical voltage that caused deflection of one division. Note the (volt/div) is called the oscilloscope sensitivity.
5. **Delay Line:** A delay line is used to delay the signal in the vertical section. The signal in the Horizontal has to pass through the "Trigger pick off" circuit, 'Time Base' Generator and Horizontal amplifier before it finally reaches the Horizontal Deflection plates (HDP). The signal therefore takes longer to reach the HDP as

compared to the vertical section, this gives an error, in which the initial portion of the input signal is lost. The distortion can be overcome by using a delay line in the vertical section.

13.3 Cathode Ray Tube (CRT) (Fig. 13.2)

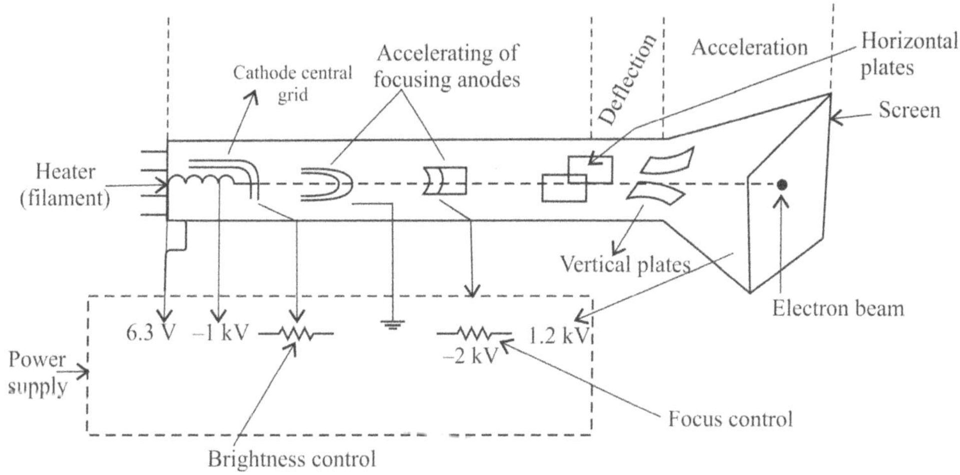

Fig. 13.2 Block diagram of the CRT

13.3.1 Idea of Operation of CRT

1. ***Electron Gun***: Electron gun is used to generate a beam of electrons.
2. ***Screen:*** The screen of a CRT is formed by depositing a coating of a phosphor material on the inner side of the tube face.

   When the electron beam strikes on the screen, electrons within the strike area are raised to a higher energy level.

   Then they loss this energy in the form of light as they return to their normal energy level.

   The glow may persist for a few milliseconds (or) even longer, depending on the material employed.

   The colour of the glow produced at the screen may be blue, red, green (or) white.

   If no deflecting fields are present, the electrons travel in a straight line from the hole in the accelerating anode to the center of the screen where they produce a bright spot.

3. ***Vertical and Horizontal Deflection Plates*:** During the way to the screen, the electron beam passes between two pairs of deflecting plates (horizontal and vertical plates).

   If there is no potential difference between both pairs of plates, the electrons strike the screen on the origin.

   A potential difference between the horizontal pair of plates deflects the electrons horizontally, and a potential difference between the vertical pair deflects them vertically.

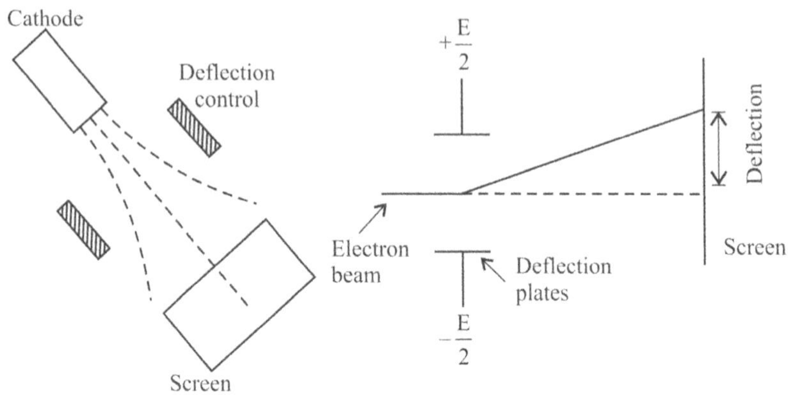

Fig. 13.3 Vertical and horizontal deflection plate

As shown in the Fig.13.3, when the upper vertical plate potential is +E/2 and the lower potential is –E/2 the (negatively charged) electrons will be attracted towards the upper (positive) plate. So the bright spot will be deflected vertically upwards. If the potential difference between the vertical plates is reversed, the bright spot will be deflected vertically downwards. The deflection height depends on potential difference.

13.3.2 The Deflection System

The Deflection system consists of two pairs of parallel plates, reffered to as vertical and horizontal deflection plates. One of the plates in each set is permanently connected to the ground (zero volt), where as the other plate of each set is connected to input signals (or) triggering signals of the CRO.

As shown in the Fig. 13.4 the electron beam passes through the deflection plates. In reference to the schematic diagram Fig. 13.5, a positive voltage applied to the Y-input terminal causes the electron beam to deflect vertically upward, due to attraction forces, while a negative voltage applied Y-input terminal causes the electron beam to deflect vertically downward, due to repulsive forces. Similarly a positive voltage applied to the

X-input terminal will cause the electron beam to deflect horizontally towards the left of the screen. The amount of horizontal and vertical deflection is directly proportional to the corresponding applied voltage. When the electron hit the screen, the phosphor emits light and a visible light spot is seen on the screen.

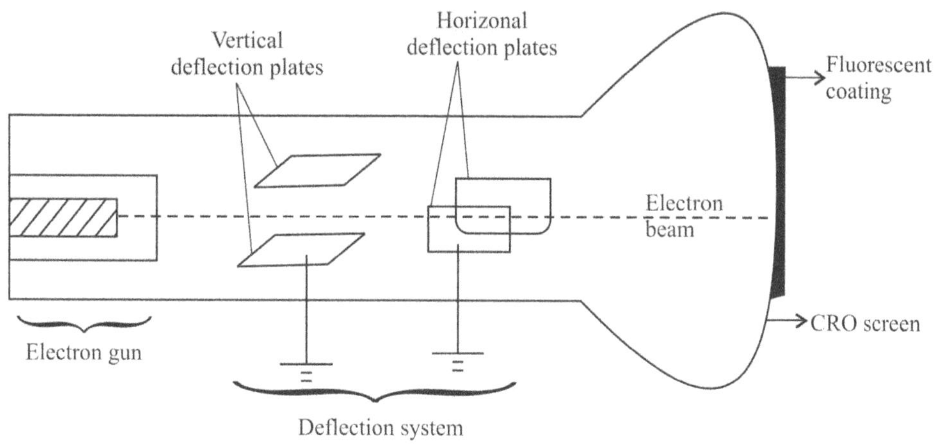

Fig. 13.4 Deflection system

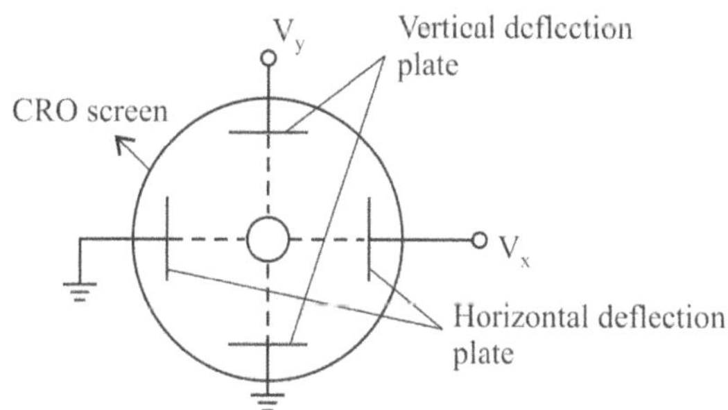

Fig. 13.5 Horizontal deflection plates

Example: Suppose v_x = sin t, v_y = cos t are applied to the horizontal and vertical deflection plates respectively then bright spot would follow a circular path on the CRO screen.

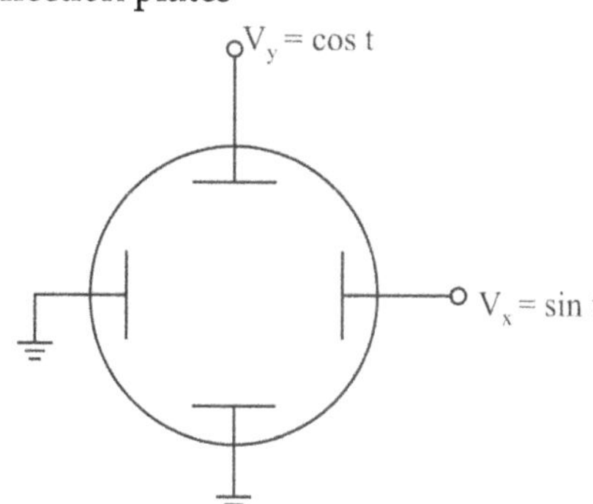

Fig. 13.6

13.3.3 Displaying a Voltage Waveform

In numerious applications it will be required to display a periodical voltage waveform as a function of time. By applying the voltage to be displayed on the CRO, to the vertical deflection plates (v_y), the vertical deflection of the beam spot will be proportional to the magnitude of this voltage. It is then necessary to convert the X-axis (horizontal deflection) into time axis. If the signal to be observed is periodic, then a periodic voltage waveform that varies linearly with time as shown in Fig.13.7 is applied to the horizontal deflection plates. This type of waveform is called sawtooth waveform.

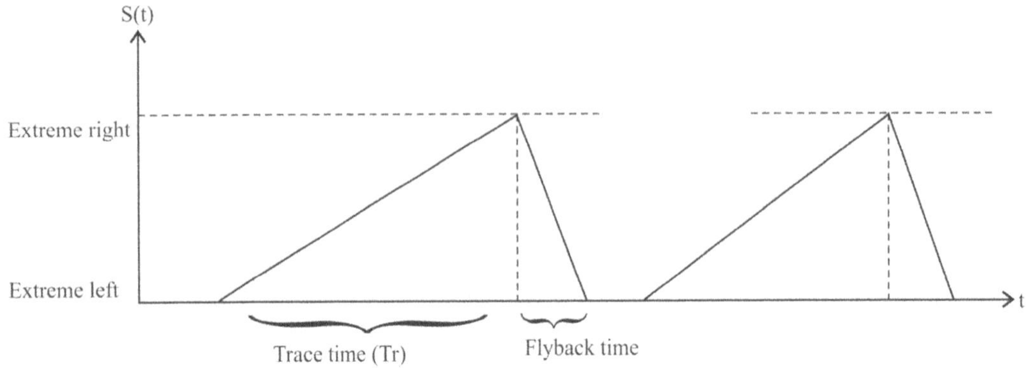

Fig. 13.7

When S(t) = OV, the bright spot is at the extreme left hand position, and when s(t) is maximum, the bright spot is at the extreme right position. Therefore, the bright spot travels from extreme left to extreme right in a time equal to the trace time, a high negative voltage pulse is applied to the control grid of the electron gun to prevent electron beams reaching the CRO Screen. This action is called blanking and prevents any reverse retrace as the beam is going back to the extreme left-hand position. The time period including the trace time and the flyback time is called the sweep period.

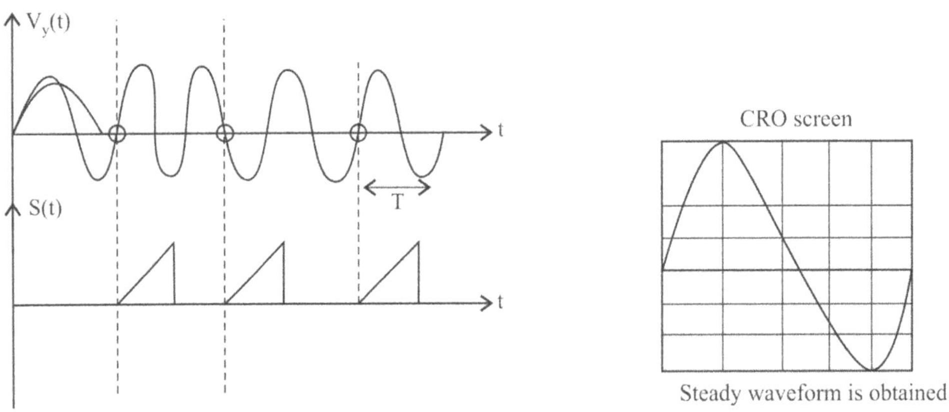

Fig. 13.8

Suppose the input $v_y(t)$ and s(t) in Fig. 13.8 are applied to the vertical and horizontal deflection plates of the CRO respectively. Note that at the beginning of each sweep cycle (i.e., when the bright spot is at extreme left) $v_y(t)$ gets exactly the same value. Therefore the bright spot is following exactly the same path in each sweep cycle. Thus we can observe a steady waveform on CRO screen. Notice that the time between the beginning of two consecutive sweep cycles is a multiple of input signal period.

13.3.4 Pulse Generator and Sweep Generator

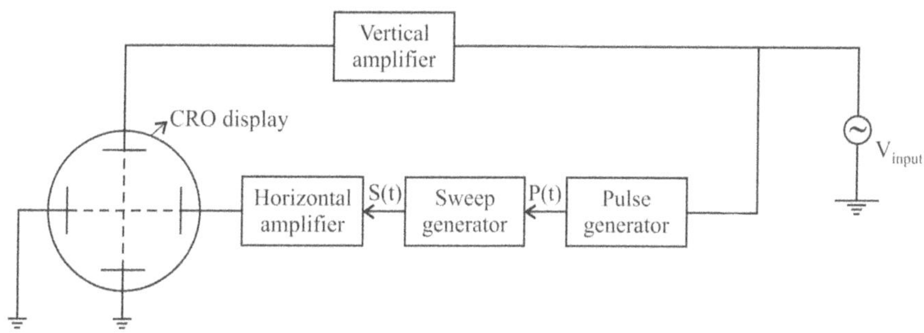

Fig. 13.9 Block diagram of the pulse generator and sweep generator

Pulse Generator: The main function of the pulse generator is to produce periodical pulses with a period of T, which is equal to the period of the input signal. For that purpose, the input signal is compared to a certain voltage level as shown in Fig. 13.10 producing pulses each time the input voltage is equal to that certain voltage level, may seem to result pulses which are periodic with period T. But this is not case, notice that the level intersect the input signal more than once in one period. Therefore one of the intersection point is neglected in each period. The decision on which intersection point to neglect made by inspecting the slope of the signal at the intersection point.

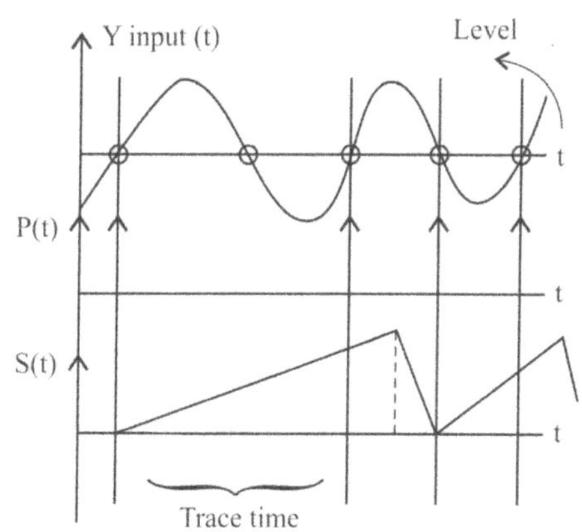

Fig. 13.10

Sweep Generator: The main function of sweep generator is to produce the one cycle of sawtooth waveform, when it receives a pulse at its input. If the sweep generator receives a trigger pulse during its sweep cycle it will simply ignore the pulse and continue with completion of its sweep cycle.

Depending on the selected level and the slope of the input signal, the output of the pulse generator will consist of narrow trigger pulses separated from each other by one period T. Each time the input signal crosses a preselected level, the pulse generator emits one narrow trigger pulse.

X-Y Operation

When the variation of one voltage form $v_y(t)$, as a function of another, $v_x(t)$, eliminating the parameter time t, is desired. x-y mode of operation is used. In x-y mode one signal is applied to the vertical deflection plates whereas the other signal is applied to Horizontal deflection plates. The xy button on the front panel of the oscilloscope disconnects the triggering signals from the horizontal deflection system, and connects the second input signal instead. This process is done by using a switch shown as x-y button on the Fig.13.11.

External Triggering

Rather than the input signal itself an external signal can also be used for triggering. For that purpose multi positional switch, which corresponds to external button of the CRO should be set to position 2 as shown in Fig. 13.11. The external signal should satisfy certain conditions in order to obtain a steady wave form on the CRO screen.

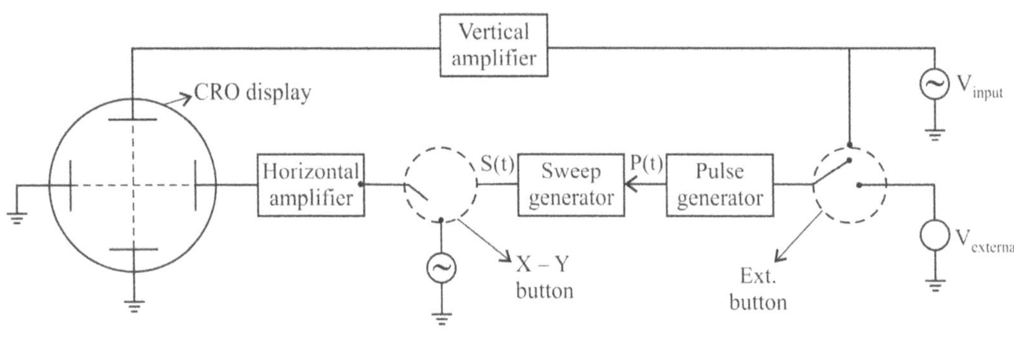

Fig. 13.11

13.4 Digital Storage Oscilloscopes (DSO)

The concept behind the digital oscilloscope is somewhat different to an analog oscilloscope. Rather than processing the signals in an analogue fashion, the DSO converts them into a digital format using an Analog to Digital converter (ADC), then it stores the digital data in the memory, and then process the signals digitally, finally it converts the resulting signals in a picture format to be displayed on the screen of the scope.

Since the waveform is stored in a digital format, that data can be processed either within the oscilloscope itself, or even by a pc connected to it. One advantage of using the DSO is that the stored data can be used to visualize or process the signal at anytime. The analog oscilloscopes do not have memory therefore the signal can be displayed only instantaneously. The transient part of the signal cannot be observed using an analog oscilloscope.

The DSO's are widely used in many applications in view of their flexibility and performance.

13.4.1 Measurement Techniques

The major concern in observing a signal, the oscilloscope screen is to make voltage and time measurement, or the circuit itself, depending on what you measure. Except for the X-Y mode of operation, the oscilloscope displays the voltage value of the waveform as a function of the time. The oscilloscope screen is partitioned into the grids, which divides both the horizontal axis and vertical axis into divisions which will be helpful in making the measurement as in Fig. 13.12. Obviously one needs to know the time (or) the voltage values corresponding to each division, in order to make accurate calculations. These values are determined by two variables namely the time/div and volt/div both of which can be adjusted from the relevant buttons available on the front panel of the oscilloscope. Also note that the time/div button controls the trace time of the sweep generator, where as the volt/div button controls the 'gain' in the vertical amplifiers in the vertical deflection system.

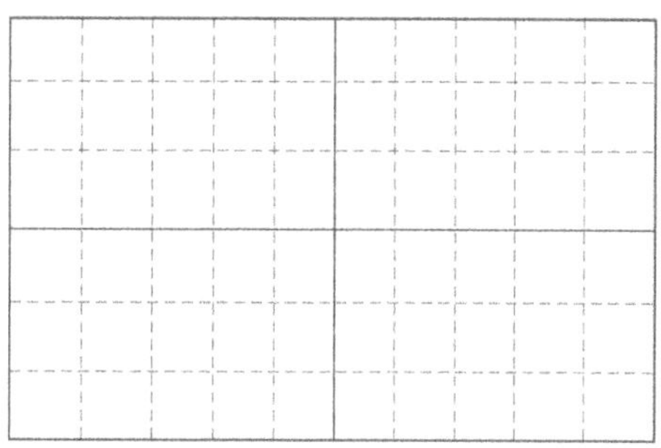

Fig. 13.12 Oscilloscope screen

The typical quantities, which are of primer interest when observing a signal with the scope are shown in Fig. 13.13.

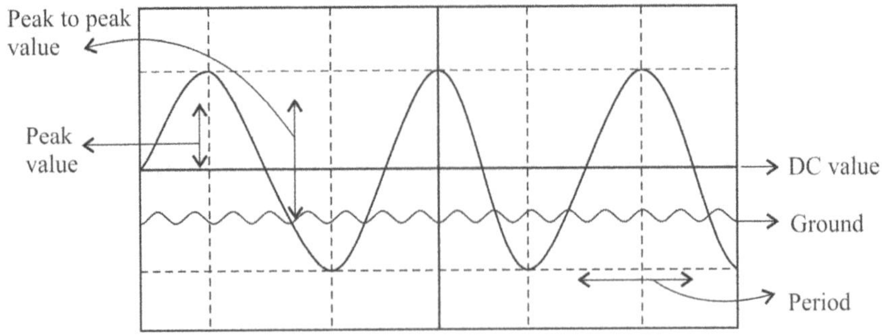

Fig. 13.13 Sinusoidal signal on oscilloscope screen

For the given Fig. 13.13 suppose the variables volt/div and time/div are set to

Volt/div = 2 volts/div

Time/div = 1 milli second/div

Then the corresponding values shown on the figure are calculated to be:

Peak value = 6 volts

Peak to peak value = 12 volts

DC value (Average value) = 2 volts

Period = 3 milliseconds

$$\text{Frequency} = \frac{1}{\text{period}} = 333 \text{ Hz}$$

Note that the signal s(t), shown on the screen can be expressed as:

$$S(t) = v_{peak} \sin(2\pi ft) + V_{DC}$$

$$= 6 \sin(2\pi ft) + 2$$

$$= 6 \sin(666\,\pi t) + 2 \text{ volts}$$

13.4.2 Phase Difference

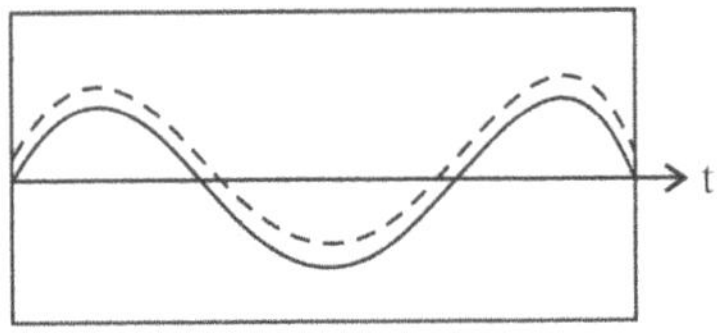

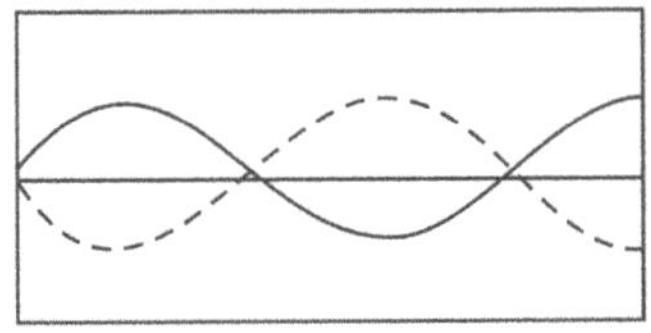

Fig. 13.14 In-phase waves

Fig. 13.15 Out-phase waves

In some applications, one may need to monitor or compare two or more signals simultaneously. A typical example can be the comparision of the input voltage with the output voltage of two-port circuit. If the signals that are being monitored have the same frequency, a time delay may occur between the signals. Two waves that have the same frequency have a phase difference that is constant (Independent t). When the phase difference that is constant (Independent t). When the phase difference (Modulo 2π) is zero, the waves are said to be "in phase" with each other, otherwise, they are out of phase with each other. If the phase difference is 180° (π radians), then two signals are said to be in "antiphase". If the peak amplitudes of two anti-phase waves are equal, then their sum is zero at all values of time, t.

Phase difference is expressed in terms of radians (or) degrees. In dual mode of the oscilloscope the difference can be calculated easily as follows:

Given the two signals having the same frequency as shown in Fig. 13.16.

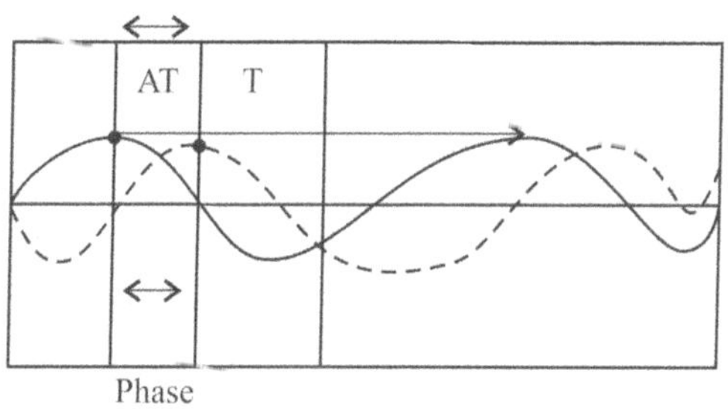

Fig. 13.16 Two signals displayed in dual mode

ΔT = horizontal spacing of the peak value of the two signals

T = horizontal spacing for one period

The phase difference θ is

$$\theta = \frac{\Delta T}{T} \times 360^\circ \text{ in degrees}$$

$$\theta = \frac{\Delta T}{T} \times 2\pi \text{ in radians}$$

13.4.3 Controls

Display system may vary between analog and digital oscilloscope. Common controls include.

An intensely control to adjust brightness of the waveform.

A focus control to adjust sharpness of the waveform.

Vertical controls are used to position and scale waveform vertically.

Horizontal control are used to position and scale the waveform horizontally.

XY-Button: Most oscilloscopes have the capability of displaying a second channel signal along the X-axis, this is called xy mode.

Dual-Button: The oscilloscopes have the capability of displaying both channel signals on the screen at the same time. This is called the Dual mode.

13.5 Applications of CRO

1. Used to investigate transient waveforms
2. Slowly varying waveforms can be monitored continuously.
3. Waveforms can be stored in memory as reference for comparision.
4. Stored waveforms can be reproduced.
5. Complex signal processing can be built into the oscilloscope.
6. The primary uses of the Cathode Ray Oscilloscope are to measure voltage, to measure frequency and to measure phase.

13.6 CRO Probes

An oscilloscope probe is a device specifically designed to minimize the capacitive load and maximize the resistive load added when the instrument is connected to the circuit. The price to pay is an attenuation of the signal that reaches the oscilloscope input.

Let's analyse the behaviour of a passive probe. The Fig. (13.17) shows the equallant circuit of a passive probe and of the input stage of an oscilloscope. The capacitance of the probe cable can be considered. Considering the voltage divider equation we have:

$$H(jw) = \frac{V_s}{V_i} = \frac{Z_s}{Z_p + z'}$$

where $$\frac{1}{Z_s} = jw_{cs} + \frac{1}{R'j} - \frac{1}{zp} = jwp + \frac{1}{R'P}$$

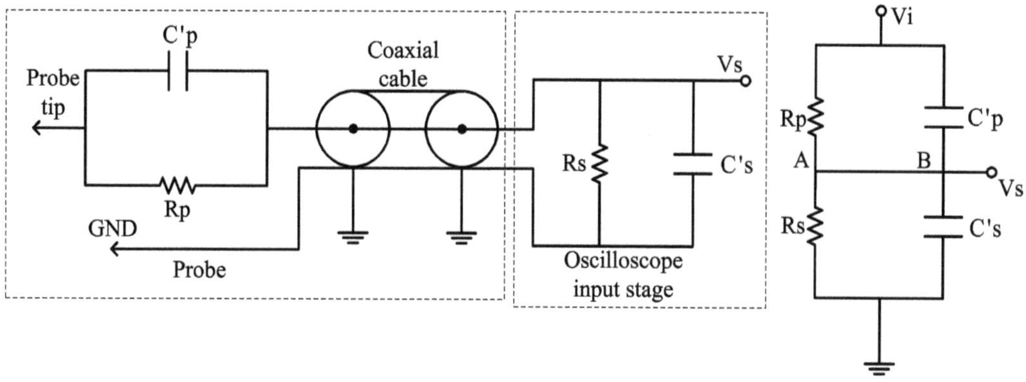

Fig. 13.17 Oscilloscope input stage and passive probe schematics and equalant circuit using ideal components

Defining the following parameters

$$\tau_P = C_P.R_P \quad \alpha = \frac{R_S}{R_S + R'P} \quad \beta = \frac{C_P}{c_s + C'P}$$

after substion all the parameters in H(jw)

$$H(j_w) = \alpha.\frac{1 + jw\tau P}{1 + jw\frac{\alpha}{\beta}\tau'_P}$$

$$Z_s = \frac{R_s}{jw\tau s + 1}; \; Z_p = \frac{RP}{jw\tau s + 1}$$

which is the transfer function from the probe input to the oscilloscope before the ideal amplification stage. The DC and high frequency gain of the transfer function H(jw) are respectively.

$$H(o) = \alpha \; H(\infty) = \beta$$

the numerator and denominator of H(jw) are respectively equal to zero (the zeros and poles of H) when

$$w = wz = j\frac{1}{\tau p}; \; w = v_p = j\frac{\beta}{\alpha}.\frac{1}{\tau p}$$

Fig.13.8 shows the qualitative behaviour of H for $\frac{\alpha}{\beta} > 1$

13.6.1 Probe Frequency Compensation

By tuning the variable capacitor 'C_P' of the probe, we can have three possible cases:

$$\frac{\alpha}{\beta} < 1 \Rightarrow \text{over} - \text{compensation}$$

$$\frac{\alpha}{\beta} = 1 \Rightarrow \text{compensation}$$

$$\frac{\alpha}{\beta} > 1 \Rightarrow \text{under compensation}$$

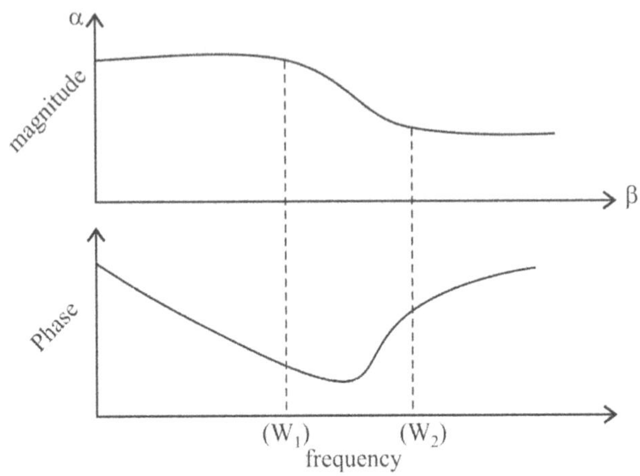

Fig. 13.18 Qualitative transfer function from the under compensated input to the oscilloscope before the ideal amplification stage

If '$\alpha < \beta$' the transfer function attenuates more at frequencies above wz, and the input signal vi is distorted.

If '$\alpha = \beta$' the transfer function is constant and the input signal vi will be undistorted and attenuated by a factor β.

If $\alpha > \beta$ the transfer function attenuates more at frequencies below wp and the input signal (vi) is distorted.

The ideal case is indeed the compensated case, the oscilloscope input impedence by a fator 'α' without distorting the signal.

The probe compensation can be tuned using a signal able to show a clear distortion when it is filtered. A square wave signal is very useful in this case because it shows a quite different distortion if the probe is under (or) over compensated. Fig.13.19 sketch is the expected square wave distortion for the two uncompensated cases.

It is important to notice that if

$$\frac{\alpha}{\beta} = 1 \Rightarrow \frac{R_s}{R_P} = \frac{C_P}{C_s}$$

and this condition implies that:

the voltage difference v_1 across R_S is equal the voltage difference V_2 across Cs i.e., $V_1 = V_2$.

The voltage difference V_3 across R_P is equal to the voltage difference V_4 across C_P, i.e., $V_3 = V_4$.

and therefore $V_1 + V_2 = V_3 + V_4$.

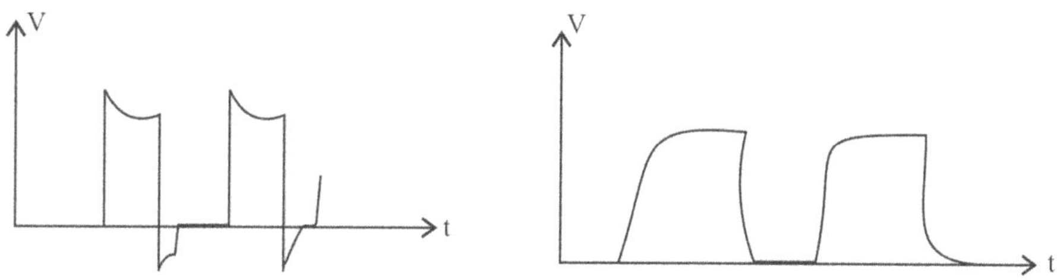

Fig. 13.19

Here $\qquad V_1 + V_2 = V_3 + V_4$

This means that no current is flowing through the branch AB when the probe is compensated and we can consider just the resistance branch of the circuit to calculate vs. Applying the voltage divider equation. We finally set

$$V_s = \left(\frac{R_S}{R_S + R}\right) Vi$$

The capacitance of the oscilloscope does not affect the oscilloscope input any more, and the oscilloscope + probe input impedence R_i becomes greater i.e.,

$$R_i = R_S + R_P$$

Review Questions

1. Draw the Block diagram of CRO and explain the operation of each block.
2. Draw the Block diagram of CRT and explain the operation of each block.
3. Explain the operation of 'Deflection System' in detail.
4. Explain the operation of sweep generator and pulse generator.
5. Explain the operation of Digital storage oscilloscope by taking measurement techniques.
6. Explain in detail about CRO probes.
7. What is the difference between Analog oscilloscope and digital storage oscilloscope.
8. What is the function of Delay line in CRO.
9. What is the function of trigger circuit.
10. Write the applications of Cathode Ray Oscilloscope.
11. Write the Advantage of Digital Storage Oscilloscope.

Quiz Questions

1. CRO displays
   (a) A.C signals (b) D.C signals
   (c) Both (a) and (b) (d) None of above
2. Digital oscilloscope is more preferable because
   (a) Digital display (b) Storage
   (c) a and b (d) None of above
3. Sawtooth voltage of CRO means?
   (a) Only sweep time (b) Flyback time + sweep time
   (c) Sweep time + flyback time (d) Only flyback time
4. Deflection sensitivity of CRO depends on
   (a) Deflection voltage, separation between the plates and plate length
   (b) Only deflection voltage
   (c) Only separation between plates
   (d) Electron density

5. The difference between CRO and Spectrum Analysis.
   (a) CRO and SA both measure the time domain signal.
   (b) CRO and SA both measure the frequency domain signal
   (c) CRO measure time domain and SA measure frequency domain
   (d) CRO measure frequency domain and SA measure time domain

Questions and Answers

1. Where do you use Cathode Ray Oscilloscope.

Ans. To calculate time period and amplitude of a signal

2. What is the Deflection Sensitivity of a CRO?

Ans. It is the deflection on the screen (meter) per volt of deflection.

3. Why pre-amplifier is required to CRO?

Ans. To get high acceleration for the beam.

4. How a CRO is used to measure period?

Ans. Period is measured in CRO by using (or) calculating time division.

5. Explain the working of delay line in CRO.

Ans. Delay line is used to delay the incoming signal vertical information and syncronize it with horizontal.

Quiz Questions Key

1. (c) 2. (c) 3. (b) 4. (a) 5. (c)